Physics FOR GEARHEADS™

Randy Beikmann
Ph.D.

**An Introduction to Vehicle Dynamics,
Energy, and Power**
with Examples from Motorsports

B BentleyPublishers®
.com

Physics FOR GEARHEADS™

Chapter 2: Kinematics Basics

Chapter 5: Forces

Chapter 7: Torque, Force
Resolution, and 2-D Vectors

Chapter 8: Angular Dynamics
Basics

Chapter 9: Angular Dynamics
Applications

Contents

Chapter 11: Dynamics in a Plane Applications

Chapter 14: Power Basics

Chapter 15: Power Applications

Chapter 16: Statics and Quasi-Statics Basics

Chapter 17: Statics and Quasi-Statics Applications

Bentley Publishers, a division of Robert Bentley, Inc.
1734 Massachusetts Avenue
Cambridge, MA 02138 USA   Information that makes
800-423-4595 / 617-547-4170   the difference®

BentleyPublishers®
.com

This book is prepared, published and distributed by Bentley Publishers, 1734 Massachusetts Avenue, Cambridge, Massachusetts 02138 USA.

Copies of this book may be purchased from selected booksellers, or directly from the publisher. The publisher encourages comments from the readers of this book. Please contact Bentley Publishers by visiting http://www.BentleyPublishers.com.

Physics for Gearheads: An Introduction to Vehicle Dynamics, Energy, and Power with Examples from Motorsports, by Randy Beikmann

Unless otherwise indicated, all artwork is courtesy of Randy Beikmann.

ISBN 978-0-8376-1615-5
Job code: GEBP-03

Library of Congress Cataloging-in-Publication Data

Beikmann, Randy, 1960-
  Physics for gearheads : an introduction to vehicle dynamics, energy, and power with examples from motorsports / Randy Beikmann, Ph.D.
     pages cm
  Includes bibliographical references and index.
  ISBN 978-0-8376-1615-5 (alk. paper)
  1. Automobiles--Maintenance and repair--Mathematics. 2. Physics--Popular works. I. Title.
  TL152.B346 2015
  629.201'53--dc23
                                    2014004773

On the Front Cover: Background line drawings by Randy Beikmann; photo of the GM/So Cal Ecotec Lakester concept courtesy General Motors Co.
On the Back Cover: Author photo by Tara Beikmann; background line drawings and figure drawings for ch. 3, 10, and 16 (along right) by Randy Beikmann; ch.12 photo of Corvette at 12 Hours of Sebring by Richard Prince; ch.4 photo of NHRA Winterfinals courtesy Ford Motor Co.

Design by Andrea Corbin.

The paper used in this publication is acid free and meets the requirements of the National Standard for Information Sciences-Permanence of Paper for Printed Library Materials.(∞)

Manufactured in the USA.
201608R001

Dedicated to my daughter Alise, and my son Aric,
both aspiring engineers, and my wife Tara,
who gets to listen to the three of us.

A Word to the Reader

There are many science books on the market in general, and no shortage of physics books in particular. But unless you consider yourself a hard-core science nut, you may not even wander down that aisle in the bookstore to see them, much less enjoy reading them.

Where's the physics book for the hard-core car-nut? The book for people who were curious at a young age about why a car worked. Who weren't satisfied with being told you turn the key, the engine starts, and the car goes. Who kept asking *why*? What *made* it go?

In early 1983, when I was finishing my Master's degree, I started asking myself where that book was. Having read the popular car magazines since I was ten years old, I had pieced together how an engine ran, the effects of gearing, etc. But until I took physics, in high school and in college, I had no way to judge whether an idea or product that an article was touting was better than a previous or competing design. Once the physics began filling the gaps in my understanding, I could usually figure out when someone was blowing smoke.

At that point, I figured the ideal physics book would combine a gearhead's knowledge of cars with the physical explanations for how they work. Familiar car examples could make the physics practically obvious, and the physics could explain why designs end up the way they do (as well as why they don't end up differently).

So this is the book for those same people who kept asking why? And who still want to know more today.

What Do You Mean, *Physics for Gearheads?*

This book is all about classical physics, which I would call a straightforward description of how things *work*. Classical physics was fully developed by the end of the 1800s, with each of its basic laws having withstood the repeated challenges that form the "scientific method." So the question is not whether they work, but rather how to use them.

The laws of physics are universal, the same whether you are tuning a car or designing a rocket bound for Mars. The title *Physics for Gearheads* simply means that the physics will be illustrated by applying it to automotive problems.[1] From tightening a nut to running the engine, everything on your machine obeys the laws of physics. It doesn't have a choice.

Even if you've never studied it, you know more physics than you might realize. As a gearhead, you have a better feel for mechanical things than most, and that's right where physics starts. We'll build on that advantage.

The book is also intended as a "stepping stone" to other automotive technical books in vehicle dynamics, suspension design, engines, etc. Most of those texts assume you've learned at least basic physics, so if you haven't, it would be a struggle to read them. The background and insight you gain here will make it easier to dig deeper.

Don't look for this to be *All You'll Ever Need to Know About Physics*. It's not. There are a lot of other valuable details and examples in traditional physics books. But after developing a good grounding in the basics here, you will get more out of those too.

Basic Physics and Ingenuity

Whether for racing, the road, or the garage, gearheads come up with a lot of cool ideas, more of them than they could ever have the time or money to carry out. But which ideas will actually work? Of those, which are worth the effort? The automotive world is full of heated debates around transmission gear ratios, tire and wheel sizes, or tuning to maximize torque vs. horsepower. But which of these mechanical changes make the most difference in vehicle performance? Trial and error is a slow and expensive way to find out. A firm grasp of physics will help you start your design closer to the best solution. Then you can fine-tune it, using physics-based intuition, and data.

Learning from experience is of course very valuable, but combining it with a strong physics background makes it much more powerful. No matter what Jim Hall, Colin Chapman, Mark Donohue, or Adrian Newey have done, there is one thing they haven't done: violate the laws of physics. They've used their understanding of it to break convention, produce the best possible cars and tune them for performance.

Note 1: "Automotive" means self-moving, so when I say "automotive," I mean cars and motorcycles. OK, mostly cars, but I promise to include some motorcycles.

Henry Ford (nearest) as he over-
takes Alexander Winton on his
way to win "The Race" held in
Grosse Pointe, Michigan, in 1901.
*Photo by permission of Ford Motor
Company and the Ford family.*

Sometimes "common knowledge" is flawed, and you have to unlearn it to make some leap in performance. The Wright brothers found out that most of the published research on flight was dead wrong, so they skillfully did their own. Go ahead and consider others' opinions, but also learn to be skeptical and think for yourself.

When tackling a new project, my personal strategy involves three crucial steps—in this order: 1) think of my own approach, 2) investigate how other people would do it, and 3) put together the best of everyone's approaches. I always like to think it through before "corrupting" my mind with someone else's ideas. Plus, working through the problem myself arms me with the best questions to ask of the other experts. Whatever improved approach we develop, at least I feel confident in using it.

You'll notice quite a few race cars in this book, partly because they illustrate ingenuity so clearly—and yeah, they're exciting. But cars have been closely tied to racing from the very start. Henry Ford made his name driving his car Sweepstakes to a win in a ten-lap race against Alexander Winton, an accomplished automotive builder of the time and heavy favorite to win. Up to then, Ford had been relatively unknown, but the demonstrated speed and durability of his car brought many potential investors to his door.

Louis Chevrolet (left) and Bob
Burman in their Buick "Bug
Racers," which were very
successful around 1910. *Photo by
permission of General Motors Co.*

Louis Chevrolet also made his name racing, most notably the Buick "Bug Racer" powered by a 4-cylinder engine displacing 622 cubic inches (10.2 liters). The company Chevrolet started with William C. Durant soon became successful enough to purchase General Motors (which then included Buick, Oldsmobile, Oakland, and Cadillac) and become part of it. Ironically, he later started a company called Frontenac that modified Fords for racing.

Keep in mind that when these early races were being run, every law of physics we cover in this book was already known. Steady progress in technology, education, and ingenuity has simply allowed us to exploit these same laws to greater and greater effect in designing race cars.

Breaking Down the Problem

I am convinced that the key to solving a problem isn't so much in your knowledge as in finding the right way to look at it. Almost every problem can be split into a few manageable parts, allowing you to see it from different viewpoints. Learning how to break a problem down also tends to make the solution appear simpler and clearer. Sometimes it becomes so obvious that you can't believe you didn't get it sooner.

Maximizing the competitiveness of a road racer like the Corvette C7.R is a good example. It requires solving for the best balance of its performance capabilities, like acceleration, braking, cornering, top speed, and fuel range. These depend on design variables like engine size, mass, and placement; transmission design and placement; brake size and materials; and vehicle shape and structure. There are tradeoffs everywhere. For example, a bigger engine may make more power, but it is heavier, consumes more fuel, and reduces range, meaning more pit stops.

In automotive design, physics is the right tool for splitting the problem up, and for solving each part. Physics is the science of how forces change

The Corvette C7.R as introduced at the North American International Auto Show (the Detroit Auto Show) in January of 2014. *Photo by Randy Beikmann.*

motion: how torque can be used to produce forces; how the kinetic energy of motion is converted to heat in braking; and how a fuel's heat is turned into power. Physics describes these processes in a simple way with clear physical laws that can be put into mathematical formulas.

And yes, physics really is a simple way to describe complex processes. As you study the laws of physics, you'll notice that each description of a particular law and its corresponding mathematical formula "say" the same thing. To me, tying the math and the physics together makes them both easier. Math-wise, you are just putting numbers from familiar physical measurements (length, pressure, force, etc.) into formulas. Most physical laws are easy to visualize, which in turn makes their formulas easy to follow. When you are able to picture what the formula is describing, the math is no longer abstract.

Don't get the wrong idea, though. This isn't trivial stuff. It took many great thinkers years of effort to develop these laws, and to then gradually combine them into a nice, neat package called physics. Don't worry when you need to read something twice. Very few people can put it all together the first time. The laws of physics are like building blocks, with one idea building on another. If you give yourself the time you need to fully understand the concepts presented in each chapter of *Physics for Gearheads*, then the blocks, or concepts, will fit together more securely.

How the Book Layout Works

Each major topic will be covered in the same sequence, with one chapter on basics and another on applications. Here's how these two chapter types will break down.

Part 1: The Basics

The first chapter on a topic will explain the basic concepts and reinforce them with automotive examples. The physical laws will be described, with some historical context of how and when they were discovered. Each law will be put into a formula (usually an equation), and variations of it may also be shown for use in different situations. Then one or more numerical examples are worked out, to solve specific automotive problems. These problems would be good to work through yourself.

Part 2: Applications

The second chapter on the topic will show several applications of the new concepts, emphasizing the insight you can gain from applying them. Some problems are there to reinforce the basic definitions, and some are there to stretch your brain, being more involved but more interesting. Some allowed me to explore concepts I've always been curious about. One resulted from my editor "challenging" me to explain a concept in more detail.[2]

Note 2: This challenge became the discussions in Section 10 of Chapter 13, and Section 6 of Chapter 15, which together describe producing power from burning a fuel.

While nothing beyond high school math is necessary to work the problems here, the number crunching is sometimes more than I would expect of the average reader. But math is not the point; it's used to show that the physics gets us to the right answer. Following and interpreting the results of these problems is enough to understand the concept. Of course you'll need the math when you want to *use* the physics.

Speaking of which, if you do want to take things further than I have in this book, you may want to check out the alternative calculus-based derivations of some of the basic equations. These are called out in the Major Formulas table at the end of each chapter, as indicated by the formula number being bold in the far-right column of the table. The derivations are found in Appendix 5.

Overall Chapter Lineup

Chapter 1 is a "warm-up lap" that quickly gets you thinking by going through examples and establishing some ground rules for the rest of the book. Then there are seven pairs of chapters on the main topics:

- Kinematics, in Chapters 2 and 3.
- Dynamics, in Chapters 4 and 6.
- Angular dynamics, in Chapters 8 and 9.
- Dynamics in a plane, in Chapters 10 and 11.
- Energy, in Chapters 12 and 13.
- Power, in Chapters 14 and 15.
- Statics and quasi-statics, in Chapters 16 and 17.

Where necessary, a "support chapter" was thrown in. Chapter 5 was added to describe many types of forces, so we could use them in Chapter 6. Chapter 7 was added to explain torque and vectors, getting us ready for angular dynamics in Chapter 8, and for dynamics in a plane in Chapter 10.

Miscellaneous

The beginning of each chapter has a list of terms describing new physical quantities, with their definitions, their symbols, and their units of measurement. At the end of each chapter is a summary, usually about a page, to pull together the new material and provide some perspective. There is also a table of major formulas that were introduced in the chapter, for easy access.

The end of the book has several appendices for quick reference. Appendix 1 covers measurement units and conversions, Appendix 2 lists Greek letters, and Appendix 3 includes a few useful math tools. Appendix 4 lists the symbols used in the book, and Appendix 5 has derivations of formulas that were too long to be in the main text, or used calculus. Appendix 6 is a glossary of common terms used. Following all of the appendices, you'll find a list of references and recommended readings.

And if you've wondered how one person puts together a book like this, the answer is that they don't. It takes lots of help. The

Acknowledgements page at the back is my attempt to thank at least some of the many people who so generously contributed their time, talent, and encouragement to make this book possible.

How Do I Read It?

How you read this book depends on what you want to take from it. It is written as an introduction to physics, so it starts at a very basic level. But if you want a challenge, that's here too.

The physics is presented so that you can absorb each concept in several ways:

- Verbal explanations. I learn best by someone explaining it in words.
- Mathematical formulas. Some people prefer to think in equations.
- Diagrams. A lot of effort has been put into helping you visualize the physics.
- Examples. Real-life images or stories are used to illustrate the concepts.

Combining all these, especially putting physical meaning into the mathematical formulas, is the most powerful way to lodge the material in your head. You learn to read the formulas just like words. The more you do this with the material, the more you "own it."

If you are new to physics, I would recommend starting from the beginning and reading straight through each chapter. You should also work some of the examples. Think the physics through. When you think you've grasped a main concept, take some time to ponder it (do people ponder anymore?). Think of how you'd use it. Refer back when necessary. But don't skip ahead, because you'll probably miss a crucial concept needed for the later material.

If you already know physics, you can pick and choose what to read. Find whatever topic you need to solve the problem at hand. If something doesn't make sense, you should be able to find an earlier section that makes it clear. Besides reinforcing the basics, some examples should challenge you. And you should see some applications that you haven't encountered before. You'll probably enjoy just looking at physics through an automotive lens.

Regardless of your background, "soak time" is important. You could breeze through a topic, but it likely won't stick unless you think about it, apply it, and come back to it. So while a lot of this book is about speed, reading it is not. Take your time, and enjoy it!

Randall S. Beikmann
Brighton, MI
March 2015

1 A Warm-Up Lap

Getting Up to Speed

Michael Schumacher's Benetton-Ford and Damon Hill's Williams-Renault lead the field warming up for the 1994 Japan Grand Prix. *Photo by permission of Ford Motor Company.*

We warm things up by examining an F1 suspension design problem, splitting it into several easy physics problems, and connecting each to the math that produces the result. Then we establish some of the ground rules for science and math.

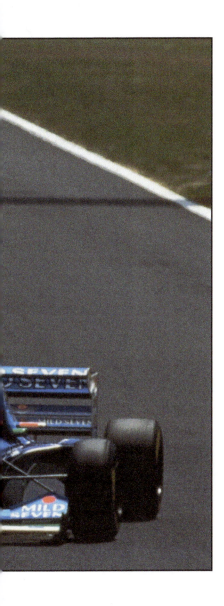

Contents

1   Introduction

No matter how capable a race car is, you don't just hit the ignition and race. The engine needs to warm up so the pistons can expand and fit the cylinders right, and so the oil can thin out and lubricate right. Perhaps most important, the tires need to "come up to temp" to increase their traction and avoid a spin at the first corner.

In the same way, if you go into a subject "cold," it takes a while to get up to speed. You won't get the most out of it without a little perspective. So let's warm up on a few concepts in physics and see how to combine them. Don't worry that they haven't been explained yet; we'll develop each one in the coming chapters.

2   Predicting a Suspension Pushrod Force

Formula 1 engineers are fanatical about reducing the weight of their cars, even though there is a minimum weight regulation. By getting below the minimum weight, they can add ballast low in the car to lower its center of gravity (the CG) and improve handling. In some years the minimum has been so tight that the drivers were asked to lose weight, to the point of losing muscle mass. Another obsession is with aerodynamics. Every ounce of drag becomes power wasted to push the car through the air, when it could be accelerating the car instead.

Both of these factors drive the design of an F1 front suspension, as in the 1997 Benetton B197 in **FIGURE 1**. The springs and shocks are kept inside the bodywork to reduce drag, exposing only the pushrods and control arms. Even then, if the suspension can be designed so the pushrods see less force, they can be made thinner and lighter to further reduce drag, as well as vehicle weight.

FIGURE 1 On recent F1 cars like this Benetton, the front suspension pieces are all tucked inside the body, except the control arm links and the (downward angled) pushrods. *Photo by Randy Beikmann with permission of owner Brian French.*

So imagine you're designing a pushrod in this suspension. You need to know how much force it carries in order to size it correctly. We'll find the pushrod force for when the car is traveling in a straight line at top speed, so the body is generating its maximum downforce. There may be a higher pushrod force in some other condition, but first things first.

How would you start? You'd start with what you know (as shown in the frontal view in **Figure 2**):[1]

1. Two downward forces act at the front axle of the car, shown by the downward arrows: the portion of the vehicle weight carried by the front wheels is 500 lb, and the maximum aerodynamic downforce acting at the front axle is 1,000 lb.

2. The pushrod is leaned over 60° from vertical.

3. The upper and lower control arms are horizontal (not exactly, but close).

4. The car is symmetric side-to-side (centered between the tire patches).

5. The car is traveling straight ahead.

Let's find the pushrod force. We have a total of 1,500 lb acting downward on the front of the car. Something must balance this out, or the front end would fall to the ground; a total of 1,500 lb must push up on the front of the car, to support it. This comes from the road surface pushing upward on the tire contact patches with a total force of 1,500 lb.[2]

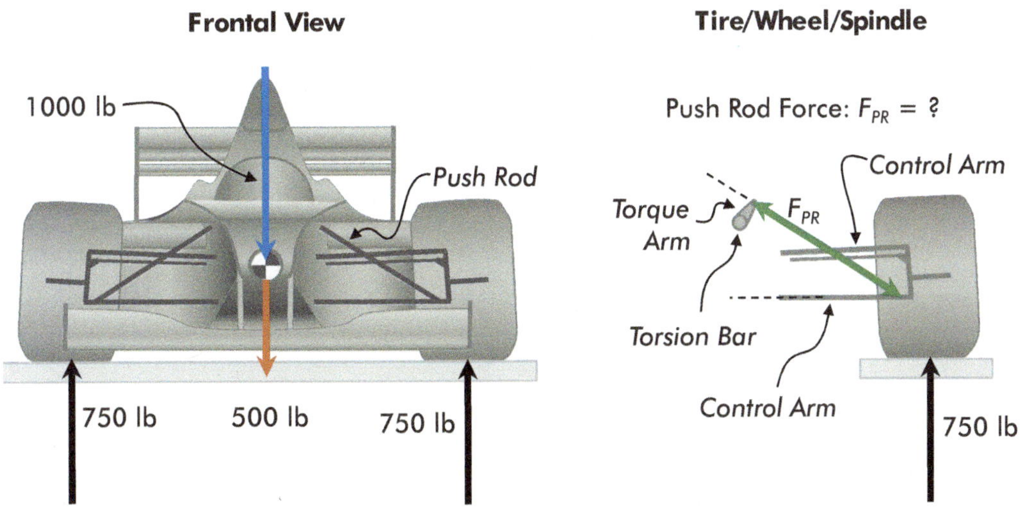

Frontal View   **Tire/Wheel/Spindle**

Since the car is symmetric side-to-side, each front tire is the same distance from center. When traveling in a straight line, this means that the upward forces on each tire are equal, so there is 750 lb pushing up on each in order to total 1,500 lb.

Figure 2 In this frontal view (left) the pushrod angle is 30° from horizontal. Focusing on the vertical forces acting on the wheel/tire/spindle (right), you can see that the pushrod force by itself must balance the tire vertical force.

Note 1: FYI, the symbol in the middle of the car means CG (center of gravity), though it could have meant BMW.

Note 2: If a road pushing up on the tires seems odd, think of what would happen if it didn't! Chapters 16–17 cover support forces like this in detail.

Now look closer at the forces on one side of the front suspension, shown on the right side of the figure. There are two horizontal control arms attached to the spindle, as well as the pushrod, which is 30° up from horizontal.[3] This leaves only the pushrod to support the vertical force, so it must balance the upward 750-lb force on the tire by itself (we're ignoring the wheel and tire weight).

The pushrod also has a hinged joint at each end. Because of this, the force it carries, which we'll call F_{PR} for "force (in the) pushrod," can only be directed along the line going through both joints (see **FIGURE 3**, left). Any force that lies at an angle can be split into vertical and horizontal parts, to form a rectangle (**FIGURE 3**, right). The downward part of the pushrod force must balance the upward force on the tire; the downward force must be 750 lb.

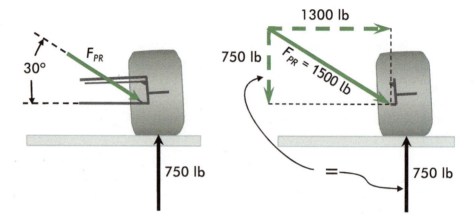

FIGURE 3 The pushrod force is found by balancing the vertical forces acting on the wheel/spindle assembly: total upward force equals total downward force.

Knowing the pushrod's angle and its vertical force component, we can now find the force along the pushrod by drawing its force diagram to scale (again, right side of **FIGURE 3**). The length of each of the sides is proportional to the force in that direction, enabling you to scale the unknown forces from the known ones. For instance, if the 750-lb force line measures 1 inch on your diagram (750 pounds/inch), the diagonal will measure 2 inches, so the diagonal force is 1,500 lb.

The pushrod force is found to be 1,500 pounds. If you follow a similar process for all the possible racing conditions, and find the one producing the highest pushrod force, you can design the pushrod to withstand it. This whole process could also be done mathematically, using techniques we'll develop.

Note 3: Note that the other end of the pushrod pushes on the torque arm (right side of the figure) that twists the torsion bar, which acts as the suspension spring. You are viewing the torsion bar end-on.

3   Divide and Conquer

I hope that seemed more interesting than difficult. We took a complicated problem that has real significance to the design of a car, and narrowed it down to a series of simpler parts. We then solved each of them, one at a time, using each result to solve the next. Each was solved using solid principles of physics. Another aspect of the solution was not adding in unnecessary factors, or jumping ahead of ourselves in the problem: we didn't "clutter" it up. For example, we never calculated the forces in the control arms, because we didn't need to know them to calculate the pushrod force. In other cases you might need to consider the control-arm forces, but you tackle problems as they come.

To make your life easier, and speed things up:

1. Isolate the problem and focus only on it (eliminate the clutter),
2. Divide the problem into a set of simple, small steps, and
3. Solve each step.

The first two parts of this may be the hardest, even through you haven't done any calculations to that point! Both require a bit of common sense, at least a gearhead's common sense, so you should already have an advantage over most people. But coupling that common sense with a good overall grasp of physics makes it much more powerful. That's what this book is for.

By the time you get to part three, the rest is just basic physics. The problem's difficulty often "falls away." [4] While this text will certainly help with the basics, reading other physics texts and working more exercises would provide valuable practice.

4   Let the Physics Drive the Math

Once, my high school physics teacher gave me a problem to work at the chalkboard. I got stuck trying to write the formula to work it, so he asked me to describe in words what would happen in the problem. That I could do. After that, he said, "Okay, now put that into an equation." [5] The light switched on in my head. I saw the connection, I could write the equation, and could solve the problem.

More than any particular physical law, this was my most important insight into physics: the physics drives the math. The math in a formula

---

Note 4: You could call it "seeing through" a problem, to simplify it and get to the answer. Charles Kettering (inventor of the electric starter and founder of General Motors Research) said, "A problem well-stated is a problem half-solved."

Note 5: Most of the formulas in this book are equations, where one quantity (like a force) is set equal to a calculation based on other quantities (like mass and acceleration). We'll see some examples shortly.

says the same thing as the physics. If you can imagine the situation in the problem, you should be able to put it to math (just like putting something to music). So don't just "find the right equation" and plug numbers into it to find the answer; think through the physics of the situation to see how the equation describes it. Then it will stick with you.

As an example, let's calculate the downward force on a piston with a 4-inch bore when there is 500 psi (pounds per square inch) of pressure in the cylinder. The pressure of 500 pounds per square inch literally means there are 500 pounds of force pushing on each square inch of the piston face. Here are the quantities we'll use in this problem:

B = bore (defined here as 4 inches)

A = area of the top face of the piston (unknown so far)

r = radius of the cylinder (½ the bore)

π = 3.1416 (pi, rounded to 4 decimal places)

F = force (unknown so far)

p = pressure (defined here as 500 psi)

On the left side of **FIGURE 4**, you see the top face of a piston that fits in a cylinder of bore B. Since we need to find the area of the piston to calculate the pressure force, let's do that first. We use the formula for the area of a circle, which is $A = \pi \times r^2$, or $A = \pi \times r \times r$, with π being about 3.1416 and r being the radius of the piston. Since the radius of the cylinder is half the bore (you can see in the diagram that r is $B \div 2$), this becomes $\pi \times (B \div 2)^2$.[6] Our 4-inch piston has a radius of 2 inches and so has an area of

$$A = \pi \times (2 \text{ inches}) \times (2 \text{ inches})$$
$$= 12.566 \text{ square inches}$$

Stop and think for a minute about that piston area. Why does the area depend on the square of the bore, $B \times B$? It's because when you increase the bore, the top surface of the piston gets wider in two directions (across the page and vertically on the page in **FIGURE 4**). If you double the bore, the area is four times as big, not two times as big (okay, that makes sense).

Note 6: Note that the formula shown in Figure 4 uses a fraction symbol "/" to signify division, rather than a "÷" sign. This is how division will typically be shown.

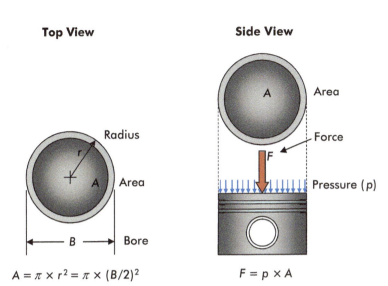

Top View   **Side View**

Radius

Area

r

A

Bore

B

A

Area

Area

Force

F

Pressure (p)

$A = \pi \times r^2 = \pi \times (B/2)^2$ $F = p \times A$

FIGURE 4 Two examples of formulas relating two or more physical quantities. On the left, the piston area A is calculated from the bore B. On the right, the cylinder pressure p pushes on the piston face area A, creating the downward force F on the piston, $F = p \times A$.

Now let's calculate the downward force on the piston, as pictured on the right side of FIGURE 4. Since each square inch of area is experiencing 500 pounds of force, we multiply the pressure times the area to get the total force:

$$F = p \times A = \left(500 \; \frac{\text{pounds}}{\text{inch}^2}\right) \times \left(12.566 \; \text{inch}^2\right)$$

$$= 6,280 \; \text{pounds}$$

So the 500 psi of pressure produces a total of 6,280 pounds on the top face of the piston, which is a lot of force! Note that the formula for the force produced by the pressure, $F = p \times A$, is something we can now use in other situations, no matter what the pressure and area are.

When you put numbers into the formulas (usually equations), don't think of them as just numbers. Think of them as measurements, or sizes. So the higher the cylinder pressure or the larger the piston area, the larger the force. This correctly suggests that a bigger bore (to get a bigger area) or supercharging (to produce higher pressure) would increase the piston force and the engine would produce more torque.[7]

If you take the small amount of extra time to think through the formulas like this, you hardly have to memorize them. They come naturally from the physical situation. Think physics first, math second.

Note 7: Of course, there's more to it than that. But it's a start.

5 | The "Fine Print"

Before buying that camera with the rebate offer, you need to read the fine print. Likewise, the laws of physics have unpleasant surprises if used outside their limitations. On one hand, Newton's laws are extremely accurate in almost any situation. But Hooke's law is only good for limited conditions. First let's demonstrate Newton's second law on the GT40 in FIGURE 5.

FIGURE 5 A Ford GT40 with mass m being accelerated by a net force F.

If you exert a net force F on the car as shown, it will accelerate (speed up). Newton's second law of motion covers this situation, saying that the force acting on a body is equal to the body's mass m multiplied by its acceleration a:

$$F = ma$$

This equation is only true if the correct units are used for m and a. This will be detailed in Chapter 4.

The relationship between force and acceleration (in g's) for the GT40, assuming it weighs 3,000 pounds, is shown in FIGURE 6. The difference between a car's weight and its mass will be explained in Chapter 4. The graph shows that a positive (forward) net force on the car produces a positive (forward) acceleration, and that a negative force produces a negative acceleration (as in braking). Furthermore, the acceleration is directly proportional to the force. This plots as a nice straight line (there is a linear relationship between the two). Newton's second law is so general that we could keep increasing the force, either forward or rearward, and it would still be linear.

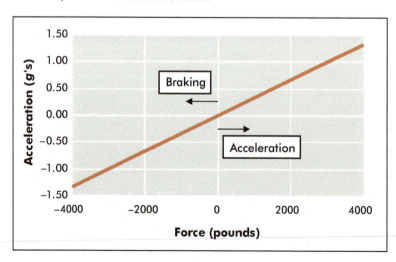

FIGURE 6 The acceleration of a 3,000-pound car (here in g's) is proportional to the net force acting on it, following Newton's second law. This is true for any circumstances you'll see in a car.

Now let's look at Hooke's law, which is used to calculate the force in a compressed or stretched spring, such as the coil springs in a suspension or a valvetrain:

$$F_{Spring} = -kx$$

Here, F_{Spring} is the force produced by the spring, k is the spring stiffness, and x is the distance the spring is compressed, as shown in **FIGURE 7**. Note that the multiplication sign is left out in the equation, as will usually be done. Again, correct units must be used for correct results.

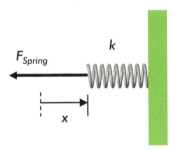

FIGURE 7 A spring that is held stationary at the right end and compressed a distance x, pushes back with a force F_{Spring}. The higher the spring stiffness k, the higher the spring force produced for the same compression.

As you can see in **FIGURE 8**, the straight line predicted by Hooke's law holds closely from 4 inches of compression to 9 inches of tension, making it very useful in practice. But outside this range, it no longer follows reality; the force changes much more quickly than Hooke's law would predict.[8] Why is this?

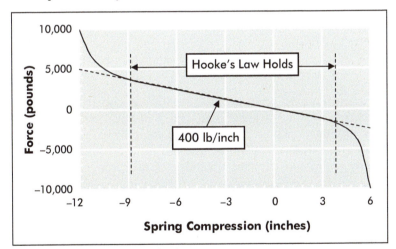

FIGURE 8 The resisting force of a coil spring is proportional to its compression for small deflections, where Hooke's law, $F = -kx$, is valid. Outside this linear range, the relationship falls apart.

It is because past 4 inches of compression, the spring stiffens as it begins to bind (the coils are pushed into each other). Past 9 inches of tension, the spring stiffens as it straightens out and becomes more of a rod than a spring. The range where Hooke's law holds, and the force plots as a straight line, is called the **LINEAR RANGE** of the spring. But

Note 8: Note that tension (stretching) is plotted here as negative compression of the spring.

note that springs are sometimes designed to use the nonlinear range, to limit travel.

KEY CONCEPT

Many laws in physics have limitations on when they can be used. That doesn't make them useless; it just means you need to know the limitations. Better yet, see if the limitations can be used to your advantage.

---

6   Formulas, Calculations, and Curves

In the F1 suspension problem, we found that for a specific loading condition and pushrod angle, the pushrod force is 1,500 pounds. But if you're designing the suspension, you probably haven't chosen the angle yet. You might try many different angles and vehicle conditions to find the *best* angle. For each possibility, you'd find the tire forces, draw the force diagram, and measure the pushrod force. That's a lot of work to repeat!

Instead of repeating the entire process, a better way is to use a little algebra and trigonometry to create a general formula for the force. In the case of the pushrod suspension, all the calculation steps could be combined into one formula:

$$F_{PR} = \frac{F_{Aero,Front} + F_{Weight,Front}}{2\sin\theta}$$

Here, $\sin\theta$ is the sine of the push rod angle θ from horizontal. This short formula is the mathematical summary of the thinking in the entire problem, using the same assumptions. If you plug in the same 1,000-lb aerodynamic force, 500-lb weight force, and 30° pushrod angle as before, you will get the same 1,500-lb pushrod force.

Finding and saving this formula has its advantages. One is that you can avoid a lot of work the next time you do a similar design. The other is that you can program it into a computer (whether in a spreadsheet or a program like MATLAB), have it calculate results for many pushrod angles, and plot the results in a curve (as in FIGURE 9).

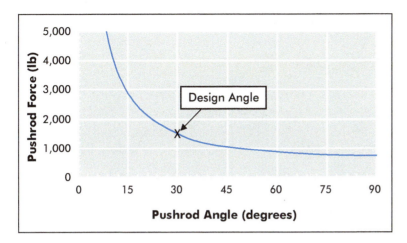

FIGURE 9 To scope out the effect of the pushrod angle on the pushrod force, you can program our formula into a spreadsheet and plot it.

Note in **FIGURE 9** that the pushrod force doesn't depend drastically on the angle when you design it to be 30°. As the angle is reduced much more, however (making it more horizontal), the required force really takes off. At 8.5°, the force is up to 5,000 lb, vs. 1,500 lb at 30°. You can probably imagine that as the pushrod is laid over, the vertical force on the tire is magnified in the pushrod, and that at some point could crush it.[9] In other words, **FIGURE 9** says that poor suspension geometry will require a stronger, heavier pushrod.

Studying the height and shape of a curve such as **FIGURE 9** is especially useful in gaining intuition about design choices, and for explaining trends to others. One look at this figure and it's obvious you'd like to keep the pushrod as vertical as possible (near 90°), keeping in mind you still want to tuck the spring and shock inside the body. That may have also been obvious to you just from looking at the design, but then the point is that the physics agrees with your gut.

Throughout this book, we will see how plotting data curves helps us to understand physical concepts, such as motion during a car's acceleration or braking, and also to see how design changes or operating conditions affect vehicle performance.

7 | Symbols

To keep things clear, every major symbol will mean the same thing throughout the book. For example, the symbol F will always represent a force, and an r will always be the radius of a circle or arc. In situations with more than one force acting, we'll use subscripts to keep them straight. So if you want to say that the weight of a stationary car

---

Note 9: If the angle were about 5° or less, the suspension would look a lot like the mechanism in a pair of locking pliers, such as VISE-GRIPS™.

is equal to the load on the front tires plus the load on the rear tires, you could write that as

$$F_{Weight} = F_{Front} + F_{Rear}$$

Note that many automotive textbooks, such as on vehicle dynamics and engines, will use several different variables to denote forces (besides F, they might use T for tension and W for weight, for example). But most of these are engineering books, written for people who have already had physics and already know calculus. In this book, physics is the subject, and calculus is not used in the main text. However, we do show calc-based derivations of some of the basic formulas in Appendix 5, as alternatives for the reader.[10]

7.1 Greek Letters

Up to now, your knowledge of Greek letters may have been that the fraternity in *Animal House* was $\Delta T X$ (Delta Tau Chi), and the one in *Revenge of the Nerds* was $\Lambda\Lambda\Lambda$ (Lambda Lambda Lambda). Learning more about Greek letters may not be your idea of fun, but there just aren't enough letters in the Roman alphabet for all of the physical quantities we need to describe. Physics books must make a choice— either use Greek letters, or reuse some Roman letters for more than one quantity, which could be confusing. The decision here is to use Greek letters where necessary. There aren't too many, and for the most part they will be used as in other physics texts. Learning Greek letters here will give you a leg up when you read other references. Not every character means the same thing in every book, however, so read their definitions carefully.

A few common Greek letters are introduced in the text now, and the entire Greek alphabet is in Appendix 2 in the back of the book. Pronunciations are given in the text when they are first used. Greek letters are often used for angles and rotational motion. Angles are often called θ (spelled "theta" and pronounced *THAY-tuh*).[11] For angular velocity, which is how fast something spins in a certain direction, we use ω (spelled "omega" and pronounced *oh-MAY-guh*). For angular acceleration, which is how quickly angular velocity increases, we use α (spelled "alpha" and pronounced *AL-fuh*).

Note 10: I had to resort to using calculus in finding some other results, but these calculations are also placed in Appendix 5, and calc isn't necessary to interpret the results.

Note 11: More likely *THAY-duh* in American English.

8   Computers and Calculators

Since mathematics makes the perfect tool set for using physics, a computer makes the perfect assistant. After you do the fun part of a problem (thinking about what to try, setting it up, etc.) the computer is happy to do the rest, no matter how boring and repetitive. The only limits are its software, its speed, and its storage.

It may surprise some readers that computers were originally used for calculations, rather than surfing the Internet. One of the first major applications was to calculate the range of artillery shells during World War II, from their muzzle velocity and barrel angle. These calculations, including the air resistance on the shell, had been done by hand by rooms full of people with pencil and paper, which couldn't have been fun, fast, or reliable. Even the early digital computers, which weren't too fast yet (and also filled a room) could outdo a set of clerks, and not make as many mistakes.

Today's computers are much faster and much cheaper than those of even 20 years ago, so we can have them not only solve for a solution for one design, but also try out many designs and help you choose the best one.

9   Precision and Accuracy

One thing that computers and calculators may be a little *too* good at is precision. Let's look more closely at the previous piston area calculation. With the 4-inch bore, and using the calculator's value for π, the actual answer is 12.566370614 square inches (stopping after "only" 9 decimal places). This answer implies you know the area to within one billionth of a square inch, which is highly doubtful.

Since the computer cannot decide for you how many of the digits are meaningful, you'll need to decide that for yourself. The following sections cover how to recognize and maintain realistic precision.

9.1 Precision vs. Accuracy

Suppose you want to calculate the stiffness of the solid antiroll bar in **Figure 10**, so you need to know its diameter. First you use a ruler (left) that is marked in tenths of an inch, and the bar measures a shade less than 9/10 of an inch. Since you know that eyeballing it will slightly underestimate the thickness, you call it 0.9 inches as a decimal.[12] Then you measure it with a micrometer, which measures it as 0.882 inches.

Note 12: Since the ruler is slightly closer to your eye than the thickest part of the bar, the ruler measurement will come out slightly smaller than the actual thickness, due to parallax.

Being careful, you then check the micrometer by turning it to its closed position, and it reads –0.012 inches (instead of zero).

FIGURE 10 Micrometers and rulers measure the same thing, but we have different expectations about their accuracy and precision.

Since the micrometer was reading low by 0.012 inches, the bar is actually 0.894 inches thick (0.882 inches + 0.012 inches). So the 0.9-inch ruler measurement is too high by 0.006 inches, and we already know the micrometer is low by 0.012 inches.

Now, is the ruler or the micrometer more accurate? Which is more precise? The accuracy of a measurement says how "correct" the measurement is; how true it is. The measurement that's closest to the actual value is the most accurate. The ruler was only off by 0.006 inches, while the micrometer was off by 0.012 inches. The ruler was more *accurate*!

On the other hand, precision states how "fine" a measurement is: how small a difference it can separate; how closely does it repeat? The ruler only reads to the nearest tenth of an inch, while the micrometer reads to the nearest thousandth. The micrometer, with resolution of 0.001 inches, is much more *precise* than the ruler, which has 0.1-inch resolution.[13]

10   Engineering Precision, Rounding, and Significant Digits

From previous examples, it's obvious you need to pay attention when measuring, doing calculations, and giving results. You don't want to lose precision by being sloppy, but you don't want to overstate it either. Because of this, we need specific rules for significant digits used when

Note 13: In this case, the precision of the micrometer allowed us to recover its accuracy, once we subtracted the reading in its "closed" position.

measuring, stating a problem, doing calculations within a problem, and giving a result.

The significant digits in a number are those that are reliably meaningful. For example, a measurement of 79,571 pounds for the weight of a truck has five digits, which probably promises too much precision. If the scale is only accurate to the nearest 100 pounds, it would be more correct to write 79,600 pounds. Only three digits are significant, with the two trailing zeroes used as place-holders (to convert the number 796 into 79,600). Leading zeroes in decimals, such as in 0.00234 inches, are also not significant. In general, the precision of a measurement must be limited to the precision of the instrument.

10.1 For Final Results

We will state results to 3 significant digits, unless the number starts with a "1," in which case we will state it to 4 significant digits. For most problems in this book, results will be given using this "3/4" significant digits rule in the tables on the next page.

10.2 For Intermediate Results

In the middle of a problem, you don't want to create calculation errors from rounding. Therefore, we'll keep two more significant digits for intermediate results (a "5/6" rule).

10.3 For Problem Statements

The remaining issue is how to treat the data given when stating the problem:

1. Large numbers will be given according to the 3/4 rule, such as 3,220 pounds, 1,256 pounds, 265,000 miles, etc.
2. Decimals will be given according to the same rule, such as 32.2 ft/sec^2, 9.81 m/sec^2, 0.01250 inches, or 12.50 kg.
3. When a value is given as a "round number," assume that it follows the above rules. If a value is given as 8 pounds, take it as 8.00 pounds, and take 10 pounds as 10.00 pounds. Multiplying 10 by π (3.1415927…) would then produce 31.416 as an intermediate result, or 31.4 as a final result.
4. If a number needs more precision, it will be stated that way. For example, the speed of light could be given as 299,792,458 meters per second, and g could be given as 32.1740 ft/sec^2.

Note that the third rule is a common-sense rule, so I can call it a 17-inch wheel, instead of a 17.00-inch wheel, which sounds, well, silly. The precision should be understood. You and I know that a wheel has to be made pretty accurately to mate with a tire. TABLE 1 summarizes these rules in a compact form.

	First Non-Zero Numeral = 2–9	
	Significant Digits	Example
Problem Statements	3	3,220 lb
Intermediate Calculations	5	2.5834 lb
Results	3	2.58 lb

	First Non-Zero Numeral = 1	
	Significant Digits	Example
Problem Statements	4	1,728 in^3
Intermediate Calculations	6	1.35583 ft
Results	4	1.356 ft

TABLE 1 Stated data and results will follow the "3/4" rule. Intermediate results use 2 more significant digits to avoid rounding errors.

10.4 Rounding

The other part of reducing calculation errors is to round numbers correctly. The rule will be to round the last kept digit up if the following digit is 5 through 9, or leave it as is if the following digit is 0 through 4. In a previous case, we rounded the value of 79,571 pounds to 3 significant digits. The fourth digit, a 7, is between 5 and 9, so the third kept digit (the 5) is rounded up to 6, giving 79,600 pounds.

11 | Getting Good Data to Work With

All the formulas and best computer tools in the world won't do you any good if you don't have good numbers to plug into them. You need realistic data, such as for a tire's friction coefficient or a car's weight distribution, to produce worthwhile answers. That's why, since the 1960s, we've seen an explosion in onboard recording of variables like engine speed and wheel speeds in race cars. But as we'll see in the coming chapters, you can use test results from popular car magazines to get numbers that are "in the ballpark," and estimate quantities like tire properties, a car's power at the wheels, etc.

12 The "Systems Approach" and Creativity

Typical problems in physics books are very straightforward. You are asked a specific question about a specific object moving in a specific way. There is only one correct answer, so you are right or wrong. That is the basic side of physics, which is critical to learn well, comparable to blocking and tackling in football. You need to understand what can and can't happen in simple situations, before tackling more involved ones. The first eight chapters mainly focus on this.

But the further we go, the more we'll focus on dynamic *systems*. A **DYNAMIC SYSTEM** is a group of components that move together and exert forces on each other, like an engine, driveline, suspension system, or the entire car. This is like learning offensive and defensive schemes in football, where the entire team works together. Most physics books don't deal much with this, but this "systems approach" is so important in automotive use that it really can't be ignored.[14]

Given a system to design, different people might get different answers for the same problem. As an example, let's take another look at Formula 1 front suspensions. Before, we examined the pushrod suspension, which is the current "state of the art" in many racing series. But of course it wasn't always done that way. Most early cars had beam front axles, which have problems with "shimmy" and also take up large amounts of room in their vertical motion. Independent front suspensions took care of these issues, and **FIGURE 11** and **FIGURE 12** show two different designs used, besides the pushrod suspension.

The 1975 March 751 seen in **FIGURE 11** mounts the springs and shocks outside the control arm pivots. While it provides good wheel control and is structurally efficient, it is not made for reducing aerodynamic drag if exposed. But it doesn't need to be, because the March's front air dam diverts airflow around the suspension.

Note 14: Of course it's also what makes cars so interesting and challenging, because they contain so many systems that interact to affect performance.

FIGURE 11 The March 751 has a coil-over double A-arm front suspension, with the spring and shock outboard of the control-arm pivots. *Photos by Randy Beikmann with permission of owner Robert Blain.*

On the other hand, the 1976 Lotus 77 in FIGURE 12 uses a rocker-arm suspension, which allows putting the spring and shock much farther inboard.[15] You can also see that the control arms are strengthened and "streamlined" by welding flat plates between the tubes.

FIGURE 12 The Lotus 77 (top) has a double A-arm front suspension, but uses the upper control arm as a rocker arm (bottom). This places the spring and shock farther inboard, leaving more unobstructed area for airflow. *Photos by Randy Beikmann with permission of owner Chris Locke.*

Note 15: In many cases, the spring and shock were completely inside the bodywork, but in this case, the driver's feet get in the way (race drivers can be such an inconvenience).

The point of this comparison is that, unlike in the most basic physics, designing a system lets you use your creativity. The March, the Lotus, and the previous Benetton suspensions were all solutions to the same problem, but done by different engineers, on different vehicles, and at different times.

Today, the pushrod suspension has the best combination of wheel control, light weight, and aerodynamics. Certainly it works well with the high nose of the 1997 Benetton and of current F1 cars. It is currently used in other road racing series, and a couple of production cars. But in the future, materials improvements and rules changes might make another design superior.

These are just examples, but you'll see the "systems" aspect come up when we deal with problems like load transfer in acceleration and cornering, energy storage in a vehicle, and power transfer through a driveline.

2 Kinematics Basics
Straight-Line Motion

Top Fuelers like Cory McClenathan's are the ultimate acceleration machines. *Photo by permission of General Motors Co.*

The very basics of straight-line motion are covered: how acceleration changes velocity, and how velocity changes displacement (the location). This is then used to find a drag racer's acceleration, and to calculate the average velocity in a land-speed-record run.

Contents

Key Symbols Introduced

Symbols are measured in either SFS or MKS units. The SFS system (slug-foot-second) is based on British units, and the MKS system (meter-kilogram-second) is based on metric. Both will be detailed as we go.

Symbol	Quantity	SFS Units	MKS Units
t	Time	seconds (sec)	seconds (sec)
d	Distance	feet (ft)	meters (m)
S	Speed	ft/sec	m/sec
s	Linear Displacement	feet (ft)	meters (m)
v	Linear Velocity	ft/sec	m/sec
a	Linear Acceleration	ft/sec^2	m/sec^2
g	Acceleration from Gravity (Downward)	32.2 ft/sec^2	9.81 m/sec^2
a_g	Linear Acceleration in g's	Not in SFS Units	Not in MKS Units

1   Introduction

Have you ever wondered how one dragster can have a higher top speed than the other, but a longer (slower) elapsed time? Why does a NASCAR engine "only" spin 9,500 rpm, but a Formula 1 engine will turn 18,000? How can you pass someone in a road race by having better brakes, when all brakes do is slow you down?

To answer these questions, you need to understand **KINEMATICS**. Kinematics (*kin uh MAT iks*) is the description of motion, without considering the forces causing it.[1] To describe a car's straight-line motion, you must define its:

1. Linear displacement (the position along its line of travel),
2. Linear velocity, (its speed with direction along this line), and
3. Linear acceleration (the rate of increase of its velocity).

A dragster accelerates hard as it leaves the start, to get it to the finish in the least time possible. If you measure the dragster's motion, you

---

Note 1: The word "kinematics" comes from the same Greek word as for "cinema," meaning motion. "Cinematography" literally means motion picture, and in German it is *Kinematographie*.

are studying its kinematics. If you include the forces applied to the car (traction, aerodynamic drag, etc.), the *causes* of the acceleration, you are studying its dynamics. That begins in Chapter 4.

> **NOTE**
>
> Kinematics is an essential foundation for dynamics, the main subject of this book.

This chapter will also introduce the very important concept of tracking measurement units, making sure the units in the calculations match up. In Chapter 4, we'll introduce consistent units for length, mass, and time, which makes the formulas used in this book as simple as possible.

While the motion of a car is used to illustrate many concepts, any other body's motion could of course be examined or predicted, such as an airplane, a person, a piston, a valve, etc.

> **REMINDER**
>
> Please remember the suggestion to visualize the physical situation when putting the physics to math, rather than blindly plugging numbers into formulas. Think of the numbers going into the formulas as "measurements" or "sizes."

There are two main types of motion problems in physics. The first is reaching a velocity or distance in a certain time, such as getting from zero to 60 mph in 5 seconds or opening a valve a half inch and closing it again in 0.0075 seconds. The second type is getting to a certain velocity in a certain distance, such as braking from 70 mph to zero in 150 feet. The first type will be covered thoroughly now, while the second will be covered in more detail in Chapter 3.

2   Straight-Line Motion at Constant Speed or Velocity

Before digging in, we need to establish ground rules for rectilinear motion, which is a continuous change in an object's position along a straight line.[2] This is the simplest kind of motion, which you've seen if you've watched a dragster race straight (hopefully) down the track. We'll see how acceleration changes the velocity of an object, and how velocity changes its position. Taking the time for this may seem like a nuisance, but this is important, fundamental stuff. Even distance, which sounds easy to define, could be interpreted in two different ways. So let's get to it by making things a bit more scientific.

Note 2: The prefix "recti-" means straight, and "linear" means a line.

2.1 Particle Motion and Rigid-Body Motion

Physics books usually kick off kinematics with the motion of a parti-cle.[3] That makes it very simple. But it's still simple, and more interest-ing, to track the motion of a body, assuming that all of its particles move the same amount (see the Dodge Challenger in **FIGURE 1**). We can measure the motion of its front or rear bumper, or its center of gravity (the CG), and get the same answer of 50 feet. (The center of gravity will be more fully described as we go, starting in Chapter 4 on Dynamics, but especially in Chapter 16 on Statics.)

In **RIGID-BODY MOTION**, we track the "gross motion" of the body while ignoring any bending, twisting, or stretching it might undergo.

In reality, every physical body flexes somewhat when it moves, but we can usually ignore it or deal with it separately. This may seem like a fine point, but there are times when we need to acknowledge a vehicle is not rigid, such as when studying suspension motion or body vibration. In rigid-body motion, we'll typically track the mo-tion of the CG, but we could use another convenient point. In races, we often track the front tires or front bumper. With that, let's get to straight-line motion.

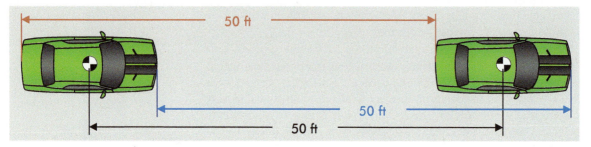

FIGURE 1 In straight-line motion, each part of a car will move the same amount, as long as it acts like a "rigid body" and stays pointed straight.

2.2 Distance and Displacement

Imagine you take a trip on a perfectly straight road, first driving 140 miles east, turning around, and driving 70 miles west, as shown in **FIGURE 2**. How far have you gone?

What sounds like a simple question isn't so simple. On one hand, you could say you've gone 210 miles, since that's how far your trip odome-ter reads. On the other hand, you ended up 70 miles from your starting point, so you could also say that you've only gone 70 miles. We need a way to distinguish between the two. We'll call distance how far the trip odometer would show.

DISTANCE (let's call it d for short) is the total path length a body travels from its starting point, regardless of which direction it goes during dif-ferent parts of its motion. When the car travels 140 miles east and then 70 miles west, the distance traveled is 210 miles.

Note 3: In basic physics, a "particle" is defined as having zero size, with all its mass concentrated at one point.

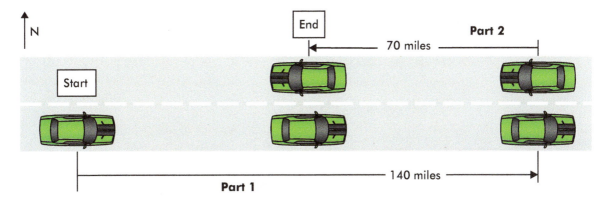

But distance alone doesn't say where the car ends up. For that we need the change in position, including the direction, from the beginning to end of the trip. The amount the car is "displaced" from some reference point is called its **DISPLACEMENT**.

LINEAR DISPLACEMENT (let's call it s) is the difference in position from some reference point, such as from a starting point, taking direction into account.[4] Traveling 140 miles east, then 70 miles west, is a total displacement of 70 miles east. If east is defined as positive displacement, then 70 miles west is a negative displacement, written as -70 miles.

By its definition, distance is always positive. But because displacement has direction, it can be positive or negative. If you define displacement as positive upward, 5 feet downward is -5 feet of displacement. Displacement could be a bit confusing because it can be zero when the distance traveled is not zero. For example, any "round trip" has zero displacement.

FIGURE 2 If you take a trip that goes 140 miles east, and then 70 miles west, have you gone 210 miles, or only 70 miles? Not drawn to scale (obviously).

KEY CONCEPT

Displacement makes motion specific. Since its positive or negative sign specifies direction, displacement carries more information with it than distance does.

So a displacement is the "net motion" of a body. We call it *linear* displacement when we have motion along a straight *line*. The term "linear" is used several ways in physics, so we need to be clear. In this case, linear refers to the straight line of the motion.

Although you don't call it that, you think in terms of displacement all the time. To plan a trip, you have to know how many miles to go, and which way. Traveling 500 miles north doesn't do you much good if you wanted to go 500 miles south (besides maybe getting to know your family a little better).

Note 4: When an engine's cylinder brings in a volume of air, we say it has "displaced" that volume from the atmosphere into the engine. But that is called a volumetric displacement, and is what gearheads usually mean by saying "displacement." Engineers also refer to it as "swept volume."

The motion in **Figure 2** is summarized in **Figure 3**. The distance is shown by the length of the looping blue arrow that extends to the far end of travel and back to the endpoint. The displacement is shown by the straight red arrow, which extends directly from the starting point to the endpoint.

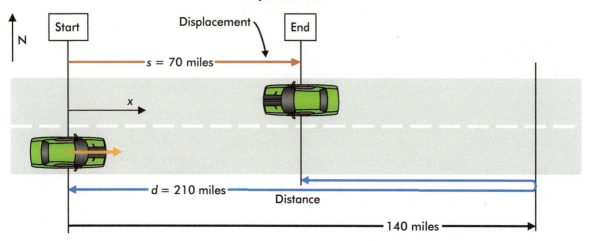

Figure 3 The net motion of 70 miles east is the car's displacement (red arrow), while the total motion of 210 miles is the distance (blue arrow).

Note that both measurements have their uses. If you want to estimate how much gas you'll need for a trip, or the wear on your car, you'll want to know the total *distance* traveled. If you want to know where you'll end up after a trip, you're more interested in the total *displacement*.[5]

Displacement from one place to another is also called **TRANSLATION**. It doesn't make much difference now, but later we'll distinguish linear motion from rotational motion by calling it **TRANSLATIONAL MOTION**.

2.2.1 Adding Distances or Displacements

Let's use the previous example to show how distances add, calling the eastbound stretch Part 1, and the westbound stretch Part 2. Following **Figure 4**, we simply add the 140 miles to the 70 miles, and end up with a total distance of 210 miles, as we said. Note that the arrows for d_1 and d_2 have arrowheads at both ends, signifying that you could travel in either direction to produce the same distance. The formula for adding two distances is simple:

General Equation 2-1

$$d_{Total} = d_1 + d_2$$

Note 5: If you tried to trade in your car and convince your dealer that your car has "zero miles" because you bought it there and are returning it there (zero total displacement), you probably wouldn't get very far.

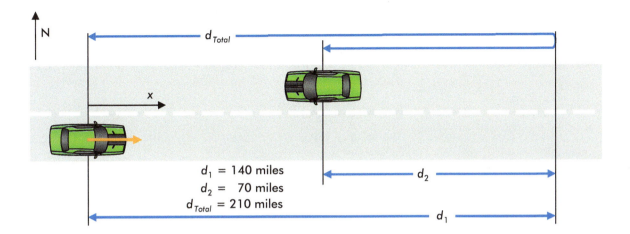

d_1 = 140 miles
d_2 = 70 miles
d_{Total} = 210 miles

Here, d_{Total} is the total distance traveled, and d_1 and d_2 are the distances for Parts 1 and 2 of the trip. Adding two distances always results in a larger total distance.

FIGURE 4 Distances are considered positive regardless of the direction traveled, so they always add to produce a larger amount.

The displacements of each part of the trip are added in **FIGURE 5**, taking their direction of travel into account. Note that here the arrowheads specify the direction of travel. The formula for adding the displacements of the two parts of the trip is:

General Equation 2-2

$$S_{Total} = s_1 + s_2$$

Of course, equation 2-2 could be adapted for a trip with three or more parts, by adding s_3, s_4, etc. The same could be done for equation 2-1 with distances.

FIGURE 5 Displacements include the direction of motion, so adding them can produce either a larger or smaller amount.

While equation 2-2 looks like the formula for distance, each displacement can be positive or negative, and may add to a smaller result.[6]

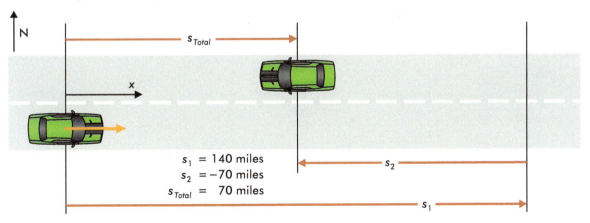

s_1 = 140 miles
s_2 = −70 miles
s_{Total} = 70 miles

Note 6: Summing numbers while including their sign ("+" or "−") is often called finding their "algebraic sum."

The distance traveled during any *constant-direction* part of the trip is equal to the *absolute value* of the displacement, where the straight brackets signify the absolute value of what's inside them:

General Equation 2-3

$$d_1 = |s_1|, \quad d_2 = |s_2|$$

The absolute value of a number is its magnitude, or "size," regardless of whether it is positive or negative. For example, 50 and −50 both have an absolute value of 50. In the case of our previous trip, the displacements convert to distances through equation 2-3 as:

$$d_1 = |s_1| = |140 \text{ miles}| = 140 \text{ miles}$$
$$d_2 = |s_2| = |-70 \text{ miles}| = 70 \text{ miles}$$

2.3 Speed and Velocity

So far we've discussed the *result* of motion (the change in location), but we haven't said anything about how *quickly* the motion happens. We'll start by covering the terms speed and velocity, and then discuss the difference between instantaneous and average velocity.

FIGURE 6 Making the previous trip with a constant 70-mph speed, the entire trip takes 3 hours. What is the average speed? What is the average velocity?

Let's take another look at our previous 210-mile (70-mile?) trip. Assuming you keep a constant speed of 70 mph, it would take 3 hours for the whole trip of 210 miles distance. **FIGURE 6** shows the time t it takes to get to each point. We indicate the speed and direction of the car with an arrow drawn from its center.

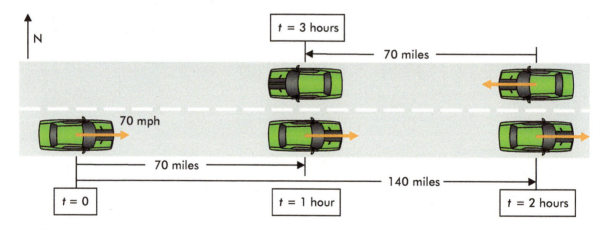

The formal definition of speed is how quickly distance is increasing:

SPEED (represented by S) is the distance a car travels, divided by the time it takes to do it.[7] Traveling 300 miles in 5 hours produces a speed of 60 mi/hr.

> **NOTE**
>
> "Miles per hour" will often be written as mi/hr, rather than mph. This indicates that speed is distance divided by time. The same is true for writing other quantities like "feet per second" as ft/sec rather than fps, "revolutions per minute" as rev/min rather than rpm, etc.

VELOCITY is the rate of change of displacement. The formal definition:

LINEAR VELOCITY (represented by v) is the increase in displacement divided by the time it takes to do it, so it is speed *with direction*. Unless otherwise noted, velocity will mean linear (translational) velocity.

So like displacement, velocity can be positive or negative. If east is positive, as in **FIGURE 6**, then 70 mph west is -70 mph. Our trip speed as plotted in **FIGURE 7** is always 70 mph. But the velocity for the first 2 hours is $+70$ mph, and for the last hour is -70 mph. Here, time is plotted as the x-variable, and speed or velocity as the y-variable.

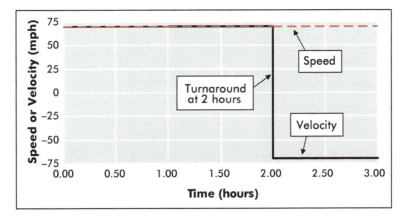

FIGURE 7 The velocity and speed of the car during our trip. Note that when the car changed directions after 2 hours, the speed stayed the same, but the velocity flipped from positive to negative.

> **NOTE**
>
> The term "x-" usually applies to the horizontal, independent variable, and the term "y-" to the vertical, dependent variable.

The y-variable is called "dependent" because it depends on changes in the x-variable. Here, velocity depends on time. It's not always easy to decide which variable is dependent on the other, but here it's clear; the passage of time certainly does not depend on your velocity!

Note 7: Hopefully the use of "S" for speed and "s" for displacement in this chapter doesn't cause too much confusion. Regardless, you won't be seeing "S" used for speed much in the rest of the book.

The effect of changing directions is not only seen in the velocities, but also in the resulting displacements (see FIGURE 8). During the first two hours, both the displacement and distance increase steadily by 70 miles each hour. But at the 2-hour mark, the displacement begins to decrease (the car is getting closer to its starting point), while the distance continues to increase (putting more miles on the car).

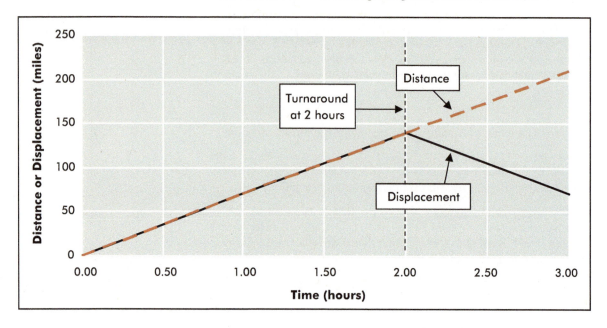

FIGURE 8 Displacement and distance during our trip. Both increase during the first 2 hours, when velocity is positive. But after 2 hours velocity is negative, so displacement decreases.

Now let's put these relationships into formulas, assuming constant speed. Mathematically, speed equals distance divided by time:

Equation 2-4a   **Constant speed**

$$S = \frac{d}{t}$$

You can rearrange this equation to solve for distance:

Equation 2-4b   **Constant speed**

$$d = St$$

or for time:

Equation 2-4c   **Constant speed**

$$t = \frac{d}{S}$$

Equation 2-4b says that distance traveled is *directly proportional* to how much time t has passed since timing started, if you keep the speed constant. If two quantities are directly proportional to each other, when one changes, the other changes by the same proportion. For example, at a constant speed, doubling the time doubles the distance. Note that the three equations 2-4a, 2-4b, and 2-4c say the same thing, and we can call them equations 2-4; they are actually the same equation.

NOTE

During the first part of this book, several versions of an important equation will be shown, as above. Later on, you may need to re-arrange some equations yourself for a specific use, using some math tips from Appendix 3.

The equations relating displacement to velocity mirror the ones for distance and speed. If the displacement $s = 0$ at time $t = 0$, and the velocity v is constant, then the equation for velocity is:

Equation 2-5a   **Constant velocity**

$$v = \frac{s}{t}$$

You can rearrange this equation to solve for displacement:

Equation 2-5b   **Constant velocity**

$$s = vt$$

This says that the displacement is directly proportional to time t when velocity v is constant. Solving for time:

Equation 2-5c   **Constant velocity**

$$t = \frac{s}{v}$$

In our example trip, speed was constant, but velocity was not. This means that the speed equations 2-4 could be used for the whole trip at once, while the velocity equations 2-5 could only be used on each part separately. But they could be used for *average* velocity.

2.3.1 Average vs. Instantaneous Velocity and Speed

In real life, your speed varies during any trip. Your speedometer reads your speed at that instant, which we'll call the instantaneous speed. Take the direction into account and you have instantaneous velocity. But you are often interested in your average (mean) speed or velocity. Calculating averages is typically easier than the instantaneous values, since we only need to know how much the distance or displacement changed, and how much time passed:

AVERAGE SPEED (let's call it S_{Ave}) is the total distance traveled divided by the time it took to do it.

AVERAGE VELOCITY (called v_{Ave}) is how much the displacement has in-creased, divided by the time it took to do it.

As formulas, these definitions become:

General Equation 2-6

$$S_{Ave} = \frac{d_{Total}}{t}$$

for average speed, and:

General Equation 2-7

$$v_{Ave} = \frac{s_{Total}}{t}$$

for average velocity.

> **NOTE**
>
> Although many equations are set up to calculate a "total" or "final" distance or displacement, etc., any point in a car's motion could be picked, regardless of whether the motion has stopped.

Our trip had an average speed of (210 miles)/(3 hours) = 70 miles/hour, but an average velocity of (+70 miles east)/(3 hours) = 23.3 miles/hour. The average velocity was lower than the average speed because much of the trip was spent reducing the displacement by traveling west. If the trip had continued for one more hour at 70 mph west, the car would have returned to its starting point, and the average velocity would have been zero.

The most general equation for average velocity is to divide a change in displacement by the change in time that occurred:

General Equation 2-8

$$v_{Ave} = \frac{\Delta s}{\Delta t} = \frac{s_2 - s_1}{t_2 - t_1}$$

The uppercase Greek letter Δ (spelled "delta" and pronounced *DEL-tuh*) is commonly used to mean "a change in" the following variable. So Δs means the change in displacement.

This equation can be used for small or large differences in time.

2.4 Interpreting the Velocity vs. Time Graph

It will help us for the next section to develop a graphical way of finding displacement from velocity and time. Let's use an example with constant velocity of 70 mph east for 1.5 hours. Using equation 2-5b, the displacement at that time is:

$$s = vt = 70 \ \frac{\text{miles}}{\text{hour}} \times 1.5 \text{ hours} = 105 \text{ miles (east)}$$

Nothing special so far, but let's take a close look at its velocity curve in **Figure 9**. If we think of the area under the velocity curve as a rectangle with "height" of 70 miles/hour and "width" of 1.5 hours, its "area" would be 105 miles, same as the displacement we just calculated.[8] This calculation technique is referred to as finding the "area under the curve." Here, our constant-velocity curve produces a straight and level line.

NOTE

In geometry, a "curve" can be any shape, including a straight line. It's different from the curve on a road, which we think of as an arc of a circle.

KEY CONCEPT

The area under the velocity vs. time curve is equal to the displacement, even when velocity isn't constant.

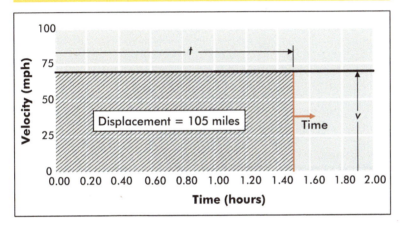

Figure 9 Plotting constant velocity vs. time produces a rectangular area under the curve. The area (cross-hatched) is equal to the displacement. As time passes, the right end of the rectangle (red) steadily moves to the right, increasing the displacement.

But we're not done. If the motion continues at 70 mph, the area keeps growing as the right end of the rectangle moves right with time (again, **Figure 9**). It is like a curtain being drawn to the right across a window, with the area of the curtain (here vt) increasing with time. Now let's step things up a notch with a more dramatic case of average velocity.

Note 8: It may seem strange to define the "length" and "height" of a rectangle in measurements other than a true length, like feet, inches, or meters. It probably should seem strange, in fact. If it seems too odd, think of it as reading the scale on a map, such as "1 inch = 50 miles," except now instead of scaling a large distance to a small distance, you are scaling velocity or time into a small distance on the figure.

2.4.1 An Extreme Example of Velocity

PROBLEM: During its second timing run for its land speed record attempt, the Thrust SSC (see **FIGURE 10**) took only 4.696 seconds (!) to travel a measured mile. What was its average velocity in mph?

SOLUTION: Using equation 2-7, you can find the average velocity as $v_{Ave} = s/t$. To use this directly for an answer in mph, put displacement in miles and time in hours:

$$s = 1.0 \text{ miles}$$

$$t = (4.696 \text{ sec}) \times \left(\frac{1 \text{ hour}}{3600 \text{ sec}} \right) = 0.00130444 \text{ hours}$$

Plugging this into equation 2-7, you divide the displacement by the time:

$$v_{Ave} = \frac{1.0 \text{ miles}}{0.00130444 \text{ hours}}$$

So (Answer):

$$v_{Ave} = 766.61 \text{ miles/hour}$$

FIGURE 10 An aerial view of the Thrust SSC record run. Note the dust to the sides of the car being stirred up by the shock wave. In this picture, s represents the displacement between hypothetical start and finish lines. *Photo by Richard Meredith-Hardy.*

This is the average velocity during the measured mile, but **Figure 11** shows the velocity wasn't constant. You can see that the Thrust SSC was still accelerating.[9]

The speed of sound at the record site was about 751.6 mph (1,102 ft/sec) at the time of the run, so its average velocity broke the sound barrier. In fact, **Figure 11** shows that the entire measured mile was run above the speed of sound. Since the speed of sound in air depends on the air temperature, the value given was calculated for the temperature at 5:05 P.M., October 15, 1997, at Black Rock Desert, Nevada.

NOTE

60 mph is 88 ft/sec, which makes it easy to remember how to convert between them.

As you can probably imagine, the forces and power required to push a 21,000-pound car to these speeds are almost unimaginable. We'll revisit this run several times to calculate those.

Figure 11 Instantaneous velocity (dark blue) and average velocity (black dashed) for the Thrust SSC during its 4.696-second measured mile. The speed of sound (orange dashed) is shown for comparison.

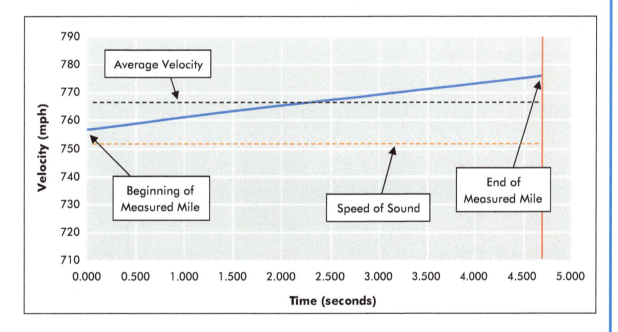

Note 9: If you look it up, you may notice that the top speed in **Figure 11** doesn't quite match the top speed of 771 mph in the record book. Due to the corrosive effects of the alkali desert on their computers' hard drives, much of the recorded data were lost. The data here were pieced together by Ron Ayers (chief aerodynamicist on the Thrust SSC), to match the overall characteristics of the run, not every detail. Many thanks to Ron, and Richard Noble for providing the data.

3 Accelerating Motion vs. Time

Now we'll examine motion where velocity varies predictably. Specifically, we'll include *constant* acceleration, so that velocity increases (or decreases) the same amount each second. Constant acceleration is a pretty good approximation for situations like braking with a constant pedal effort, or when braking or acceleration is limited by the available traction. Traction limitations will be discussed in Chapters 15 (on Power) and 16 (on Statics).

3.1 Acceleration and Velocity

During a Friday night cruise, you start out from a stoplight and want to get to 35 mph (about 51.3 ft/sec) and then stay there (see **FIGURE 12**). You accelerate at the moderate rate of one quarter of a *g*, which means your velocity will increase about 5.5 mph (8.05 ft/sec) each second. A simple explanation of *g*'s is that an object falling in Earth's gravity accelerates downward at 1 *g* (increasing its speed by 32.2 ft/sec each second), if drag forces on it are small enough. So 0.25 *g*'s is speeding up at one quarter of that rate.

How long will it take to get to 35 mph? What does the plot of your velocity look like, and what does it tell you? How far will you go in 8 seconds? We'll answer these questions in the following sections.

> **NOTE**
>
> From here on, most calculations will be done with distances in feet (rather than miles) and time in seconds (rather than hours). This will make it easy to work with acceleration over short time periods.

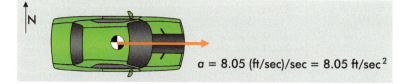

FIGURE 12 The Challenger with acceleration *a* of 8.05 ft/sec per second, that is, 8.05 ft/sec².

$$a = 8.05 \text{ (ft/sec)/sec} = 8.05 \text{ ft/sec}^2$$

Before we get too far, let's formally define straight-line acceleration:

LINEAR ACCELERATION (called *a*) is how quickly linear velocity is increasing. Acceleration is positive in the same direction as positive velocity. Unless otherwise stated, **ACCELERATION** will mean linear acceleration.

Our acceleration of "8.05 feet/second every second" eastward is written as 8.05 feet/second², and read as "8.05 feet per second-squared." East being defined positive here, 8.05 ft/sec² westward would be −8.05 ft/sec². Whether acceleration is positive *or* negative, the velocity is changing. So if acceleration is zero, then velocity is constant.

> **NOTE**
>
> Displacement, velocity, and acceleration are all defined positive in the same direction, usually the direction the car is supposed to travel. On a drag strip running from south to north (start to finish), I would set north as positive.

If velocity starts out as zero, and acceleration (the rate of change in velocity) is constant, then it makes sense that the acceleration is the velocity divided by time:

Equation 2-9a   **Standing start**

$$a = \frac{v}{t}$$

Putting values into equation 2-9a emphasizes that the units for acceleration are length divided by time squared. You can also rearrange this equation to solve for velocity:

Equation 2-9b   **Standing start**

$$v = at$$

Or it could be solved for time:

Equation 2-9c   **Standing start**

$$t = \frac{v}{a}$$

If a vehicle accelerates at a constant rate (such as 8.05 ft/sec²) from a standing start, equation 2-9b says that velocity is directly proportional to how much time t has passed; the longer you accelerate, the faster you go.

We can now answer the first part of our example problem, the time to get to 35 mph (51.3 ft/sec) with 8.05 ft/sec² acceleration. Using equation 2-9c, this is straightforward:

$$t = \frac{v}{a} = \frac{51.333 \text{ ft/sec}}{8.05 \text{ ft/sec}^2} = 6.38 \text{ sec}$$

As most things are in physics, this is very logical. If it takes one second to increase velocity by 8.05 ft/sec, it will take 6.38 times that long to go 6.38 times that fast. You probably could have done that without any formula.

The line graph in **FIGURE 13** makes the steady increase in velocity up to 51.3 ft/sec obvious. Notice how it increases in a straight line (it is said to increase "linearly"). Here "linear" is meant in the mathematical sense, as in the velocity curve plots as a straight *line*. You can also see the rate of increase of velocity per unit time (the slope of the line). The slope of a line is an increase of the *y*-variable divided by an increase in the *x*-variable. In this case it is the increase in speed divided by the increase in time. The slope is 8.05 ft/sec² until velocity reaches 51.3 ft/sec, and then is 0 ft/sec², in both cases equal to the acceleration.

KEY CONCEPT

By its definition, instantaneous acceleration is the slope of the velocity vs. time curve. For that matter, instantaneous velocity is the slope of the displacement vs. time curve.

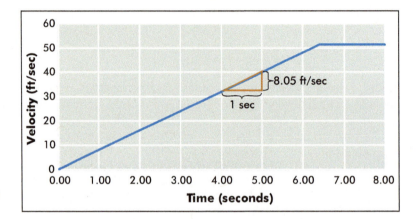

FIGURE 13 A plot of the velocity for the Challenger, which accelerates at 8.05 ft/sec² until it gets to 51.3 ft/sec (35 mph), then holds a constant velocity.

What if the velocity is not zero at $t = 0$, but some other value v_0? With constant acceleration, the increase in velocity $(v - v_0)$ is directly proportional to how much time has passed, so $(v - v_0) = at$. To find the velocity any time after the start, isolate v on the left side of the equation:

General Equation 2-10   **Constant acceleration**

$$v = v_0 + at$$

If you start off at 10 ft/sec and accelerate at 20 ft/sec² for 2 seconds, you'll end up at 50 ft/sec.

What if the acceleration isn't necessarily constant? You can write a more general equation to find the average, or mean, acceleration:

General Equation 2-11

$$a_{Ave} = \frac{\Delta v}{\Delta t} = \frac{v - v_0}{t - t_0}$$

Equations 2-9 and 2-10 for constant acceleration can all be gotten from equation 2-11, by setting some variables equal to zero, and setting a_{Ave} equal to the constant acceleration a.

> **REMINDER**
>
> Naming of the beginning and ending velocities is up to you. In equation 2-11, the initial velocity is v_0, but could have been called v_1. Likewise, the final velocity was simply called v, but could have been called v_F (for "final") or v_2.

You will see equations rewritten like this for new situations, so be aware. This renaming doesn't change the meaning of the equation.

3.1.1 Acceleration in g's

When dealing with measurements, you want to have a feel for their size. That's easy for displacement and velocity, because you can visualize them. Acceleration may be tougher to relate to, but fortunately we live with a convenient acceleration to compare to.[10] A body falling in Earth's gravity, neglecting air resistance, will increase its downward velocity by 32.2 feet per second (9.81 meters per second) each second. An acceleration of 32.2 ft/sec², or 9.81 m/sec², is called 1 g. To put acceleration rates into g's, we simply divide the actual acceleration by the acceleration of gravity, in either British (here SFS) units:

General Equation 2-12a

$$a_g \text{ (in } g\text{'s)} = a \text{ (in ft/sec}^2) \times \frac{1 \ g}{32.2 \text{ ft/sec}^2}$$

or in metric (here MKS) units:

General Equation 2-12b

$$a_g \text{ (in } g\text{'s)} = a \text{ (in m/sec}^2) \times \frac{1 \ g}{9.81 \text{ m/sec}^2}$$

So from equation 2-12a, an acceleration of 40 ft/sec² is 1.24 g's, and from equation 2-12b, an acceleration of 12.19 m/sec² is also 1.24 g's.

> **NOTE**
>
> To avoid confusion, the "g" for acceleration will be italicized, while the "g" for gram is not. But you will also be able to tell by how it's used.

Note 10: Acceleration is hard to see, but much easier to feel. We'll see why in Chapters 4 and 5.

3.2 Displacement During Constant Acceleration

Before describing the equations for displacement from constant acceleration, let's visualize the motion. **FIGURE 14** shows "snapshots" of the Challenger while accelerating at 8.05 ft/sec², at 1-second intervals. Note how the displacement traveled after 1 second is quite small (about 4 feet). But after 2 seconds, it has gone about 16 feet total. More ground is covered in each second, because the velocity is increasing. The smooth curve joining the CG positions at each second shows the displacement between snapshots.

FIGURE 14 The displacement from a stoplight after each of the first 5 seconds of travel, accelerating at a constant 0.25 g's.

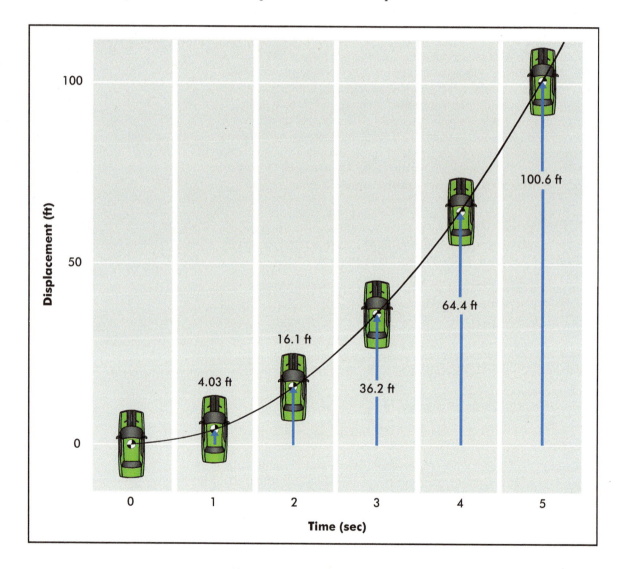

The velocity curve is plotted in **Figure 15**. Let's use it to calculate the displacement after 5 seconds.

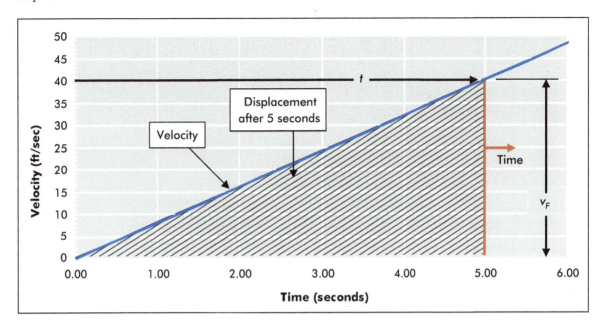

As before, the displacement accumulated up to a certain time is found by finding the area under the velocity vs. time curve. The triangle under the velocity curve is 5 seconds long and 40.25 ft/sec high (the velocity after 5 seconds). Since the area of a triangle is one half its length times its height, the displacement after 5 seconds is:

$$s = \frac{1}{2} t \times v_F = \frac{1}{2} \times 5 \text{ sec} \times 40.25 \ \frac{\text{ft}}{\text{sec}} = 100.6 \text{ ft}$$

Figure 15 The first 5 seconds of the Challenger's acceleration. Here, the velocity (in blue) starts off at zero, increasing by 8.05 ft/sec per second (or 8.05 ft/sec²). How far does it go in that time?

as was shown in **Figure 14.** So finding displacement from constant acceleration is about as easy as for constant velocity.

> **CAUTION**
>
> With a negative velocity, the "area under the curve" is actually *above* the velocity curve, producing a *negative* area, and thus a negative displacement.[11]

The other four displacements in **Figure 14** could also be found this way, but instead let's put it into a formula, so we don't need to keep drawing diagrams.

Note 11: Negative areas aren't physically meaningful, but work fine mathematically. Remember that math is just a tool, and we get to define the rules, as long as they work.

3.2.1 Constant Acceleration from a Standing Start

You may have noticed that the smooth curve in Figure 14 showed displacement increasing with the square of time traveled. In the previous calculation, we multiplied the final velocity v_F by half the final time t. Since the final velocity is equal to the acceleration multiplied by time (at), we can rewrite the area $\frac{1}{2}v_Ft$ as $\frac{1}{2}(at) \times t$, which is $\frac{1}{2}at^2$. So from a standing start, the displacement is:

Equation 2-13   **Standing start**

$$s = \frac{1}{2}at^2$$

Now we plot in Figure 16 the displacement for our Challenger for the first 8 seconds. Notice how it curves upward more steeply as the velocity increases with time. But the slope gets no steeper after the 6.38-second mark (signified by a blue dot), where acceleration stops. From there on, the displacement curve is a straight line with a slope of 51.3 ft/sec, the (constant) velocity.

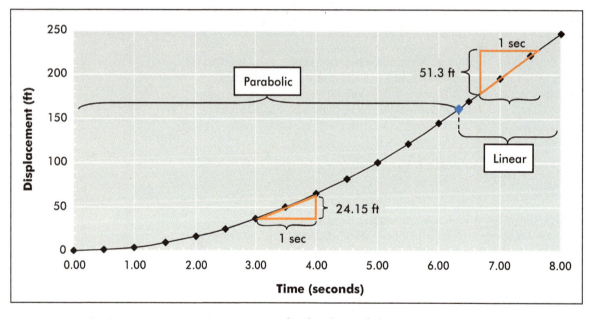

Figure 16 The displacement traveled from the stoplight, which increases very slowly at first, but then quickens. The curve becomes a straight line once you hit a constant 51.3 ft/sec (at 6.38 seconds, marked by the blue dot).

But you can take the slope of the curve at any point in time to find the instantaneous velocity. At the 3-second mark, a right triangle is drawn with its upper edge parallel to the curve. Doing this precisely, and drawing the horizontal edge 1 second "long," the length of the vertical edge comes out to be 24.15 feet. Dividing 24.15 feet by 1 second gives a slope of 24.15 ft/sec, the velocity at 3 seconds. You could have easily calculated that using equation 2-9b.

3.2.2 An Example of Extreme Acceleration

PROBLEM: In the finals at Phoenix in 2010 (which he won), Cory Mc-Clenathan's top-fueler accelerated so that its velocity increased vs. time according to the graph in **FIGURE 17**. During the first 3 seconds, what was its average acceleration in g's?

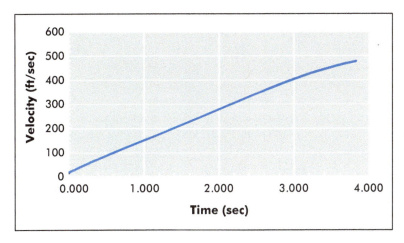

FIGURE 17 Velocity of McClenathan's top-fueler as it accelerates through a 1,000-ft run that took 3.813 seconds. The velocity increases steadily for the first three seconds, in nearly a straight line (i.e., constant acceleration).

SOLUTION: The initial velocity v_0 when leaving the starting line is 10 mph. (We'll discuss why it's not zero in Chapter 3.) After 3 seconds, it is 275 mph. This gives us an initial velocity of 14.667 ft/sec and a final velocity of 403.33 ft/sec. Using equation 2-11, the average acceleration is:

$$a_{Ave} = \frac{v_F - v_0}{t} = \frac{388.66 \text{ ft/sec}}{3 \text{ sec}} = 129.55 \frac{\text{ft}}{\text{sec}^2}$$

Using equation 2-12a, the acceleration in g's is:

$$a_g = 129.55 \frac{\text{ft}}{\text{sec}^2} \times \frac{1 \, g}{32.2 \text{ ft/sec}^2} = 4.02 \, g\text{'s}$$

So during the first 3 seconds, it accelerates about four times faster than a rock falling off a cliff! We'll calculate the traction force required to do this in Chapter 6.

3.2.3 Constant Acceleration from a Rolling Start

Equation 2-13 assumes that a vehicle is stationary at $t = 0$. But what's the displacement if you're already moving when you begin accelerating, like when passing another car? With an initial velocity v_0 at $t = 0$, the displacement grows from both the initial velocity, as in equation 2-5b, and the increasing velocity caused by the acceleration. Also, you may not choose to call the displacement s_0 at $t=0$ equal to zero, but some other value.[12] Including all of these factors produces:

Equation 2-14   **Rolling start**

$$s = s_0 + v_0 t + \frac{1}{2} at^2$$

You can think of it as s_0 added to equations 2-5b and 2-13. This is shown graphically in **FIGURE 18**, which combines the rectangular area of **FIGURE 9** with the triangular area of **FIGURE 15**. The rectangular area, $s_1 = v_0 t$, would be the displacement if velocity were held constant at v_0. The triangular area, $s_2 = \frac{1}{2} at^2$, is the additional displacement from increased velocity.

The total displacement s equals $s_1 + s_2$. In **FIGURE 18**, v_0 is 10 ft/sec, and $a = 5$ ft/sec². After 2 seconds, $s_1 = v_0 t = 10$ ft/sec × 2 seconds = 20 feet, and $s_2 = \frac{1}{2} at^2 = \frac{1}{2} \times 5$ ft/sec² × (2 sec)² = 10 feet. Adding these gives a total displacement $s = s_1 + s_2 = 30$ feet, after 2 seconds.

FIGURE 18 An illustration of equation 2-14. The area in blue cross-hatch is the displacement s_1 (20 feet) from initial velocity v_0. In the red cross-hatch is the additional displacement s_2 (10 feet) from the acceleration, for a total of 30 feet, after 2 seconds.

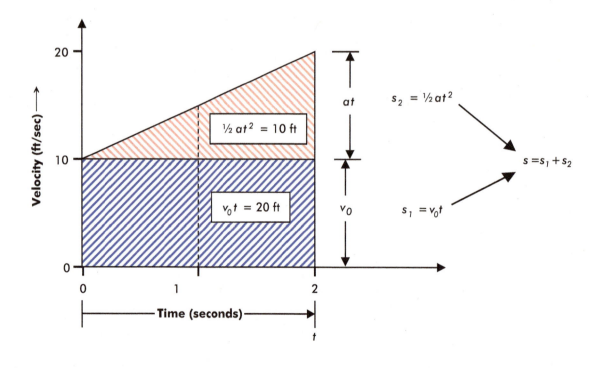

Note 12: After all, if you're behind another car, you can't call both your displacements $s = 0$!

Finally, you might want to calculate the displacement starting from a time t_0 different from zero. For example, you might split a trip into several consecutive parts, and only one could start at $t = 0$. Then you can let the starting time equal t_0 instead:

General Equation 2-15   **General Case**

$$s = s_0 + v_0(t - t_0) + \frac{1}{2}a(t - t_0)^2$$

This is the most general formula for constant acceleration vs. time, and is really the only one you need; you can always zero out any unnecessary parts.

4 Accelerating Motion vs. Displacement

The discussion to this point involved how much time a vehicle accelerates. But you may want to know how far a vehicle must accelerate to get to a certain velocity, as in braking distance from 70 mph to zero. Using the previous equations and doing a little bit of algebra, you can write an equation for the required increase in displacement to reach a desired velocity v (the derivation of this equation is in Appendix 5):

General Equation 2-16   **General Case**

$$s = s_0 + \frac{v^2 - v_0^2}{2a}$$

Here, s_0 and s are the initial and current displacements, and v_0 and v are the initial and current velocities. Let's take a look at some special cases. For acceleration from a standing start, the initial velocity v_0 and displacement s_0 are zero, and equation 2-16 becomes:

Equation 2-17a   **Standing start**

$$s = \frac{v^2}{2a}$$

You can see that to get to twice the velocity v with a given rate of acceleration, you must go four times as far. You can rework this to find the acceleration needed to reach a certain velocity in a specified displacement s:

Equation 2-17b   **Standing start**

$$a = \frac{v^2}{2s}$$

You can just as easily calculate stopping distance from a given velocity v_0, where constant acceleration is usually valid. Since the velocity decreases, the acceleration will be negative. Now the ending velocity v in equation 2-16 is zero. Then you get:

Equation 2-18a $$s = \frac{-v_0{}^2}{2a}$$ **Braking to zero**

Here, s is the stopping "distance" (really displacement), given the braking acceleration a. But you might find the stopping distance from 60 or 70 mph in a magazine, and want the braking acceleration. Reworking equation 2-18a:

Equation 2-18b $$a = \frac{-v_0{}^2}{2s}$$ **Braking to zero**

Many other situations can be covered by equation 2-16, like passing distances, or braking from one speed to another, as long as acceleration is constant.

4.1 Velocity Changes vs. Displacement

Let's revisit the Challenger acceleration problem from before, where we started off at rest and accelerated at ¼ g, and see how the velocity changes vs. displacement. We'll use equation 2-17a and solve for the velocity:

$$v = \sqrt{2as}$$

For our problem, this becomes:

$$v = \sqrt{2 \times 8.05 \frac{\text{ft}}{\text{sec}^2} \times s} = 4.01\sqrt{s}\ \frac{\text{ft}}{\text{sec}}$$

FIGURE 19 uses this to plot velocity vs. displacement during the constant acceleration part, which lasts from $s = 0$ to about 163.5 ft. Note how, with constant acceleration, you need more and more room to increase the velocity. Getting up to 70 mph instead of 35 mph would take four times the distance (654 ft instead of 163.5 ft).

CAUTION

Note that you need to be careful here taking the square root in solving for v, because there are two possible answers, the positive or negative square root. Usually only one makes physical sense.

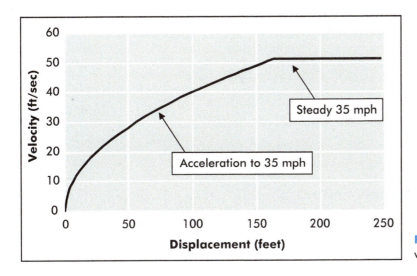

FIGURE 19 The Challenger's velocity vs. its displacement.

5 Dimensions and Tracking of Units

It's time to start paying closer attention to dimensions and units, because standardized units are critical to science, engineering, and manufacturing. Imagine if every person used a different standard for measuring weight (I guess then there wouldn't be a standard!). It would cause total confusion. But if someone tells you that a rear axle weighs 140 pounds, you know what it means, and you can measure it to check. Even if you're more familiar with metric units, there is a standard way to convert pounds into kilograms (or grams).

We'll be working with two standard systems of units in this book. Americans will be most familiar with the SFS system, which is based on British units. The second, the MKS system, is based on the metric system, which the rest of the world uses (including, ironically, the British). These systems of units will be detailed in Chapter 4.[13]

By dimension we mean the type of physical quantity. For instance, distance and linear displacement are each a type of length, so "length" is their dimension. For each type of dimension, there are standard measurement units (or simply units) to measure them in. For example, the dimension "length" could be measured in units of feet, meters, kilometers, miles, etc., each of which have a standard size.

The **DIMENSION** for a physical quantity is its general type of measurement, regardless of the units it is given in. **UNITS** are standard-sized amounts of the dimensions of a physical quantity.

A good knowledge of measurement units is essential to understanding and using the rest of the book, as well as in checking for errors.

Note 13: In the U.S., wrench sets are often called either "standard" or "metric," standard being in inches and metric being in millimeters. There's more irony here, since metric *is* the standard internationally.

5.1 Derived Dimensions and Units

When we've multiplied two quantities together, (like velocity and time to get distance), we haven't just multiplied the numbers. We've also multiplied the units (like ft/sec by seconds to get feet). The same was true for division. Combining the **BASIC UNITS** of feet and seconds produced the **DERIVED UNIT** of ft/sec.

The basic (or fundamental) units are mass, length, and time. From these, we'll be able to handle all the types of quantities needed in this book. Since mass isn't introduced until Chapter 4, we'll go over the basic units then. But with only length and time units, we've already described displacement (and distance), velocity (and speed), and acceleration:

Displacement has the dimension length, and is measured in units like feet, meters, kilometers, miles, etc.

Velocity has the dimensions length/time, and is measured in units like feet/sec, meters/sec, kilometers/hour, miles/hour, etc.

Acceleration has the dimensions length/time^2, and is commonly measured in units like feet/sec^2 or meters/sec^2.

Note that velocity and acceleration are derived dimensions, both combining length and time. Using the basic dimension of length, we can also derive the dimensions for area and volume.

AREA has the dimensions length^2, and is measured in units like ft^2, inch^2, meter^2, centimeter^2, etc.

VOLUME has the dimensions length^3, and is measured in units like ft^3, inch^3, meter^3, centimeter^3, etc.

Examples of length, area, and volume are shown in **FIGURE 20**. The first (upper left) is a straight 24-inch length of wire. The second (lower left) is a piece of sheet metal 24 inches by 20 inches, giving it an area A of 480 in^2 (square inches). The third is a sheet metal box 24 inches by 20 inches by 8 inches, giving it a volume V of 3,840 in^3 (cubic inches).

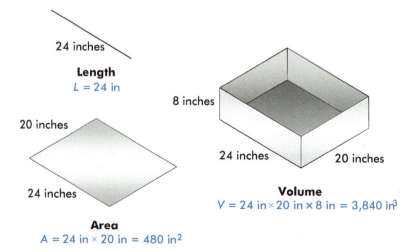

FIGURE 20 An example of dimensions derived from length. Length, area, and volume have dimensions length, length^2, and length^3, respectively. Their units here are in, in^2, and in^3.

24 inches

Length
$L = 24$ in

8 inches

20 inches

24 inches

24 inches

20 inches

Volume
$V = 24$ in $\times 20$ in $\times 8$ in $= 3{,}840$ in^3

24 inches

Area
$A = 24$ in $\times 20$ in $= 480$ in^2

Let's illustrate how to track the units and dimensions by calculating how many gallons a fuel tank that is 24 inches long and 20 inches wide can hold, if the fluid is level and 8 inches deep (the volume shown in **FIGURE 20**). Note that you wouldn't really fill it quite full, to leave some air space at the top, for expansion, etc.

To put it into more reasonable numbers, let's calculate the volume in cubic feet instead of cubic inches. The volume V of a rectangular box is equal to its length L times its width w times its height h:

$$V = Lwh = \left(24 \text{ in} \times \frac{1 \text{ ft}}{12 \text{ in}}\right) \times \left(20 \text{ in} \times \frac{1 \text{ ft}}{12 \text{ in}}\right) \times \left(8 \text{ in} \times \frac{1 \text{ ft}}{12 \text{ in}}\right)$$

$$= 2 \text{ ft} \times 1.66667 \text{ ft} \times 0.66667 \text{ ft} = 2.2222 \text{ ft}^3$$

Notice how the length, width, and height are converted from inches to feet in the top line, by multiplying each by 1 ft/12 inches. In the second line, the numbers are multiplied out, as well as the units. Now we have the volume in cubic feet, but we want it in gallons. There are 7.4805 gallons in 1 cubic foot (from Appendix 1), so:

$$V = \left(2.2222 \text{ ft}^3\right) \times \left(\frac{7.4805 \text{ gallons}}{1 \text{ ft}^3}\right) = 16.62 \text{ gallons}$$

To convert, we took the volume of 2.2222 cubic feet, and then multiplied and divided it by two identical things (7.4805 gallons is the same as one cubic foot). Notice that in multiplying the units, the "cubic feet" in the left term cancel with the "cubic feet" in the denominator of the right term. So all the units cancel out, except for "gallons."

KEY CONCEPT

To keep conversions straight, view it as multiplying and dividing by equal quantities. This is like restating ½ as ⁵⁄₁₀, done by multiplying ½ by ⁵⁄₅.

When you don't do it this way, it can be hard to remember whether to multiply or divide by a conversion factor.[14] The time you spend using this check will save you time in mistakes and rework, much like the saying, "measure twice, cut once."

Note 14: Of course, when I have hurried and skipped this step is when I have made a stupid mistake.

6 Summary

Now we know how to talk about straight-line motion, and are set to apply it in the next chapter. To wrap this up, let's review the major points on kinematics:

- Displacement is distance with direction.
- Velocity is speed with direction.
- Acceleration is the rate of change of velocity.
- To change velocity, a body must accelerate.
- Displacement, velocity, and acceleration can be positive or negative.
- Velocity is the slope of the displacement vs. time graph.
- Acceleration is the slope of the velocity vs. time graph.
- Displacement is the area under the velocity vs. time curve.
- Velocity is the area under the acceleration vs. time curve.

Let's summarize this with one last look at the Challenger problem, where we accelerated from rest to 51.3 ft/sec at 8.05 ft/sec², and then held at 51.3 ft/sec. **FIGURE 21** shows velocity vs. time.

Let's concentrate on what has happened up to the 4-second mark. First off, the velocity at that point is 32.2 ft/sec, which is read directly off the graph. To find the acceleration at that point, we take the *slope* of the velocity, which is 8.05 ft/sec². To find the displacement, we take the *area* under the velocity curve up to 4 seconds, getting 64.4 feet.

FIGURE 21 The acceleration, velocity, and displacement at *t* = 4 seconds are highlighted on the velocity vs. time graph.

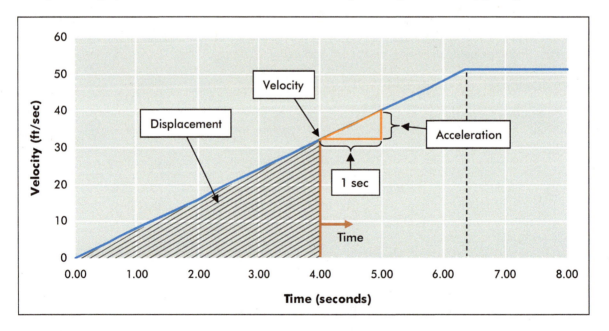

This one plot summarizes how acceleration, velocity, and displacement are related. These are important concepts to store away.

Major Formulas

Definition	Equation	Equation Number
Total Distance = First Distance + Second Distance	$$d_{Total} = d_1 + d_2$$	2-1
Total Displacement = First Displacement + Second Displacement	$$s_{Total} = s_1 + s_2$$	2-2
Distance is the absolute value of the displacement (only when the velocity is all in the same direction)	$$d = |s|$$	2-3
Velocity = Displacement ÷ Time (Assumes Constant Velocity)	$$v = \frac{s}{t}$$	2-5
Average Velocity = Change in Displacement ÷ Change in Time	$$v_{Ave} = \frac{\Delta s}{\Delta t} = \frac{s_2 - s_1}{t_2 - t_1}$$	2-8
Constant Acceleration: Velocity = Initial Velocity + (Acceleration × Time)	$$v = v_0 + at$$	**2-10**
Average Acceleration = Change in Velocity ÷ Change in Time	$$a_{Ave} = \frac{\Delta v}{\Delta t} = \frac{v - v_0}{t - t_0}$$	2-11
Acceleration in g's = Acceleration × 1 g per 32.2 ft/sec²	$$a_g = a \ (\text{in ft/sec}^2) \times \frac{1 \ g}{32.2 \ \text{ft/sec}^2}$$	2-12a
Displacement = Initial Displacement + Initial Velocity × Time Change + ½ × Acceleration × Time Change²	$$s = s_0 + v_0(t - t_0) + \frac{1}{2}a(t - t_0)^2$$	**2-15**
Change in Displacement = Initial Displacement + ((Final Velocity)² − (Initial Velocity)²) ÷ (2 × Acceleration)	$$s = s_0 + \frac{v^2 - v_0^2}{2a}$$	**2-16**

Equation numbers shown in bold above have an alternative calculus-based derivation demonstrated in Appendix 5.

3

Kinematics Applications

Straight-Line Motion in Vehicles and Mechanisms, and Racing Strategy

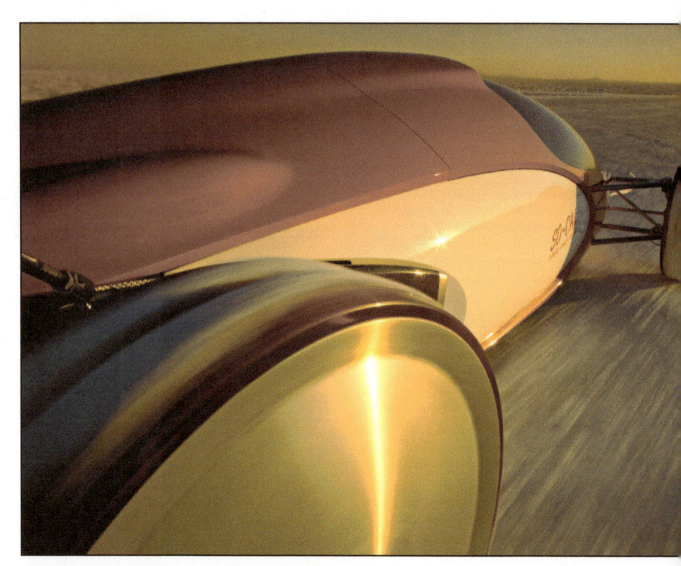

The 2003 GM/So-Cal Ecotec Lakester concept vehicle, shown on the Bonneville Salt Flats. Don Sherman drove the 2005 version to a G/BGL (G Class/Blown Gas Lakester) record speed of 189.205 mph.
Photo by permission of General Motors Co.

Engine speeds and mean piston speeds are compared between a NASCAR engine and an F1 engine. The distance required to pass a car is calculated. Kinematics is applied to find a drag racer's velocity when it leaves the starting line, and to calculate launch acceleration from its 60-foot time. The effects of braking acceleration are examined from 60 mph to zero, and in a road race.

Contents

Key Symbols Introduced

Symbol	Quantity	SFS Units	MKS Units
d_{Stroke}	Crankshaft Stroke	feet (ft)	meters (m)
$s_{Rollout}$	Rollout Distance	feet (ft)	meters (m)
a_{Launch}	Launch Acceleration	ft/sec^2	m/sec^2
v_{Launch}	Launch Velocity	ft/sec	m/sec
t_{60}	60-Foot Time	seconds (sec)	seconds (sec)
t/s	Slowness, or Time Consumption	sec/ft	sec/m

1   Introduction

Having only been through the basics of kinematics, what insight can you possibly gain into designing, driving, or racing a car? Maybe more than you think. Now we can explain the difference in redlines between a NASCAR engine and an F1 engine, how better brakes "speed you up" in a road race, and how a dragster can have a quicker ET (elapsed time) than its competitor, but have a lower top speed.

In this book, the main purpose of each "Applications" chapter is to show what you can do with the concepts and equations from the matching "Basics" chapter. All the problems in this chapter were selected to make specific points about using kinematics, only using concepts and formulas from Chapter 2, plus a bit of algebra. Some of the conclusions are almost obvious, while some of them are more subtle.

This chapter will also add more emphasis on how velocity changes vs. displacement, whereas Chapter 2 dealt mainly with velocity vs. time. This is especially important in racing, where reaching a certain displacement in the least possible time (such as ¼ mile in drag races), is what matters, not reaching a certain velocity in the least time.

> **REMINDER**
>
> Remember that some of the math in the "Applications" chapters is more involved than I would expect many readers to perform at this point, so please don't be put off by it. The main idea of the problems is to show how to use the concepts, and how to make the data easier to understand and interpret. The discussions should still be clear, and you can always brush up on the math. With some thought, you can learn a lot from a little bit of data.

2 Time-Averaged Speed vs. Distance-Averaged Speed

What is your average speed if you travel 80 mph for 60 minutes, and then 40 mph for 60 minutes? How about if you drive 80 mph for 60 miles, then 40 mph for 60 miles? Why are they different?

2.1 Traveling 80 MPH for 1 Hour, Then 40 MPH for 1 Hour

Using equation 2-1 and adapting 2-4b, you can add the distances from each part of the trip, d_1 and d_2, to find the total distance traveled:

$$d_{Total} = d_1 + d_2 = S_1 t_1 + S_2 t_2 = 80\frac{mi}{hr} \times 1\,hr + 40\frac{mi}{hr} \times 1\,hr$$

$$= 120\,mi$$

Now we can calculate the average speed from equation 2-6:

$$S_{Ave} = \frac{d_{Total}}{t} = \frac{120\,mi}{2\,hr} = 60\,mi/hr$$

2.2 Traveling 80 MPH for 60 Miles, Then 40 MPH for 60 Miles

We begin with equation 2-1 immediately, since we know the distance in each part of the trip:

$$d_{Total} = 60\,mi + 60\,mi = 120\,mi$$

Now we need to find the time it takes to cover each leg of the trip. Adapting equation 2-4c:

$$t_1 = \frac{d_1}{S_1} = \frac{60\,mi}{80\,mi/hr} = 0.75\,hr$$

$$t_2 = \frac{d_2}{S_2} = \frac{60\,mi}{40\,mi/hr} = 1.5\,hr$$

So the total time t is 2.25 hours (2 hours and 15 minutes). Again we calculate the average speed from equation 2-6:

$$S_{Ave} = \frac{d_{Total}}{t} = \frac{120 \text{ mi}}{2.25 \text{ hr}} = 53.3 \text{ mi/hr}$$

The average is slower, though not as slow as it probably seemed during those last 60 miles.

2.3 Why the Answers Are Different

The two answers are different because speed is correctly averaged over time, not distance. In the first trip, you spend *equal times* at 80 mph and 40 mph, and the average speed comes out to be the simple average of the two speeds (60 mph). In the second trip, you travel *equal distances* at 80 mph and 40 mph, meaning you spend twice as much time at 40 mph as you do at 80 mph. The result is that the slower speed counts twice as heavily in the average.

This is shown graphically in **Figure 1**. In either case, the distance (the area of the dashed box) is 120 miles, and is the average speed multiplied by the total trip time. But in the first case, the distance is covered in two hours, making the box two hours "long" and 60 mph "tall." The average speed is midway between the high and low speeds. In the second case, the total time is longer (2.25 hours), so the average speed must be lower (53.3 mph) to still have the same 120-mile rectangular area. Note that the average speed is twice as far away from the higher speed as from the lower speed, because twice as much time was spent at the lower speed.

Figure 1 The instantaneous speeds (blue lines) and average speeds (dashed black) for the cases of "equal time" and "equal distance" at 80 mph and 40 mph. Note that twice as much time is spent at the lower speed (40 mph) in the second case, making the lower speed count twice as heavily in the average speed.

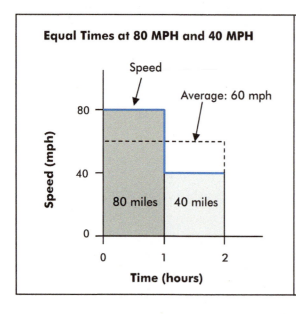

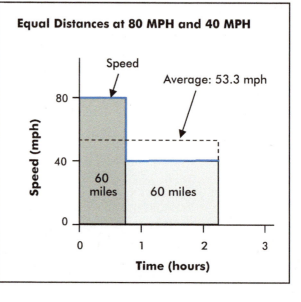

You may have noticed this effect when watching an oval track race. As long as there are no caution flags, the average speed stays very high, but cautions seem to make the average speed drop more than they should. Suppose a 5-lap caution is called 15 laps in, and the lap speed is 60 mph for cautions and 180 mph for race laps (one lap is 2 miles). **FIGURE 2** shows that even though only one quarter of the first 20 laps are run under caution, the caution laps take half the total time!

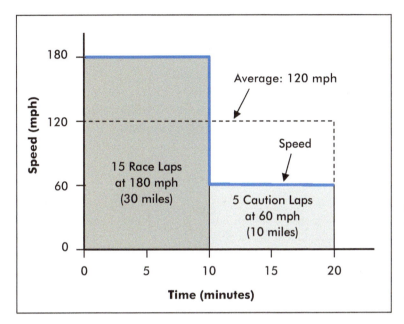

FIGURE 2 Instantaneous (blue line) and average (dashed) speeds for the first 20 laps of an oval track race with a caution called 15 laps in, with 2-mile laps. The 5 caution laps take the same amount of time as the 15 race laps, killing the average speed.

The result is that running ¼ of the laps under caution doesn't drop the average speed just ¼ of the way from 180 mph to 60 mph; it drops it *halfway* there, to 120 mph! (Meanwhile, you've spent half your time watching caution laps.)

3 | Different Strokes

Why can Formula 1 engines turn 18,000 rpm but NASCAR engines are kept to about 9,500 rpm? A common measure used to compare engines is called "mean piston speed" at redline (the maximum engine speed). "Mean" is another word for average, so you don't need to know the speed at all times to calculate it. Since the piston moves downward the distance of the stroke and then back up each crankshaft revolution, the distance traveled per crank revolution is twice the stroke. Knowing this, we can find the mean piston speed by multiplying the distance per revolution by the crank's rotational speed:

Equation 3-1

$$S_{Mean} = 2d_{Stroke}\omega_{Crank}$$

where S_{Mean} is the mean piston speed, d_{Stroke} is the stroke, and ω_{Crank} is the crank rotation speed in revolutions/minute (or rpm). Rotation speed is symbolized by the Greek letter ω. Rotational motion will be more fully covered beginning in Chapter 8, where we'll measure it in different units.

Take a 2010 NASCAR 5.9-liter V-8 engine with a bore of 4.185 inches and a stroke of 3.25 inches (0.27083 feet), as shown in **Figure 3**. At its redline of 9,500 rpm, the mean piston speed is:

$$S_{Mean} = \left(2 \times 0.27083 \frac{\text{ft}}{\text{rev}}\right) \times 9,500 \frac{\text{rev}}{\text{min}} = 5,150 \text{ ft/min}$$

This piston has an average speed of almost 60 mph in the cylinder, despite having to stop and change directions twice each crank revolution (19,000 times each minute)!

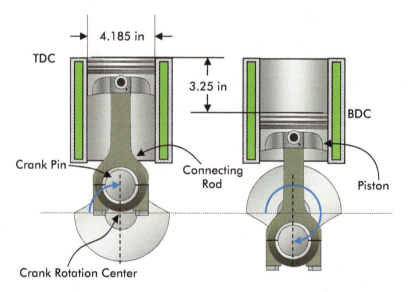

Figure 3 The piston in a typical NASCAR engine moves downward by its stroke of 3.25 inches between TDC (top dead center) and BDC (bottom dead center) during 180° of crank rotation.

By contrast, a 2010 2.4-liter V8 Formula 1 engine has a bore of 98 millimeters and a stroke of only 39.772 millimeters (that's 1.56583 inches, or 0.130486 feet), as shown in **Figure 4**. Its redline is (by 2010 rules) 18,000 rpm, so its mean piston speed is:

$$S_{mean} = \left(2 \times 0.130486 \frac{\text{ft}}{\text{rev}}\right) \times 18,000 \frac{\text{rev}}{\text{min}} = 4,700 \text{ ft/min}$$

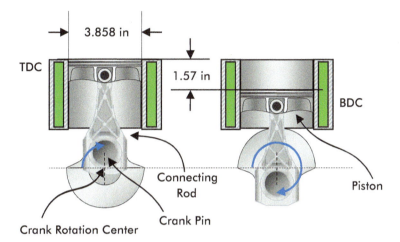

FIGURE 4 The typical 2010 F1 engine has a stroke of only 1.566 inches (39.8 millimeters).

So even though the F1 engine turns nearly twice as fast as the NASCAR engine, its mean piston speed is actually a bit lower, due to its very short 1.566-inch stroke. Of course this is no coincidence. Engines made for similar use often have similar mean piston speeds, since air can only flow into the cylinder so fast.[1] Turning the engine much faster makes it difficult to fill the cylinders and produce torque.

Speaking of filling the cylinders, notice that the 2.4-liter F1 engine's bore (98 millimeters or 3.86 inches) is nearly as large as the 5.9-liter NASCAR engine's (4.185 inches), even though the engine is only 40% as large in swept volume. This allows the F1 engine nearly as much space in the cylinder head for valves and ports, for good airflow.

> **NOTE**
>
> Swept volume of a cylinder is the amount the cylinder volume increases from the top of piston travel to the bottom. The formula is $V_{Swept} = A_{Piston} \times d_{Stroke}$. Most people call this the cylinder displacement.

Note 1: The main limiting factor is what fraction of the speed of sound the incoming air has to travel at as it moves through the intake ports and follows the piston down the cylinder.

4   Velocity vs. Speed in Reciprocating Motion

Repetitive back-and-forth motion, such as the piston motion in the previous example, is called **RECIPROCATING MOTION**. Imagine an engine spinning a whole number (an integer number) of revolutions, so that the piston begins and ends its motion at top dead center. Its total linear displacement is zero, since the motion ends where it starts. This means the average piston velocity for any engine is zero, so "mean piston velocity" would tell you nothing useful in comparing engines to each other.

So while speed may not be as technical sounding as velocity, it is much more useful in this case. Since velocity is a "speed with direction," it is most useful when you care about the details of motion from one point to another. We use speed when it makes sense, and velocity when it makes sense.

5   Haulin' It Down

Road & Track tested a Ferrari Enzo, and measured its stopping distance from 60 mph as 109 feet (see **FIGURE 5**). What was its braking acceleration in *g*'s? How much time does it take to stop? What does its velocity look like vs. displacement?

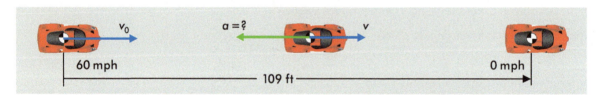

60 mph $a = ?$ v 0 mph

|← 109 ft →|

FIGURE 5 The Ferrari Enzo brakes from 60 mph to zero in 109 feet. What is its average braking acceleration?

To make this easy, let's convert the initial velocity from mph to ft/sec:

$$v_0 = 60\,\frac{\text{mi}}{\text{hr}} \times \frac{5{,}280\ \text{ft}}{1\ \text{mi}} \times \frac{1\ \text{hr}}{3{,}600\ \text{sec}} = 88\,\frac{\text{ft}}{\text{sec}}$$

NOTE

Note how the conversion from mph to ft/sec was done using two exact conversions (5,280 ft/mile and 3,600 sec/hr), rather than the approximate conversion 0.68182 ft/sec per mph. There were two advantages to it, the first being that you don't need to store the approximate conversion in your head. The other is that there is no rounding in the conversion factor. Remember that you can use the fact that 60 mph is 88 ft/sec to convert any velocity from mph to ft/sec by multiplying it by 88/60.

Now let's lay out the problem by putting the beginning and ending conditions in **TABLE 1**. The beginning and ending displacements are known, as are the beginning and ending velocities. The braking acceleration so far is unknown; we'll assume it's constant during the stop, at $a_{Braking}$.

	Start of Braking	**Vehicle Stopped**
Displacement	$s_0 = 0$	$s_F = 109$ ft
Velocity	$v_0 = 88$ ft/sec	$v_F = 0$
Acceleration	$a_{Braking}$	$a_{Braking}$

TABLE 1 The beginning and final conditions for the Ferrari braking.

Now let's work through it.

5.1 Braking Acceleration in g's

We can use equation 2-18b directly to calculate the average braking acceleration:

$$a_{Braking} = \frac{-v_0^2}{2s_F} = \frac{-(88 \text{ ft/sec})^2}{2 \times 109 \text{ ft}} = -35.523 \frac{\text{ft}}{\text{sec}^2}$$

REMINDER

As shown in Chapter 2, when we multiply or divide quantities we do the same with the units. Here, we divided ft^2/sec^2 by feet to get ft/sec^2.

Using equation 2-12a to convert this to g's:

$$a_{Braking}\left(g\text{'s}\right) = -35.523\,\frac{\text{ft}}{\text{sec}^2} \times \frac{1\,g}{32.2\,\dfrac{\text{ft}}{\text{sec}^2}} = -1.10\,g$$

So during braking, it pulls -1.1 g's acceleration, or a deceleration of 1.1 g's. Pretty good braking for a road car.

5.2 Stopping Time

We can now calculate the time to stop from the change in velocity (-88 ft/sec) and the constant acceleration rate (-35.53 ft/sec²). Rearranging equation 2-11:

$$\Delta t = \frac{\Delta v}{a_{Braking}} = \frac{-88\text{ ft/sec}}{-35.523\text{ ft/sec}^2} = 2.48\text{ sec}$$

For the record, that's eight-tenths of a second less than the 3.3 seconds the Enzo takes to go zero to 60 mph. Braking (negative acceleration) capability is almost always greater than for positive acceleration, mainly because the traction from all four tires is being used—all-wheel-drive vehicles would have an advantage here.

5.3 Velocity vs. Displacement

To plot velocity vs. displacement, we first use equation 2-16 and solve it for the velocity v as a function of displacement s:

Equation 3-2

$$v = \sqrt{v_0^{\,2} + 2a_{Braking}\left(s - s_0\right)}$$

We set $s_0 = 0$ and substitute our other known values into the equation:

$$v = \sqrt{\left(\frac{88\text{ ft}}{\text{sec}}\right)^2 + 2\times\left(-35.53\,\frac{\text{ft}}{\text{sec}^2}\right)\times s}$$

where s is in feet.[2] The resulting velocity is plotted vs. displacement s in FIGURE 6.

Note 2: This equation is only valid until s gets to the braking distance of 109 feet, where v gets to zero. And we only want the positive square root, which is probably obvious.

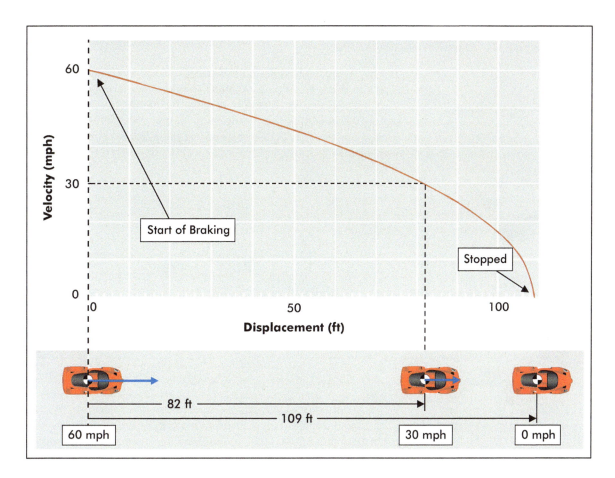

Notice how the velocity changes slowly vs. displacement at first, then drops like a rock. This is because at its high initial velocity, the car is covering a lot of ground each second, but near the end it's practically crawling. For reference, the distance to brake to half the original speed is also shown (82 feet). This shows that ¾ of the braking distance is used to cut the speed in half. It also says that if you drove 60 mph where 30 is safe, your required stopping distance would be 4 times what is safe.

FIGURE 6 A plot of the velocity of the Ferrari Enzo vs. its displacement along the road, assuming constant braking acceleration.

FUEL FOR THOUGHT

If you were doing 60 mph and had to brake for a stopped truck, and you could have *just* stopped safely from 30 mph, how fast would you be going when you hit the truck? You should get about 52 mph!

6 | Calculating Passing Distance

You're driving the Dodge Challenger in FIGURE 7 at 60 mph, following a pickup that's also doing 60 mph. You then accelerate at 11 ft/sec² (about 0.342 g's) for 5 seconds to pass. At the start, your front bumper is 40 feet behind the pickup's rear bumper (that's 60 feet behind its front bumper). Use the positions of their front bumpers to mark their displacements.

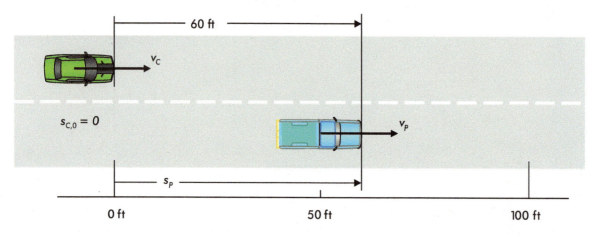

FIGURE 7 Setting out to pass a pickup traveling your speed of 60 mph, starting off 40 feet behind the pickup (60 feet behind its front bumper).

What is your final velocity (in miles per hour) after 5 seconds? What is your displacement after 5 seconds? What is the pickup's displacement? Is it safe to pull in?

6.1 Final Velocity of the Car

First off, we'll use a subscript "C" for the car's variables (like velocity v_C) and "P" for the pickup. The car's initial velocity $v_{C,0}$ is 60 mph, which is 88 ft/sec. [3] Using this in equation 2-10, and knowing the car's acceleration a_C:

$$v_{C,F} = v_{C,0} + a_C t = 88\,\frac{\text{ft}}{\text{sec}} + \left(11\,\frac{\text{ft}}{\text{sec}^2} \times 5\ \text{sec}\right) = 143\,\frac{\text{ft}}{\text{sec}}$$

Converting this to mph:

$$v_{C,F} = 143\,\frac{\text{ft}}{\text{sec}} \times \frac{60\ \text{mph}}{88\ \text{ft/sec}} = 97.5\ \text{mph}$$

Ending up at 97.5 mph, you might get a ticket.

Note 3: When designating the initial velocity for the car, $v_{C,0}$, we use a comma to separate the "C" for "car" from the "0" for "initial," to make it clear. The same is done for the other displacements and velocities. It's important to reduce the confusion factor when doing these problems.

6.2 Final Displacement of the Car

We start with equation 2-16, using the initial and final displacements ($s_{C,0}$ and $s_{C,F}$) of the car:

$$s_{C,F} - s_{C,0} = \frac{\left(v_{C,F}{}^2 - v_{C,0}{}^2\right)}{2a_C}$$

The initial displacement $s_{C,0}$ is zero, so then substituting in our values for the rest:

$$s_{C,F} = \frac{\left(143\,\dfrac{\text{ft}}{\text{sec}}\right)^2 - \left(88\,\dfrac{\text{ft}}{\text{sec}}\right)^2}{2\times 11\ \text{ft}\,/\,\text{sec}^2} = 577.5\ \text{ft}$$

The car's front bumper is 577.5 ft farther along than when it began.

6.3 Final Displacement of the Pickup

The pickup stays at the initial velocity of 60 mph (we'll assume its driver isn't being a jerk), so its acceleration is zero. We use equation 2-14, and remember that its initial displacement $s_{P,0}$ was 60 feet (it started out 60 feet ahead of the car):

$$s_{P,F} = s_{P,0} + v_P t = 60\ \text{ft} + 88\,\frac{\text{ft}}{\text{sec}}\times 5\ \text{sec} = 500\ \text{ft}$$

So the pickup's front bumper is at the 500-foot mark after 5 seconds.

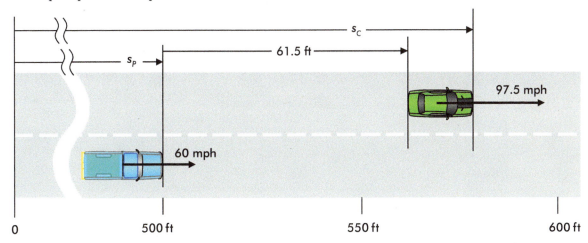

Figure 8 After 5 seconds, the displacement of your car is 577.5 feet, and the pickup's displacement is 500 feet. Now your front bumper is 77.5 feet ahead of his.

6.4 Is It Safe to Pull Back in After 5 Seconds?

After 5 seconds, your displacement (or position) is 577.5 feet, vs. 500 feet for the pickup, so your front bumper is 77.5 feet ahead of his (see Figure 8). Since your Challenger is 16.5 feet long, your rear bumper is now 61.0 feet ahead of his front bumper (about 3 ½ car lengths). You should probably still wait to pull back in.

Note how 5 seconds of acceleration at 11 ft/sec², a pretty stout acceleration for a road car above 60 mph, didn't buy you that much space. Plus, you end up going 97.5 mph! It's a bit less attention-grabbing to start from a little farther behind, accelerate hard earlier and more briefly, hold a reasonable passing speed, and take a little more time.

7   Drag Racing "Staging"

Early drag races were started much like other car races, by waving a flag. But as the cars got quicker and the stakes got higher, more accurate timing was needed to make sure the start was fair. The solution was to use timing lights and photocells that would detect when the front tires were on the starting line, and when they left (see Figure 9).

Figure 9 Dick Moroso's 1966 Corvette staged at the starting line, its position tracked from the three lights at the edge of the track, aimed cross-track. Their light is detected by the three sensors in the white box between lanes. *Photo by permission of General Motors Co.*

Although the light beam has long since been replaced by an infrared beam, the idea is still the same. As you nudge your car forward toward the line, your front tire "breaks" the first beam (the rightmost in Figure 9), and your car is "pre-staged." The second beam senses when your car is *on* the starting line (positioned here, your car is "staged"). Your time starts once your tire rolls out of this beam. If you roll out before you get the green light (not shown), you've "red-lighted" and are disqualified. The distance your car rolls before leaving the staging beam is called the rollout distance.

Most racers agree you should "stage shallow," so that your tire is barely into the staging beam (see **FIGURE 10**, dashed circle). I've read that it is to avoid "red-lighting," but there's more to it than that. Staging shallow reduces your ET (elapsed time) for the quarter mile because your car is moving faster when it leaves the line; you get a rolling start.[4] How fast are you going when the timing starts if you stage as shallow as possible?

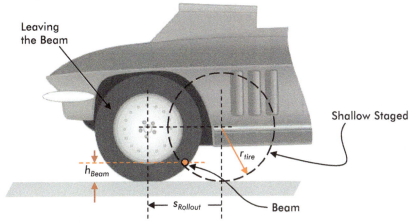

FIGURE 10 The rollout of a drag racer when staging at the beam, which is at height h_{Beam}. The car had staged shallow (dashed tire outline), and the tire is about to roll out of the beam.

First let's estimate the rollout distance, assuming a tire radius r_{Tire} of 12 inches, and a beam height h_{Beam} of 1.1 inches. Referring to **FIGURE 11**, the rollout is twice as long as the lower edge of the right triangle, drawn from even with the wheel center to the beam. The hypotenuse of the triangle is r_{Tire}, and the vertical side is $r_{Tire} - h_{Beam}$. From the Pythagorean Theorem, $(r_{Tire} - h_{Beam})^2 + (s_{Rollout}/2)^2 = r_{Tire}^2$. Massaging this a bit, the rollout displacement is:

$$s_{Rollout} = 2 \times \sqrt{2 r_{Tire} h_{Beam} - \left(h_{Beam}\right)^2}$$

$$= 2 \times \sqrt{2 \times 12.0 \text{ in} \times 1.10 \text{ in} - \left(1.10 \text{ in}\right)^2} = 10.0 \text{ in}$$

While this is an approximation because we neglected the width of the beam, it does point out that the rollout grows with a higher beam and larger diameter tires.[5]

Note 4: Of course to take advantage of this, you need to launch slightly *before* the green, so that you roll out of the staging beam *just after* the green.

Note 5: Using the largest permitted front tire diameter has long been a way to maximize the rollout distance.

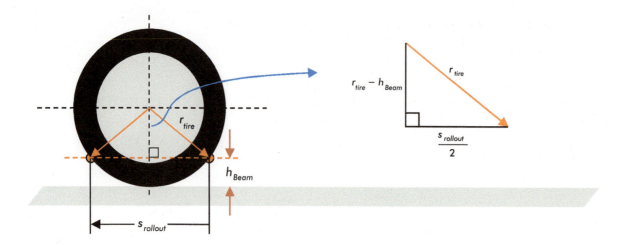

FIGURE 11 Calculating the rollout distance from a shallow stage, neglecting the width of the beam.

Now we can find the launch velocity using equation (2-16), by setting the initial velocity v_0 to zero, and the ending velocity v to v_{Launch} (the velocity when the car leaves the beam):

Equation 3-3

$$v_{Launch} = \sqrt{2\,a_{Launch}\,s_{Rollout}}$$

If the car rolls 10 inches before it leaves the beam (which is typical), and its launch acceleration is 1.5 g's (48.3 ft/sec²), then by the time the car rolls out of the beam its velocity v_{Launch} is:

$$v_{Launch} = \sqrt{2 \times 48.3\,\frac{ft}{sec^2} \times 0.83333\ ft}$$

$$= 8.9722\,\frac{ft}{sec} = 6.12\,\frac{mi}{hr}$$

So with 1.5-g acceleration, the timing won't start until you are traveling 6.11 mph, from a shallow stage! Of course the harder your car accelerates off the line, the faster you'll be going when timing starts. A Top Fueler launching at 4.0 g's would get up to 14.7 ft/sec, or about 10 mph in those same 10 inches!

This rolling start has confused at least one magazine publisher, who had to figure out why their new electronic equipment, based on a truly stationary start, didn't agree with previous tests. Their fix was to start timing at 5 mph, not too far off what we calculated here.

8 | 60-Foot Time and Launch Acceleration

A standard judgment of a dragster's launch is its 60-foot time, the time to cover the first 60 feet (see FIGURE 12). Why is it so important?

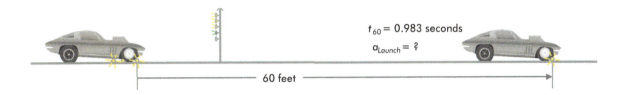

$t_{60} = 0.983$ seconds

$a_{Launch} = ?$

60 feet

At first glance, the 60-foot time is just a way of measuring an average speed for the first 60 feet. But what you're really after is the acceleration a_{Launch} during those 60 feet. If we assume the acceleration is constant, we can use equation 2-14, with $s_0 = s_{Rollout}$ and $v_0 = v_{Launch}$, to get:

FIGURE 12 The "60-foot time" starts when the car leaves the staging beam (second light, left) and stops when it blocks the 60-foot beam (right).

Equation 3-4

$$s_{60} = s_{Rollout} + v_{Launch} t_{60} + \frac{1}{2} a_{Launch} \left(t_{60} \right)^2$$

where s_{60} is 60 feet, t_{60} is the 60-foot time, and v_{Launch} is the velocity leaving the start beam, from equation 3-3. Using this, we can substitute equation 3-3 into equation 3-4, so it now becomes:

$$s_{60} = s_{Rollout} + \sqrt{2 a_{Launch} s_{Rollout}} \; t_{60} + \frac{1}{2} a_{Launch} \left(t_{60} \right)^2$$

With some work, you can solve this for the launch acceleration: [6]

Equation 3-5

$$a_{Launch} = \frac{2}{t_{60}^2} \left(\sqrt{s_{60}} - \sqrt{s_{Rollout}} \right)^2$$

Now let's assume a 60-foot time of 0.983 seconds, and the same 10-inch rollout as before. The average acceleration would then be:

$$a_{Launch} = \frac{2}{\left(0.983 \text{ sec} \right)^2} \left(\sqrt{60 \text{ ft}} - \sqrt{0.83333 \text{ ft}} \right)^2 = 96.64 \frac{\text{ft}}{\text{sec}^2} = 3.00 \text{ g's}$$

Note 6: The solution to this is shown in Appendix 5, and uses the quadratic equation.

Just for grins, let's see how far off you'd be if you ignored the rollout, and incorrectly assumed the velocity is zero when the timing starts. Putting zero in for the rollout distance in equation 3-5:

$$a_{Launch} = \frac{2}{(0.983 \text{ sec})^2}\left(\sqrt{60 \text{ ft}}\right)^2 = 124.2 \frac{\text{ft}}{\text{sec}^2} = 3.86 \text{ g's}$$

So the launch acceleration is overstated by about 28%, just from neglecting the 10-inch rollout. Of course, a good 60-foot time is still a good 60-foot time, whether you calculate the acceleration correctly from it or not. Just make sure you stage the same depth each time.

9 | Going Faster by Stopping Faster

I don't think this is obvious. In fact, I still remember hearing for the first time how someone covered laps faster in a road race by having better brakes. I couldn't figure out how slowing down harder could make you go faster. The explanation was that harder braking allows *later* braking for a curve, keeping your speed up for more of the straight. Okay, but how much does it help?

FIGURE 13 You accelerate out of a 60-mph curve and must brake for a 45-mph curve at the other end of the 1,000-foot straight. When do you need to brake?

The problem could be put something like this (see **FIGURE 13**): You exit a turn at 60 mph (88 ft/sec) and then accelerate as long as possible before braking for a 45-mph turn. To make it simple, we'll assume you accelerate at a constant 0.6 g, and brake at a constant −0.9 g.[7] The distance between the turns is 1,000 feet.

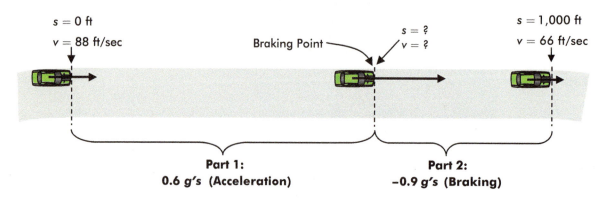

$s = 0$ ft
$v = 88$ ft/sec

Braking Point

$s = ?$
$v = ?$

$s = 1,000$ ft
$v = 66$ ft/sec

Part 1:
0.6 g's (Acceleration)

Part 2:
−0.9 g's (Braking)

Note 7: Note that braking with −0.9-g acceleration would normally be called "0.9-g braking," the negative sign being left out by assuming braking is negative acceleration.

Where do you have to brake? How much later could you brake if you could brake at 1.0 g instead? How much time would it save per lap?

You can solve this just as if you were the driver. As you accelerate, your velocity is steadily increasing, and you are mentally judging the last moment you can switch from gas to brake and still get down to 45 mph by the 1,000-foot mark. FIGURE 14 graphs the physics going on in your head. The 0.6-g acceleration is shown by the velocity trace starting at 60 mph at the left end of the graph, which climbs as you travel down the straight. The 0.9-g braking is shown by the decreasing velocity which reaches 45 mph at the right end of the graph. Where the two traces cross (at v_{Peak}), it's time to brake (note that the velocity traces aren't drawn to scale, since we don't know the answer yet).

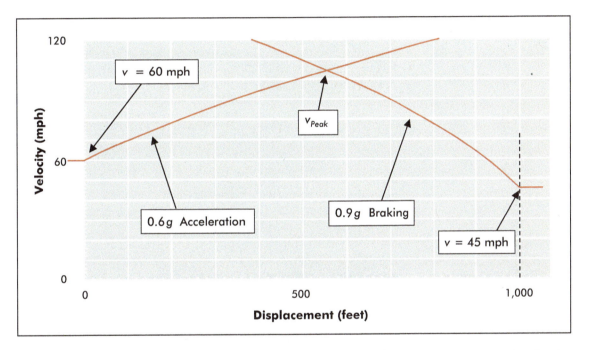

Since this involves the intersection of two motion traces, we need to be very organized. TABLE 2 summarizes FIGURE 14 in numbers. It splits the straight into Part 1 (acceleration) and Part 2 (braking). The two parts have their own displacements (s_1 and s_2), velocities (v_1 and v_2), and accelerations (a_1 and a_2). The "0" subscript stands for beginning and "F" for final.

FIGURE 14 You can visualize the problem as one velocity trace accelerating from 60 mph out of the first curve (left) and a second trace decelerating to 45 mph for the next curve. Where the two traces intersect, you need to brake.

	Part 1 (Acceleration)		Part 2 (Braking)	
	Beginning	**End**	**Beginning**	**End**
Displacement	$s_{1,0} = 0$	$s_{1,F} = s_{2,0}$	$s_{2,0} = s_{1,F}$	$s_{2,F} = 1{,}000$ ft
Velocity	$v_{1,0} = 88$ ft/sec	$v_{1,F} = v_{2,0}$	$v_{2,0} = v_{1,F}$	$v_{2,F} = 66$ ft/sec
Acceleration	$a_1 = 19.32$ ft/sec^2	$a_1 = 19.32$ ft/sec^2	$a_2 = -28.98$ ft/sec^2	$a_2 = -28.98$ ft/sec^2

TABLE 2 The beginning and final displacements and velocities for each part of the straight, along with the acceleration values.

As **TABLE 2** shows, we know the displacement and velocity at the beginning and end of the straight, and the acceleration rates during acceleration and braking. The peak velocity and its location aren't yet known, but we do know that the displacement and velocity at the end of Part 1 are equal to those at the beginning of Part 2:

$$s_{1,F} = s_{2,0}$$

$$v_{1,F} = v_{2,0} = v_{Peak}$$

So instead of having two unknown displacements and two unknown velocities, we are down to one unknown displacement ($s_{1,F}$, the braking location) and one unknown velocity ($v_{1,F}$, the peak velocity on the straight).

9.1 Where Do You Need to Brake?

Now that we're set up, let's solve for the braking location $s_{1,F}$. Adapting equation 2-16 to indicate it is for Part 1, we have

Equation 3-6a

$$s_{1,F} - s_{1,0} = \frac{v_{1,F}^{\,2} - v_{1,0}^{\,2}}{2a_1}$$

This says that the increase in displacement from the start of the straight during Part 1 depends on the beginning velocity $v_{1,0}$, the ending velocity $v_{1,F}$, and the acceleration rate a_1 during Part 1. It is similar for Part 2:

Equation 3-6b

$$s_{2,F} - s_{2,0} = \frac{v_{2,F}^{\,2} - v_{2,0}^{\,2}}{2a_2}$$

with beginning velocity $v_{2,0}$, ending velocity $v_{2,F}$, and the acceleration rate a_2. Using the conditions from the previous section, we can eliminate the unknown variables in equations 3-6, and solve for the displacement at the braking point: [8]

Equation 3-7

$$S_{1,F} = \frac{-2a_2 S_{2,F} + v_{2,F}^2 - v_{1,0}^2}{2a_1 - 2a_2}$$

Plugging our numbers into this says you need to brake when you hit the 564.93-foot mark. So with 0.9-g braking, you can accelerate for the first 565 feet of the straight, at which point you must brake to make the 45-mph curve (see **Figure 15**, bottom curve).

Figure 15 Acceleration and braking on the 1,000-foot straight. We'll see the 1.0-g braking point is 27 feet later than for 0.9-g, and saves about 0.1 seconds/lap.

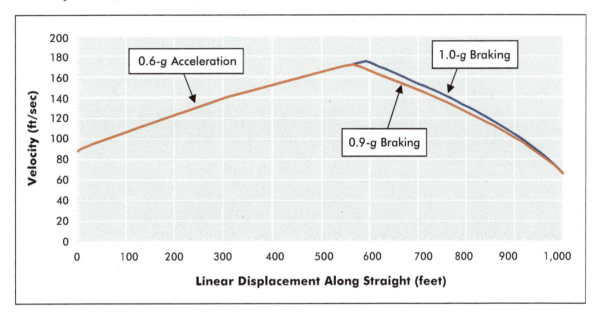

9.2 How Much Later Could You Brake with 1.0-g Braking?

For 1.0-g braking, we can use equation 3-7 again, to find that the braking point is at 592 feet, 27 feet later than for 0.9-g braking. **Figure 15** shows the velocity during acceleration and braking, for both braking rates. You can see a speed advantage with 1.0-g braking, from the 0.9-g braking point until the end of the straight.

Note 8: The derivation of equation 3-7 is in Appendix 5, if you'd like to see all the details.

9.3 Time Saved by 1.0-g Braking

To find how much time is saved by braking at 1.0 g, we first need to calculate the time on the straight in the baseline condition. We'll calculate the peak velocity, and then calculate the acceleration and braking times from that. We solve the displacement equation for Part 1, equation 3-6a, for the final (peak) velocity:

Equation 3-8

$$v_{1,F} = \sqrt{2a_1 s_{1,F} + v_{1,0}^2}$$

Substituting in our known values for Part 1:

$$v_{1,F} = \sqrt{2 \times \left(19.32 \frac{\text{ft}}{\text{sec}^2}\right) \times 564.93 \text{ ft} + \left(88 \frac{\text{ft}}{\text{sec}}\right)^2}$$

$$= 171.97 \frac{\text{ft}}{\text{sec}}$$

This is a peak velocity of 117.25 mph. We can calculate the elapsed time for Part 1 by using equation 2-11, knowing that $v_{1,0} = 88$ ft/sec and $a_1 = 19.32$ ft/sec²:

$$t_1 = \frac{v_{1,F} - v_{1,0}}{a_1} = \frac{171.97 \text{ ft/sec} - 88 \text{ ft/sec}}{19.32 \text{ ft/sec}^2} = 4.3463 \text{ sec}$$

Doing the same calculation for Part 2, with $v_{2,F} = 66$ ft/sec and $a_2 = -28.98$ ft/sec²:

$$t_2 = \frac{v_{2,F} - v_{2,0}}{a_2} = \frac{66 \frac{\text{ft}}{\text{sec}} - 171.97 \text{ ft/sec}}{-28.98 \text{ ft/sec}^2} = 3.6567 \text{ sec}$$

Adding both of these times gives the total time down the straight as 8.003 seconds. When braking is changed to 1.0 g, the top speed increases to 122.7 mph, and the total time is reduced to 7.888 seconds, a 0.115 second reduction. That's from *one straight on one lap*. Add that up over a race!

Mark Donohue used braking strategy to win a NASCAR road race at Riverside in 1973, where his AMC Matador could brake lap after lap at −0.5 g, while his competitors could only sustain −0.25 g without fading! Especially with the long braking distances of that time, cutting it in half was his huge ("unfair," as he put it) advantage.

A tenth of a second advantage per lap is impressive enough, but what about passing? Let's say you could brake at 1.0 *g* and you were trying to pass a car that could only brake at 0.9 *g*. You could brake 27 feet later than that car on this straight, and accelerate during that time instead. At the entrance to the 66-ft/sec corner you would have gained 7.5 feet of track position (a good calculation to verify), which would help you nose inside your rival into the curve.

10 Don't Be Slow

Sometimes the effect of a measurement isn't obvious if you look at it in a conventional way. It may be clearer if you "flip it on its head." For example, Americans typically calculate "fuel economy" in miles/gallon (which is distance/volume), to find out the car's range on a tank. In many other countries, they calculate "fuel consumption" in liters/100 kilometers (which is volume/distance) instead, to estimate their fuel cost on a trip of a certain distance.[9] Same data, different interpretation.

It's similar in drag racing, where the idea is to minimize the elapsed time for the run. Top speed (velocity) is impressive, but it doesn't win the race.[10] It's common in drag racing to use "short" gearing, which hurts the car's top speed, in favor of better launch acceleration, thereby spending less time at low speed. A car that's improperly geared may reach a higher top speed by the finish line, while its competitor gets there first. We'll see that "going fast" isn't nearly as important as "not going slow."

Take a look at the velocities of two drag racers in FIGURE 16. Car A has twice the initial acceleration because of better traction, but because it only has about half the power, its acceleration drops off sooner than for Car B. The calculations neglect air drag. Traction forces and power are topics that will be covered in Chapters 5 and 14, respectively. For now, just note that Car A accelerates harder than Car B at first, but later accelerates less hard.

Car A has a higher velocity for the first 254 feet of the track, during which time it opens up a lead. Car B has a higher velocity for the last 1,066 feet of the track, and at the end of the quarter-mile, it is moving 20 mph (29 ft/sec) faster! It's clear that Car B is a lot faster for a lot of the time.

Note 9: Fuel consumption in the U.S. would be in gallons/100 miles. For example, 25 miles/gallon would be 4 gallons/100 miles.

Note 10: For this problem, "speed" and "velocity" mean the same thing, as long as we can agree that the cars are always moving forward—and in a straight line!

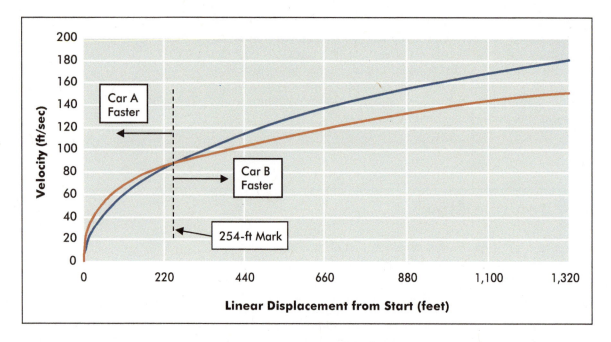

FIGURE 16 The velocities of two drag racers. Car A (red) has twice the traction of Car B (blue), but Car B has twice the power. For most of the race, Car B is faster. Who wins? A little more analysis is needed to find out.

Now let's take the opposite view. Instead of plotting velocity v, which is $\Delta s/\Delta t$ (displacement per unit time), let's plot its reciprocal, $1/v$, or $\Delta t/\Delta s$ (the time per unit displacement). In other words, we track how long the car takes to cover a short distance. For lack of a better term, we'll call t/s "time consumption," or "slowness":[11]

General Equation 3-9   **Slowness, t/s**

$$\frac{\Delta t}{\Delta s} = \frac{1}{v}$$

As an example, if a car has a velocity of 50 ft/sec, then it takes two hundredths of a second to go 1 foot, so its "time consumption" is 0.02 seconds per foot. The greater the rate of time consumption, the more time it takes to cover a certain distance: greater "slowness." The slowness of the two cars is plotted in **FIGURE 17**.

What seemed like a small difference in velocity at the start of the race (in **FIGURE 16**) makes a large difference in how long it takes to move each foot (**FIGURE 17**). For example, at the 100-foot mark, Car A covers 1 foot in 0.0158 seconds, while Car B takes about 0.0183 seconds. On the other hand, the large difference in velocity during the later part of the race makes little difference in the slowness (in seconds/foot), both cars taking very little time anyway.

Note 11: There seems to be no better word for this. "Slothfulness" and "lethargy" didn't quite work, so "time consumption" and "slowness" will have to do.

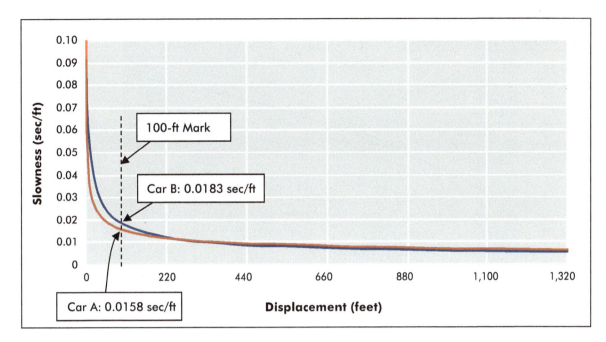

Figure 17 The slowness of the two drag racers, plotting how long it takes them to cover each foot of the track. Car B puts itself way behind in the first 200 feet, as you'll see in **Figure 18**.

Car B, with its slow start, eats up a lot of time covering the first part of the track, and reaches the 254-foot mark well behind Car A (see **Figure 18**). To find out how much, we calculate the time it takes to get to that point. Since the slowness $\Delta t / \Delta s$ is the amount of time it takes to move the car 1 foot forward, adding the slowness for each foot traveled gives us the elapsed time (ET) up to that point.[12] The ET vs. displacement is plotted in **Figure 18**. Note that this figure plots displacement s on the x-axis and time t on the y-axis, opposite of "normal."

Note 12: This was done with a spreadsheet program (Microsoft Excel). What we are doing is multiplying the slowness t/s by the s of 1 foot to find the time to travel each 1-foot segment. We then add these times up to find the total time up to a certain distance. In calculus, finding the total time in a similar way is called *integration*. Then instead of breaking the distance into 1-foot increments, it is broken up into *infinitely small* distances.

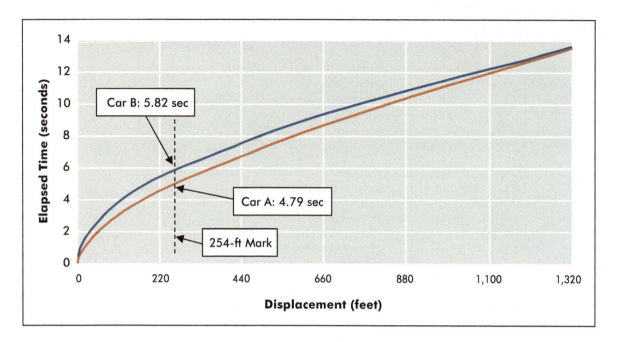

FIGURE 18 The elapsed time up to each point on the track. Car A (in red) reaches the 254-foot mark in 4.79 seconds, compared to 5.82 seconds for Car B (blue). By the end of the quarter-mile (1,320 feet), Car B catches up—just about.

Car A covers the same ground in much less time for the first few seconds. In fact, it reaches the 60-foot mark 0.77 seconds sooner (2.06 vs. 2.83 seconds). By the 254-foot mark, its quick start gives it a 1.03-second edge (4.79 vs. 5.82 seconds). Despite its higher velocity during the last 80% of the distance, Car B can't quite catch up, and loses by 0.01 seconds, with an ET of 13.52 seconds compared to 13.51 seconds.[13]

Because of the huge hole a poor start will put you in, off-the-line acceleration (the hole shot) is recognized as the most important part of a drag race. This doesn't mean that improvements that only help at higher speeds (like aerodynamics or power) don't play a role, but acceleration off the line is the top priority.

KEY CONCEPT

"Not being slow" is more important than being fast.

Note 13: Note that these comparisons were made ignoring the "running start" from staging, described earlier. This would give Car A, with its higher initial acceleration, still more of an advantage!

11 Summary

You may have noticed a common thread in the problems in this chapter. Most of them involved the fact that reducing your lap time or quarter-mile ET is about increasing *average* speed. Keeping a high average speed is more about increasing speed in the slow parts of the track than increasing speed in the fast part of the track. This point resurfaces several times in the book.

Another major point is that maximizing acceleration is always the key to doing this. Increasing *positive* acceleration gets you away from your minimum velocity more quickly. Increasing negative acceleration (braking) delays your braking point, increasing the time spent at higher velocities.

This chapter also shows that even though "physics is physics," taking a viewpoint that is just slightly unconventional can make issues that were hidden, or at least not obvious, pop right out at you. And often, a close examination of the rules of the game (such as with the staging problem) can help you take full advantage of the rules (legally!). Speaking of which, let's take one last look at the 60-foot time and the rollout issue.

Is there any reason not to stage shallow in a drag race? One theory is that you might want to stage deep because you could start off closer to the finish line. Let's examine the velocity vs. displacement for the "shallow stage" condition first (see **FIGURE 19**). Assuming our 10-inch (0.833-ft) rollout and a 3-*g* (96.6-ft/sec^2) launch acceleration, the velocity v_{Launch} when timing starts is about 12.7 ft/sec (8.65 mph). With that acceleration, it takes 0.131 seconds to get to the point where the timer starts. In other words, the first 0.131 seconds (which is just about forever in a drag race) are "off the clock."

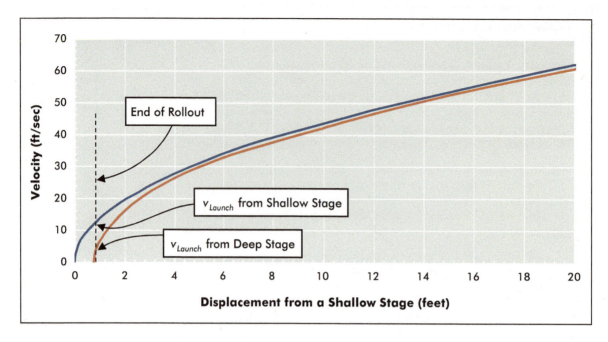

Displacement from a Shallow Stage (feet)

FIGURE 19 Velocity vs. displacement over the first 20 feet for shallow (blue) and deep (red) staging positions. A 3-g launch acceleration is assumed. Notice how the velocity traces for the accelerating motion here look just like FIGURE 6 for braking, except swapped right-for-left.

Now let's look at the deep stage (the lower curve in FIGURE 19), assuming you stage 1 inch short of rolling out of the timer. With the same launch acceleration, you roll out at about 4 ft/sec and accelerate for only 0.0415 seconds before the timer starts.

Staging shallow allows you to have an extra 0.09 seconds of your run before timing starts, which is a pretty big advantage to throw away. By staging deep, your only advantage is taking ¾ foot off the end of the run, which at your top speed of 150 ft/sec is only 0.005 seconds. So even considering the longer distance to cover, staging shallow is still quicker by 0.085 seconds. For me, it's hard to see any advantage in staging deep, as long as you're not caught napping at the line!

Major Formulas

Definition	Equation	Equation Number
Mean Piston Speed (in ft/min)	$S_{Mean} = 2d_{Stroke}\omega_{Crank}$ (in rpm)	3-1
Velocity vs. Displacement, Constant Braking ($a_{Braking}$ negative)	$v = \sqrt{v_0^2 + 2a_{Braking}(s - s_0)}$ for $v_0^2 + 2a_{Braking}(s - s_0) \geq 0$	3-2
Launch Velocity at End of Rollout, Constant Acceleration	$v_{Launch} = \sqrt{2a_{Launch}\, s_{Rollout}}$	3-3
Launch Acceleration Calculated from 60-Foot Time	$a_{Launch} = \dfrac{2}{t_{60}^2}\left(\sqrt{s_{60}} - \sqrt{s_{Rollout}}\right)^2$	3-5
Slowness, $\dfrac{t}{s}$	$\dfrac{\Delta t}{\Delta s} = \dfrac{1}{v}$	3-9

4 Dynamics Basics

Forces, Mass, and Acceleration

You can see the rear tires distort on John Force's Mustang funny car (near lane), as they develop the immense forces needed for a quick launch. Force beat his teammate Tony Pedregon here, in the semifinals of the 1997 NHRA Winternationals in Pomona, California, to advance to the finals and then win the meet. *Photo by permission of Ford Motor Company.*

Chapter 4

Now we get to the "meat of the book," Newton's laws of motion. Every object has mass that tends to keep it at a constant velocity. To change this, and cause the object to accelerate, a net force must act on it. The greater the mass, the more force it takes to produce a given acceleration. Also, every force on a body has an equal and opposite twin acting on another body.

Contents

Key Symbols Introduced

Symbol	Quantity	SFS Units	MKS Units
m	Mass	slug	kg
$\mathbf{F}$	Force	pound (lbf)	newton (N)
$\mathbf{F}_{Res}$	Resultant Force	pound (lbf)	newton (N)
$\mathbf{F}_{Action}$	"Action" Force	pound (lbf)	newton (N)
$\mathbf{F}_{Reaction}$	"Reaction" Force	pound (lbf)	newton (N)

1   Introduction

Whether a 50-mile/gallon econo-car, a Top Fuel dragster, or a Grand Prix motorcycle, one thing is common: each vehicle needs to move its own mass by its own power.[1] So despite completely different ultimate goals, their designers must each use their expertise in vehicle motion to eke out that last 0.1 mpg or cut that last 0.01 second from a race or a lap time. To out-perform their competition, they need to know how to move the vehicle as effectively as possible. They need to know *dynamics*.

Dynamics is the "why" of motion. In **DYNAMICS**, we study how forces cause the motion that we examined in Chapters 2 and 3. Forces are what start motion, stop motion, speed you up, and slow you down. In the same way, some forces push your car down, some push it up, and some propel it up a hill. Forces push your engine's pistons to propel your car, and cam and valve-spring forces push on your engine's valves to open and close them.

So while kinematics showed that maximizing acceleration (positive *or* negative) is the key to getting somewhere fast, this chapter will show how to use Newton's laws of motion to find the traction force to do it.

You might also say that dynamics is the "why not" of motion, as in why your street-legal car doesn't accelerate zero to 60 mph in 2 seconds, why it won't corner at 2 *g*'s, why it doesn't gain speed coasting uphill, and why it won't break the sound barrier. In each case, your car won't produce the forces necessary to do it.

You must know your car's performance limitations to avoid exceeding them (i.e., driving like an idiot), and to know how to increase them.

Note 1: The term "mass" will be defined shortly. For now, just know that a heavier vehicle has more mass.

Only through identifying its "dynamics bottlenecks" can you efficiently improve your vehicle's performance; an engine making more power won't help you accelerate if your tires can't apply that power to the ground. Likewise, making the powertrain more efficient may not improve fuel economy if it makes the car much heavier.

There are many types of forces that affect vehicle motion, such as gravity, rolling resistance, aerodynamic drag and lift, spring forces, and tire traction forces. Some forces will be introduced here, but Chapter 5 will cover the more important ones in detail.

> **NOTE**
>
> There will be a lot of attention paid to units and terminology in this chapter, especially in distinguishing mass from weight. In this area, our everyday language about cars isn't always specific enough for science. But automotive tradition has too much momentum to turn it around, so we will "translate" the common terms into the necessary scientific framework.

2   Physical Concepts

It's been said that the subject of physics can be summed up by "force = mass × acceleration; all the rest is details." That's not far off. Force causing acceleration of a mass is the focus of this chapter, probably the most important in the book; the rest build on this as they add in details. It all starts with that simple law, discovered by Sir Isaac Newton:

$$F \sim ma$$

This is read "Force is proportional to mass times acceleration," where F is force, m is mass, a is (still) acceleration, and "$\sim$" means "is proportional to." How and why objects move wasn't well understood before Newton made this profound observation in the late 1600s.

The ancient Greek philosopher Aristotle had taught that heavier objects fall faster than lighter objects, and for 2,000 years, no one seriously questioned this! In retrospect, this seems ridiculous, but no one probably realized the importance of this error in the big picture (science wasn't even defined yet). Also, early philosophers thought that physical truth came from the mind alone, and they didn't necessarily let reality get in the way of a "good" theory. Plus, Aristotle tended to be right about many things, and people just didn't question authority (or might do so at their own peril).

Finally Galileo Galilei, around 1604, had the open mind (and open schedule) to do experiments rolling balls down an incline, to show that heavy and light objects fall at the same rate.[2] He also theorized that if he could reduce the rolling friction of a ball to zero, it would roll forever if on a level surface. But thinkers of the day *still* didn't like proving their theories, and largely ignored Galileo's results.[3]

He had set the stage, though, for Newton, who took the ball and ran with it. After years of developing the theory and mathematics, he published his major work on physics, the *Principia,* in 1687.[4] It revolutionized the way we look at the world, and the universe, by cutting through the complexity that was hiding the simplicity of $F \sim ma$. It instantly made him an international celebrity, after which he became the first scientist to be knighted.[5]

2.1 Newton's Discoveries

Before Newton, philosophers had deduced that the "natural state" of an object is at rest, because when you throw something, like a rock, it rolls, bounces, or slides until it stops. Once it stops, it usually stays there.

Newton's breakthrough, building on Galileo's work, was realizing that the earlier philosophers weren't seeing the whole picture. Yes, a stationary object is in a natural state, but that is just a special case of motion at *constant velocity* (a constant speed with constant direction).

All objects have INERTIA, which means they resist any "changes in their motion." Newton realized this specifically meant "changes in their velocity," and that the only time an object's velocity changes is when unbalanced forces act on it.[6] The thrown rock comes to rest after

---

Note 2: Most people of his time were probably more concerned with where their next meal would come from than with how balls roll downhill.

Note 3: Although he proved that heavier bodies do not naturally fall faster than lighter ones, he probably never actually dropped objects off the leaning tower of Pisa, as legend suggests. Rolling balls down a ramp made an experiment slow enough for him to time.

Note 4: The full title was *Philosophiae Naturalis Principia Mathematica,* Latin for "Mathematical Principals of Natural Philosophy." At the time, physics was called natural philosophy.

Note 5: Newton was as much gearhead as egghead, though. Having no cars to obsess over, he was obsessed with machines like the local windmill being built when he was a kid, even making a working model of it.

Note 6: It's important to remember that velocity has a magnitude (speed) and a direction. Changing speed requires a force, and changing direction also requires a (sideways) force. The sideways force causes lateral acceleration, as we'll see in Chapter 10.

bouncing and sliding because friction forces act on it to slow it down, not because it's the object's "natural state" to be at rest (see **FIGURE 1**). Once it is at rest, friction forces balance the forces from gravity and support from the ground. Friction holds it there (at its constant velocity of zero) unless another force acts to upset this balance.

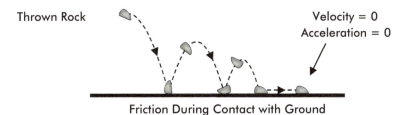

Thrown Rock

Velocity = 0
Acceleration = 0

Friction During Contact with Ground

FIGURE 1 A thrown rock flies until it hits the ground, then bounces a few times before sliding to a stop. Ancient philosophers thought the rock was seeking its "natural state" of being at rest. Now we understand that friction forces are what make it stop, and keep it there.

Newton also realized that the stronger the net force pushing or pulling on an object, the quicker the object will change its velocity, i.e., accelerate. He further noted that if something has more mass, it takes a larger force to give it a certain amount of acceleration. Anyone pushing a wheelbarrow full of dirt knows that even on a smooth, level surface, it takes a good push to get it going. Once it's going, it takes a strong pull to stop it quickly. When the wheelbarrow is empty, it takes much less force to accelerate it, forward or rearward.

People had also believed that the reason an object thrown into the air fell back to Earth was that it was returning to be with other "heavy" things, rather than being up with the lighter air. Newton correctly explained that it falls because a force, caused by Earth's gravity, pulls it downward. We call this force the object's weight. Earth and the object, say a falling apple, actually pull on each other with gravitational forces that are equal in strength, but opposite in direction.[7] As you might expect, a force this small hardly affects the massive Earth, but it changes the motion of the apple quite a bit.

With this theory, Newton could also explain why not everything that went up came down. For example, a bird flies by producing an upward force on its body with its wings (by contact with the air), which works against the downward acting force of its weight.

But for us, the greatest part of Newton's discoveries was the clear, uncluttered picture of how motion occurs, describing it with a set of three simple laws. The next few sections will show how powerful these laws are, especially when put into basic equations.

Note 7: Newton may have observed falling apples, but his being hit in the head by one is probably just a legend.

3   Newton's First Law and Static Equilibrium

Newton's first law describes a body's motion in very simple condition, where there is no net force acting on it. Galileo had discovered this law, and called it the law of inertia:[8]

FUNDAMENTAL LAW

Newton's First Law: A body that is not acted on by a resultant force has a constant velocity.

That's pretty simple. A body with no resultant force (meaning net force) acting on it would continue to travel along the same straight line, at constant speed, forever. Once the object is put into this motion, it won't change *speed* or *direction* until it is acted on by a net force. A very clean, simple idea, but thinking of it took an incredible leap of imagination for someone living on Earth and always being pulled by its gravitational force!

FIGURE 2 shows an example of Newton's first law. In it we have an object traveling at some initial velocity v_0, under no net influences of forces. This object could be an asteroid traveling through space, far away from any stars or planets. A good example on Earth would be an "air hockey" puck floating on a level cushion of air; whichever way you start it, it travels in that direction, at that speed, until it hits a wall (which does apply a net force to the puck).

FIGURE 2 Newton's first law applied to a body with no net force on it. It travels along a straight line at constant speed. The speed *could* be zero (as for a stationary object), but it doesn't have to be.

Initial Velocity

v_0

Future Path Without a Net Force

This also means that if you started a car coasting at a constant speed, with no resultant force acting on it, it would roll forever (how's that for gas mileage?). But don't get your hopes up. We know from experience that a coasting car doesn't coast forever. Why not? We'll answer that in the next few pages.

Note 8: An "inert" object is one that is unaffected by outside influences. Inertia then is the tendency of an object to resist changes in its motion.

3.1 Forces, Their Magnitude, and Their Direction

Before going any further, let's define forces a little more fully:

A **FORCE** (usually labeled *F*) is an *action* that pushes or pulls on an object. A force is a *vector*, being defined by its size (called its magnitude) and its direction.

It's hard to describe a force any better in words; you know it better "by feel." When you push on a wrench, you are producing a force that acts on it. When you pull on a rope, you are exerting a force on it. When you stand on a scale, you are exerting a force on it, equal to your weight.

> **NOTE**
>
> **VECTORS** are quantities that have magnitude and direction, like force, displacement, and velocity, and they are typically shown as arrows scaled to their magnitude (see **FIGURE 3**). Quantities with no specified direction, like speed and distance, are called **SCALARS**.

When you pull hard, you are producing a force with a large magnitude; when pulling softly, a force with a small magnitude. We measure its magnitude in pounds-force (abbreviated "lbf") when using SFS units, and in newtons (abbreviated "N") when using MKS units.[9] The pound is familiar to everyone in the U.S., and the newton is familiar everywhere else (or at least it should be).[10] Even in the U.S., we do see newton-meters on torque wrenches, and automotive engineers commonly use newtons for force.

The direction a force acts in is just as important as its size. Pushing with a force of 50 pounds has the opposite effect of pulling with 50 pounds. Putting your car in reverse produces a rearward traction force, while first gear produces a forward force.

> **KEY CONCEPT**
>
> If we define a forward force as positive, then a rearward force is negative. A reverse traction force of 350 pounds is a forward force of −350 pounds. The same is true for left- and right-acting forces, and for upward and downward forces.

Note 9: Normally, pounds are abbreviated as lb, but in this book (and many engineering texts), it is abbreviated as "lbf" to emphasize it is used as a unit of force: a "pound-force."

Note 10: Unfortunately, people often use the kilogram as a unit for force instead of the newton.

3.2 Resultant Forces

Considering positive and negative forces, how do we calculate the *resultant* force mentioned in Newton's first law? We find the sum of the forces acting on an object, taking their direction into account. For example, if two forces act on an object, and they are equal in magnitude but opposite in direction, the resultant force is zero. We experience this "stalemate" every day.

For example, **Figure 3** shows a car traveling on a level road, the rear tires pushing forward on it with a 100-pound traction force, and aerodynamic drag pushing rearward with an 80-pound force. There is also rolling friction (in the tires, bearings, etc.) that pushes rearward with 20 pounds. In this case, the three forces add to a resultant force of zero, so from Newton's first law, the car's velocity (whatever it may be) will be constant (no acceleration).

x

Aero Drag, 80 pounds

Rolling Resistance (20 pounds)   Traction Force (100 pounds)

Figure 3 Newton's first law demonstrated on a Ford GT40. The sum of the forward forces is zero (+100 lbf traction, −80 lbf aerodynamic drag, and −20 lbf rolling friction). The car will continue at its current velocity until one of the forces changes.

The equation for the resultant of two or more forces is straightforward:

Physical Law Eq. 4-1

$$\mathbf{F}_{Res} = \mathbf{F}_1 + \mathbf{F}_2 + ...$$

Here, $\mathbf{F}_1$ could be the traction force and $\mathbf{F}_2$ could be the aerodynamic force. If there are three forces, $\mathbf{F}_{Res} = \mathbf{F}_1 + \mathbf{F}_2 + \mathbf{F}_3$, and so on. We used this in our pushrod problem in Chapter 1.

> **NOTE**
>
> For now, forces and other vectors will be signified by bold characters, to emphasize their directionality. This will be relaxed in future chapters.

3.3 Static Equilibrium

Newton's first law, describing a body subjected to a resultant force of zero, is really a statement of **STATIC EQUILIBRIUM**. The word "equilibrium" comes from the Latin *aequi* (meaning equal) and *libra* (meaning balance): the forces on the object are "equally balanced" by each other.[11] For **Figure 3**, the traction force is balanced by the total drag force. Another way to state Newton's first law is that a body in static equilibrium has zero acceleration.

KEY CONCEPT

Equilibrium is an important general concept in several areas of physics. Notice that an object being in static equilibrium (no net force) doesn't mean there are no forces acting on it; they just cancel each other out! A tug of war where neither team is gaining ground is a good example.

So why can't we shut off a car's engine on a straight and level road and let it roll forever? If we let the car in **Figure 3** coast, the (rearward) drag forces would still be there, with no forward force to offset them. We'd have a resultant force of 100 pounds rearward. The car is no longer in static equilibrium, and it will no longer travel at a constant velocity. Hold that thought.

4   Newton's Second Law: Resultant Force, Mass, and Acceleration

Exactly what happens when a net force pushes rearward on a car going forward? Anytime we have a net force acting, we need to use Newton's *second* law. But first, we need to be more specific about inertia, the property of a body that resists changes in its velocity. After defining this inertia as *mass*, we'll use it in Newton's second law.

4.1 Mass

If you leaned into a car's rear bumper and pushed as hard as you could, the car wouldn't just shoot forward (even without friction in the wheel bearings). It would speed up, but gradually, because it has a large amount of mass. Let's say the car has a mass of 1,000 kilograms (it weighs about 2,200 pounds), and after 5 seconds of pushing, it goes 5 km/hr. If you pushed just as hard on a car having twice the mass, all else being equal, the same 5-second push would only get it up to 2.5 km/hr.

Note 11: Since weight is often measured on a balance, *libra* was also used as a unit for weight in Roman times, and abbreviated "lb." When we started using the word "pound," we kept the abbreviation for *libra*.

Mass is often confused with weight, but it's not the same thing. As mentioned, weight is a *force* that pulls downward on an object when it is in a gravitational field (like Earth's). The same object's weight would only be ⅙th as much if it were on the Moon, even though the object itself hasn't changed.

MASS is the inertial resistance of a body to acceleration. It is a property of the body, so it does not change with the body's location. It is equally resistant to motion in *any* direction. This makes mass a scalar quantity, while weight, being a force, is a vector.

So an object's mass is the same on Earth, on the moon, or in space.[12]

4.2 Newton's Second Law

Newton's second law describes what happens when a body *is* acted on by a resultant force, including when it's *not* zero:

FUNDAMENTAL LAW

Newton's Second Law: The acceleration of a body is directly proportional to the resultant force acting on it, inversely proportional to the mass of the body, and in the same direction as the resultant force.

That's a mouthful, but it is intuitive. It says that the larger the force on an object, the more it accelerates (that's directly proportional). However, the more mass the object has, the less it accelerates from that force (that's inversely proportional). So "force is proportional to mass times acceleration." This is written mathematically as:

Fundamental Law Eq. 4-2   **Any units**

$$\mathbf{F}_{Res} \sim m\mathbf{a}$$

Again, the "squiggle" means "is proportional to." This is always valid, regardless of the measurement units used. But to get the numerical answers we want, we need equations, and preferably simple ones. If (and only if) we use consistent units that are selected and defined to work together, we can rewrite Newton's second law in this classic equation:[13]

Fundamental Law Eq. 4-3a $\mathbf{F}_{Res} = m\mathbf{a}$ **Consistent units**

Note 12: It only makes sense that there should be some measure of an object's material that "belongs" to it, and isn't affected by gravity or other factors. Just because something becomes "weightless" by traveling far away from Earth's gravity, that doesn't mean its substance disappears.

Note 13: Consistent units will be explained shortly. The main point is that Newton's second law is not an equality unless you use a matched set of units for force, mass, and acceleration.

Here, **a** is the acceleration of the object (in either ft/sec² or m/sec²), $\mathbf{F}_{Res}$ is the resultant force acting on it (in pounds or newtons), and m is its mass, in slugs or kilograms (a one-slug mass weighs about 32.2 lbf). Notice that both $\mathbf{F}_{Res}$ and **a** are vectors. The equation says the resultant force and acceleration vectors point in exactly the same direction (in agreement with the last part of the above law).

This equation allows you to calculate the resultant force $\mathbf{F}_{Res}$ required to give a mass m an acceleration **a**. Equation 4-3a can then be re-written as:

Equation 4-3b

$$\mathbf{a} = \frac{\mathbf{F}_{Res}}{m}$$

This is "worded" the same as the above definition, saying acceleration is directly proportional to the resultant force and inversely proportional to the mass. Rearranging the equation again, the mass that will undergo an acceleration a from a given force $\mathbf{F}_{Res}$ is:[14]

Equation 4-3c

$$m = \frac{\mathbf{F}_{Res}}{\mathbf{a}}$$

CAUTION

Please remember not to confuse the "m" for mass with the "m" for meters when working in MKS units. To help, the variable m for mass is always in *italics*, but be careful anyway.

NOTE

Equations 4-3 tell you mathematically what every drag racer knows: to accelerate quicker (larger **a**), maximize the traction force while minimizing opposing forces such as aero drag (larger $\mathbf{F}_{Res}$), and keep the car as light as possible (lower m).

FIGURE 4 shows three examples of applying Newton's second law to a 1-slug mass or a 2-slug mass (weighing 32.2 lbf and 64.4 lbf, respectively). On the left, a 2-lbf force is applied to each. Using equation 4-3b, we find that the lighter mass accelerates at 2 ft/sec², while the heavier mass accelerates at 1 ft/sec². If we double the force on the larger mass to 4 lbf (middle of figure), it accelerates at 2 ft/sec², same as the smaller mass.

On the right side of the figure, the forces have the same magnitude as on the far left, but they are applied in the opposite (negative) direction.

Note 14: Care must be taken when dividing two vectors. It's okay here, because we know they point in the same direction.

The resulting accelerations are the same magnitudes as on the left, but in the negative direction.

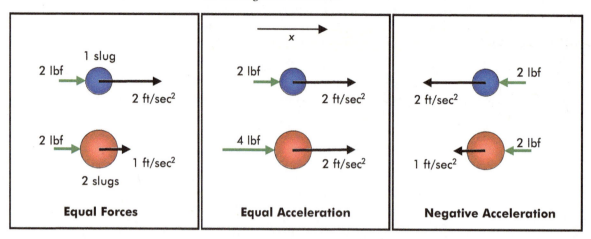

FIGURE 4 Masses of 1 and 2 slugs, with applied forces and resulting accelerations. With equal forces (left), the 1-slug mass accelerates twice as much. Applying twice the force to the 2-slug mass (middle) produces equal accelerations. Applying negative forces (right) produces negative accelerations. A mass of 1 slug has a weight of 32.2 lbf.

The units needed to use Newton's second law as an equation are summarized in TABLE 1.

Physical Quantity	SFS System	MKS System
Acceleration	ft/sec^2	m/sec^2
Mass	slug	kilogram
Force	lbf	newton

TABLE 1 Consistent units we'll use in the equation for Newton's second law.

Usually you will see equation 4-3a written as $F = ma$, which does roll off the tongue better than $F_{Res} = ma$. But early on, I want to emphasize that the acceleration is from $\mathbf{F}_{Res}$, the *sum* of the forces acting on the body. Quoting "$F = ma$" can lead to including only one force, and forgetting other important forces in the problem.

NOTE

After this chapter, Newton's second law will typically be written as "$F = ma$," assuming that you will use the resultant force in the equation.

Again, to form the *resultant force*, just sum the forces acting on the body, remembering that each force has direction. For instance, the tires might be pushing a car forward with 1,000 pounds, but aerodynamic drag is pushing it rearward with 600 pounds (-600 pounds forward). The resultant forward force is 400 pounds, and the resulting acceleration will be forward.

Revisiting our coasting car, the 100-lbf rearward resultant force from aerodynamic drag and rolling resistance will give the car a negative acceleration (a deceleration), slowing it down (see **FIGURE 5**).

KEY CONCEPT

Just as in Chapters 2 and 3, we'll assume the vehicle is rigid, so that all its parts move together. Then it behaves as if its mass were concentrated at its CG.

To find the car's acceleration, let's plug in some numbers. The car in **FIGURE 5** weighs 3,220 lbf, so its mass is 100 slugs. Our resultant force from the rolling friction and aerodynamic drag is equal to −100 lbf. Putting this into equation 4-3b gives the magnitude of the acceleration:

$$\mathbf{a} = \frac{\mathbf{F}_{Res}}{m} = \frac{-100 \text{ lbf}}{100 \text{ slugs}} = -1.0 \text{ ft/sec}^2$$

The resultant force gives the car's 100-slug mass an acceleration of −1 ft/sec² (about ⅓₀th of a g); the car will slow down at the rate of 1 ft/sec each second it travels. Note that this doesn't mean it will keep slowing down this quickly. Both rolling friction and aerodynamic drag will be reduced as the car slows down, so it won't slow down as quickly later.

CAUTION

Don't confuse negative acceleration with going backwards. Negative acceleration means the *velocity* is decreasing. Going backwards means the velocity is negative, and the *displacement* is decreasing. Braking hard is severe negative acceleration, but it won't make you go backwards!

FIGURE 5 A demonstration of Newton's second law on a 3,220-pound coasting GT40. Without a traction force, the sum of the forward forces F_{Res} is −100 pounds (from drag). This force acts on the car's mass m and produces rearward acceleration, slowing it by 1 ft/sec each second.

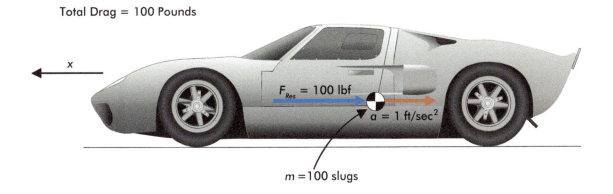

Total Drag = 100 Pounds

F_{Res} = 100 lbf

a = 1 ft/sec²

m = 100 slugs

Enough of this cruising and coasting; let's deal with some positive acceleration. What traction force would we need to accelerate the same car at 0.7 g's, assuming the same drag force? First of all, 0.7 g's is 22.54 ft/sec² (as shown in FIGURE 6). With the car's mass of 100 slugs, we would need a resultant force of 100 slugs × 22.54 ft/sec² = 2,254 lbf. With our rearward drag force of −100 lbf, we would need to increase the traction force to 2,354 lbf, to get the resultant force of 2,254 lbf.

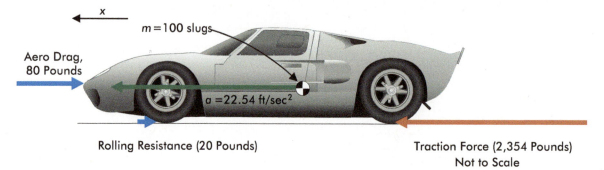

Rolling Resistance (20 Pounds)

Traction Force (2,354 Pounds)
Not to Scale

FIGURE 6 To accelerate at 0.7 g's (22.54 ft/sec²), we need enough traction force to overcome the total drag on the car, plus enough to produce the 0.7 g's acceleration.

We could have also written an equation for the resultant force ($F_{Traction}$ + F_{Drag}), plugged it into equation 4-3a, and solved it for the traction force. That's a more "mechanical" algebraic way, but it's best to start out thinking through the physics step by step.

4.3 Inertial Reference Frames

So far we've used Newton's second law to calculate acceleration forces without questioning whether acceleration was measured correctly. But we do need to be a bit careful. Just as we need to measure length or distance relative to some reference (like an edge on a part, or a starting line), we must measure acceleration relative to the right reference. But what is it?

Without saying so, we've been measuring acceleration relative to "stationary" references, such as a racetrack or a street. But Earth travels around the sun at about 67,000 mph, so the track is hardly stationary. Earth also rotates, so its surface moves 1,037 mph at the equator (767 mph in Detroit), relative to Earth's center. What on Earth makes it okay to consider a road to be stationary? To use Newton's second law, we must only use acceleration measured relative to an *inertial frame of reference*.

An **INERTIAL REFERENCE FRAME** is a reference coordinate system that isn't accelerating. It also can't be rotating. Newton's laws of motion are only valid with acceleration measured relative to an inertial reference frame.

We know Earth's surface isn't stationary, and that it rotates. But its acceleration is very small (less than $\frac{1}{100}$ of a g), and its rotation is slow. So for motions in the automotive field, we can ignore a road's acceleration and use it as an inertial reference.[15]

An example of a good inertial reference frame would be you sitting in the stands and watching a drag racer accelerate (see **FIGURE 7**, top). Since you aren't accelerating, you could measure the acceleration of the dragster *by careful observations* from your position, and use it to calculate the force acting on the car, or on the driver. Assuming the driver weighs 150 pounds (making his mass 4.6584 slugs), you could calculate the amount of force from the seat needed to accelerate him (600 lbf for the 4-g acceleration shown).

You could instead use the track and safety barrier as the inertial frame, since they do not accelerate. Either way, you correctly conclude that the dragster and driver will soon be several feet forward, but the track and safety barrier remain where they are (**FIGURE 7**, bottom).

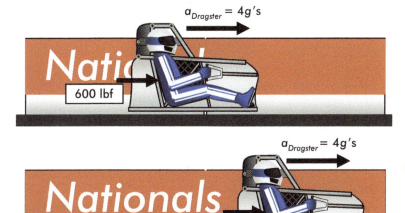

FIGURE 7 A cutaway of a drag racer that is accelerating at 4 g's relative to the track. The track, since it's not accelerating, makes a good inertial reference frame to correctly predict the 600-pound force on the driver's back.

An example of a *non-inertial* reference frame is the dragster itself. Let's contrast the previous (correct) view that the dragster is accelerating forward (**FIGURE 8**, upper), with the interpretation that the dragster and driver are not accelerating forward, but that the track is accelerating rearward (**FIGURE 8**, lower).

If the driver measures his acceleration relative to the dragster, he will conclude he's not accelerating, since he's moving with it. If he knows that $F = ma$, he should wonder why he feels a 600-lbf force from the seat pushing him, but it's not accelerating him!

Note 15: You can't always ignore Earth's motion. It can affect motion in large systems or where other effects are small. It causes low pressure systems, including hurricanes, to rotate counterclockwise in the Northern Hemisphere.

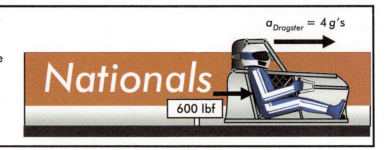

Inertial Frame:

Track (Correctly) Used as Reference

Car Accelerates Forward

Forward Force on Driver Explained

$a_{Dragster} = 4\,g's$

600 lbf

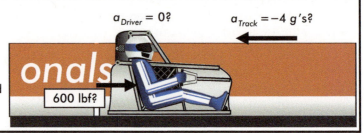

Non-Inertial Frame:

Car (Incorrectly) Used as Reference

Track "Accelerates Rearward"

Forward Force on Driver Unexplained

$a_{Driver} = 0?$

$a_{Track} = -4\,g's?$

600 lbf?

FIGURE 8 At top, the driver correctly uses the track as an inertial frame, concluding that the dragster is accelerating. At bottom, he incorrectly uses the accelerating car as reference, and concludes he is not accelerating. This leaves no explanation for the 600 pounds pushing him forward.

It may seem ludicrous to even consider the track (and the ground that it's attached to) to be accelerating, rather than the dragster. But measuring speeds between track and car only gives you their *relative* velocity and acceleration. The only way to prove this is incorrect is to show that the force F on the driver does not equal his mass m times his (incorrectly measured) acceleration; Newton's second law fails to agree. The reference frame must not be valid.

However, a moving reference frame *can* be an inertial frame if it's not accelerating, like a bus moving in a straight line at a constant speed. If you're sitting in the bus and toss a ball straight up, it comes back down to you just as if you were sitting in your living room.[16]

But not if the bus is braking or speeding up. If braking, the ball will appear to "fly forward" in the bus; if speeding up, it will seem to "fly rearward." This would look odd to you in the bus, because there was no force acting forward or rearward on the ball.[17] But it would make perfect sense to someone outside the bus, who would see the ball moving with a constant forward speed the whole time!

Note 16: I was tempted more than once to try this in the school bus with my lunch box, but with my luck, the driver would have hit the brakes at the same time.

Note 17: While there are no horizontal forces acting on the ball here, there is gravity acting downward and changing its vertical velocity. But gravity does not affect its horizontal velocity. In Chapter 7 we will begin working with forces acting in two directions, and in Chapter 10 with motions in two directions (dynamics in a plane).

5   Newton's Third Law: Forces Always Come in Equal and Opposite Pairs

Newton further realized that forces acting on a body don't appear "out of thin air." You've probably heard it spelled out: "For every action, there is an equal and opposite reaction." What it really means is that whenever one body is pushed on by a second body, the first body simultaneously pushes back on the second, with an equal strength:

FUNDAMENTAL LAW

Newton's Third Law: If body A exerts a force on body B, then body B exerts an equal and opposite force on body A, along the same line of action. This is true regardless of the type of force.

Let's say two spheres have opposite electrical charges, as shown in **FIGURE 9**. Each of the two bodies is acted on with 5 lbf, along the (dashed) line of action, in opposite directions. Note that these are two separate forces with the same cause, and the same magnitude. Here they are attractive forces, pulling the spheres toward each other.

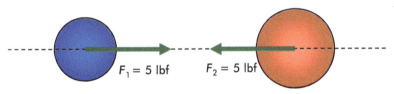

$F_1 = 5$ lbf $F_2 = 5$ lbf

FIGURE 9 When one body is acted on by a force (electrical, mechanical, or otherwise), there is always an equal and opposite force acting on another body. These are called the *action* and *reaction* forces. Which is which?

So when you push on the head of a wrench, it pushes back on you. When you pull on a rope, it pulls back on you. When a truck pulls forward on a trailer, the trailer pulls rearward on the truck. A force never acts alone. Newton's third law as an equation is:

Fundamental Law Eq. 4-4

$$\mathbf{F}_{Action} = -\mathbf{F}_{Reaction}$$

KEY CONCEPT

Note that although we call one force the action and the other the reaction, there is actually no order implied in this. Both forces occur simultaneously, so either force could be called the action, and the other the reaction. It's best to say that *each* force is the other's reaction.

FIGURE 10 takes FIGURE 3 and separates the car (mentally) from the road, making these force pairs clearer. Now we see that each of the forces acting on the car is half of a pair of forces. The 100-lbf traction force pushing forward on the car is paired up with a 100-lbf rearward force acting on the road surface. Same for the 20-lbf rolling resistance force. The 80 lbf of aero drag acts rearward on the car, while the car pushes forward with 80 lbf on the air.

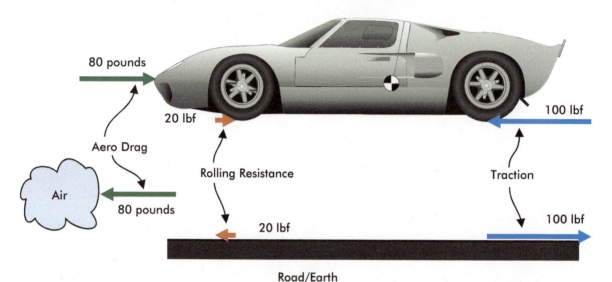

FIGURE 10 Per Newton's third law, each force acting on the GT40 has a twin acting in the opposite direction, on the road surface or on the air. These actions and reactions are paired by color in this figure.

Often, you don't need to show both forces in the pair, but you should realize that the second one is always there, because sometimes it does matter.

You could say that Newton's third law is one of the "big picture" laws that broaden your view of a situation when it's needed, such as where the rubber meets the road.

5.1 Free-Body Diagrams (FBDs)

FIGURE 10 above is a good example of what's called a free-body diagram, an excellent way to clearly define the forces acting on each body. With all the reaction force pairs, you need a way to keep them straight. As in FIGURE 10, you draw each interacting body separately, with the forces acting *on* it (not from it), according to Newton's third law.[18] You may only be interested in the motion of one of the bodies, but seeing the forces acting between them may be necessary to make their interaction clear. Usually both the horizontal and vertical forces are all shown, but FIGURE 10 only shows the horizontal.

Note 18: It's a lot like an exploded diagram of a transmission, with the parts moved slightly apart so each can be seen clearly.

> **KEY CONCEPT**
>
> Because of their importance, free-body diagrams (FBDs) are con-
> stantly hammered into physics students' minds. It's tempting to
> skip them, but they often show aspects of the problem you hadn't
> thought of. Besides, if you can't draw the free-body diagram, you
> probably don't understand the problem well enough to solve it.
> You haven't seen the last of these.

It's worth having a long look at **Figure 10**, because it may not be intuitive immediately. For instance, it might seem like the traction force
arrow should be pointed rearward from the rear tire. But remember
the traction force pushes rearward on the road, and forward on the car.

6   Dimensions and Units: Length, Mass, Time, Etc.

So far we've dabbled with dimensions and units, just enough to use the
formulas being discussed at the time. Now we're at the point where we
can see the bigger picture. The next few pages will detail basic dimensions and units, derived units, and how to track units to improve the
accuracy and reliability of your calculations.

To review, every measurement in physics — velocity, mass, etc. — has
certain dimensions, regardless of the measurement system you use. So
a tire's width could be measured in different units, like 255 millimeters,
10.04 inches, or 0.837 feet, but always has the dimension "length."

6.1 The Basic Dimensions

We've introduced the three basic (fundamental) dimensions you will
need in mechanics:

1. Length (in units like feet, meters, inches, miles, kilometers, etc.),
2. Mass (in units like grams, kilograms, and slugs), and
3. Time (in units like seconds, minutes, hours, etc.).

You'll see that the other dimensions and units used, including force,
energy, and power, are created from just these three. Recall from Chapter 2 that these other units are called *derived units*.

6.2 A Gearhead's Dilemma

Units are a real problem in a physics book geared toward gearheads and written in the U.S. We use customary (British) measurement units, which are not well suited to physics. On the other hand, many U.S. readers do not have a feel for metric units. That's why I've decided to do most of the problems in British Engineering Units (the slug-foot-second, or SFS, system). After all, drag racers quote "60-foot" times, not "18.288-meter" times!

As a result, I've already taken a lot of flak from fellow automotive engineers, since we only use metric units.[19] But engineers will already know that they can use the same equations with SFS or MKS units, and hold their noses (if necessary) at the SFS units. For most other readers, SFS units will be more familiar, and the few metric problems done will show how easy they are to use.

I have to admit that even though everything in my engineering job is metric, I still "think" in pounds and feet. A few simple conversions make it bearable, though. Since there are 25.4 millimeters in one inch, I mentally divide lengths in mm by 25 and think of them in inches. Another crutch is remembering that 1 mm, about $\frac{1}{25}$ of an inch, is 0.040 inches, or 40 thousandths of an inch. So when you bore an engine out by 0.040 inches, you increase the bore by about 1 mm. For "weight," remember that 1 kilogram is about 2.2 pounds. Just remember that a kilogram is not really weight, but mass.

6.3 Customary U.S. Units of Measure

As mentioned, customary U.S. measurement units originated in Britain. They were picked for convenience, or to compare to some familiar object. The "foot" is said to have been based on measuring the length of twelve men's feet as they left church one particular day, and taking the average. The "cup" is about as big as a coffee cup. But confusingly, you could also measure volume in ounces, pints, quarts, gallons, or bushels.[20] And however units like gills, yards, or stones were decided on, it wasn't to make physics calculations easy.[21]

Note 19: The Society of Automotive Engineers (SAE) Standards bulletin TSB003 refers to anything other than MKS units as "Old Units." So metric wrenches conform to SAE Standards, but wrenches sized in inches (often called SAE wrenches) do not!

Note 20: Even though it's not a consistent unit, I don't object to a "pint" now and then.

Note 21: The stone is a British unit of weight equal to 14 pounds, which never caught on in the U.S. However the 'Stones, also from Britain, were introduced to the U.S. in the 1960s and became extremely popular.

In the late 1700s, the French government commissioned a group of scientists to define a more logical measurement set.[22] This produced the *metric system* (which simply means "measurement system"), which is now the international standard. The U.S. began converting to it in the 1970s, but the effort stalled out.

6.4 Consistent Units of Force and Mass

In dealing with speed and distance, we didn't worry too much about unit sets. It was fine to have time in seconds or hours, and distance in feet, meters, miles, or kilometers. But once acceleration was introduced, it would have been awkward to use time in hours (measuring acceleration in miles/hour² would produce huge numbers that are difficult to relate to). So we used *seconds*.

Now that we've introduced force and mass, we need to be even more particular. Within a consistent set of units, one (1) unit of force applied to one (1) unit of mass must produce one (1) unit of acceleration. Only this turns the proportionality "$F \sim ma$" into the equality "$F = ma$."

> **NOTE**
>
> A consistent set of units has one unit for length, one for mass, and one for time, with the force unit defined so that force = mass × acceleration. This also sets the dimension of force as "mass × length/time²."

Now let's get down to brass tacks and discuss some other details.

6.4.1 The Meter-Kilogram-Second (MKS) System

The metric system has produced two consistent unit sets, the MKS (meter-kilogram-second) system and the earlier CGS (centimeter-gram-second) system. The CGS system worked fine, but has been officially "decommissioned," so we use MKS.

The MKS system was established in 1960, and is officially the SI system (from the French *Systeme International des Unites*), or System International. Being metric, it uses prefixes to multiply units to form larger or smaller units, where convenient. For example, "kilo-" means the size of the unit is multiplied by 1,000, so 1 kilogram is 1,000 grams. Also, "milli-" means dividing the unit by 1,000, so a millimeter is $\frac{1}{1,000}$ of a meter.[23]

Note 22: King Louis XVI approved the project on June 19, 1791, the day before he attempted to escape from the country at the beginning of the French Revolution. He was instead arrested and imprisoned.

Note 23: Other metric prefixes are conveniently listed in Appendix 1, on page 535.

The metric system definitions were originally based on natural quantities that were not likely to change (or if they did, measurement systems wouldn't be our main concern). For example, the length of one **METER** (abbreviated as "m"), was defined to be 1 ten-millionth of the distance from Earth's North Pole to the equator (making that distance 10,000 kilometers).[24] A standard "meter bar" was made of platinum, with two fine scratches on it one meter apart, and was kept in a vault for reference.[25]

Once the meter was defined, area in square meters (m²) and volume in cubic meters (m³) were also defined (remember the section in Chapter 2 on "derived units"). Since a cubic meter is often too large to be convenient, we typically give metric volumes in liters (such as a 5.7-liter engine). The liter is the volume in a cube $\frac{1}{10}$ of a meter on a side, and is $\frac{1}{1,000}$ of a cubic meter (see **FIGURE 11**).

Likewise, the **KILOGRAM** was originally based on nature. It was defined as the *mass* of one liter of pure water at 4° Celsius (39.2° Fahrenheit).[26] In the SI system, this is no longer exactly true, but it's very close. So one liter of diet pop (mostly water) has a mass of about 1 kg.

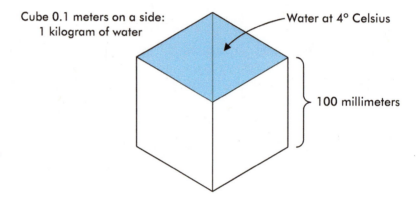

Cube 0.1 meters on a side: 1 kilogram of water — Water at 4° Celsius

100 millimeters

FIGURE 11 A cube that is $\frac{1}{10}$ of a meter on each side, a volume of 1 liter. If it's filled with water at 4° Celsius, the mass of the water is about 1 kilogram.

One **SECOND** of time was originally defined as $\frac{1}{3,600}$ of $\frac{1}{24}$ of the average day on Earth (the time between "high noons").[27]

The size of a newton follows from knowing that each newton of force applied to a mass of 1 kilogram results in 1 m/sec² acceleration. Dropping an object in Earth's gravity, where the applied force is the object's weight, produces a free-fall acceleration of 9.80665 m/s². So the weight of a 1-kilogram body must be 9.80665 newtons:

Note 24: King Louis XVI of France gave the order to measure this distance in 1792, from his jail cell. The survey was delayed by several wars along the surveyed path, and was finished six years later—five years after Louis XVI was executed.

Note 25: One meter is now officially defined as the distance light travels in $\frac{1}{299,792,458}$ seconds.

Note 26: This also means that a gram is the mass of a cubic centimeter of water.

Note 27: One second is now defined more precisely, from the duration of certain energy fluctuations in a cesium atom.

The *force* of one **NEWTON** is equal to 1 kg-m/s^2, and a mass of one kilogram has a weight of 9.80665 newtons in standard gravity. This is very important to remember when checking units in equations, which will sometimes require converting newtons to kg-m/sec^2, or vice versa.

6.4.2 The Slug-Foot-Second (SFS) System

The SFS (slug-foot-second) system is a consistent unit set created by specifying that length be in feet, time in seconds, force in pounds, and then introducing a new mass unit elegantly named the slug. This makes British units as easy to work with as MKS. Since 1893, British units have been defined in terms of metric units. Let's go through the basic SFS units for length, mass, and force.

For length, one **FOOT** is defined as exactly 0.3048 meters, so one meter is about 3.2808 feet. Accordingly, one square foot is exactly 0.09290304 square meters, and 1 cubic foot is exactly 28.316846592 liters. Following this through, 1 inch is exactly 25.4 millimeters, one cubic inch is exactly 0.016387064 liters, and one liter is about 61.0237 cubic inches.

Now for force and mass. We've said a couple of times that a mass of 1 slug weighs 32.2 lbf. But let's see why. It'll take a couple of steps.

First off, 1 pound is defined as exactly 0.453593237 kilograms, making 1 kg about 2.2046 pounds. But the kilogram is a unit of mass, not force. So we define 1 **POUND-MASS** (1 lbm) to be 0.453593237 kg. Just remember that the pound-mass is not a consistent mass unit; it *cannot* be used in the equations in this book. With that, defining the pound-force is easy. The *force* of one **POUND-FORCE** (1 lbf) is the weight in standard gravity of a body of 1 pound-mass. It's handy to know that 1 pound-force is exactly 4.4482216152605 newtons, but for us, 4.4482 newtons/lbf will do.

Now to define the slug. We know that applying a force of 1 lbf to a mass of 1 slug produces an acceleration of 1 ft/sec^2. To produce the free-fall acceleration of 32.174 ft/sec^2, the applied force (here, weight) on 1 slug must be 32.174 lbf. So we define one **SLUG** as an amount of *mass* that weighs 32.174 lbf in standard gravity. This also makes 1 slug equal to 32.174 lbm.

Having defined these fundamental SFS units, we'll derive most others from these. For example, per Newton's second law, 1 lbf is equal to 1 slug-ft/sec^2. And now we're fully justified in dividing a vehicle's "curb weight" in pounds by 32.174 to produce its mass in slugs.[28]

6.5 Other Derived Dimensions and Units

With the dimensions length, mass, and time, you can describe every other physical quantity you will need in mechanics. The most common

---

Note 28: The curb weight is the weight of the car as it sits with a full tank of fuel and no passengers or luggage.

are summarized in **TABLE 2**. A more detailed table, breaking each unit into its most basic units, can be found on the inside cover of this book.

Physical Quantity	Dimensions	Customary Units	Consistent Units	
			MKS	SFS
Time	Time	seconds, hours	seconds (sec)	seconds (sec)
Linear Displacement	Length	inches, feet, miles	meters (m)	feet (ft)
Speed or Velocity	Length/Time	miles/hr	m/sec	ft/sec
Acceleration	Length/Time^2	g's	m/sec^2	ft/sec^2
Mass	Mass	pounds, kg	kg	slug
Force	Mass × Length/Time^2	pounds	newton (N)	pound-force (lbf)
Torque	Mass × Length^2/Time^2	ft-lbf, N-m	N-m	ft-lbf
Pressure	Mass/(Length × Time$^{2)}$)	psi, kPa	pascal (Pa) 1 N/m^2	lbf/ft^2
Energy	Mass × Length^2/Time^2	BTU, calories	joule (J) 1 N-m	ft-lbf
Power	Mass × Length^2/Time^3	horsepower, kilowatt	watt (W) 1 J/sec	ft-lbf/sec

TABLE 2 Common physical quantities used in mechanics, in customary and consistent units.

Note that the MKS unit of pressure, 1 newton/meter^2, is named the pascal (abbreviated Pa) for convenience.[29] One newton acting on one square meter is a pretty small pressure, so pressure is usually given in kilopascals (abbreviated kPa). The label on the side of your tire that says "max inflation 350 kPa" means 350 kilopascals, or 350,000 newtons per square meter. This is about 50.75 psi or 7,308 lbf/ft^2.

7   Working the Problems Reliably

When you're driving, you need to keep your eyes on the road, or bad things happen. When using formulas, you need to keep your eyes on units. Depending on how you approach problems, the units can be either a stumbling block or a way to keep things straight. Let's demonstrate

---

Note 29: The pascal is named after Blaise Pascal, who was a scientist, mathematician and philosopher from the mid-1600s. Using his learnings in fluid mechanics, he invented the hydraulic press and the syringe.

good practice while calculating the acceleration of a 2,000-pound car from a forward 1,200-pound traction force, using $F = ma$.

7.1 Step 1: Put All Quantities into Consistent Units

The small step of converting to consistent units allows using the simple equations in this book. Note that the traction force, given as 1,200 pounds, is already in consistent SFS force units. But we have a "2,000-pound car," which is not in consistent mass units. To convert its mass from pounds of mass (lbm) to slugs, we know that there are 32.174 lbm in a slug:

$$m = 2{,}000 \text{ lbm} = 2{,}000 \text{ lbm} \times \frac{1 \text{ slug}}{32.174 \text{ lbm}} = 62.162 \text{ slugs}$$

Note that this unit conversion was done by multiplying and dividing by equal quantities, since 32.174 lbm equals 1 slug. Multiplying and dividing by the same quantity doesn't change the quantity we're looking at — only the numbers and units. This is the safest way to convert from one set of units to another (otherwise, it can be easy to forget whether to multiply or divide).

7.2 Step 2: Plug the Consistent Values into the Equations

Now we can calculate the acceleration using equation 4-3b:

$$\mathbf{a} = \frac{\mathbf{F}_{Res}}{m} = \frac{1{,}200 \text{ lbf}}{62.162 \text{ slug}} = 19.30 \text{ lbf/slug}$$

So, the acceleration is 19.30 lbf/slug? That doesn't sound like acceleration at all! We've got one more step to do to make sense of this.

7.3 Step 3: Convert Answers to Basic Units and Reduce Down

It was easy to get the numerical answer of 19.30, but now we need to work the units out. We'll convert the lbf to its basic units, which are slug·ft/sec² (use the table in the inside cover). We again multiply and divide by equal quantities to convert:

$$\mathbf{a} = 19.30 \frac{\text{lbf}}{\text{slug}} \times \frac{1 \text{ slug} \cdot \text{ft/sec}^2}{1 \text{ lbf}} = 19.30 \text{ ft/sec}^2$$

Note how once the lbf was converted to slug·ft/sec², most units can-celed out, "reducing" the units to their simplest form: ft/sec². At each step in this problem, we multiplied or divided the number out, as well as the units. When the units didn't neatly combine, we converted to the basic consistent units so they would.

If you follow this procedure when using the equations, you'll find yourself catching mistakes that you would have otherwise missed, and you'll develop more confidence in your results.

8   Summary

Nothing in physics is more important than Newton's second law:

"The resultant force is proportional to mass times acceleration."

Only a net force can cause acceleration. This is obviously important when your car accelerates, as in a drag race, when leaving a stoplight, or stopping for an intersection. In each case, your tires must be able to produce the required traction force. One way to increase accelera-tion is to increase traction, which is what stickier tires are all about. Another (often better) option is to reduce the vehicle mass, so a given traction force will produce more acceleration.[30] You also need to keep an eye on aerodynamic drag and rolling resistance.

But the second law is just as important when there is *no* acceleration, because then you know the net force on your car is zero; the traction force exactly offsets drag forces when driving a constant speed, and your parking brake must oppose other forces to keep your car from rolling downhill.

This isn't just useful to engineers and designers. The club racer knows that a bigger wheel/tire set may help traction, but any extra mass will partially offset the gain in performance — look for the ones with the lightest construction. Tires with a grippier rubber compound are always good, but will usually have more rolling resistance to overcome (not that it stops me from buying them). Bigger brakes are cool, but if you're not doing repeated heavy braking, maybe they're just extra mass.

Another critical factor in designing or driving a vehicle is Newton's third law:

"Every action has an equal and opposite reaction."

Note 30: Whether the lightened car can still produce as much traction depends on how it's lightened. On a rear-drive car, you can put on a lighter hood and front fenders, and still keep the same load on the rear tires to maintain drive traction.

In other words, the forces you need to accelerate your car aren't going to happen, unless you have something else to push against. To push forward on your car, your tires push rearward on the road. Likewise, when your cam pushes down on a valve to accelerate and open it, the valve pushes upward on the cam lobe.

But Newton's laws often mean other reaction pairs must develop. If you draw a free-body diagram of an overhead cam valvetrain and cylinder head, you'll see that while opening the valve, the cam must push upward on the cylinder head (through the bearing) while the cylinder head pushes down on it. The cam bearing and the cam lobe and follower driving the valve must be strong enough, and well lubricated enough, to withstand these forces.

Finally, working problems consistently is essential to trusting the results of your calculations. To do it, the step-by-step procedure outlined is essential:

1. Put all quantities into consistent units (SFS or MKS).
2. Plug these consistent values into the equations.
3. Convert your answers to basic units and reduce down.

By doing this, you can catch problems in your work, and therefore increase your confidence that not only will you get "the right number," but also that the answer will make good physical sense. Not doing so is like walking a high wire without a net.

Major Formulas

Definition	Equation	Equation Number
Resultant Force = the Sum of the Individual Forces	$\mathbf{F}_{Res} = \mathbf{F}_1 + \mathbf{F}_2 + ...$	4-1
Newton's Second Law: Resultant Force Is Proportional to Mass × Acceleration (Using Any Units)	$\mathbf{F}_{Res} \sim ma$	4-2
Newton's Second Law: Resultant Force = Mass × Acceleration (Consistent Units Only)	$\mathbf{F}_{Res} = ma$	4-3a
Newton's Third Law: An Action on One Body Always Has an Equal and Opposite Reaction on Another Body	$\mathbf{F}_{Action} = -\mathbf{F}_{Reaction}$	4-4

5 Forces

Agents of Change

Many forces are in action as Juan Pablo Montoya's crew lifts his Chevy SS and replaces the tires during a pit stop at Auto Club Speedway in Fontana, CA. *Photo by permission of Jan R. Wagner, AutoMatters.*

Chapter 5

The major types of forces important in the automotive field are described, such as gravity, friction, springs, and aerodynamic drag and lift.

Contents

Key Symbols Introduced

Symbol	Quantity	SFS Units	MKS Units
G	Universal Gravitational Constant	$lbf\text{-}ft^2/slug^2$	$N\text{-}m^2/kg^2$
F_{Weight}	Weight	pound-force (lbf)	newton (N)
mg	Weight	pound-force (lbf)	newton (N)
$g_{Standard}$	Standard Gravitational Acceleration	ft/sec^2	m/sec^2
$F_{Applied}$	Applied Force	pound-force (lbf)	newton (N)
k	Spring Stiffness	lbf/ft	N/m
ρ	Density	$slug/ft^3$	kg/m^3
p	Pressure	lbf/ft^2	N/m^2
$A_{Frontal}$	Frontal Area	ft^2	m^2
C_D	Aerodynamic Drag Coefficient	dimensionless	dimensionless
C_L	Aerodynamic Lift Coefficient	dimensionless	dimensionless
q	Electrical Charge	No SFS Units	coulomb (C)
μ_{Static}	Static Coefficient of Friction	dimensionless	dimensionless
$\mu_{Kinetic}$	Kinetic Coefficient of Friction	dimensionless	dimensionless

1   Introduction

Now we flesh out the forces that make things happen in the automotive world, such as vehicle weight, rolling resistance, aerodynamic drag and lift, spring forces, and traction forces. Most can be loosely grouped into these seven types:

1. Gravitation
2. Friction between surfaces
3. Elastic deformation and internal friction
4. Fluid pressure
5. Fluid dynamics
6. Adhesion
7. Electromagnetism

Some of these groups actually overlap. For instance, aerodynamic forces are fluid dynamics forces, and are mostly caused by differences in air pressure (a fluid pressure).

Let's get right to it.

2   Gravitational Forces, Weight, and Freefall

As much as we would like to defeat or get rid of gravity at times, it is very important in automotive physics. Gravity is the main force pressing the tires against the ground to produce traction. Without gravity, tires wouldn't get us anywhere!

Gravity is also one of the most important forces in the universe because there is a gravitational attraction force between every pair of objects in existence. This includes the forces between stars, planets, moons, and even between particles in the same body.[1] There is even a gravitational force between your hands! However, unless at least one of the objects is *very* large, the gravitational force between them is too weak to worry about. For example, when a magnet picks up a piece of steel, it is overcoming the gravitational pull on the steel from the entire Earth!

Note 1: As a solid planet forms, each part of it is pulled by gravity toward the other parts, so it becomes almost perfectly round.

Newton (who else?) proposed the universal gravitational law that describes this.

FUNDAMENTAL LAW

> The gravitational force between two bodies is proportional to the product of their masses, and inversely proportional to the square of the distance between their centers of gravity.

This law is true for any pair of masses. As an equation, it is:

Fundamental Law Eq. 5-1

$$F_{Gravitation} = \frac{Gm_1 m_2}{d^2}$$

where G is the universal gravitational constant, m_1 and m_2 are the masses of the two bodies, and d is the distance between their centers of gravity. The larger each mass is, and the closer they are to each other, the larger the force. The gravitational constant is equal to the amount of gravitational force between unit masses that are a unit distance apart, and is a very small value. The constant G is 6.673×10^{-11} N/(kg^2/m^2) in the MKS system, and 2.967×10^{-10} lbf/(slug^2/ft^2) in SFS.[2] This formula gives you the force of gravity between any two masses, such as the sun and Jupiter, or Earth and your car. In fact, once the constant G was found, the mass of Earth was calculated from equation 5-1!

Of course on the surface of Earth we call the gravitational force on a body its **WEIGHT**, which we could call F_{Weight}. To calculate the weight of a body, you could use equation 5-1, plug in the mass of the body and of Earth, and the distance between their centers. But if you do, you will just get the relationship between weight and force that we hinted at in Chapter 4. Letting m_1 in equation 5-1 be Earth's mass, and replacing m_2 by the mass m of the object on Earth's surface, its weight is:

Physical Law Eq. 5-2   **On Earth's surface**

$$F_{Weight} = F_{Gravitation} = mg$$

NOTE

> You rarely see weight written as F_{Weight}. Usually it is simply mg.

The value of g varies slightly with your location on earth. Because earth spins, it bulges out at the Equator. Due to this and other effects, the gravitational pull is slightly less on the Equator, and is a maximum at the North and South Poles.

Note 2: This is equal to 0.00000000006673 N-m^2/kg^2, or 0.0000000002967 lbf-ft^2/slug^2. Pretty small.

To put everybody on the same page, the scientific community has agreed on a standard ratio of weight to mass at Earth's surface:[3]

$$g_{Standard} = \frac{F_{Weight}}{m} = 32.1741 \ \frac{\text{lbf}}{\text{slug}} = 32.1741 \ \frac{\text{ft}}{\text{sec}^2} \quad \text{(SFS System)}$$

or:

$$g_{Standard} = \frac{F_{Weight}}{m} = 9.80665 \ \frac{\text{N}}{\text{kg}} = 9.80665 \ \frac{\text{m}}{\text{sec}^2} \quad \text{(MKS System)}$$

In this text, we can usually round off these values:

Standard gravity (approx.)

$$g \approx 32.2 \ \frac{\text{lbf}}{\text{slug}} \approx 9.81 \frac{\text{N}}{\text{kg}}$$

Earth's mass attracts us downward toward its center, so when we say downward, we always mean toward the center of Earth (see **Figure 1**). Weight *always* pulls the same amount, and straight down, even when you crest a hill or round a banked curve. Notice that the attraction is mutual, per Newton's third law; we pull up on Earth as hard as Earth pulls down on us. This is shown in **Figure 1**.

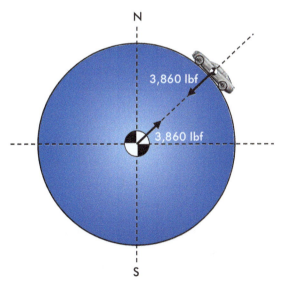

Figure 1 The gravitational forces between a 2012 Camaro SS and Earth, shown near Detroit (about 42 degrees north latitude). We call the force on the Camaro its *weight*, *mg*. The direction down is always toward the center of Earth.

Notice that the numbers we just used in relating a body's weight to its mass near Earth's surface (32.2 lbf/slug and 9.81 N/kg) and for its free-fall acceleration (32.2 ft/sec² and 9.81 m/sec²) are the same. They were called 1 *g* in both cases. It's no coincidence. Galileo had shown that all objects fall with the same acceleration at Earth's surface (neglecting air resistance). But Newton, using his laws of gravitation and motion, could show why.

Note 3: Remember that in Chapter 4 we used unit conversions to show that 1 lbf/slug equals 1 ft/sec², and 1 N/kg equals 1 m/sec².

The downward force on an object in freefall is its weight, *mg*, from the gravitational equation 5-2. If we use this in Newton's second law of motion (equation 4-3b) to calculate the body's downward acceleration this weight causes, we get:

Freefall

$$a = \frac{mg}{m} = g$$

So a "heavier" object (more mass *m*) has a greater downward gravitational force acting on it (more weight *mg*), but it also has more inertia *m* resisting acceleration. The two effects cancel out, and all objects fall with the same acceleration *g*. When dropped in air, this is only true until aerodynamic forces act on the objects—some more than others.[4]

Logically enough, the force from gravity (the weight) acts on an object's center of gravity, as shown in **FIGURE 1**.

2.1 A Falling Rock and Its Zero-to-Sixty Time

In freefall with no air resistance, gravity causes an object to accelerate downward at 1 *g*; its *downward* velocity increases by 32.2 ft/sec each second. **TABLE 1** shows the acceleration, velocity, and displacement for the first 6 seconds of freefall, in SFS units and MKS units.

Here, "up" is considered positive for displacement, so the downward acceleration is negative. So are the resulting displacement and velocity. That's not a problem; we just need to remember which way is up.

TABLE 1 is a good review of the difference between acceleration, velocity, and displacement. During the fall, the acceleration stays the same all the way through, weight being a constant force. Velocity changes by the same amount each second. Displacement changes more and more the longer the rock falls.

Note 4: Some will remember astronaut David Scott dropping a hammer and a feather on the moon, where there is no atmosphere, during the *Apollo 15* mission. They fell side by side to the moon's surface. Afterward he said, "Mister Galileo was correct in his findings!"

Time	SFS Units			MKS Units		
	Acceleration	Velocity	Displacement	Acceleration	Velocity	Displacement
(sec)	ft/sec²	ft/sec	ft	m/sec²	m/sec	m
0.0	−32.2	0.0	0.0	−9.81	0.00	0.00
1.0	−32.2	−32.2	−16.1	−9.81	−9.81	−4.91
2.0	−32.2	−64.4	−64.4	−9.81	−19.62	−19.62
3.0	−32.2	−96.6	−144.9	−9.81	−29.43	−44.15
4.0	−32.2	−128.8	−257.6	−9.81	−39.24	−78.48
5.0	−32.2	−161.0	−402.5	−9.81	−49.05	−122.63
6.0	−32.2	−193.2	−579.6	−9.81	−58.86	−176.58

Now, what about the rock's 0-to-60 mph time? **TABLE 1** (column 3) shows that the rock reaches 88 ft/sec (60 mph) in just a bit under 3 seconds. Using equation 2-9c, the exact time is:

TABLE 1 Downward motion of a rock falling from the pull of gravity, starting from rest at time = 0, until 6 seconds later.

$$t = \frac{v}{a} = \frac{88 \text{ ft/sec}}{32.2 \text{ ft/sec}^2} = 2.73 \text{ sec}$$

So the next time you read about a car going from 0 to 60 in 3.1 seconds, remember it isn't too far behind what it would do in freefall!

2.2 Using MKS Units

PROBLEM: Since we've been using mostly SFS units, let's plot how far the rock falls after each of the first 3 seconds, using MKS units.

SOLUTION: The acceleration from gravity is 9.81 m/sec² downward. Starting with $s = 0$, displacement positive upward, and using equation 2-13, gives:

$$s = \frac{1}{2}at^2 = \frac{1}{2}\times\left(-9.81 \text{ m/sec}^2\right)\times t^2$$

If you calculate the results every second, you can compare your results to those in TABLE 1. At $t = 1$ second, $s = -4.905$ meters; at $t = 2$ seconds, $s = -19.62$ meters; at $t = 3$ seconds, $s = -44.2$ meters.

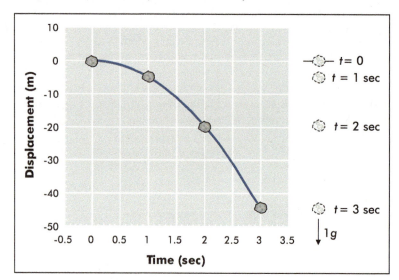

FIGURE 2 The downward motion of a rock in freefall, starting from rest. You can match the graphed motion with the pictures on the right, "frozen" at once per second.

KEY CONCEPT

Note that it is just as easy to use MKS units in the equations as SFS, even if you've never done it before. Same quantities, same equations, different numbers. Again, this is the advantage of using equations made for consistent units.

3 Friction Forces

When I took Driver's Education, I remember they said a tire's traction comes from friction. I only thought of friction as something that occurred between sliding surfaces (like when using sandpaper), so that didn't make much sense to me. But there are two types of friction: 1) **STATIC FRICTION** between contacting surfaces not sliding against each other, and 2) **KINETIC FRICTION** between surfaces that are.

Tires work mostly on the static friction in the contact patch between tire and pavement (**FIGURE 3**, left). When you talk about how much traction is available, it's mainly the amount of *static* friction the tires could produce. Spinning the tires of a Top Fueler slides the tread, and hurts acceleration. In slowing your car, your brakes work by *kinetic* friction (**FIGURE 3**, right). The brake pads each produce a friction force $F_{Friction}$ on the moving rotor as they are pressed against it by a force F_N. But when stopped, your brakes hold your car in place with static friction.

In either case, friction is a contact force that opposes sliding motion between two surfaces that are pressed together by a force perpendicular to the surfaces (called the **NORMAL FORCE**). "Normal" may sound odd here. In mathematics, a "normal" vector at a point is perpendicular to the surface. So our "normal force" is the force acting perpendicular to the contact surface. In **FIGURE 3**, the tire is pressed to the pavement by the load on it (from the car's weight, downforce, etc.), and the brake pads are pressed against the rotor by hydraulic pressure.

Static Friction   **Kinetic Friction**

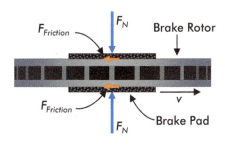

FIGURE 3 Static friction between the tire and pavement provides most of a tire's traction (left), while kinetic friction creates the braking forces on the brake rotor to slow your car (right). Once you've stopped, your brakes work by static friction. *Photo by permission of General Motors Co.*

Friction is arguably the most important type of force controlling how a vehicle operates, and for holding it together. Tires, brakes, clutches, nuts, bolts, drive belts, interference fits, and many other parts rely on friction to function. On the other hand, friction works against some vehicle functions, as in bearings and between piston rings and cylinders.

The amount of friction acting between two surfaces depends on their physical properties and surface finish, and on the normal force. Rubber on concrete has a lot more friction than rubber on ice, and metal on oily metal has a lot less friction than metal on dry metal.

Static friction is almost always greater than kinetic friction, because the small "bumps" in the surfaces (called asperities) can "lock together" more. No matter how smooth two mating surfaces may seem, there is roughness if you look closely enough. On the left side of **FIGURE 4**, you can see small friction forces developing between the bumps on the two surfaces. If you push on an object with an applied force $F_{Applied}$, and the friction keeps it from moving, the friction force acts in the opposite direction of your push, with *just enough* force to prevent the motion; it is equal and opposite to the applied force.

Kinetic friction is shown on the right side of **FIGURE 4**. Because the surfaces are moving past each other, they can't interlock as deeply, there are fewer contact points, and the forces at these points are directed less against the direction of motion; thus the friction force is smaller. One exception is when rubbing between two surfaces causes enough heat to produce micro-welds (as in aluminum on aluminum). In that case, kinetic friction can be higher than static.

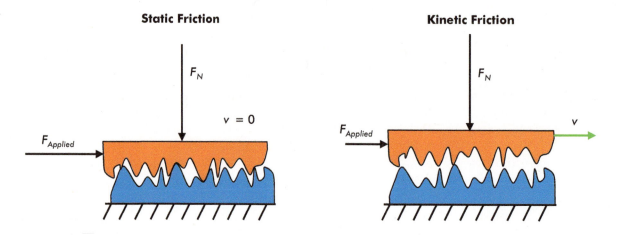

Static Friction   **Kinetic Friction**

F_N F_N

$v = 0$ v

$F_{Applied}$ $F_{Applied}$

FIGURE 4 The two main mechanisms of friction. Static friction allows more "interlocking" between the surfaces than kinetic friction does, so static friction forces are usually greater.

Many factors affect friction, like surface finish, corrosion, dirt, lubrication, and temperature. Some of these traits are intentional, and some are contaminants that get between the surfaces accidentally or due to aging. They are sometimes hard to know, control, or describe. The description here gives you a *rough* idea, though.[5] Some forms of friction actually include significant adhesion, which is described later.

One reason rubber is so good at producing traction on pavement is that, being "soft," it can *deform* to get down into the cracks and crevices in the road, as pictured in **FIGURE 4** (left). It's no coincidence that auto-crossing tires are made of very soft rubber, to get good grip without needing to be heated up first.

Note 5: Sorry, but I had to slide that in there.

4 Deformation Forces: Elastic and Frictional

Two types of forces are created as a flexible body is deformed from its original shape, that is, when it gets stretched or compressed, bent, or twisted. One is a stiffness force resisting the deformation, and the other is internal friction.

4.1 Elastic Deformation

Coil springs and torsion bars work by resisting twist in the spring wire, and leaf springs work by resisting bending in the metal leaves. Whole car bodies even bend and twist when you hit bumps, especially older convertibles. Seat belts and shoulder belts also stretch during a collision to cushion your stop (yes, they *do* stretch). In each of these cases, **RESTORING FORCES** develop within the body, which attempt to return it to its original shape. If the body does return to its original shape when the force is removed, the deformation is said to be **ELASTIC**.

Most elastic forces can be described by some sort of "spring" stiffness, whether or not you'd actually call the object a spring. The force exerted by a spring is proportional to its deflection:

Physical Law Eq. 5-3   **Hooke's Law**

$$F_{Spring} = -kx$$

where k is the spring's stiffness in either lbf/ft or N/m, and x is the deflection of the spring from its unstressed length (also called its "free length") in feet or meters. **FIGURE 5** shows a spring at its free length and at a deflected length. The negative sign means that the force from the spring pushes in the direction opposite its deflection. The applied force *on* the spring is the opposite of the force *from* the spring, so $F_{Applied} = kx$. Equation 5-3 is known as Hooke's Law, because Sir Robert Hooke proposed the law (he lived at the same time as Newton).

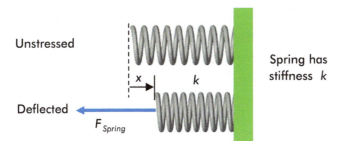

FIGURE 5 Springs push (or pull) back when compressed (or stretched). The spring's stiffness constant is k, and the force produced is $F_{Spring} = -kx$.

Hooke's "law," isn't a *fundamental law* like Newton's laws of motion or universal gravitation. Hooke's law only works under certain conditions, as seen in **Figure 6**.

A spring that acts according to equation 5-3 is called a "linear" spring, because if you plot the spring force vs. the deflection, you get a straight line over its working range (see **Figure 6**, repeated from Chapter 1). This spring pushes back with a force of 400 pounds for every inch of deflection, so its spring stiffness k is 400 lbf/in (4,800 lbf/ft in SFS units). But note how Hooke's law breaks down past a certain range in each direction, as the coils contact (under compression) or straighten out (under tension).

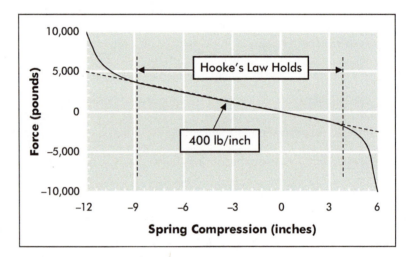

Figure 6 Hooke's law is valid over a spring's linear range, for this spring from 9 inches tension to 3.5 inches compression. There, the spring's stiffness k is 400 lbf/inch.

Coil springs with evenly spaced coils are good examples of linear springs. Multileaf springs aren't so linear because of friction between the leaves, which causes "binding." The spring doesn't move until it breaks the friction, so it acts very stiff for small motions, but then softens. The static friction causing the leaves to stick together is often called "stiction."

Tires act like stiff springs when they run over a bump, and the front of your car acts like a spring (fortunately) as it crumples when you run into another vehicle. The front end of your car does not deflect linearly, though, and it won't spring back much after being crushed. Deformation that doesn't fully spring back is not fully elastic, but is partially **PLASTIC**.

4.2 Internal Friction

Friction doesn't just happen between surfaces of two objects. All materials have **INTERNAL FRICTION** between the very particles they're made of. When you pull on a belt, most of its opposing force is from its stiffness (as in **FIGURE 5**). But an internal friction force also pulls back against you. The difference is that when you let go, the stiffness force is still working to shorten (contract) the belt. But the internal friction force is *resisting* the contraction.

Internal friction is relatively small in metals, but not so small in rubber and fabrics (like in the carcass of a tire). It is what causes tires to heat up from rolling; the tread is repeatedly straightened when contacting the road and curved again when picked back up. Even in metals, when you bend a wire back and forth far enough and often enough, you can feel the heat from its internal friction.

5   Fluid Pressure Forces

There are two types of fluids: liquids (like water, oil, and gasoline) and gases (like air and nitrous oxide). Pressure in either type of fluid exerts a force on all the surfaces the fluid touches, and vice versa. Pressure is defined as the force per unit area (force divided by area), so force is equal to pressure multiplied by area:

Physical Law Eq. 5-4   **Uniform pressure**

$$F_{Fluid\ Pressure} = pA$$

The force produced is directly proportional to the pressure and the area; the greater the pressure, the greater the force, and the greater the area, the greater the force.

Pressure acting on a flat surface like a piston face produces a force perpendicular to it. What if you curve the face by "doming" or "dishing" it? Although "side forces" are produced on each part of the curved surface, they all cancel out, and produce the same total force, in the same direction, as if it were flat.

Fluid pressure forces are important in driving pistons down an engine's cylinders, in hydraulic brakes, in hydraulic lifters, in supercharging, and in vacuum motors. For example, a brake master cylinder "converts" force to pressure in the brake fluid, and then a piston in the caliper "converts" that pressure to a force acting on the brake pad and caliper (see **FIGURE 7**). Because the brake fluid is nearly incompressible, the brake force is produced almost instantly when the pedal is depressed. By using a large piston in the caliper, a large force can be applied to the pads when a small force is applied at the master cylinder.

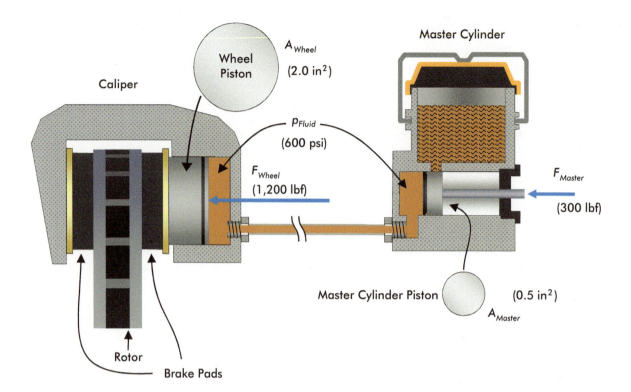

Caliper

Wheel
Piston

A_{Wheel}

(2.0 in^2)

Master Cylinder

p_{Fluid}

(600 psi)

F_{Wheel}

$(1,200 \text{ lbf})$

F_{Master}

(300 lbf)

Master Cylinder Piston

(0.5 in^2)

A_{Master}

Rotor

Brake Pads

FIGURE 7 Forces in a hydraulic brake system. The pressure in the brake fluid trapped between the pistons is the same throughout, so a small force F_{Master} at the master cylinder becomes a large clamping force F_{Wheel} on the brake pads.

Airbags in supplemental restraint systems provide examples of how fluid pressure can be used, both in deploying them and in producing their cushioning force. Airbags are quickly inflated by a high-pressure gas, for example by a chemical reaction from a solid mixing with a liquid. The gas starts off at a high pressure to fill the bag quickly, getting it into position in thousandths of a second. But to produce the appropriate restraining force, it quickly vents off some of the pressure to provide a softer "air spring" to cushion you while slowing your forward motion.

6   Fluid Dynamics Forces

Fluid motion causes forces from two main sources: 1) internal friction from viscosity, and 2) pressure from fluid inertia. The forces act on both the fluid and whatever it touches. One example is viscosity in liquid flow in a pipe, causing a pressure drop downstream; or between two plates (as in **FIGURE 8**, left). You can see that the viscosity causes the flow to "stick" at the walls, where it transmits the drag force to the wall in the direction of the flow (**FIGURE 8**, right). Flow velocity is greatest midway between the plates. Other examples are the drag and lift forces a car experiences moving through the air, which are mostly from pressure differences from the fluid's inertia.

Fluid Velocity Profile   **Free-Body Diagram of Fluid**

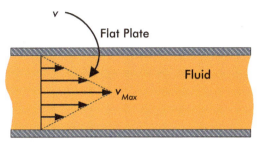

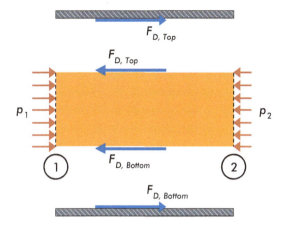

We won't fully describe aerodynamic forces here, since that would involve a deeper dive into fluid mechanics. But since aerodynamic forces are very important to us, we'll summarize the formulas. The force F_D from aerodynamic drag is:

Physical Law Eq. 5-5

$$F_D = \frac{1}{2}\rho v^2 C_D A_{Frontal}$$

FIGURE 8 Laminar (layered) flow between two plates. Viscous friction causes the layers of fluid to have different velocities (left). They "rub" against each other and create a drag force that tries to pull the boundary walls with it (right). It also causes a pressure drop (p_1 is larger than p_2).

Here, the Greek letter ρ (spelled "rho" and pronounced *row*) is the density of the surrounding air, and v is the velocity of the vehicle relative to the air. The vehicle's design is summarized by its drag coefficient C_D, and by its frontal area $A_{Frontal}$ (the size of the hole it punches through the air, in **FIGURE 9**). If you have a headwind, you have to add the wind velocity to the car's velocity before squaring it. Since aerodynamic forces grow with the square of velocity, a car that has 100 lbf of drag at 60 mph would have 400 lbf at 120 mph (this has a profound effect on drag horsepower, as we'll see in Chapter 14).

FIGURE 9 The frontal area of this CTS-V can be estimated by taking its picture from a long distance (to reduce parallax), drawing a border around it, and splitting that into triangular and rectangular sections. Scale the figure with some known distance, like the length of the trim piece over the grille.

The frontal area of the CTS-V in **Figure 9** is 26 ft², and its drag coefficient is 0.355. Using that, and the fact that the density of air at 70 degrees F is 0.0747 lbm/ft³ (which is 0.00232 slugs/ft³), the drag force at 70 mph is 112.9 lbf. Most production cars have drag coefficients in the 0.25 to 0.4 range, depending on the purpose and shape of the car.

Vertical aerodynamic forces F_L are classified as **LIFT**. Lift is usually split up into components acting at the front and rear axles. Downforce is negative lift. The lift forces $F_{L,F}$ and $F_{L,R}$ are calculated as:

Physical Law Eq. 5-6a   **Front lift**

$$F_{L,F} = \frac{1}{2}\rho v^2 C_{L,F} A_{Frontal}$$

and:

Physical Law Eq. 5-6b   **Rear lift**

$$F_{L,R} = \frac{1}{2}\rho v^2 C_{L,R} A_{Frontal}$$

where $C_{L,F}$ and $C_{L,R}$ are the front and rear lift coefficients, respectively. If a car has downforce on the front or rear, it has a negative lift coefficient on that axle. **Figure 10** shows the aerodynamic forces, with lift exaggerated relative to the drag. Lift coefficients for a production car might be in the 0.05 range, much smaller than the drag coefficient, and are sometimes slightly negative.

Figure 10 The aerodynamic forces on an (admittedly stationary) CTS-V: the drag F_D, the front axle lift $F_{L,F}$, and the rear axle lift $F_{L,R}$. Lift is shown oversized for clarity.

Aerodynamic forces are of course critical in racing, where downforce is used to press the tires harder to the ground (increasing their normal force) to increase traction, especially for high-speed cornering. The downside of downforce is that it comes at the cost of increased drag. The combined lift coefficients for a Formula 1 car can be around −3.00, but the resulting drag coefficient can be around 1.30. At a high enough speed, these cars produce more downforce than their weight, which would allow them to drive on an upside-down road!

7   Adhesion Forces

Adhesion is the property that makes one surface bond (stick) to another one, strongly as with glue, or weakly, like with a Post-It™ note. There are many instances of adhesives in gaskets, plastic parts, and more recently as structural adhesives, especially on aluminum bodies. Soldering and brazing could also be considered as adhesion, in which bonding occurs between two pieces of metal (as opposed to welding, where the metals are melted into one piece).

Drag slicks are one example of adhesion that is less obvious. They have a soft and gummy rubber tread, and do a "burn-out" to heat the tread and make it sticky. While other tires use adhesion to some extent for grip, drag slicks take it to another level.[6]

8   Electricity and Magnetism

Electricity and magnetism are closely linked to each other; in fact, magnetic fields are caused by electrical currents, which are nothing more than moving electrical charge. Electrical charge is measured in coulombs, which is the charge of 6.24×10^{18} electrons. One ampere is an electrical current of one coulomb of electrons per second.

There are two main types of mechanical forces caused by electromagnetism. ELECTROSTATIC forces are produced between electrical charges that are stationary (static). ELECTROMAGNETIC forces (or simply MAGNETIC forces) are produced by electrical current.

8.1 Electrostatic Forces

Every atom of every material is composed of protons, neutrons, and electrons (hydrogen alone has no neutrons). Protons have a positive (+) electrical charge, electrons have a negative (−) charge, and neutrons have no charge. Charge is represented by q. Atoms usually have an equal number of protons and electrons, so their charges cancel each other; most objects have no significant net charge.

But if an object is made to shed electrons, it has more protons than electrons, and it has a positive charge. Likewise, an object that acquires extra electrons has a negative charge. Rubbing the soles of your shoes on dry carpet is a good way to build up a negative charge on yourself by picking up electrons. Touching a metal door knob (or a nearby cat) is a good way to send them back to the room, through a spark of electric current. This equalizes your charge with the room's. If we have two

---

Note 6: Also, drag strips are sprayed with a traction compound that is essentially an adhesive. I've heard that it will pull your shoes right off if they're not tied tight enough!

objects with opposite charges q_1 and q_2, electrostatic forces will attract them (see **FIGURE 11**, top). If the charges both have the same sign (positive or negative), the forces will repel them (**FIGURE 11**, bottom).

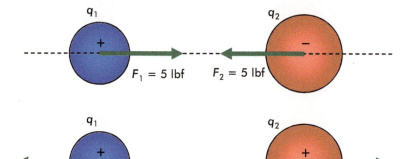

FIGURE 11 When two objects have opposite electrical charges (top), it produces attracting forces between them. With the same sign, the forces are repulsive (bottom).

The electrostatic forces between the two bodies are proportional to each of the two charges, and inversely proportional to the square of the distance between them:

Physical Law Eq. 5-7

$$F_{Electrostatic} \sim \frac{q_1 q_2}{d^2}$$

With opposite charge, this produces attracting forces. Notice how this is very similar in form to equation 5-1 for the gravitational force between two bodies. Just remember that while gravity is always attractive, electrostatic forces can be either attractive or repulsive. Since this is only an introduction to electrostatics, we'll leave this as a proportionality, rather than an equation.

8.2 Electromagnetic Forces

All magnetic fields are produced by electric currents, which are charged particles in motion (usually moving electrons). In permanent magnets, the electric current originates from lining up many atoms so that their electrons orbit in the same direction. In electromagnets, also called solenoids, the current is pushed by a battery's voltage through many turns of wire, with the result that many electrons circle in the same direction (see **FIGURE 12**). The result is a magnetic field in the solenoid, with its north pole at the left end, south at the right. It also induces a magnetic field in the nearby steel bar, forming a south pole at its right end. The north pole of the solenoid and the south pole of the bar are attracted to each other by equal and opposite (there's that phrase again) magnetic forces.

In a solenoid, the more turns in the windings, the stronger its magnetic field. Also, a greater electrical current (more amperes) produces a stronger magnetic field. Using an iron core in a solenoid is not absolutely necessary, but it is good practice. Depending on the type of iron, or perhaps steel, that is used, it can increase the magnetic field strength by a factor of more than 1,000!

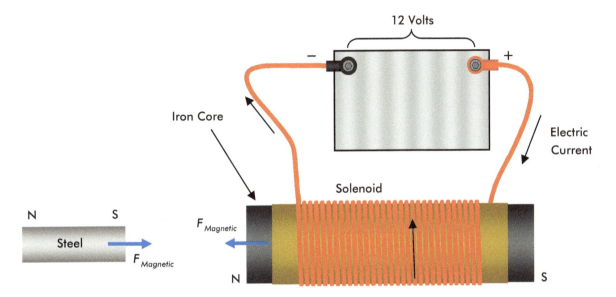

Although electricity and magnetism are closely related, they are not exactly the same. Electric charges come in two kinds: positive and negative, but they do not have to be paired together. Magnetic fields, on the other hand, always have two poles, one called north, and one south. You'll never find a north pole by itself, but you can find a negative electrical charge (like an electron) by itself. Magnetic poles do act similar to electrostatic charges, in that a north pole from one magnetic field (magnet) attracts the south pole of another magnet. A north pole repels another north pole, and south poles repel each other. Like many other things in life, opposites attract, and likes repel.

FIGURE 12 An electromagnet (solenoid). A battery creates a voltage difference, which forces a current of electrons through the wire as shown. The electric current flowing around the iron core creates a magnetic field with its north pole to the left, south pole to the right, attracting the steel bar (seen at the left) with a magnetic force.

Electric motors have magnets in the (stationary) stator and the (rotating) rotor, some of which are electromagnets. For example, **FIGURE 13** shows part of a permanent magnet motor, similar to those in many hybrid-electric vehicles. The motor works by changing the current direction in each electromagnet in the stator back and forth, to alternate its inner pole between north and south (currently, the current is going clockwise through the wire loop and producing a north pole at its inner end). By timing the current correctly, each electromagnet attracts an approaching magnet in the rotor, and then repels it once it passes, to propel it in the desired direction. Because of the alternating direction of the voltage and current necessary to run it, this type of motor is called an alternating current (AC) motor.

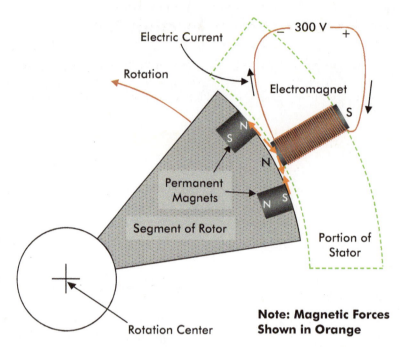

FIGURE 13 A permanent magnet electric motor uses magnetic forces (in orange) between the magnets in the stator and rotor to produce torque. Currently, the north pole of the electromagnet is attracting the incoming south pole and repelling the north pole that just passed.

Note how the outward-facing polarity of the rotor's permanent magnets must be arranged in an alternating pattern (N-S-N-S...) so that each electromagnet in the stator will attract the approaching magnet and repel the receding magnet, to produce a positive torque with both.

Electromagnetic forces are important in the solenoids in power door locks, relays, and fuel injectors, as well as motors that drive fuel pumps, start engines, and propel vehicles.

9   A Closer Look at Friction Forces

Since friction is so important, let's look at it in more detail. As mentioned before, there are two main types of friction between two surfaces: static (stationary contact) and kinetic (sliding contact). Let's look at static friction first.

> **CAUTION**
>
> The following gives a first approximation of friction, assuming that friction force capability is directly proportional to the normal force pressing the two surfaces together. This is often good enough, but in Chapter 6 we'll see some situations where it's not.

9.1 Static Friction

FIGURE **14** shows a block lying on a surface (left), pulled down by its weight mg, and a free-body diagram of the block and floor (right). An applied force $F_{Applied}$ pushes on the block to the right, but not hard enough to break it loose, so the block remains stationary. To have no acceleration, the forces on the block must add to zero, so:

Physical Law Eq. 5-8   **Static friction**

$$F_{Friction} = -F_{Applied}$$

Static friction is always equal and opposite the resultant applied force. Equation 5-8 holds as long as the magnitude of the friction force doesn't exceed the maximum available static friction. Assuming it is proportional to the normal force, that maximum is:[7]

Physical Law Eq. 5-9a   **Static friction**

$$\left| F_{Friction,\,Max} \right| = \mu_{Static} F_N$$

In this equation, μ_{Static} is called the static coefficient of friction, which is defined as the ratio of the maximum achievable static friction force to the normal force:[8]

Equation 5-9b   **Static friction**

$$\mu_{Static} = \frac{\left| F_{Friction,Max} \right|}{F_N}$$

Note 7: Again, the straight brackets signify the absolute value of what is inside them (remember that the absolute value of a number is its "size": 5 and −5 both have an absolute value of 5).

Note 8: The Greek letter μ is spelled "mu" and pronounced mew, like a kitten says.

Equation 5-9a not only sets the limit to the static friction, but also defines a threshold for when the friction changes from static to kinetic.

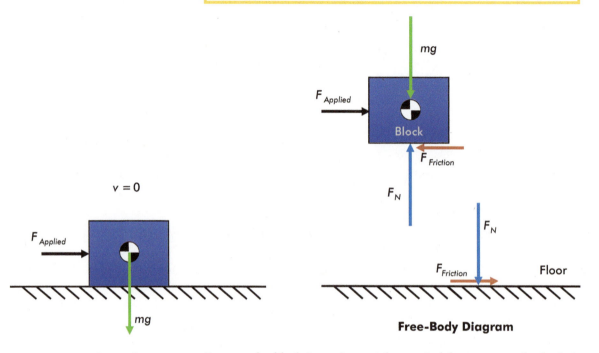

Free-Body Diagram

FIGURE 14 Static friction between a stationary block and floor, with an applied force $F_{Applied}$ on the block. The friction force $F_{Friction}$ just cancels the applied force. The normal force here is equal to the weight mg of the block.

Because the block is stationary, the vertical forces must also be balanced. This tells you that the magnitude of the normal force is mg. Putting this all together, if the block weighed 100 lbf, and the coefficient of friction was 0.70, the maximum static friction force would be 70 lbf. If you push on it to the right with 30 lbf, friction will push on it to the left with 30 lbf; it doesn't budge. With 60 or 65 lbf, same thing. Only when the applied force exceeds 70 lbf will the block move.

9.2 Kinetic Friction

Kinetic friction, or sliding friction, is shown in **FIGURE 15**. Notice that there is a larger applied force acting on the block than in **FIGURE 14** (static friction), but the friction force is smaller. In fact, the friction force no longer depends on the applied force at all; it only depends on the kinetic coefficient of friction and on the normal force:

Physical Law Eq. 5-10   **Kinetic friction**

$$\left| F_{Friction} \right| = \mu_{Kinetic} F_N$$

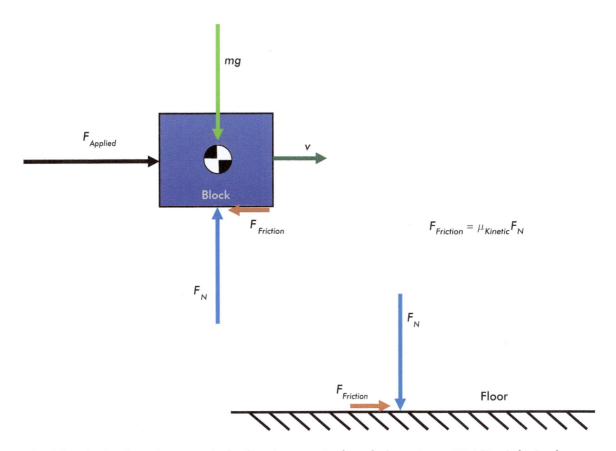

$$F_{Friction} = \mu_{Kinetic} F_N$$

The sliding friction force always acts in the direction opposite the velocity.

Since kinetic friction force only depends on the friction coefficient, the normal force, and the direction of the sliding, it wouldn't matter if there were no applied force at all! Imagine the block was initially given a push to start it sliding, and then it slides on its own. There is no applied force, but the kinetic friction force is still acting on the block with the same magnitude, opposing its motion. It would continue to do so until the block stopped.

Conversely, the object could be accelerating, but the friction force would still be the same. In **FIGURE 15**, the applied force is larger than the friction force, so the block will accelerate with $a = (F_{Applied} - \mu_{Kinetic}F_N)/m$. As the block accelerates the friction force does not change, because while its direction depends on the velocity, we assume its magnitude does not (in reality, the friction force will vary somewhat with velocity).

FIGURE 15 A kinetic friction force is equal to the kinetic coefficient of friction multiplied by the normal force, and acts in the direction opposite the velocity.

9.3   Coefficients of Friction

The coefficient of friction is what determines whether we would call a surface "slick" or say it has a good grip. Putting your weight on a slick surface like ice doesn't produce much friction between your shoe and the ice, so it has a low coefficient of friction. Putting water on the ice reduces the friction even further, and so gives an even lower coefficient of friction.

A sample of coefficients of friction is shown in **TABLE 2**. As you can see, the range is huge, from 0.04 for Teflon® on steel, up to about 1.2 for the rubber in performance street tires. Tire traction depends a lot on the compound and aging of the rubber compound, the condition of the pavement, whether there is gravel or "marbles," temperature, and a number of other factors.

Contacting Materials	μ_{Static}	$\mu_{Kinetic}$
Teflon on Teflon or Steel	0.04	0.04
Clutch Material on Steel (in Oil)	0.08–0.15	0.06–0.12
Clutch Material on Steel (Dry)	0.20–0.50	0.15–0.40
Steel on Steel (Screw Threads)	0.12–0.18	-
Steel on Steel (Dry)	0.6–0.74	0.4–0.57
Steel on Steel (Greased)	0.1	0.05
Brake Material on Cast Iron (Wet)	0.2	-
Brake Material on Cast Iron (Dry)	0.4–0.8	0.2–0.3
Rubber on Ice (Dry)	0.1–0.2	0.1–0.2
Rubber on Ice (Wet)	0.03	0.01
Rubber on Concrete (Wet)	0.5–0.8	0.4–0.7
Rubber on Concrete (Dry)	0.7–1.2	0.6–0.9
Aluminum on Aluminum	1.05–1.35	1.4

TABLE 2 Approximate values for the coefficients of static and kinetic friction for various material pairs.

There is a large uncertainty in each coefficient of friction. This is reason for caution, especially when it comes to tire traction. Unexpected changes in the road surface, like gravel, can take you from driving at 90% of your car's cornering capability to needing 110%. Not good.

But friction can let you down in other places, like in clutches and brakes. Too little friction in a clutch may let it slip and overheat, while too much might make it touchy to engage. In brakes, too little could require you to push on the pedal with both legs to stop fast, while too much will make them hard to modulate. For brake and clutch linings, higher friction materials are called more "aggressive."

When a tire is rolling, but not sliding, the tread in the contact patch is primarily in static (stationary) contact with the road at any given instant, so its traction is limited by the static coefficient of friction. When most of the tread is sliding, the coefficient of friction drops quite a bit, which is one reason you typically don't want to lock up your tires and skid when braking, and why spinning your wheels to accelerate isn't normally the fastest way to go.

9.4 Units and Dimensions of Coefficients of Friction

Being picky about units, I should mention why there are none listed for coefficients of friction. In equation 5-9b, the coefficient of static friction is the ratio of the maximum friction force to the normal force. If a normal force of 100 newtons could produce a maximum friction force of 70 newtons, the coefficient of friction for the two surfaces is:

$$\mu_{Static} = \frac{\left|F_{Friction,Max}\right|}{F_{Normal}} = \frac{\left|70\ N\right|}{100\ N} = 0.70$$

The answer of 0.70 is just a pure number. What happened to the units? The newtons in the top (numerator) were divided out by the newtons in the bottom (denominator). The coefficient of friction is said to be dimensionless, or to have a dimension of 1 (anything divided by itself is 1, even units). Whether the forces are in newtons or lbf, the ratio is the same. If you prefer, you can think of the friction coefficient units as "newtons per newton" or "pounds per pound."

Of course, this means that any ratio between two measurements with the same dimensions will be dimensionless. This includes gear ratios, as well as drag and lift coefficients.

10 | Summary

In Chapter 4, we summarized physics as "Force = Mass × Acceleration, all the rest is details." Here we started filling in the details about the forces causing the acceleration.

The basic forces described here form a fairly complete list of those determining the physics of your car (and most other things, for that matter):

1. Gravitation
2. Friction between surfaces
3. Elastic deformation and internal friction
4. Fluid pressure
5. Fluid dynamics
6. Adhesion
7. Electromagnetism

These types of forces stock the gearhead's toolbox; they are what you can use to make the car do what you want. It's your job to use them the best you can to propel and control your car, hold it together, and keep you and your passengers comfortable and safe.

Much of the challenge in building and tuning cars is that so many forces depend on others. For example, increasing downforce usually increases drag. Picking the best suspension stiffness and damping to keep the tires following the road may not be the best to produce a good ride, or to keep the body at the best ride height for aerodynamic downforce.

The first step in improving performance is to know what forces are necessary to meet your goals. In-car data is indispensible to measure the actual effect of each change you make. But if you have problems matching your performance calculations with actual results, don't blame Newton:

KEY CONCEPT

Newton's laws are fundamental, and for our purposes are exact. Any difficulty we have in accurately predicting motion is not from his laws, but from our inability to correctly predict and/or measure the forces that go into them.

Major Formulas

Definition	Equation	Equation Number
Gravitational Force between Two Masses Separated by Distance d	$F_{Gravitation} = \dfrac{Gm_1 m_2}{d^2}$	5-1
Weight = Mass × Acceleration of Gravity	$F_{Weight} = mg$	5-2
Spring Force = − Spring Constant × Spring Deflection	$F_{Spring} = -kx$	5-3
Fluid Pressure Force = Pressure × Area	$F_{Fluid\ Pressure} = pA$	5-4
Drag Force in Terms of Air Density, Drag Coefficient, Frontal Area, and Velocity	$F_D = \dfrac{1}{2}\rho v^2 C_D A_{Frontal}$	5-5
Lift Forces in Terms of Air Density, Lift Coefficients, Frontal Area, and Velocity ("F" for Front Axle, "R" for Rear Axle)	$F_{L,F} = \dfrac{1}{2}\rho v^2 C_{L,F} A_{Frontal}$	5-6a
	$F_{L,R} = \dfrac{1}{2}\rho v^2 C_{L,R} A_{Frontal}$	5-6b
Electrostatic Force between Two Charges Separated by Distance d (Repulsive for Same Sign)	$F_{Electrostatic} \sim \dfrac{q_1 q_2}{d^2}$	5-7
A Static Friction Force Is Equal and Opposite to the Resultant Applied Force	$F_{Friction} = -F_{Applied}$	5-8
Maximum Static Friction Force in Terms of the Static Coefficient of Friction and Normal Force	$\left| F_{Friction,\ Max} \right| = \mu_{Static} F_N$	5-9a
Kinetic Friction Force in Terms of the Kinetic Coefficient of Friction and Normal Force	$\left| F_{Friction} \right| = \mu_{Kinetic} F_N$	5-10

6 Dynamics Applications

Forces and Motion in Vehicles and Mechanisms

The Ford GT40 of Jacky Ickx and Paul Hawkins digs in, charging hard exiting one of the many curves of the Nürburgring's infamous Nordschleife (North Loop). Despite the new rules reducing its engine size from 7.0 liters to 5.0 for the 1968 season, it finished this 1,000-km race in third place. *Photo by permission of Ford Motor Company.*

Chapter 6

Dynamics is applied to a variety of automotive problems, including traction forces in drag racing and braking. Cam forces in a rigid valvetrain are calculated, and thrust and drag forces are estimated for the Thrust SSC land speed record run from its velocity trace.

Contents

1   Introduction

In Chapter 4 we covered the *equation of motion* for a rigid body in straight-line motion:

$$F_{Res} = ma$$

which means "resultant force equals mass × acceleration."[1] This is simply Newton's second law of motion. Because it is so short and simple, it may not be obvious how powerful it is. But using it correctly can straighten out many situations that seem complicated. To begin with, using this law can give you control over a situation by applying the necessary forces:

1. To accelerate something (to change its velocity) you must apply a net force to it.
2. To maintain its velocity, you must have zero net force on it.
3. To keep it stationary, you must have zero net force on it.

Of course, stationary is just a special case of constant velocity. Conversely, if you know a body's motion, you can figure out what's causing it:

1. If a body is accelerating, there must be a net force acting on it.
2. If a body moves at a constant velocity, the net force on it must be zero.
3. If a body is stationary, the net force on it must be zero.

In each case, if you know two of the three (resultant force, mass, or acceleration), you can calculate the third. We'll be demonstrating that for different situations in this chapter.

KEY CONCEPT

The kinematics we have developed so far is very basic; we can only predict motion from constant acceleration. But today we are often given the acceleration and/or velocity from some measurement, such as in a car magazine or from an in-car data recorder. In that case it's easy to calculate the forces necessary to cause that acceleration, using $F_{Res} = ma$ directly. We'll be doing just that.

Note 1: An equation relating the forces acting on a body to its acceleration is often called an "equation of motion."

2   What's Pushing Me Back in the Seat?

Well, that's easy: nothing is. When your car accelerates, you need to be accelerated, too, or you'll be left behind (remember **Figure 7** and **Figure 8** in Chapter 4). Since your seatback is right behind you, that's the first thing to catch up to you as the car moves forward. Your body has inertia (mass), so the seat has to push you to make you accelerate with the car. Nothing is pushing you into the seat; it's the seat that's being pushed into you!

3   Braking Forces in a Performance Road Car

PROBLEM: The Ferrari Enzo in Chapter 3 braked with an acceleration of -35.523 ft/sec^2, or -1.10 *g*'s. If the curb weight of the Enzo is 3,230 lbf, and the driver weighs 200 lbf, what total tire force is required to cause this level of braking performance?

SOLUTION: The total weight is $3,230 + 200 = 3,430$ lbf. Converting weight to mass by using equation 5-2, $m = 106.61$ slugs. To find the braking force, we use equation 4-3a:

$$F_{Res} = ma = 106.61 \text{ slugs} \times \left(-35.523 \text{ ft/sec}^2\right) = -3,787 \text{ lbf}$$

For the units in this calculation, remember that 1 slug·ft/sec^2 is 1 lbf.

Note that the braking force is greater than the weight of the car. Is that reasonable for a road car? Read on.

4 A Sense-Check on Tire Coefficients of Friction

PROBLEM: An Internet website I found said the coefficient of static friction for rubber on dry pavement ranged from 0.7 to 0.9. How does that square with the braking performance of the Enzo in the previous problem?

SOLUTION: Remember that the total weight with driver was 3,430 lbf. This is how much force is available to press the tires against the road (producing the combined normal force, F_N). We're assuming that there is no aerodynamic downforce at such low speeds, and that drag forces are small compared to the traction forces. Let's also assume all the tires have the same friction coefficient, and are being braked to their maximum traction capability. Using equation 5-9b:

$$\mu_{Static} = \left| \frac{F_{Friction,Max}}{F_N} \right| = \frac{3,785 \text{ lbf}}{3,430 \text{ lbf}}$$

so:

$$\mu_{Static} = 1.10$$

The tires on a road car certainly *can* have a coefficient of friction on pavement greater than 0.9, because the Enzo's is 1.10 at the very least.

Rather than relying on some μ_{Static} number off the Internet to characterize tires, you are better off using 60–0 or 70–0 braking data out of a reputable car magazine, and back-calculating the tire numbers you're after. This way, you know the rubber was part of a tire used on the vehicle it was designed for, and that load transfer from rear to front was involved. Note that load transfer is usually called "weight transfer." (More on that beginning with Chapter 16.)

4.1 A Shortcut to Finding μ from a in Braking

The calculations here showed that when the braking deceleration was 1.10 g's, the minimum coefficient of friction for the tires was also 1.10. This is no coincidence. During braking, all the car's weight is on tires that are braking (we're assuming four-wheel brakes, of course). If each tire's maximum traction is being used, the total braking force is equal to the coefficient of friction times the car's weight, mg:

$$F_{Braking} = -\mu_{Static} mg$$

This force results in the (negative) braking acceleration, so using $F_{Res} = ma$:

$$-\mu_{Static} mg = ma$$

Dividing both sides of this by the mass and by g:

$$\mu_{Static} = \frac{-a}{g}$$

From equation 2-12a, you'll recognize the right-hand side as just the acceleration in g's. Substituting that in,

Equation 6-1   **Max Braking**
$$\mu_{Static} = -a_g$$

Remember we assumed that all tires were at their maximum traction level. If they weren't, it means they could have produced a little more traction, and the friction coefficient must be somewhat higher.

> **NOTE**
>
> Finding μ from forward acceleration isn't so simple, because the load on the driving tires is less than the car's weight (unless it's all-wheel drive). But you do know that the acceleration in g's can't be *more* than μ_{Static}, at least not without aerodynamic downforce.

5   Braking Distance on the Moon

PROBLEM: If we took the same Ferrari to the moon, it would weigh about ⅙ as much. Relieved of all that weight, what would be the stopping distance on an identical road surface?

SOLUTION: The gravitational acceleration at the moon's surface is 5.31 ft/sec², compared to 32.2 ft/sec² on Earth. The Ferrari's mass is still the same as on Earth, at 106.61 slugs. Its weight on the moon can be calculated using equation 5-2:

$$F_{Weight} = mg = 106.61 \text{ slugs} \times 5.31 \text{ ft/sec}^2 = 566.10 \text{ lbf}$$

So the Ferrari only weighs 566.10 lbf on the moon. Now we need to find the maximum braking force, based on this new weight. Remember that the car's weight is the normal force pressing the tires to the road. Assuming the same 1.10 static coefficient of friction for the tires, the maximum braking force is:

$$F_{Friction} = \mu_{Static} F_{Weight} = 1.10 \times 566.10 \text{ lbf} = 622.71 \text{ lbf}$$

Now that we have the maximum braking force and already know the car's mass, we can calculate the braking acceleration:

$$a = \frac{-622.71 \text{ lbf}}{106.61 \text{ slugs}} = -5.841 \text{ ft/sec}^2$$

Remember that its braking acceleration on Earth was −35.53 ft/sec², which produced a stopping distance from 60 mph of 109 ft. With an acceleration of −5.841 ft/sec² on the moon, the braking distance would be, using equation 2-18a:

$$s_F = \frac{-v_0^2}{2a} = \frac{-(88 \text{ ft/sec})^2}{2 \times (-5.841 \text{ ft/sec}^2)} = 663 \text{ ft}$$

So reducing the car's weight by 87% *increased* our stopping distance by 500%! (This was sort of a trick question). By moving the car to the moon, we reduced its weight, and with it most of the load on the tires, and their traction. This reduced traction force had to decelerate just as much mass as on Earth, so the braking distance ballooned.

I've actually seen it suggested (incorrectly, of course) that you should use lift to reduce the weight of the car (making it "lighter"), and make it easier to accelerate!

Why talk about this "moon scenario" in a physics book about cars on Earth? Besides recommending longer following distances on the moon, we can make a couple of more important points:

1. Mass is the enemy of performance; weight is not.
2. Weight produces the normal force on the tires, producing traction.
3. Increasing the normal force on the tires without increasing vehicle mass (as with aerodynamic downforce) gives you the best of both worlds.

Incidentally, this explains why everything involving traction happens more slowly on the moon. You see in NASA footage that astronauts walk slowly, and the lunar rover has to stop and turn slowly. It's easier to lift things, but harder to accelerate their mass using friction for traction.

6   Real Tire Properties

As a gearhead, have you noticed something missing in our description of traction? Tire size! To find traction, we've multiplied the coefficient of friction for rubber on pavement by the normal force pressing it to the road. That's it! There was no mention of anything to do with tread width, inflation pressure, carcass construction, or tread block shape. According to what is written so far, you could take a performance tire (size and all) that produces 1.10-g braking on a Honda Civic, put it on a (much heavier) BMW 7 Series, and get the same performance. Are all us gearheads who put wider tires on for better traction wasting our money? Not likely!

Using the friction coefficient to find the available friction (traction) is an example of a first-order (or linear) *approximation*, because if you plot the predicted friction force vs. the normal load, you get a straight line. That's a good first shot with small loads on the tire, but this doesn't last forever. The grip begins to tail off at a certain load.

The vertical **LOAD** on a tire, as we've been referring to it, is the downward force on the tire from the suspension. Since it provides the normal force F_N between the tire and the road, it largely determines how much traction that tire can produce on a given surface.

To illustrate this, **FIGURE 1** shows the traction capability of two tires that are identical in construction, but the larger one is built for a car that weighs 50% more. At loads below 800 lbf, each tire can produce a friction force about 1.10 times the load it's carrying (same as for the braking Enzo), so its traction capability is the straight line $F_{Friction} = \mu_{Static} F_N$, shown on the chart. Note that after 800 lbf load, the smaller tire can't produce 1.10 times the load any more. But the larger tire still does, up to about 1,200 lbf load.

Let's say you're accelerating with 1,500 lbf load (vertical line in **Figure 1**) on each of the two driving tires. If you used the smaller tires, each one could produce 1,360 lbf of forward traction. But if you used the larger ones, each could produce 1,620 lbf. If the car weighed 4,000 lbf (its mass is 124.22 slugs), the maximum acceleration with the smaller tires would be 22.0 ft/sec^2 (0.683 g's), while the larger tires could accelerate the car at 26.1 ft/sec^2 (0.811 g's).

For the record, the friction coefficient of 1.1 would predict 1,650 lbf per tire, and produce 26.6 ft/sec^2 (0.825 g's). The larger tires come pretty close to this. We'll get into more detail on tires as we need it.

Figure 1 A comparison of the linear tire friction capability based on the friction coefficient (green dashed line) vs. actual (solid lines), for tires of different sizes.

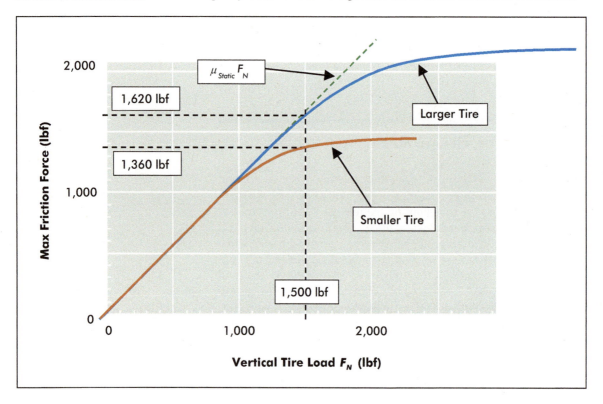

7   Extreme Acceleration, Extreme Force

PROBLEM: In Chapter 2 we found that Cory McLenathan's dragster had an average acceleration of 129.55 ft/sec² for the first 3 seconds. During this time, what was the resultant traction force required to cause this acceleration, ignoring aerodynamic drag? The dragster weighs 2,400 lbf, including the driver. What is the coefficient of friction for the rear tires, assuming the load on them is 95% of the car's weight?

SOLUTION: Convert the weight of 2,400 lbf to mass, using equation 5-2. Use $m = 74.53$ slugs in equation 4-3a to find the tractive force:

$$F_{Res} = ma = 74.53 \text{ slugs} \times 129.55 \text{ ft/sec}^2 = 9,660 \text{ lbf}$$

which is a lot of force on such a light car. As for the coefficient of friction of the rear tires, we're assuming that the load on them (the normal force) is 95% of the car's weight of 2,400 lbf, or 2,280 lbf, assuming no downforce. So equation 5-9b would indicate the minimum μ_{Static} is:

$$\mu_{Static} = \frac{9,660 \text{ lbf}}{2,280 \text{ lbf}} = 4.24$$

Those would be some sticky tires! Now to be fair, I have read that the headers shooting the exhaust gases upward can provide 800 lbf of downforce for a Top Fueler (a great use of Newton's third law!). Assuming this is in the ballpark, the normal force on the rear tires would be about 3,080 lbf, giving a coefficient of friction of "only" 3.17. Either way, these estimates are still way over the typical values given for rubber.

7.1 A Theoretical Limit to Friction?

Is there a theoretical limit to friction? Forty-some years ago someone got the idea that a coefficient of friction of more than 1.0 was "theoretically impossible," which would say that tire traction couldn't accelerate your car at more than 1.0 *g*'s. Using that, they calculated a maximum theoretical top speed (199 miles/hour) for a quarter-mile. (You could easily check the ET and top speed for a 1-*g* run using equations 2-13 and 2-10, remembering that ¼ mile is 1,320 ft.)

Shortly afterward, Don Garlits broke through this "barrier." Now the acceleration is over four times this "theoretical" limit! Even though most physics books list μ_{static} for rubber to be below 1.0, none I've read said it *couldn't* be greater than 1.0.

The claim that car acceleration over 1 *g* is impossible still shows up on Internet forums. Call it an automotive urban legend.

8   F = ma: It Isn't Just a Good Idea, It's the Law

PROBLEM: You are riding in a car without a seat belt at 30 mph, and the car hits a tree. The car comes to a stop, and *after the car stops* your head hits the dashboard, denting it in 4 inches before coming to a stop. Using Newton's second law, estimate the average force on your head during its collision with the dash, if your head weighs 15 lbm.

SOLUTION: Newton's first law says you won't slow down until a resultant force makes you. With no seat belt, this doesn't happen until you hit something (in this case, the dashboard), so your velocity v_0 at the beginning of your 4-inch stop is 30 mph (44 ft/sec). If the acceleration were constant over the 4-inch stop (⅓ foot), then using equation 2-18b:

$$a = \frac{-v_0^{\,2}}{2s_F} = \frac{-\left(44\ \text{ft/sec}\right)^2}{2 \times 0.3333\ \text{ft}} = -2{,}904\ \text{ft/sec}^2$$

That's about 90 g's! Your head has a mass of 15 lbm/(32.2 lbm/slug) = 0.466 slugs. So, using equation 4-3a, the force on your head is:

$$F_{Res} = ma = 0.466\ \text{slugs} \times \left(-2{,}904\ \text{ft/sec}^2\right) = -1{,}353\ \text{lbf}$$

That's way more force on your skull than I'd want! Stopping your head in 4 inches from 30 mph just isn't fun. On the other hand, wearing a seat belt gives you about 4 feet to stop in: the belt stretches about 2 feet, while the front end of the car is crushing about 2 more feet. Stopping in 4 feet is still sudden, but the average force could go down from 1,353 lbf to 113 lbf (your neck pulls back on your head to slow it). You may still get injured, but most likely it will be much less severe.

I'm not a crash expert, and these calculations are just estimates, but I'd rather stop in 4 feet than 4 inches any day!

It's disturbing to think of yourself as a missile hurtling toward an obstacle, but in a crash, that's what you become. The forces involved are too large for your muscles to have any effect. You need to rely on seat belts for protection. Engineers put a lot of design work into protecting you during a collision. Don't waste it. Wear the belt! Even with an airbag, the seat belt is your first line of defense.

Full Disclosure: I must admit I never wore my belt until I did some calculations like this for a college report in 1982; I had thought seat belts just kept you from flying out of the car (which is important in itself)! Once I realized that they decelerate you (relatively) slowly while the front end of the car crushes, I got the idea. I have *never* gone without them since. I don't need any law to tell me to wear seat belts, besides the laws of physics.

9 A Crude Estimate of Piston Pin Force

PROBLEM: Chapter 3 had an example of a NASCAR engine turning 9,500 rpm, or 158.3 rev/sec, with a stroke of 3.25 inches (0.27083 feet). Estimate the maximum force on the piston pin, assuming the piston mass is 1 lbm (0.031056 slugs). Calculate the results at 9,600 rpm, to make the numbers work out cleaner (making it 160 revolutions/second).

SOLUTION: Before calculating pin force, we need to ballpark the piston's acceleration. For lack of better knowledge, let's assume that when the piston accelerates, downward or upward, the acceleration is constant. Then the velocity will increase (or decrease) in a straight line during each part of its motion (see **FIGURE 2**). Since the piston stops at TDC and at BDC, the velocity is zero at both times, and the piston's top speed must occur halfway down the cylinder (90° after TDC, or 90° ATDC). Remember that TDC stands for "Top Dead Center," and BDC stands for "Bottom Dead Center," the top and bottom of the piston's travel.

At 9,600 rpm, each piston stroke takes 1/320 of a second (0.003125 seconds) and covers 0.27083 ft, so the average speed is 86.667 ft/sec. The peak speed must be twice that, or 173.33 ft/sec.

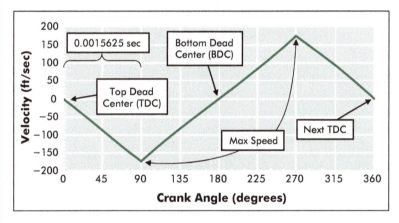

FIGVURE 2 The velocity profile of the piston at 9,600 rpm, assuming its acceleration is constant and downward from TDC to 90° after, constant upward from there to 90° after BDC, and constant downward from there to the next TDC.

The acceleration to maximum speed at 90° after TDC occurs in 0.0015625 seconds (not much time to accelerate!). Using equation 2-13 and solving for the acceleration a:

Equation 6-2

$$a = \frac{2s_{90°}}{t_{90°}^2}$$

where $s_{90°}$ is the displacement at 90 crank degrees (half the stroke), and $t_{90°}$ is the time to get there (0.0015625 seconds). Putting our displacement into consistent units, $s_{90°}$ is −3.25/2 inches, or −0.13542 ft. With this, equation 6-2 becomes:

$$a = \frac{2 \times (-0.13542 \text{ ft})}{(0.0015625 \text{ sec})^2} = -110{,}936 \text{ ft/sec}^2$$

PROBLEM CONTINUES

Continued

This is about $-3,445$ g's! Remember we assumed it was constant for the first 90° of crank rotation, as shown in **Figure 3**. By the same reasoning, the acceleration would be $+110,936$ ft/sec² (upward) from 90° to 270°, then downward again from 270° to 360°.

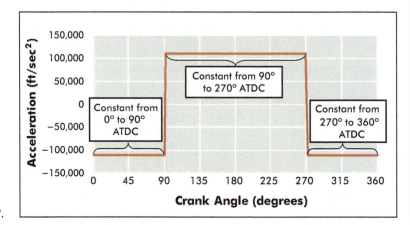

Figure 3 Crudely estimated acceleration of the piston at 9,600 rpm. Note that it is positive (upward) from 90° to 270°, and negative (downward) from 270° back to 90°.

Now that we know the acceleration and the mass, we can find the piston pin force (ignoring the forces of cylinder pressure):

$$F_{Pin} = ma = (0.031056) \text{ slugs} \times \left(-110,936 \text{ ft/sec}^2\right) = -3,445 \text{ lbf}$$

So the piston pin acts with about the weight of the car on the 1-lbm piston to push it up or down at 9,600 rpm. This is why speed takes money—standard rods and pins wouldn't survive the repetitive forces from accelerating a heavy piston at high speeds. A lot of effort and expensive materials go into making the parts not only strong, but light!

Just in case you had any doubt the piston ends up moving the right amount (couldn't blame you really), the displacement from TDC is shown in **Figure 4**. Sure enough, the piston moves downward by its stroke at BDC.

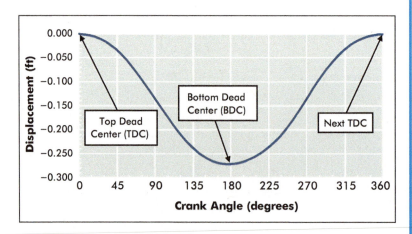

Figure 4 The piston displacement from TDC, resulting from our constant-acceleration assumption.

NOTE

This ballpark calculation produces a believable displacement trace, but if we calculated the actual piston motion, we'd find the velocity and acceleration traces are considerably off. The peak acceleration and piston pin force are actually about 55% higher, and the peak piston speed happens at about 75° ATDC, not 90°.

10 Thinking Like a Cam Designer

This is an opportunity to look at motion where acceleration isn't constant, but is *almost always changing*. The key is to remember that instantaneous velocity is the slope of the displacement-time curve at that instant, and likewise that instantaneous acceleration is the slope of the velocity-time curve.

FIGURE 5 shows a sketch of the simplest valvetrain mechanism: a cam lobe acting directly on a bucket follower. As the cam rotates, the lobe pushes down on the follower, which pushes down on the valve. The cam shown has just rotated past its base circle (the no-lift portion) and has just lifted the valve off its seat. The valve is headed downward to open the intake port and allow air to flow into the cylinder. When the nose of the cam lobe is pointed straight down, the valve is at its maximum opening (max lift).[2]

To make things simple, we'll assume the bucket follower and the valve are rigid bodies, and that the follower never loses contact with the cam.

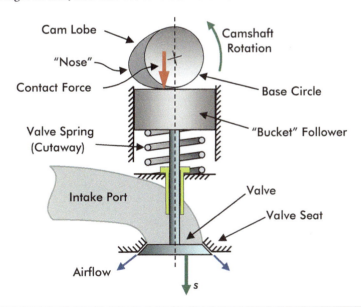

FIGURE 5 A simple overhead cam bucket-follower valvetrain, shown just after the lift event starts. Note that downward motion s is positive.

Note 2: Note how the situation changes the meaning of terms. In aerodynamics, lift is upward, while in valve motion, lift is away from the valve seat, which here is straight down!

Valvetrain forces are one of the limiting factors in engine performance and operating speed. Opening a valve 50 times a second (at 6,000 rpm) or more is serious business. The ideal intake-valve mechanism, from a flow standpoint, would jam the valve fully open instantly (limited by valve-to-piston interference) when you want to start the inlet airflow, hold it there, and then slam it closed instantly when you want the flow to stop.

But the faster you open and close the valve, the greater the acceleration. By now you know that more acceleration means more force.[3] If the accelerating forces were too great, the cam lobe would hammer the follower, and the follower would hammer the valve. The cam bearing would also take a beating, and the engine would sound like trash (but only until it broke). The only question is what would break first.

So, how can you design a good lift curve? You need to hold the valve open as far as possible, for as long as possible, while avoiding excessive acceleration of the hardware. **Figure 6** is what a typical cam lift profile might look like (although it was fabricated by the author). Lift is really the valve displacement s, with $s = 0$ at the seat. Lift increases slowly at first, ramps up to peak in the middle, and then decreases as the valve seats itself.

What part of the lift profile puts the most force on the valve stem, cam lobe, and follower? Hmmm, look at that steep part of the ramp where the lift is increasing really fast (about 70 crank degrees before peak lift). Must be there, right? All right, let's look.

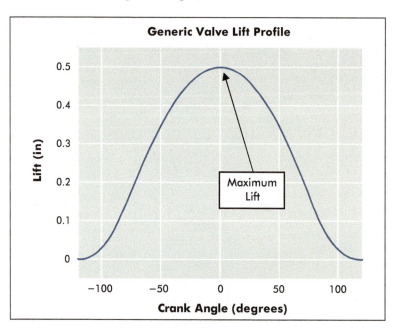

Figure 6 Valve lift (displacement) profile for a typical cam, with 240° duration and 0.5 inches peak lift. Crank angle is shown relative to peak lift.

If we look at the velocity trace for 3,000 rpm (**Figure 7**), it starts off at zero, reaches its peak at the steepest part of the lift profile (about −70°

Note 3: Using $F = ma$, you can show that you can't even cause an instant change in velocity, much less an instant change in position!

in **FIGURE 6**, remember), is zero at peak lift, then becomes negative as the valve changes direction and returns to its seat.

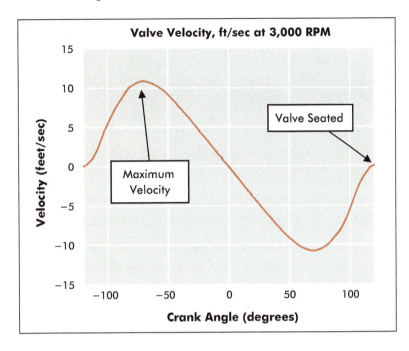

FIGURE 7 Valve velocity for the lift profile shown in FIGURE 6, with the engine running at 3,000 rpm.

The acceleration profile (**FIGURE 8**) may surprise you. It starts off at zero, quickly builds to a maximum (105 degrees before max lift), and becomes *zero* at the steepest part of the lift profile (70 degrees before max lift). It goes negative for a long time, until it's almost back at the seat. At this point, the acceleration goes large positive to slow the valve quickly so it doesn't hit the seat hard.

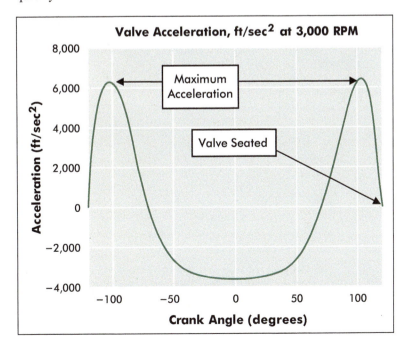

FIGURE 8 Valve acceleration for the lift profile shown in FIGURE 6, at 3,000 rpm. Note how early the maximum acceleration happens.

Let's dissect these plots as a group. The steepest part of the lift profile is at $-70°$. Because velocity is at its maximum (**Figure 7**), it isn't changing fast, and acceleration is about zero (**Figure 8**). In other words, the steepest, nastiest looking part of the lift profile is where the smallest acceleration and accelerating force are produced!

The maximum acceleration actually happens at the steepest part of the *velocity trace* (about $-103°$). If you really look closely, the maximum upward acceleration happens where the displacement curves upward the tightest (**Figure 6**).[4] Since $F_{Res} = ma$, the maximum forces occur here too. This part of the lift profile looks innocent, but it is not.

Overall, the acceleration profile jams the valve open fairly hard, and then waits until the last millisecond to slow it before it seats. This holds the valve open as long as possible without hammering the hardware. The small levels of acceleration in the middle of the profile are also important because the cam lobe can only push on the follower—it can't pull.[5] When the valve is accelerating toward the seat, only the valve spring provides the force to do so. Low acceleration during this part of the motion allows lower valve spring preload and/or spring stiffness, reducing cam lobe friction and wear.

10.1 High-Speed Effects

One more question: Since good airflow is most important at high engine speeds, what does the valve acceleration look like at 6,000 rpm vs. 3,000? Maximum lift is the same, but the valve velocity must double, since the valve has half the time to reach the same max lift.

Think carefully about the acceleration, though (see **Figure 9**). The shape of the trace is the same. It's just much larger. At 6,000 rpm vs. 3,000, the valve needs to reach twice the velocity in half the time, so the acceleration isn't twice as high; it's four times as high! The forces required to accelerate the valvetrain parts grow with the *square* of engine speed.

Note 4: This is very similar to driving through a dip in the road (you feel like you're pushed down into the seat when you drive through the curved part of the trough).

Note 5: Some cam mechanisms, called desmodronic cams, can pull the valves closed. These were used on some Mercedes engines years ago, and have long been used on Ducati motorcycles.

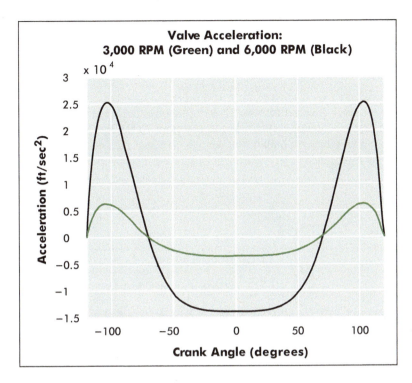

FIGURE 9 Comparing the valve acceleration at 3,000 rpm (green) vs. 6,000 rpm (black), for the same cam profile. Doubling the speed *quadruples* the acceleration.

Let's take a second to estimate the maximum force on the cam lobe (neglecting the spring force) for a valve and bucket-follower weighing 5 ounces (mass is then 0.0097050 slugs). At 3,000 rpm, the peak acceleration is 6,250 ft/sec^2, making the force 60.7 lbf. But at 6,000 rpm, it's 25,000 ft/sec^2, so the force is 242.6 lbf.

How do you make the system survive at high speeds? You can build the cam and follower surfaces tougher against wear, or make the valvetrain lighter (reducing the forces). The two main ways to accomplish this are to use 1) "exotic" materials (like titanium, which is less dense than steel) for the same-sized valves, or 2) smaller valves. A smaller valve doesn't flow as well, but you could use two small intake valves instead of one big one. This is one reason to use a four-valve head. A four-valve head with titanium valves would be even better—and pricier.

11 | Forces in a Supersonic Record Run

We've seen what happens when you accelerate hard for about 4 seconds in a top-fueler. What happens when you don't accelerate quite so hard, but you stay on it for a minute and 20 seconds or so? In 1997 Andy Green drove Richard Noble's Thrust SSC on a landspeed record run that looked something like the velocity trace in **Figure 10**.[6] Though the record mile took only 4.696 seconds to cover, the entire run lasted about 160 seconds and covered 13 miles.[7]

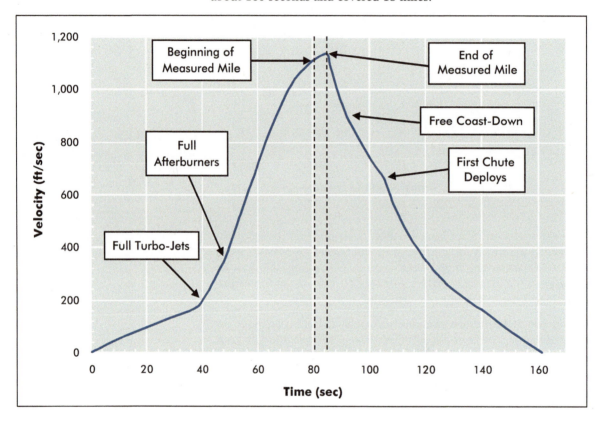

Figure 10 The velocity of the Thrust SSC during its entire record run. The peak velocity, reached at the end of the measured mile, is 1,138 ft/sec (about 778 mph). *Data courtesy of Ron Ayers and Richard Noble.*

What kind of thrust from the jet engines does it take to accelerate this 21,000-pound car to supersonic speeds? To find it, we need to account for all the forces involved. During the run up to top speed, thrust pushes the car forward, and drag force (aerodynamic and rolling resistance) acts rearward (see **Figure 11**).

Note 6: Ron Ayers, the chief aerodynamicist for the Thrust SSC, provided "velocity vs. distance" and "acceleration vs. distance" traces for the runs. The time trace was calculated from these by the author.

Note 7: If you calculate the average speed over the entire 13-mile run (161.7 seconds long), it comes out to an average of nearly 290 mph, stopped at both ends: not too shabby in itself!

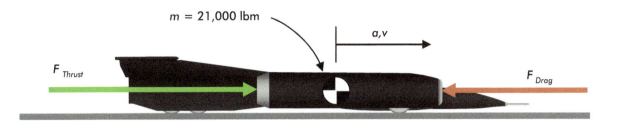

$m = 21{,}000$ lbm

a, v

F_{Thrust}

F_{Drag}

These two forces combine to produce the resultant force causing the acceleration:

Equation 6-3

$$F_{Res} = F_{Thrust} + F_{Drag}$$

So our equation of motion is

Equation 6-4

$$F_{Thrust} + F_{Drag} = ma$$

These equations assume that forward forces are positive, making the drag negative.

FIGURE 11 The Thrust SSC is pushed forward by jet engine thrust and rearward by drag. Its mass is 21,000 lbm, or about 650 slugs.

11.1 A Huge Drag

Now, how do we find that drag force? Fortunately for us, after the car went through the measured mile and shut off its jet engines, there was no thrust force acting on it, so the only force acting on the car was drag. During the coast-down, the equation of motion 6-4 is just

Equation 6-5   **Coast-down**

$$F_{Drag} = ma$$

The power-off coast-down acceleration is plotted vs. the velocity in **FIGURE 12**. The coast-down starts in the lower right portion of the plot, where the car is going 1,080 ft/sec (about 735 mph) and the acceleration is -45.7 ft/sec^2 (about -1.42 g's). As the car coasts down to lower speeds (proceeding upward and to the left in **FIGURE 12**), the acceleration becomes smaller in magnitude, since the drag is reduced.

But note the two portions in the curve: 1) a steep portion above 890 ft/sec (about 610 mph) and 2) a more gradual portion at lower speeds. The steep portion is due to a phenomenon Ayers named "spray drag," where some of the airflow around the car becomes supersonic, and the resulting shock wave knocks the sand up from the ground, which then impacts the car and slows it.[8] This is called the **TRANS-SONIC** speed region. The more gradual portion is typical aerodynamic drag as described in Chapter 5, proportional to the square of speed.

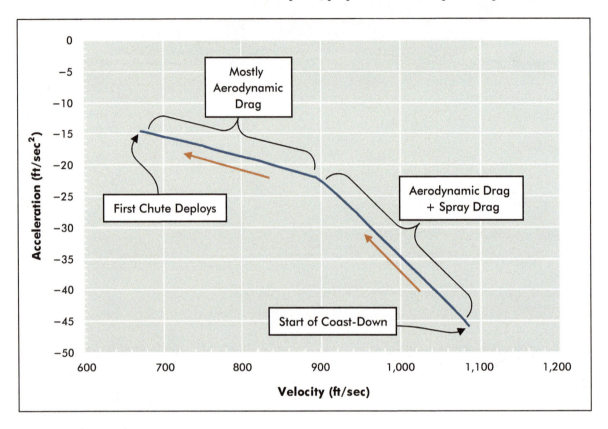

FIGURE 12 Acceleration vs. velocity during coast-down, from right after engine shutdown (1,080 ft/sec) to just before the first drag-chute deploys (670 ft/sec).

Just how big is the drag force then? We know the coast-down acceleration from **FIGURE 12**, and that the car's mass is 650 slugs. Using equation 6-5, we multiply acceleration by mass to get drag. The plot of known drag vs. velocity is shown in the solid-line portion of **FIGURE 13**. Outside this known range we estimate drag by extending the known curves with "best fit" curves.[9]

Note that the peak drag force is truly huge, at over 30,000 pounds. This is about 260 times the drag we calculated for the CTS-V at 70 mph in Chapter 5!

Note 8: Just one of the complications of "flying low" near the speed of sound. This previously unexpected source of drag has been extensively studied by Mr. Ayers, who also produced ways to minimize it. Dust being tossed up from the soft desert surface by the shock wave can be seen in Chapter 2, **FIGURE 10**.

Note 9: Below trans-sonic speeds, it boiled down to finding the best combination of a parabola (aero drag varies with speed squared) and an offset (rolling resistance is assumed constant) that blends well into the known data. The trans-sonic portion was done the same way, but also involved speed to the fourth power.

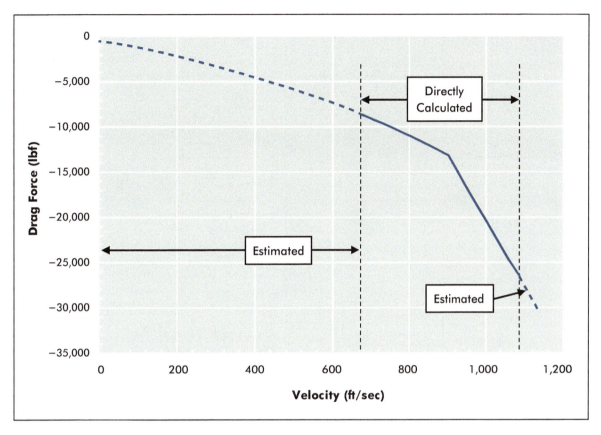

Knowing the drag force on the car vs. velocity during its coast-down, we can assume that the drag force during the run up will be the same (pretty close, anyway).

11.2 Thrust Indeed!

Now let's dig in and find the thrust applied by those engines. FIGURE 14 plots the acceleration for the entire run. During the first 35 seconds, the jets are not run hard to avoiding sucking in dust, so the acceleration is only 5 ft/sec^2 or so (around 0.2 g's). After this, the car has gone 0.6 miles and is going only 110 mph. From there until about the 49-second mark, the engines are gradually brought up to full thrust, and then the afterburners are lit. To this point, the car has gone 1.25 miles and is going 265 mph.

The acceleration is pretty stout from there until about the 67-second mark, at which point the velocity is about 890 ft/sec (610 mph). After that spray drag sets in, and the acceleration drops off rapidly. During the measured mile, the acceleration is only about 0.2 g's.

FIGURE 13 The total of aerodynamic, spray, and rolling drag forces acting vs. speed. Using the data from the free coast-down (solid line), a best-fit estimate of drag was produced for the rest of the run (dashed).

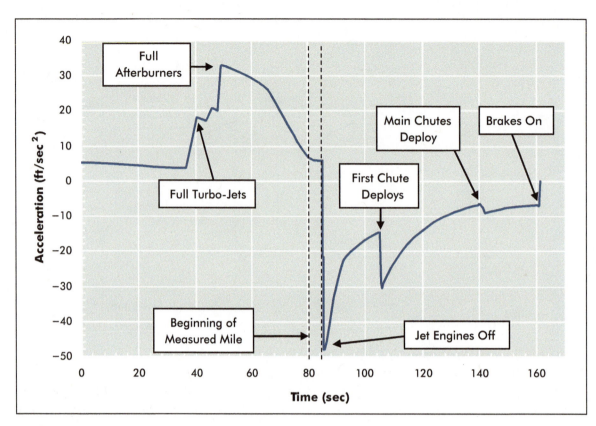

Figure 14 The acceleration of the Thrust SSC during its entire run. Note how little acceleration there is when approaching the measured mile, and the huge change in acceleration when the engines are shut down at the 85-second mark.

Now let's take the drag profile from the previous section to find the drag and thrust forces for the entire run, by reworking equation 6-4 as

Equation 6-6

$$F_{Thrust} = ma - F_{Drag}$$

Remember, the drag force is assumed negative. We know the mass, and the acceleration vs. time, so that takes care of the "*ma*" part. The drag is found by taking the velocity vs. time from **Figure 10**, and then using **Figure 13** to look up drag vs. velocity. **Figure 15** shows the resultant force (equal to *ma*), the drag force, and the thrust (their difference).

As the speed increases, more and more of the thrust goes into overcoming drag, rather than accelerating the vehicle. This is most obvious after the 49-second mark; the thrust increases, while the net force decreases. At 82 seconds, the car is going 1,107 ft/sec (755 mph), and the thrust is 32,300 lbf, but the drag force is 28,200 lbf. This only leaves 12.7% of the thrust (4,100 lbf) available for acceleration.

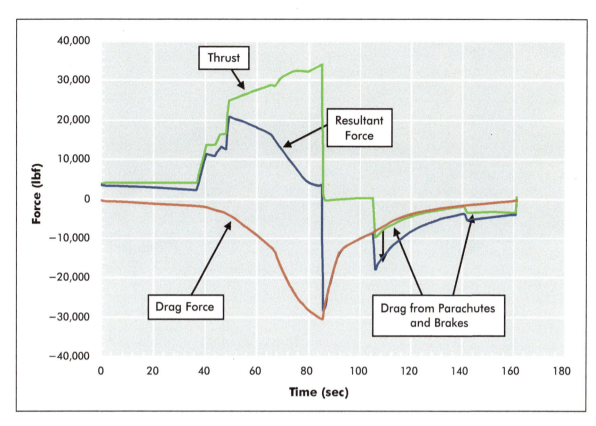

All in all, the lesson here is that incredible speeds are only produced by overcoming incredible drag forces, when you are plowing through air. The power required to produce the required propulsion forces at these speeds will be taken up in Chapter 14.

FIGURE 15 The thrust (green), drag (red), and the resultant force causing acceleration (blue) during the entire run. Note that the drag chutes and brakes show up here as "negative thrust."

12 Inappropriate Language? Applied Forces and Reaction Forces

Even though Newton's third law is straightforward, it doesn't keep engineers (including the author) from misquoting it. We often apply a force to a body (like a car, piston, or rocker arm) and then want to see what other forces develop. It's common to call the first force the **APPLIED FORCE** and any resulting forces in the system as **REACTION FORCES**. We use these terms so routinely that it's easy to forget that that's not the way Newton used the word "reaction."

Take the free-body diagram in **FIGURE 16** of a car sitting on the road. You can see the equal and opposite force pairs from 1) gravity, 2) the front tire contact patch, and 3) the rear tire contact patch. Each pair of forces acts along a common line of action.

Newton would say that each of the force pairs in **Figure 16** form an **action** and **reaction**, per his third law. Each force acting on one of the two bodies has its matching equal and opposite force acting on the other body.

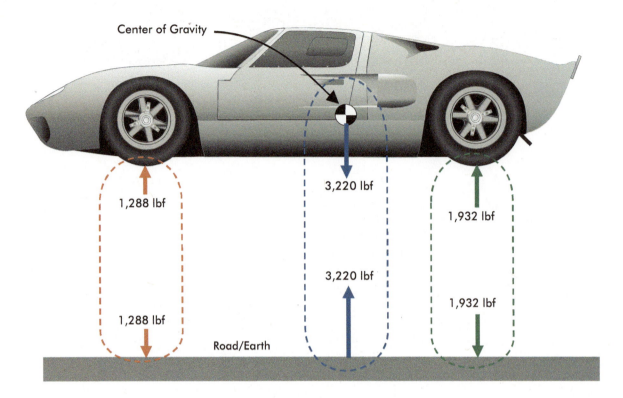

Center of Gravity

1,288 lbf

1,288 lbf

3,220 lbf

3,220 lbf

1,932 lbf

1,932 lbf

Road/Earth

Figure 16 Free-body diagram of the forces on a Ford GT40 and on the road/Earth, the way Newton meant it. Reaction force pairs are color-matched, and each act along their (shared) line of action.

Now let's look at the same situation using an engineer's terminology. **Figure 17** shows only the car and the forces acting on it (described above). Since we tend to name things based on our intentions, an engineer might call the car's weight an "applied" force, because it is the force that causes the others (without it pushing down on the car, no contact forces would exist between the tires and the road).

So far, so good, but the next step is the issue. An engineer would call the upward forces on the tires *reaction* forces, and say they *react* the downward gravitational force. Is this okay?

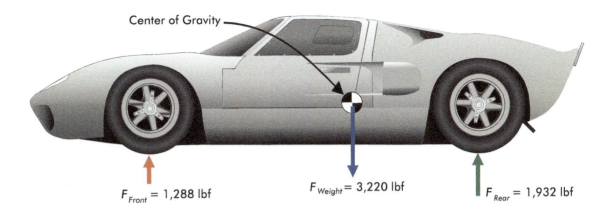

Center of Gravity

$F_{Front} = 1,288$ lbf

$F_{Weight} = 3,220$ lbf

$F_{Rear} = 1,932$ lbf

Well, no and yes. By Newton's definition, the tire forces fail to be reaction forces to the weight on four counts: 1) an action and its reaction must act on two different bodies, 2) each must be along the same line of action, 3) each action must have a single reaction (not two), and 4) action/reaction pairs are equal in magnitude whether the affected bodies are accelerating or not. All these things are true in **Figure 16**, but not in **Figure 17**.

So why bring up such heresy? Because this terminology is common in engineering texts, so it's not likely to change, and the physics police probably have greater infractions to enforce. As far as I can see, it's not a practical problem to call what are really **CONTACT FORCES** by the name "reaction forces"; just don't say it's from Newton's third law. Rather than reaction forces, these contact forces are more properly called **BEARING FORCES**, because they develop to "bear" the applied force.

Figure 17 In common engineering usage, the 3,220 lbf weight would be called the "applied" force, and the tire forces would be called reaction forces—not the way Newton would have used the word "reaction."

13 | Summary

The problems in this chapter illustrate several important points in physics. For one, they show why car guys are so obsessed with tires. Whether in a quarter-mile run or in braking to a stop, the tires largely determine the car's maximum performance capability—they are the link between car and road. Often traction is a limitation.

For example, traction is the main reason that the top land-speed record cars have not been driven by their tires since the 1960s. There simply isn't enough traction to get to record speed. For a good laugh, calculate the coefficient of friction necessary to push the Thrust SSC at top speed (see **Figure 18**), remembering that it was running on broken rock.[10]

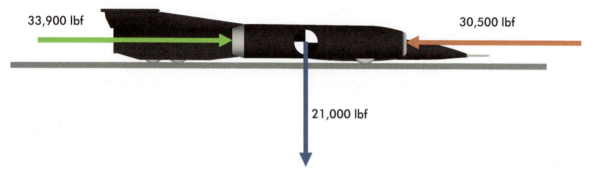

33,900 lbf   30,500 lbf

21,000 lbf

Figure 18 With the Thrust SSC weighing 21,000 lbf, what coefficient of friction would the tires need to replace the 32,300 lbf of jet engine thrust?

But while high tire forces are desirable in increasing performance, we spend a fair amount of time trying to reduce forces elsewhere. I'm thinking of the force on your head in a collision, on a piston pin at high speed, or on a cam follower. In these cases, the idea was to reduce acceleration and/or mass, in order to produce smaller forces and treat the "hardware" involved with care (especially your head).

This chapter also shows how much can be learned when you have real data to work with. Using published braking distance data, we got a more realistic tire friction coefficient than from any table in a textbook. By examining the lift profile of a camshaft, we were able to identify some critical areas in cam design. By following the acceleration and velocity traces of the Thrust SSC, we were able to estimate the drag and thrust forces acting on it during its run.

Note that these later problems were worked using a spreadsheet to make the calculations for each instant in time (or cam rotation angle), rather than looking at one point in time. For me, the ability to perform calculations on an entire trace at once is one of the best parts of using computers. Besides being faster and more accurate than plotting by hand, it makes it easy to change the values in a problem and see what will happen. And with some practice, you can read the data plots like an open book.

Note 10: Better traction could be had by producing downforce, but that comes with extra drag, requiring more traction to propel the car!

Major Formulas

Definition	Equation	Equation Number
Static Tire Friction Coefficient = − Max Braking in g's	$\mu_{Static} = -a_g$	6-1
Resultant Force on a Car = Thrust + Drag (Drag Assumed Negative)	$F_{Res} = F_{Thrust} + F_{Drag}$	6-3
Equation of Motion with Propulsion Thrust	$F_{Thrust} + F_{Drag} = ma$	6-4
Coast-Down: Drag Force = Mass × Acceleration	$F_{Drag} = ma$	6-5

7 Torque, Force Resolution, and 2-D Vectors

Basic Use of Vectors in a Plane

Tightening a single wheel nut takes torque—lots of it. Here, a crew member uses a 6-foot torque wrench on Jim Hall's Chaparral 2F, driven by Phil Hill and Mike Spence at the 24 Hours of Daytona in 1967. *Photo from the collections of The Henry Ford.*

Torque is introduced, and then vectors are used in balancing a tire. The geometry and trigonometry of working with vectors are presented, and then used to calculate straight-line distance in a plane, and in calculating torque. After explaining force resolution, the stage is set for rotational dynamics and dynamics in a plane in the following chapters.

Contents

Key Symbols Introduced

Symbol	Quantity	SFS Units	MKS Units
T	Torque	ft-lbf	N-m
r	Radius (Here a Lever-Arm Length)	ft	m
s	Displacement Vector	ft	m
$\mathbf{s}_x, \mathbf{s}_y$	Component Vectors of Displacement Vector **s**	ft	m
s_x, s_y	Vector Components of Displacement Vector **s**	ft	m
s	Displacement Vector Magnitude	ft	m
F_x, F_y	Vector Components of Force **F**	lbf	N
θ_s	Displacement Direction Angle	degrees	degrees

1 Introduction

Most books on physics take a few chapters to get to rigid-body motion, because they need to "dot the i's and cross the t's" on particle motion first. We wasted no time, going straight to it in Chapter 2. But in the straight-line dynamics of Chapters 4 and 6, you may have noticed that the forces on a body weren't always acting along the same line. They had different *lines of action*.

Take the street rod in **FIGURE 1** as an example. Here, the drag force acts on a line aimed directly at the car's CG, but the traction force acts on a horizontal line at pavement level, 1.5 feet lower. When you're only interested in the fore-aft motion, you can ignore this fact. But you know that the offset between the traction force and the CG would tend to rotate the car's nose up, which would change the loads on the tires, and might affect the aerodynamic forces and airflow. We need a way to include this influence.

FIGURE 1 The combined effects of the drag and thrust forces would be to nose the car up, in addition to accelerating it. We'll "resolve" this situation later in the chapter.

Drag: 200 lbf

1.5 ft

Thrust: 500 lbf

So first off, we'll resolve each force to the CG. The offset traction force does produce a force on the CG. But it also produces a **TORQUE** on it, which is a turning action, just as a force is a pushing or pulling action. Resolving the traction force to the CG is just replacing it with a parallel force acting on the CG, along with a couple (a "pure" torque) that the force produces on the CG. These produce the same effect on the car, as long as it's rigid. The drag force (in this example) can be resolved to the CG simply by shifting it there, because it acts directly at the CG and produces no torque on it.

Second, we simply add up the resulting resolved forces and couples, giving us their total influence, a resultant force and couple. This process greatly simplifies complex problems, making the problem easier to visualize and analyze.

Since force resolution and torque will involve the use of vectors in a plane, it's time to introduce them too. For instance, we'll see the combined effect of the lever arm length, the applied force, and the angle between them, when producing torque with a wrench or just about anything else.

We'll introduce vectors in a plane (i.e., two dimensions) for problems like motion on a level, curvy road. Here we are using the word "dimension" as most people would: a line has one dimension, a square has two, and a cube has three. In geometry, a plane is an imaginary surface that is 1) flat and 2) extends forever in all directions. We'll be interested in surfaces that are "almost flat" (like a parking lot, or the Bonneville Salt Flats), and won't worry about it being infinite.

The most important vector concepts to take from this chapter are:
1. Anything that has magnitude and direction is a vector.
2. Any vector can be split into components that are perpendicular to each other.
3. The difference in direction between vectors is important in combining their effects.

To avoid most of the vector math, we'll split vectors into their components and do the calculations part by part; again, divide and conquer.

2   Torque

Just as force is pushing or pulling action, torque is turning action. The word "action" is used here as Newton used it to describe a force *acting on* something. In the same way, a torque *acts on* something. Neither one necessarily produces motion. Torque is familiar to anyone who has used a wrench. You pull up on the wrench with a force roughly perpendicular to its handle (see **FIGURE 2**), in order to turn a bolt. Pull with a force F of 110 lbf at a radius r of 0.917 feet (11 inches) from the pivot point (the center of the bolt), and the wrench applies a torque of 100.8 ft-lbf to the bolt, in a counterclockwise (CCW) direction. (I'll often abbreviate clockwise as "CW," and counter-clockwise as "CCW.")

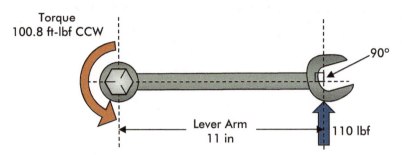

FIGURE 2 Torque is a turning action produced when a force acts perpendicular to (at a 90° angle to) a lever arm like a wrench. Here, a force of 110 lbf applied 11 inches from the center of the bolt produces a 100.8 ft-lbf CCW torque on the bolt.

When the applied force is perpendicular to the lever arm, the torque T is equal to the force F times the lever-arm length r.[1] Consequently, the equation for the torque T is simple:

Physical Law Eq. 7-1   **Torque from perpendicular force**

$$T = Fr$$

The dimensions of torque are force times length, so torque is measured in foot-pounds (ft-lbf) in SFS units, and newton-meters (N-m) in MKS. Most torque wrenches in the U.S. read both.

FIGURE 3 Two ways to get 150 ft-lbf of torque: increase the force on our wrench to 163.6 lbf (top), or use a breaker bar that is twice as long, with half that force (bottom).

Equation 7-1 agrees with experience; if you need more torque to break a bolt loose, you can either pull harder on the wrench (more force) or get a longer wrench like a breaker bar (more lever arm). This is shown in **FIGURE 3**.

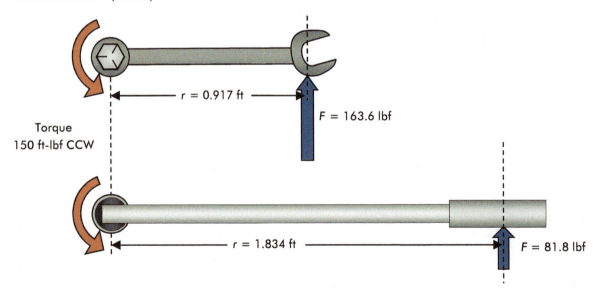

Note 1: If the force isn't applied perpendicular to the wrench, less torque will be produced. Later in the chapter we'll use vectors to see by how much.

A wrench is probably the most familiar example of producing torque, but you use the same idea for a piston turning a crankshaft by pushing on it through a connecting rod and for suspension links holding the spindle and tires in position while cornering or braking.

2.1 Direction of Torque

Knowing the direction of a force is essential to predict its effects. It's the same with torque. Imagine the 110-lbf force in **FIGURE 2** being applied *downward* instead of upward. The torque would still have a magnitude of 100.8 ft-lbf, but it would be in the clockwise (CW) direction. Obviously this would change a loosening torque (assuming right-hand threads) into a tightening torque. The same is true for engine torque; if CCW engine torque speeds the car up, CW engine torque slows it down. Note that this makes torque itself a vector.

3   Vectors in Two Dimensions

Up to this point, we've recognized that forces and velocities are vector quantities because they have a magnitude and direction. But since we only dealt with straight-line motion, the only effect on "doing the math" was a negative sign here and there. Now we'll look at vectors that are *not* along the same line.

KEY CONCEPT

Vectors are a great example of thinking in pictures, but calculating with numbers. Get a clear picture of vectors in your mind, and the battle of understanding forces and motion is half done.

3.1 Balancing a Tire Using Vectors

Let's start off with a vector example I learned working in a gas station: balancing a tire. One of my coworkers taught me this technique, but we didn't think of it as "adding vectors." Say the wheel/tire assembly in **FIGURE 4** has a 0.9-ounce (oz) "light spot" at the rim, at the 12:00 position.[2] To balance it exactly, it would need a 0.9-ounce weight at 12:00, as shown on the left.

Let's suppose that weights are usually sold in 0.25-ounce steps. A single weight at 12:00 would either underbalance the wheel (with 0.75 ounces) or overbalance it (with 1.00 ounces). What if you want to balance it more exactly? You could take two equal weights (0.5 ounces each)

Note 2: Or a 0.9-oz "heavy spot" at 6:00—either way.

that add up to a bit more than what's needed, but split them apart so they're centered about 12:00 (**Figure 4**, right). If their "effective weight" is too great, you spread them farther apart, and if too small, closer together. Once the weights are adjusted correctly, the wheel is balanced exactly. In this example, the weights need to be about 52° apart (each 26° from 12:00) to balance the tire.

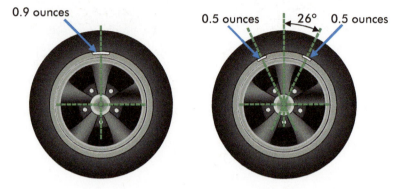

Figure 4 If you don't have the right weight to balance a wheel (0.9 ounces, left), you can use two weights of just over half that much (two 0.5-ounce weights, right), in this case each placed 26° from the light spot. This is an example of "vector addition."

Now let's think of this in terms of "imbalance vectors," as shown in **Figure 5**. Let's assume it is a 16-inch wheel (the weights mount 8 inches from the wheel center), and define the amount of imbalance as weight times distance from center. If we had a 0.9-ounce weight then, we would multiply 0.9 ounces by 8 inches, for an imbalance of 7.2 in-oz. Similarly, the imbalances of the 0.5-ounce weights are 4.0 in-oz each. **Figure 5** scales the vectors so the 7.2 in-oz imbalance just touches the center of the 0.9-ounce weight.

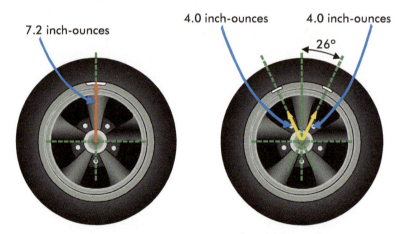

Figure 5 The imbalance effect of each weight can be shown as a vector, with each length proportional to the product of its weight and the distance from center.

How do we know that the imbalances from the smaller weights "add up" to the same as the larger one? Two vectors can be added head-to-tail to form a triangle (**Figure 6**, left), where the third side is the resultant (sound familiar?). A parallelogram with opposite sides equal (**Figure 6**, right) can also be formed to do the same thing. Using either method, **Figure 6** shows them to add to the desired resultant imbalance of 7.2 in-oz.

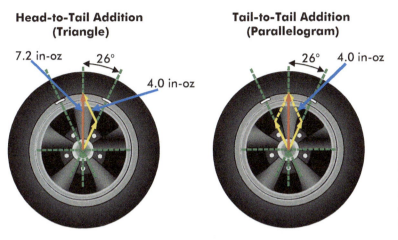

**Head-to-Tail Addition
(Triangle)**

7.2 in-oz   26°

4.0 in-oz

**Tail-to-Tail Addition
(Parallelogram)**

26°   4.0 in-oz

Figure 6 The imbalances from the two smaller weights (the two yellow vectors) add head-to-tail or tail-to-tail to equal the imbalance from the larger weight (red).

You can imagine what happens as you change the "split angle" between the 0.5-ounce weights. If they are both moved to 12:00, they act like one 1.00-ounce weight. If they are placed opposite each other (at 9:00 and 3:00), their imbalances cancel out, and you might as well not put the weights on. The vector addition for splits of 20°, 52°, 120°, and 160° are shown in **Figure 7**, along with their resultant imbalance.

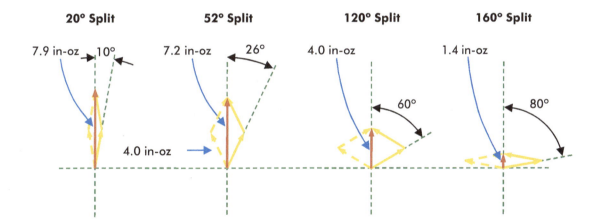

20° Split

7.9 in-oz   10°

4.0 in-oz

52° Split

7.2 in-oz   26°

120° Split

4.0 in-oz

60°

160° Split

1.4 in-oz

80°

Figure 7 The resultant imbalance vector from the two smaller weights, for several split angles. Note that the resultant doesn't change much for smaller split angles, meaning you can set the imbalance very precisely by adjusting the angle.

KEY CONCEPT

You can add any two vectors of the same type (displacement, force, etc.) graphically by placing them head to tail, and then drawing their resultant from the beginning of the first to the end of the second. Although the vectors in this example were equal in size, you can also use this method when they aren't. For example, you *could* use unequal weights.

3.2 A Vector's Magnitude and Direction

Figure 8 is a picture of a situation I was curious about when I was a kid. The old road from my home town of Linn, Kansas (point A), goes 2 miles straight north (to point B), and then 2 miles straight east to the intersection of Highways 148 and 15 (point C). They were building a new road that cut across country and took a shorter route between the same two points. Even though the new road was actually curved, I wondered how long it would be if it were the shortest possible (straight).[3] What is the straight-line distance s, and what direction is the displacement vector **s** from point A to point C?

First off, notice the map is drawn in a **coordinate system** that tracks the position on the map. Point A is picked as the **origin**, where $x = 0$ and $y = 0$. The x-direction is due east, and the y-direction is due north. So in moving from point A to point B, we move 2 miles in the y-direction. In moving from point B to point C, we move 2 miles in the x-direction. We can write these displacements as two vectors:

$$\mathbf{s}_x = 2 \text{ miles east}$$

$$\mathbf{s}_y = 2 \text{ miles north}$$

Both of these are displacement vectors, and are given by how much motion there is in each of the coordinate directions.

NOTE

In print, vectors are usually written in bold print, so displacement would be **s**, and velocity would be **v**. That doesn't work when writing by hand, so then they're written with an arrow over the symbol; **s** would be $\vec{s}$, and **v** would be $\vec{v}$.

The total displacement **s** (the dashed vector in **Figure 8**) is also given as a vector:

$$\mathbf{s} = \mathbf{s}_x + \mathbf{s}_y$$

$$= 2 \text{ miles east} + 2 \text{ miles north}$$

Note that we just wrote the displacement vector **s** in terms of its **component vectors**, $\mathbf{s}_x$ and $\mathbf{s}_y$. The fact that we wrote them in the reverse order of our travel (it was traveled y first) doesn't matter; we still would end up at point C by going 2 miles east, then 2 miles north. There may not be a road to go east first, then north. But the displacement vector only tells you where you end up, not how you got there (no different than in the straight-line motion from Chapters 2–6).

Note 3: As I remember, I asked my Aunt Dorothy how far it would be, but I don't remember the answer. Yes, I grew up in Kansas, and Dorothy is my aunt....

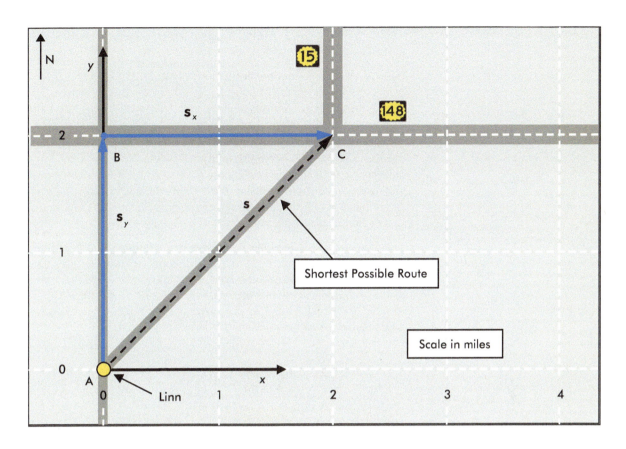

Another way to give a vector is in terms of its **VECTOR COMPONENTS**. Each vector component is the amount of motion in that coordinate direction. Here, we would say that:

$$s_x = +2 \text{ miles}$$

$$s_y = +2 \text{ miles}$$

Note that the vector components are not themselves vectors, but just numbers with units that tell how far and in what direction each component vector points. For example, an x-component of -2 miles would mean a component vector $\mathbf{s}_x$ of 2 miles west, and a y-component of -2 miles would mean a component vector $\mathbf{s}_y$ of 2 miles south.

We can use the vector components to find the length of the vector $\mathbf{s}$, called its **MAGNITUDE** s. Since the component vectors $\mathbf{s}_x$ and $\mathbf{s}_y$ are perpendicular to each other, we can use the Pythagorean theorem to calculate the magnitude of the displacement vector $\mathbf{s}$:

General Equation 7-2

$$s = |\mathbf{s}| = \sqrt{s_x^{\,2} + s_y^{\,2}}$$

FIGURE 8 In this case, we know that $\mathbf{s}_x$ is 2.0 miles east, and $\mathbf{s}_y$ is 2.0 miles north. What is the straight-line distance from A to C?

Here, we always take the positive square root, because the answer is a distance. For our example:

$$s = |\mathbf{s}| = \sqrt{s_x^2 + s_y^2} = \sqrt{(2 \text{ miles})^2 + (2 \text{ miles})^2} = \sqrt{8 \text{ miles}^2}$$
$$= 2.83 \text{ miles}$$

For displacement, the magnitude s of the vector is the straight-line distance between its beginning and end points. So if the road were built perfectly straight, it would only have to be 2.83 miles long, vs. 4 miles for the old route, and save more than a mile of pavement. Every vector has a magnitude and **DIRECTION**. Now we know the magnitude, so let's find its direction.

Unlike in compass directions, in math and physics we usually measure direction angle θ_s from due east, in the counterclockwise direction. From trigonometry, we can see that the tangent of the angle θ_s is equal to the y-component of the vector divided by the x-component. (Remember, the Greek letter θ, is often used as a symbol for angles.) So we take the inverse tangent of their ratio to find the angle:

General Equation 7-3

$$\theta_s = \tan^{-1}\left(\frac{s_y}{s_x}\right)$$

For our case, this is:

$$\theta_s = \tan^{-1}\left(\frac{s_y}{s_x}\right) = \tan^{-1}\left(\frac{2 \text{ miles}}{2 \text{ miles}}\right) = \tan^{-1}(1) = 45°$$

Its also obvious from **FIGURE 9** that the direction is 45° CCW from due east. Now that we have its magnitude and direction, we can write the displacement vector **s** as 2.83 mi $\angle$ 45°.

Getting the direction from the tangent can be a little more involved than this, because the direction angle could be anywhere from 0° to 360°, but the inverse tangent only produces angles from −90° to 90°. There is a special function in computer programs to overcome this ambiguity, described in Appendix 5.

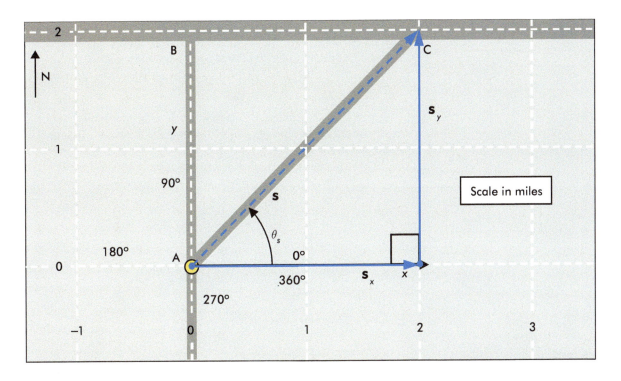

CAUTION

The terms *component vector* and *vector component* are very similar, so don't confuse them. Component vectors are the vectors that make up the full vector (two are needed in a plane). Vector components are the *x*- and *y*-quantities describing the magnitude and direction of the component vectors.

FIGURE 9 Interchanging the order of the vectors s_x and s_y, we have a right triangle that makes it easier to visualize the direction angle θ_s relative to east.

3.3 Motion Vectors Within a Coordinate System

We can use vectors to track an object's motion anywhere in a plane. **FIGURE 10** shows a car on a flat oval track with a coordinate system originating at the start/finish line. As the car moves, we specify its displacement in terms of its *x*- and *y*-positions. As shown, its *x*-displacement is 1,000 feet, and its *y*-displacement is 1,800 feet. We could write this as the coordinate point $(x,y) = (1,000 \text{ ft}, 1,800 \text{ ft})$.[4] But as in the previous problem, we could also give the vector components $s_x = 1,000$ ft, $s_y = 1,800$ ft.

Coordinate points would work fine to show displacement, but not velocity or acceleration. Vectors work much better. **FIGURE 10** shows the dynamics of the car as it speeds up on the back straight: its velocity **v** (say, 170 ft/sec west) and acceleration **a** (say, 20 ft/sec² west) vectors.

Note 4: A rectangular coordinate system like this is also called a **CARTESIAN** coordinate system, after René Descartes, who devised it.

As in straight-line motion, to calculate the force to accelerate a mass, the acceleration must be measured relative to an inertial frame. The coordinate system in **Figure 10** is stationary, so it will do.

> **NOTE**
>
> For motion in a plane, an inertial frame is a coordinate system that is either stationary or moving at constant velocity. It cannot accelerate or rotate.

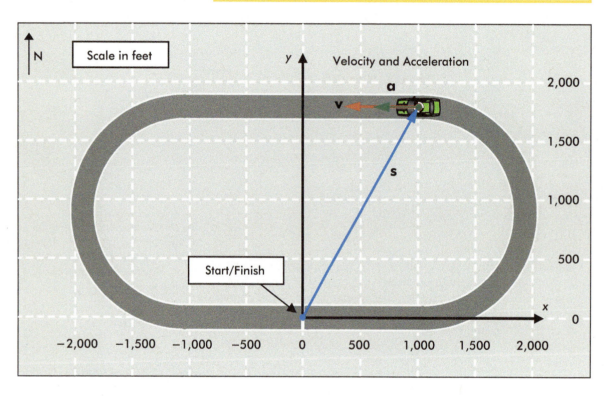

Figure 10 Displacement **s**, velocity **v**, and acceleration **a** vectors for a car on an oval track.

Figure 10 shows that you can use vectors to visualize the physics shown (whether motion, forces, etc.), before seeing any actual numbers. Their magnitude and direction say a lot in a very direct way. It's a bit like reading analog gauges. After a while, you read a gauge from the direction the needle points, without reading the numbers. Vectors have the additional advantage that their lengths also mean something.

> **KEY CONCEPT**
>
> Each vector has an origin. Displacement vectors typically originate from the coordinate origin (0, 0), while velocity and acceleration vectors originate from a body's CG. But their *directions* still refer to the coordinate axis directions.

3.4 Addition and Subtraction of Vectors

We did vector addition graphically in the wheel-balance problem. Now let's do it numerically, for the motion in **FIGURE 11**. Our car has moved 1,500 feet west from its position in **FIGURE 10**. Its new displacement vector **s** (here called **s₃**) is the original one (now called **s₁**) plus its change in position (called **s₂**). The component vectors are added head to tail, and their sum is **s₃**. The vector **s₁** has components $s_{1,x} = 1,000$ ft, and $s_{1,y} = 1,800$ ft, while **s₂** has components $s_{2,x} = -1,500$ ft, and $s_{2,y} = 0$.

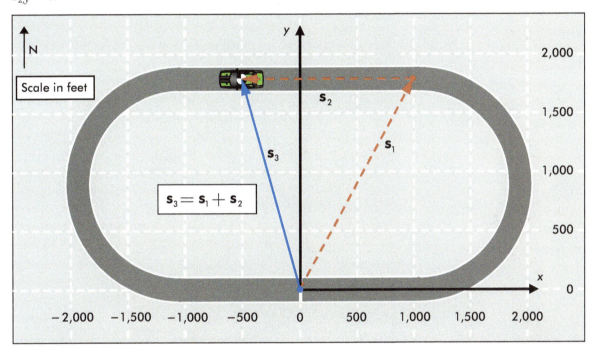

FIGURE 11 Two displacement vectors **s₁** and **s₂**, added by placing them head to tail, to get the vector **s₃**.

To find the components of **s₃**, which are $s_{3,x}$ and $s_{3,y}$, we simply add the x-components, and then the y-components:

General Equation 7-4a   **Adding vector components**

$$s_{3,x} = s_{1,x} + s_{2,x}$$

General Equation 7-4b

$$s_{3,y} = s_{1,y} + s_{2,y}$$

For the vectors in **FIGURE 11**, equations 7-4 produce:

$$s_{3,x} = s_{1,x} + s_{2,x} = 1,000 \text{ ft} + \left(-1,500 \text{ ft}\right)$$
$$= -500 \text{ ft}$$

and:

$$s_{3,y} = s_{1,y} + s_{2,y} = 1,800 \text{ ft} + \left(0 \text{ ft}\right)$$
$$= 1,800 \text{ ft}$$

Vector subtraction is just the opposite of addition, and is done by subtracting their *x*-components, then their *y*-components. For example, we could "undo" the second displacement vector $\mathbf{s}_2$ in **Figure 11** by subtracting it from the total $\mathbf{s}_3$:

General Equation 7-5a   **Subtracting vector components**

$$s_{1,x} = s_{3,x} - s_{2,x}$$

General Equation 7-5b

$$s_{1,y} = s_{3,y} - s_{2,y}$$

So we would subtract $-1{,}500$ ft from -500 ft to get $1{,}000$ ft, and subtract 0 ft from $1{,}800$ ft to get $1{,}800$ ft.

3.5 Vectors vs. Scalars

We need to distinguish between vectors, which have direction, and quantities that don't. Remember that quantities without direction are called scalars. As an example, a velocity "60 mph east" is a vector, a speed of "60 mph" is a scalar. **Table 1** lists examples of scalars and vectors, which should help to distinguish the two.

> **NOTE**
>
> From here on, we'll drop the use of S for speed. We'll simply use the magnitude v of the velocity vector $\mathbf{v}$.

Examples of Scalars	Examples of Vectors
Distance	Displacement
Speed	Velocity
Mass	Acceleration
Energy	Force
Vector Components	Torque
Vector Magnitudes	Rotational Velocity
Temperature	Momentum
	Component Vectors

Table 1 Some examples of scalars, which are quantities that don't have direction, and vectors, quantities that do.

> **CAUTION**
>
> There is sometimes confusion over whether scalars can be negative, since many quantities given as examples (speed, distance, mass, volume, area, etc.) are magnitudes, which will always be positive. But scalars *can* be negative, as you have seen with vector components.

3.6 Converting from Magnitude and Direction to Component Vectors

In a plane, every vector carries two pieces of information, regardless of how it's specified. We've done it two ways:

1. Magnitude and direction.
2. Vector components.

> **KEY CONCEPT**
>
> Specifying a vector either by its magnitude and direction, or by its vector components, will fully describe it. Each one can be converted to the other. Sometimes one way is easier to use or visualize than the other.

Take the car in **FIGURE 12**, which is going 60 mph on a heading of 30° CW from east, or 60 mph $\angle -30°$. If you'd rather use a positive angle, you can call it 60 mph $\angle 330°$. Since the magnitude v and direction angle θ_v are known, the vector components can be calculated by using a little bit of trigonometry. You would find the components as

General Equation 7-6a

$$v_x = v\cos\left(\theta_v\right)$$

General Equation 7-6b

$$v_y = v\sin\left(\theta_v\right)$$

where again θ_v is measured CCW from the x-axis. For our example, this works out to:

$$v_x = v\cos\theta_v = 60 \text{ mph} \times \cos\left(-30°\right) = 52.0 \text{ mph}$$

$$v_y = v\sin\theta_v = 60 \text{ mph} \times \sin\left(-30°\right) = -30.0 \text{ mph}$$

NOTE

Converting a vector from magnitude and direction into its perpendicular vector components is called **VECTOR RESOLUTION**.

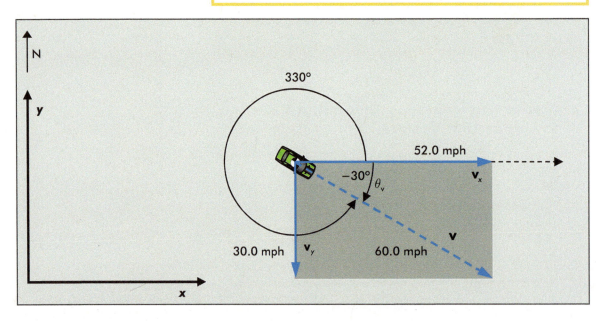

FIGURE 12 The direction of a vector can be given as a positive (CCW) angle or a negative (CW) angle. Here, 330° is the same as −30°. Either way will work.

It might be worth proving to yourself that you can get the same velocity components using an angle of 330°.

4   Calculating Torque with Vectors

When we applied a force perpendicular to a lever, the torque T was simply Fr. But if it's not perpendicular, how much different is it? Vectors are a great help in calculating the torque in this case. We'll show three ways to calculate the torque, to illustrate a few points about forces in a plane. While a wrench is used as an example, the same would hold true for torque on any lever, like a crank throw.

4.1 Using Force Components Relative to the Lever Arm

Imagine that when you pull up on the wrench back in **FIGURE 2**, it is angled 35° down from horizontal (see **FIGURE 13**), so that there is a 55° angle between the force and the lever arm. The 110-lbf force isn't "square" to the handle. How do you calculate the torque?

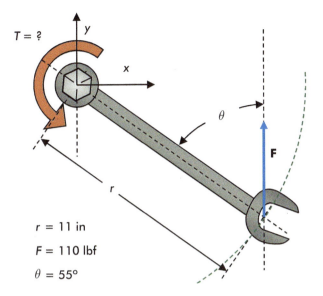

$T = ?$

y

x

θ

F

r

$r = 11$ in

$F = 110$ lbf

$\theta = 55°$

FIGURE 13 If the force isn't perpendicular to the wrench, the torque on the bolt will be less than Fr. But we can still calculate it knowing the angle θ between the force and lever arm, here 55°.

A vector can be split into components that aren't parallel to the x- and y-directions, as long as they are perpendicular to each other. Here, let's split the force vector **F** into $\mathbf{F}_{Radial}$, parallel to the wrench handle, and $\mathbf{F}_{Tangential}$, perpendicular to it (see **FIGURE 14**). The tangential force produces torque, since it is perpendicular to the lever.[5] But the radial component points straight at the center of the bolt, parallel to the lever.[6] Therefore it makes no torque, and we can ignore it here. The tangential force is $F \sin\theta$, making the torque:

Physical Law Eq. 7-7

$$T = Fr \sin \theta$$

So the tangential force is 90.107 lbf, and the torque is 82.6 ft-lbf CCW. Note that if you take equation 7-7 and use a 90° angle, it simplifies to $T = Fr$, since the sine of 90° is 1.

KEY CONCEPT

Who says we can calculate torque from an angled force this way? Varignon's theorem states that the total torque applied by a force is equal to the sum of the torques produced by its perpendicular components.

Note 5: The term "tangent" is from the Latin *tangere*, which means "to touch"; a tangent vector barely touches (it grazes) an arc, such as the green dashed arc in **FIGURE 14**.

Note 6: The term "radial" comes from acting along a radius drawn from the center of rotation, which here is the bolt center.

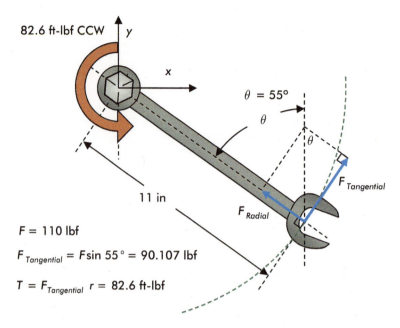

FIGURE 14 One way to calculate the torque on the bolt is to split the force into its tangential and radial components. Only the tangential force component (**F**$_{Tangential}$) produces torque.

$F = 110$ lbf

$F_{Tangential} = F\sin 55° = 90.107$ lbf

$T = F_{Tangential}\; r = 82.6$ ft-lbf

To summarize, we found the force component that acts perpendicular to the wrench, and then multiplied that part by the lever arm. Plus, for maximum torque, pull perpendicular to the wrench.

4.2 Using Lever-Arm Components Relative to the Force Direction

A second approach is to call the lever arm a vector **r**, and split it into a component **r**$_{Perp}$ that is perpendicular to the force, and a component **r**$_{Par}$ that is parallel to the force (see **FIGURE 15**). The length of **r**$_{Perp}$ is $r \sin\theta$, which comes to 0.75089 feet (just a tad over 9 inches). Since **r**$_{Perp}$ is perpendicular to the force, we can multiply the entire 110-lbf force F by r_{Perp} to get the torque on the bolt head, again 82.6 ft-lbf. As an equation, this would be the same as equation 7-7, of course.

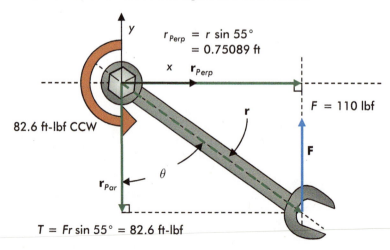

FIGURE 15 A second way to calculate the torque is to multiply the force times the component of the lever-arm radius (**r**$_{Perp}$) that is perpendicular to the force.

$T = Fr \sin 55° = 82.6$ ft-lbf

You can see that vectors are a good tool to convert one problem that may seem difficult into two easy problems. But it may not always be easy to use either of these first two methods. However, the next one works for any force and lever arm in a plane.

4.3 Using Components Relative to Coordinate-Axis Directions

A third way to calculate torque is shown in FIGURE 16, which is the wrench and force from FIGURE 2 both rotated CW by 35° (so we already know it creates 100.8 ft-lbf of torque). This method will split both the lever arm and the applied force into component vectors in the x- and y-directions. To start, we calculate the x- and y-force components, using equations 7-6, but for a force. Then $F_x = 63.1$ lbf and $F_y = 90.1$ lbf (the vectors shown by dashed arrows in FIGURE 16).

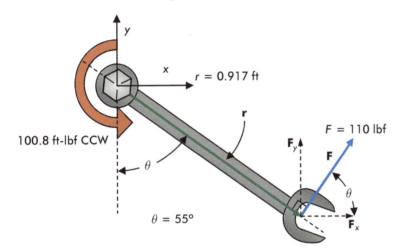

FIGURE 16 For our third method, we examine FIGURE 2 rotated downward by 35° from horizontal. The torque of 100.8 ft-lbf is the same, of course.

Next we split the lever-arm vector **r** into its x- and y-components. This is shown in FIGURE 17. On each side of the figure, we calculate the torque each force component produces, by multiplying it by the component of the lever arm perpendicular to it.

On the left side, we multiply F_x by the length r_y to get $T_1 = 33.173$ ft-lbf CCW; on the right, we multiply F_y by r_x to get $T_2 = 67.660$ ft-lbf CCW.

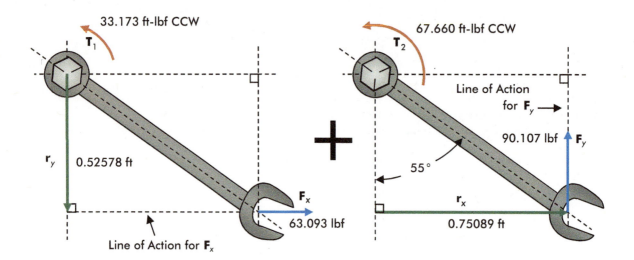

FIGURE 17 Here we calculate torque by splitting the lever arm and the force into their x- and y-components, multiplying each force component by the lever-arm component perpendicular to it, and adding up the resulting torque components.

Adding these two torque components together, we get the expected 100.8 ft-lbf CCW, same as in **FIGURE 2**. A couple of points:

1. In this case, both torque components were CCW, so their effects added. If the two were acting in opposite directions, they would partially cancel.

2. To be proper, the direction of a positive torque should be CCW, just as the direction angle θ is. We won't always keep to that convention, though.

When we do use the convention that a CCW angle is positive, the torque calculation can be put into an equation:

General Equation 7-8

$$T = r_x F_y - r_y F_x$$

Just to be clear, although the force and lever arm were perpendicular in this example, equation 7-8 works for any angle between force and arm. In vector math, this formula is called the **CROSS PRODUCT** of the lever and force vectors. So now you know where it comes from.

To use equation 7-8, you must include the sign of both the force and lever-arm vector components. For example, in **FIGURE 17** the y-component of the lever arm was negative ($r_x = 0.75089$ ft, $r_y = -0.52578$ ft). Whereas you or I can catch whether a torque applied to a lever arm will produce a CW or CCW torque, a computer or a canned equation only knows by the mathematical signs it's given. Garbage in, garbage out.

By the way, why was it valid to calculate torque components from the force and lever-arm vectors in **FIGURE 17** that didn't seem to touch? The next section answers that.

4.4 A Force's Line of Action

Every force has a **LINE OF ACTION** along which it acts. A force can be moved anywhere along its line of action and still have the same effect on a rigid body. This is called the principle of **TRANSMISSIBILITY**.

To show this more strongly, let's replace the wrench in **FIGURE 15** with a piece of plate metal (**FIGURE 18**). When the force **F** is applied at point A, it is at the same point as previously on the wrench handle. The same 110-lbf force is also shown applied to two other points, B and C. At point B, the distance between the force and the bolt is smallest, at $r \sin\theta$, which here is 0.75089 feet. Again, the torque produced is 82.6 ft-lbf. If we moved the force to point C, the torque would still be the same.

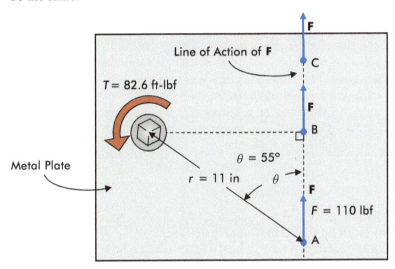

FIGURE 18 The torque produced by the force **F** on the metal plate is the same when applied anywhere along its line of action.

Now imagine the plate rotates 5° CCW, and points A, B, and C are holes drilled in the plate, so they travel with it. Point A would move up and right, while B and C would move up and left. If the force **F** were applied (still vertically) to point A, it would produce more torque than before, 87.2 ft-lbf. But the line of action of **F** through point A would no longer run through points B and C. So applying **F** at either of those points would produce a different torque than at A (each somewhat lower).

4.5 Couples

The wrench in **FIGURE 2** didn't just produce a torque on the bolt head. We'll see that pulling on a wrench with a single upward force produces a torque *and* a force on the bolt head (upward at 110 lbf in this case). To produce a *pure* torque, like what comes out of a crankshaft or an axle, it takes two forces (equal and opposite, and therefore parallel), called a "force couple" or simply a **COUPLE** (see **FIGURE 19**). On the other hand, the torque on a body from a single offset force is usually called the moment of a force, or simply a **MOMENT**.

FIGURE 19 A four-way lug wrench makes it possible to apply two equal and opposite forces (downward on the right, upward on the left), producing a pure torque called a "couple."

To calculate the torque from a couple, you can either add the torque produced by each force, or simply multiply one of the forces (they are the same size) times the perpendicular distance d between their lines of action, which is $2r$:

General Equation 7-9   **Couple**

$$T = F_{Perp}d$$

This is diagrammed in **FIGURE 20**. You can imagine you are tightening a lug nut on your wheel, your right hand pushing down on the right side of the wrench with a force of 55 lbf, and your left hand pulling up with 55 lbf on the left side of the wrench.

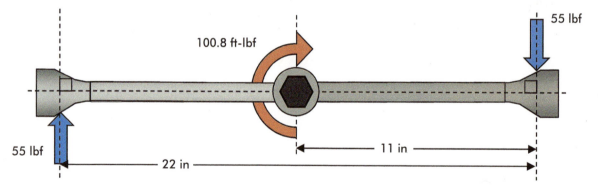

100.8 ft-lbf

55 lbf

55 lbf

11 in

22 in

FIGURE 20 This lug wrench is producing a couple of 100.8 ft-lbf, the same sized torque as the wrench in **FIGURE 2**, but from two forces of only half the size. In addition, it produces no net force on the lug, so it's less likely to slip off.

By doing this, you don't create any vertical force on the lug nut, because the two forces cancel out. Not only can you make the same torque by applying a smaller force at two points, the lack of (downward) vertical force prevents the wrench slipping off the lug nut.[7]

Note 7: What would make the wrench slip off is actually an unwanted torque from the net force being applied outward from the face of the wheel. No need to get bogged down here, though.

5 Force and Torque Resolution

Now that we have the required tools, let's revisit **Figure 1**, where we have two horizontal forces acting on our street rod at different heights. Let's find their combined action by *resolving* the forces to the CG.

First we look at each force separately (see **Figure 21**). The 200-lbf drag force happens to be pointed directly at the CG; its line of action goes through the CG. But the line of action of the thrust force is at pavement level, 1.5 feet below the CG.

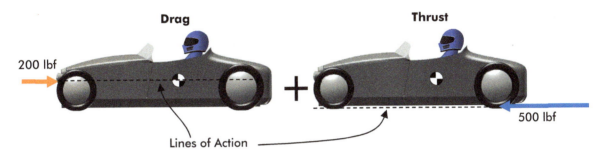

Drag

200 lbf

Thrust

500 lbf

Lines of Action

Let's **RESOLVE** each force to the CG. That's easy with the drag force; when acting on a rigid body, a force can be moved to any point along its line of action and have the same effect (remember transmissibility?). So we can just slide the force over to the CG, and we're done (**Figure 22**, left).

Figure 21 The drag and thrust forces each act along their own line of action. The drag here acts directly at the CG, and the thrust acts 1.5 feet below it.

The thrust force requires only a bit more work. Since its line of action misses the CG by 1.5 feet, it creates a clockwise torque (a moment) on it, of 750 ft-lbf. We can shift the thrust force up to the CG if we add back in the torque effect, by adding a 750 ft-lbf couple (which remember, is a pure torque). This is shown on the right side of **Figure 22**.

KEY CONCEPT

We can use **FORCE RESOLUTION** to transfer any force acting on any point of a *rigid body*, to any other point on the body, so that it will still have the same effect. The force is replaced by a force of the same magnitude and direction, plus a couple equal to the force multiplied by the distance between the new point and the line of action of the original force.

Each Force Resolved to CG

750 ft-lbf

500 lbf

200 lbf

FIGURE 22 Resolving the drag force to the CG produces no torque, but the thrust force is replaced by both a force and a torque (a couple) acting on the CG.

We have now resolved both forces to the CG, and we can add their results. First we add any fore-aft forces (of which we have two), any vertical forces (none shown), and then any couples (we have one). After summing them, we get the net result in FIGURE 23: a 300-lbf forward force and a 750 ft-lbf torque which tends to pitch the nose up. This is called a FORCE-COUPLE SYSTEM.

FIGURE 23 Resolving the two horizontal forces shows their total action on the CG: a 300-lbf forward force and a 750 ft-lbf torque tending to nose the car up.

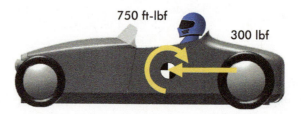

750 ft-lbf

300 lbf

How do we interpret and use this result? One part is obvious: there is a resultant force of 300 lbf (thrust minus drag) acting on this 1,600-pound car, which would result in an acceleration of 6.04 ft/sec² (0.1875 *g*'s).

What effect will the torque from the thrust force have? FIGURE 23 suggests that the torque would rotate the body nose up, but also that there's nothing to limit it. But that's not the whole picture. FIGURE 24 adds in the vertical forces on the car when stationary (left), and accelerating (right). With the car's weight of 1,600 lbf, an 8-foot wheelbase, and the CG 3 feet forward of the rear axle, the front tires carry 600 lbf and the rear 1,000 lbf when stationary (Chapter 16 will show how these all fit together).

But under acceleration, the 750 ft-lbf torque acts on the car, and the tire loads must change to counteract it. As a result, the load on the front tires must be reduced by 94 lbf, and the load on the rears must increase by 94 lbf.

Overall, we've taken the resolved forces and moments from FIGURE 23 (from traction and drag), and used them to change the vertical forces from the left side of FIGURE 24 (the weight and stationary tire loads) and produce the right side of FIGURE 24 (the weight and altered tire loads).

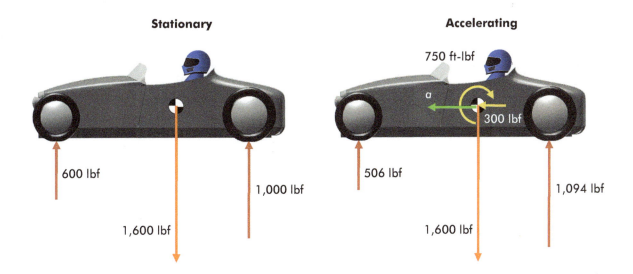

Stationary

600 lbf

1,000 lbf

1,600 lbf

Accelerating

750 ft-lbf

a

300 lbf

506 lbf

1,094 lbf

1,600 lbf

This is the first case of load transfer that we've seen, but we'll hit it pretty hard later. Chapters 16 and 17 on statics and quasi-statics will cover load transfer during acceleration, braking, and cornering.

Already though, we can feel good about being able to take any set of forces and moments in a plane and resolve them into a single force and couple.

You can actually resolve any set of forces and moments to any point on the body you want, not just the CG. But be careful. When there is a net force or couple accelerating the body, we need to resolve the forces to the CG in order to use $F = ma$.

FIGURE 24 What happens as a result of the force and torque from **FIGURE 23**? The car accelerates at about 0.2 g's (green arrow), and the load on the front tires is reduced by 94 lbf, but increased by 94 lbf at the rear.

6   Summary

What do you have to show for having read through several wrenching examples of torque, using vectors for tire balancing and navigation, and learning some other things about vectors you didn't think you needed to know?

First off, this chapter will be a good one to come back to when we use vector concepts throughout the book, and some of the forgotten details suddenly become important (it's inevitable). Remember that while we may have demonstrated certain tasks on one type of vector, most can be used on other types. For example, vector addition and subtraction were used on displacement vectors, but force vectors, velocity vectors, etc., add and subtract in the same way.

Remember that you can always think of a vector as magnitude and direction or as x- and y-components. Its magnitude and direction are often easier to visualize. Its x- and y-components are often easier to do calculations with. Same information, different purposes.

If this is your first exposure to vectors, they may seem abstract. But vectors should seem familiar to us—they are another tool we use everyday. Increasing torque by pushing *more perpendicular* to a wrench, or adjusting the angle between wheel weights are just a couple of ways we can use vectors. We also know that to push harder on something that's rolling, we need to push more parallel to its motion (*less perpendicular*).

Torque itself shouldn't seem abstract at all; if you're reading this book, you know what it is. Using a wrench, opening a twist-top, or turning a steering wheel are all good examples. But we've seen that torque shows up in more subtle places, like traction forces pitching a car's nose up and changing the loads on the tires.

We've also seen that using vector methods gives you some options on how to calculate torque. With the techniques presented here, you should be able to calculate torque for any force and lever arm in a plane. Graphically, it's all summed up in **FIGURE 18**; the torque about any point is the force times the perpendicular distance between the force's line of action and that point. You may use different ways to get there, but that's what it boils down to.

The summation of resolved forces into a force-couple system is another concept not to be taken lightly. It's vital to know that no matter how many forces act on a rigid body, their effect can be summarized by a single force and a torque (a couple) about its CG (not only in 2-D, but also in 3-D). This can make a seemingly complicated situation relatively easy—or at least give you the patience to slog through it, confident you can get there in the end.

Having added torque and 2-D vectors to straight-line dynamics and forces, we're in good shape to venture into rotating motion, and motion in a plane.

Major Formulas

Definition	Equation[8]	Equation Number
Torque from a Force F Acting Perpendicular to a Lever Arm of Length r	$T = Fr$	7-1
Magnitude of a Vector (Its Absolute Value)	$s = \lvert \mathbf{s} \rvert = \sqrt{s_x^{\,2} + s_y^{\,2}}$	7-2
Direction Angle of a Vector $\mathbf{s}$	$\theta_{\mathbf{s}} = \tan^{-1}\left(\dfrac{s_y}{s_x}\right)$	7-3
Addition of Two Vectors by Vector Components	$s_{3,x} = s_{1,x} + s_{2,x}$	7-4a
	$s_{3,y} = s_{1,y} + s_{2,y}$	7-4b
Subtraction of Two Vectors by Vector Components	$s_{1,x} = s_{3,x} - s_{2,x}$	7-5a
	$s_{1,y} = s_{3,y} - s_{2,y}$	7-5b
Vector x- and y-components in Terms of Its Magnitude and Direction Angle	$v_x = v\cos\left(\theta_v\right)$	7-6a
	$v_y = v\sin\left(\theta_v\right)$	7-6b
Torque from a Force F Acting on a Lever Arm of Length r, at an Angle θ	$T = Fr\sin\theta$	7-7
Torque in Terms of a Lever Arm and Force, in Vector Components	$T = r_x F_y - r_y F_x$	7-8
Equation for a Couple, in Terms of Two Equal and Opposite Forces and the Distance Between Their Lines of Action	$T = F_{Perp}\, d$	7-9

Note 8: Note that displacement and velocity vectors are used as examples for magnitude, direction, components, and addition and subtraction. Other vectors follow the same rules.

8 Angular Dynamics Basics

Torques, Rotational Inertia, and Angular Acceleration

Able to produce huge amounts of torque and horsepower, Chrysler's 426 Hemi was a dominant force in drag racing and, when allowed, in stock car racing. *Photo by Randy Beikmann with permission of owner John Ottino.*

Chapter 8

The kinematics and Newton's laws for rotational motion are presented, and rotational inertia is defined. Gear sets are analyzed for speed ratios, torque ratios, and tooth forces. Angular motion is connected to straight-line motion, in belt drives and rolling motion.

Contents

Key Symbols Introduced

Symbol	Quantity	SFS Units	MKS Units
θ	Angular Displacement	radians (rad)	radians (rad)
ω	Angular Velocity	rad/sec	rad/sec
α	Angular Acceleration	rad/sec^2	rad/sec^2
I	Rotational Inertia	slug-ft^2	kg-m^2

1 Introduction

You'll notice that in the automotive world, rotating parts are every-where. Crankshafts and flywheels rotate, fans rotate, shafts and tires rotate. To understand powertrains and drivelines, we need to understand angular dynamics—the dynamics of rotating bodies.

The first major purpose of this chapter is to apply kinematics and dynamics to *pure* rotational motion, as we did for straight-line motion. See, for example, the clutch/flywheel assembly in **FIGURE 1**. If the clutch is disengaged so the engine can free-rev, the engine's torque would only need to accelerate the inertia of the crankshaft, flywheel, and pressure plate.

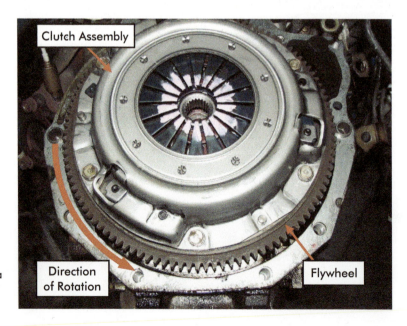

FIGURE 1 A flywheel and clutch assembly from a 1991 Miata. Together, they add rotational inertia to the crankshaft and "smooth out" its rotational speed.

In this chapter we'll see that the dynamics of rotation is very similar to that of straight-line motion. Its kinematic, inertial, and forcing action quantities are listed in **TABLE 1**. Note that torque T is the forcing action, and *rotational inertia I* is the inertia. Rotational motion uses Greek letters for symbols: θ for angular displacement, ω for angular velocity, and α (spelled "alpha" and pronounced *AL-fuh*) for angular acceleration.

	Translation	**Rotation**
Displacement	s	θ
Velocity	v	ω
Acceleration	a	α
Forcing Action	F	T
Inertia	m	I

TABLE 1 For formulas in rotational dynamics, we will simply replace the quantities for translation with the quantities for rotation.

The next purpose of this chapter is to link rotation and translation (see **FIGURE 2**). If you've ever gotten a sleeve caught in a pulley, your hair in a drill, or the thought of either makes you cringe, you know how rotation causes straight-line motion. You also know that anyone working with rotating machinery should be very careful. These examples also help explain why and how for physics, we measure angles in radians. Section 2.4 will cover this in detail.

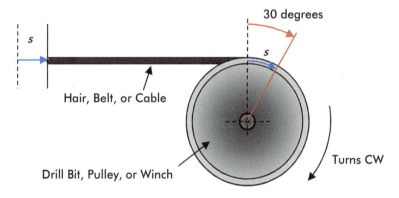

FIGURE 2 When we have rotation, we're usually using it to cause straight-line motion, or vice versa.

Fortunately, there are more productive (and less painful) ways to pair rotation and straight-line motion, such as a pulley pulling a belt, or a tire propelling a vehicle.

But let's not get ahead of ourselves. The next few sections will cover these concepts.

2 Rotation Basics

When I was in first grade, an older student showed me how to use a ruler (no problem) and then tried to explain using a protractor.[1] I wasn't getting it. Then my teacher tried to explain what an angle was. I still didn't have a clue. After a couple of minutes, they gave up. I think they, like most people, knew what an angle was, but had a hard time explaining it (at least to a first grader).

2.1 Angles and Angular Displacement

Assuming you've thought about it, you probably think of angles as a *difference in direction* between two lines or edges, or as a "misalignment" between them. For example, the left side of **FIGURE 3** shows a 25° angle between two different vectors; their directions are 25° apart.

But in angular dynamics, we will think of an angle as the amount of *rotation* of an object (see the right side of **FIGURE 3**). Here, the vector is rotated from position A to a new position A′. A weathervane pointing east can be turned (rotated) 90° counterclockwise to point north. Turn the minute hand on a clock 30° forward (clockwise, naturally), and it reads 5 minutes later.

Difference in Direction Between Vectors

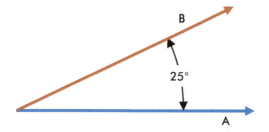

Rotation of a Vector in CCW Direction

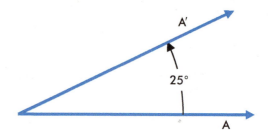

FIGURE 3 Two ways to interpret an angle. On the left, the 25° angle is thought of as a difference in direction; on the right as a rotation of the vector A to A′.

On the right side of **FIGURE 3**, the rotation angle has a definite direction. In the same way, if a piston is at top dead center (TDC), turning the crankshaft 180° CCW will put it at bottom dead center (BDC), as in **FIGURE 4**. Rotating the crankshaft 360° CCW would put it back at TDC, and the crankshaft has gone through one revolution.[2]

Note 1: For one year I was in a two-room schoolhouse with first-to-fourth grade in one room—no kidding.

Note 2: One of my co-worker's favorite quotes is from a college basketball player who said, "We have to turn our game around three-hundred and sixty-five degrees," which would only result in a 5° change.

Note: Viewed from Rear of Engine

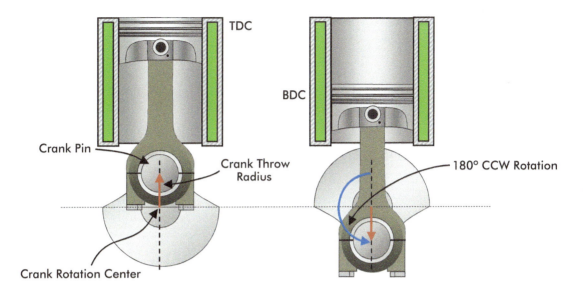

Crank Pin

Crank Throw
Radius

Crank Rotation Center

TDC

BDC

180° CCW Rotation

Since an angle is an amount of rotation, angular displacement is the rotation of an object through an angle. More precisely, **ANGULAR DISPLACEMENT** is the amount of angular motion from some reference position, with its direction specified (i.e., CW or CCW). The angle is symbolized by the Greek letter θ. In **FIGURE 5**, right, the angular displacement is 30° CCW.

Modern engine controls need to know the crank position, to trigger fuel injection and spark at the right time. A toothed wheel (see **FIGURE 5**) is mounted to the crankshaft, and the crank sensor (a Hall effect sensor, aka magnetic pickup) signals each passing steel tooth by creating a voltage pulse between the attached wires. By counting the number of pulses since the "missing tooth," and knowing the angle between teeth, the position of the crank is calculated (for example, TDC for cylinder 1 might be at the 10th pulse after the missing tooth). The system would know when the missing tooth goes by from looking at the pattern of voltage pulses from the magnetic pickup, and by noting when the time between pulses is twice as long as normal.

If not for the missing tooth, the crank wheel pictured would have 24 evenly spaced teeth, so there are 360°/24 = 15° between them. Said another way, rotating the crank by one tooth is 15°. On the right side of **FIGURE 5**, the crank has rotated CCW by two teeth, or 30°.

FIGURE 4 When the piston moves from TDC (top dead center) to BDC (bottom dead center), the crankshaft rotates CCW through 180°. The crank throw (red arrow) rotates from a straight upward position to straight downward.

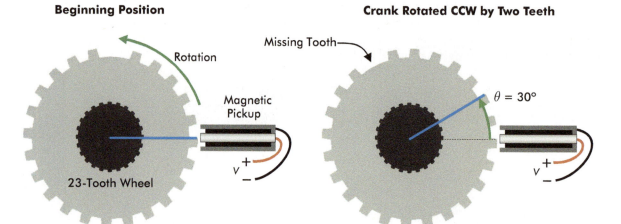

Beginning Position   **Crank Rotated CCW by Two Teeth**

FIGURE 5 Crank position is often measured by sensing the passing teeth on a crank-mounted wheel. Here, the crank rotates 30° CCW (two teeth).

In everyday use, we measure rotation angles in degrees or revolutions. We'll soon see that there is one more way, which works more naturally in physics than either of these.

2.2 Angular Speed and Angular Velocity

Just as with straight-line motion, we have angular speed and angular velocity. Angular speed is simply how fast something is rotating, such as 6,000 revolutions per minute (rpm). Angular velocity is speed with direction, specifying CW or CCW, for example.

ANGULAR VELOCITY is defined as how quickly the angular displacement is increasing, and is called ω (omega). So it's positive in the same rotational direction as the angular displacement.

You might think we would measure angular speed in revolutions/second (rps). But engineers began measuring speeds when shafts rotated more slowly than today, so rpm is more common. Sometimes we *will* talk about rotation speed in revolutions/second, because our consistent time unit is the second.

For example, a crankshaft in an engine running 8,000 rpm is revolving 133 times/second, and an F1 engine at 18,000 rpm is revolving 300 times/second! It still amazes me that some street bikes redline at 15,000 rpm (or 250 rev/sec); each cylinder brings an air/fuel mixture in, compresses it, burns and expands it, and exhausts it, in $\frac{1}{125}$ of a second (2 revs).

In **Figure 6** we have our crank sensor hooked up to a voltmeter, and to a display you might see on an engine analyzer. This shows the individual peaks and valleys of the voltage pulses from the sensor. The time between these pulses is used to calculate engine speed.

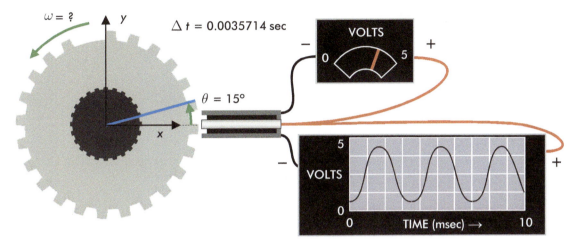

Let's calculate the crank speed from **Figure 6** using the same approach an engine controller might use.

Figure 6 As the crank wheel rotates, the magnetic pickup creates a series of voltage pulses (the "blips" seen on the voltage display at bottom). Rotation speed is calculated from the time between them.

202 Chapter 8: Angular Dynamics Basics

EXAMPLE

PROBLEM: If the sensor in **FIGURE 6** measures 3.5714 milliseconds (0.0035714 seconds) between pulses, what speed is the crank turning?

SOLUTION 1: Since the crank-wheel teeth are evenly spaced at 24 teeth per revolution (aside from the missing one), it would take 24 times as long to rotate one revolution as to rotate from one tooth to the next. The time for one revolution is then:

$$t_{Rev} = 24 \times t_{Tooth} = 24 \times (0.0035714 \text{ sec}) = 0.085714 \text{ sec}$$

If it takes 0.085714 sec/rev, then the angular speed ω (in rev/sec) is the reciprocal of that:

$$\omega = 1/t_{Rev} = \frac{1}{0.085714 \text{ sec/rev}} = 11.667 \text{ rev/sec}$$

which is 700 rev/min (rpm). The engine is at idle.

SOLUTION 2: We know there are 15 degrees between pulses, so the angular speed is the change in angle divided by the change in time:

$$\omega = \frac{\theta_{Tooth}}{t_{Tooth}} = \frac{15 \text{ deg}}{0.0035714 \text{ sec}} = 4,200 \text{ deg/sec}$$

Converting degrees/sec to revolutions/sec:

$$\omega = \left(\frac{4,200 \text{ deg}}{1 \text{ sec}}\right) \times \left(\frac{1 \text{ rev}}{360 \text{ deg}}\right) = 11.667 \text{ rev/sec}$$

which, as we've seen, is 700 rpm.

Notice that the second solution, calculating angular velocity by dividing the angular displacement by the time it took to do it, was more direct. After one calculation, we already had angular velocity—all we did afterward was convert to rpm.

NOTE

Toothed wheels and magnetic pickups aren't just used on crankshafts. They're used to measure wheel speeds for traction control and ABS (antilock brakes), and on transmission shafts to run the speedometer.

So that's our quick introduction to rotation. In the next section we'll not only show how rotation can be related to translational motion, but also find a different way to measure angles, thereby killing two birds with one stone. [3]

2.3 Connecting Rotation to Translation

Let's look closer at the link between rotation and translation. FIGURE 7 shows a CW-rotating pulley driven by a belt (most engines turn CW from a front view). In the figure, a pulley with a 1-inch radius rotates by 60°, and about 1.047 inches of belt is drawn in. [4] Rotating the pulley twice as much pulls twice as much belt in. So the belt motion s and the pulley rotation θ are linked by a simple ratio. What is that ratio, s/θ?

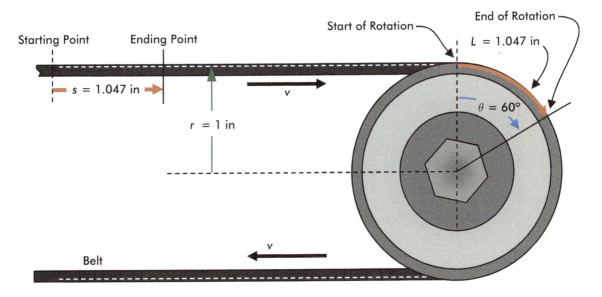

You *could* use the fact that that during one revolution of the pulley, a length of belt equal to the circumference of the pulley is pulled in—the circumference is 2 times π (spelled "pi" and pronounced *pie*) times the radius r, written as $2\pi r$. [5] Then set up a proportion: dividing the actual rotation angle θ by 360° is equal to dividing the actual belt translation s by the circumference.

FIGURE 7 A 1-inch radius pulley rotating 60° pulls in about 1.047 inches of belt. The pulley rotation and belt translation are linked by a set ratio.

Note 3: No birds were killed in the preparation of this manuscript, whether by stones or other means.

Note 4: The pulley's physical outside radius would be less than an inch, but the distance out to where the cord rides is what matters to the belt motion.

Note 5: When we moved into GM's new Noise and Vibration Building some years ago, I was confused by a staircase with the label "STAIR PI." I could *not* figure out why they would use π to number a staircase, even in an engineering building. It turned out to be "STAIR P-one."

The number π is the ratio of a circle's circumference to its diameter. It is approximately 3.14159.

We can use this proportion to get the belt's translational displacement:

Equation 8-1   **Angle in degrees**

$$s = \frac{\theta}{360°}\left(2\pi r\right)$$

For example, if the pulley radius is 1.5 inches, and it rotates 45°, the belt's displacement is:

$$s = \frac{45°}{360°} \times \left(2 \times \pi \times 1.5 \text{ in}\right) = 1.178 \text{ in}$$

We get the right answer, but equation 8-1 is unnecessarily complicated and clunky. There is a more natural way of doing this than taking ratios of angles.

2.4 Measuring Angles in Radians

In physics, a better way to measure a rotation angle is to compare its arc length L to its radius r. We'll show this by tracking the belt material in the first 3 inches of the upper span in **FIGURE 8** onto the 1-inch pulley. As we rotate the pulley clockwise, it draws in those 3 inches (3 times the radius, note). Each inch of belt is marked off on the pulley, along with the angle it spans. Each of these 1-inch arc lengths is equal to the pulley radius. We say that each of these angles is one radian.

An angle whose arc length is equal to the arc's radius is one **RADIAN**.

So here, for a rotation of 1 radian we have a 1-inch arc, for 2 radians 2 inches, and for 3 radians 3 inches (pretty simple, huh?).

FIGURE 8 A rotating pulley with radius 1 inch, with belt material divided into arcs of 1 inch each. In rotating 3 radians (171.9°), the pulley pulls in 3 inches of belt.

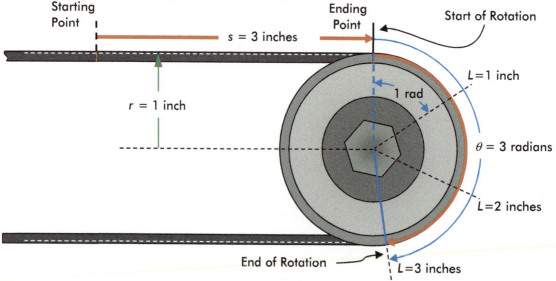

Starting Point

Ending Point

Start of Rotation

$s = 3$ inches

$L = 1$ inch

$r = 1$ inch

1 rad

$\theta = 3$ radians

$L = 2$ inches

End of Rotation

$L = 3$ inches

Of course the radius isn't always 1 inch. But regardless of the radius, if the angle's arc length L equals the radius of the arc, the angle is 1 radian. Then, if the arc length does not equal the radius, the angle is found using a ratio:

General Equation 8-2   **Angle in radians**

$$\theta = \frac{L}{r}$$

This is shown in **FIGURE 9**. When an angle has a measure of 1 radian, the ratio of the circular arc length L to the radius r is 1:1 (the fraction ⅟₁). It follows that if $L/r = $ ½, then the angle is ½ radian, and if $L/r = $ ¼, the angle is ¼ radian, etc.

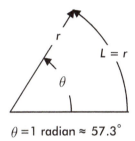

$\theta = 1$ radian $\approx 57.3°$

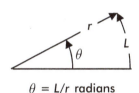

$\theta = L/r$ radians

FIGURE 9 An angle is equal to 1 radian (about 57.3°) when its arc length L is equal to its radius r (left). Otherwise, the angle in radians equals the ratio L/r (right).

When you use radians, there is a simple relationship between the pulley's angular displacement and the belt's translational displacement. Starting with equation 8-2, then referring back to **FIGURE 8**, noting that $L = s$, we get

General Equation 8-3   **Using radians**

$$s = r\theta$$

> **NOTE**
>
> Compare this to equation 8-1; this is *a lot easier* to remember! And the ratio s/θ of belt motion to pulley rotation, that we set out to find in Section 2.3, is also very simple: it's just r!

How many radians are in a revolution? In one revolution, the pulley would draw in "one circumference" of belt, so s is equal to $2\pi r$. Dividing this by the radius r, the angle turned in one revolution is 2π radians.

> **NOTE**
>
> One revolution is equal to 360 degrees and 2π (about 6.2832) radians. One radian is then about 57.3 degrees, and in exact form is $180/\pi$ degrees.

So if you know an angle in revolutions, you can turn it into radians by multiplying it by 2π:

Equation 8-4a

$$\theta_{Radians} = 2\pi\theta_{Revolutions}$$

If the angle is given in degrees, you can convert it to radians by dividing by 57.3:

Equation 8-4b

$$\theta_{Radians} \cong \frac{\theta_{Degrees}}{57.3°}$$

KEY CONCEPT

An angle in its most basic form is simply the ratio of its circular arc length L to the arc's radius r.

If you're new to it, using radians will almost certainly seem like a pain. But this is another time where getting the idea down pat pays dividends. It's best to store the simplest possible concepts and formulas in your head, so you can pull each one out as you need it.[6] Instead of memorizing a complicated equation like 8-1, remember two simpler equations: one like equation 8-3, and one to convert the angle into radians, as in equations 8-4.

You don't need (or want) to use radians all the time, of course. Asking for a "$\pi/2$-radian" brace at the hardware store (instead of 90°) would get a blank stare at best, and using radians for ignition timing would be equally ridiculous. But in physics, using angles in radians is essential.

So now that we've introduced radians (rad), we need to define the consistent units for angular motion.

NOTE

The consistent unit for angular displacement θ is the radian, and the consistent units for angular velocity θ are radians/second, or rad/sec.

Note 6: Why strain your brain?

2.5 Some Common Angles in Radians

To get a feel for radians, have a look at **TABLE 2** to compare some commonly used angles. Each is listed in degrees, revolutions, and radians.

Degrees	Revolutions	Radians (Exact Form)	Radians (Decimal Form)
0	0	0	0
57.2958	0.159155	1	1
60	⅙	$\pi/3$	1.04720
90	¼	$\pi/2$	1.57080
120	⅓	$2\pi/3$	2.09440
180	½	π	3.14159
270	¾	$3\pi/2$	4.71239
360	1	2π	6.28319

Better yet, have a look at **FIGURE 10**, which shows multiples of 45° and 30°, in both degrees and radians.

TABLE 2 Common angles in degrees, revolutions, and radians. Radian measures are shown in exact form, as well as rounded off to working precision.

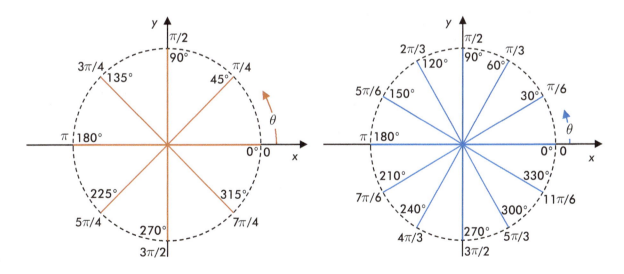

FIGURE 10 Common angles in degrees and radians. Angles in multiples of 45° ($\pi/4$ rad) are shown on the left, and in multiples of 30° ($\pi/6$ rad) on the right.

REMINDER

FIGURE 10 is set in the standard way: x to the right, y upward, and CCW for rotation positive. Sometimes we don't follow this.

3   Angular Kinematics

With the groundwork in place, let's cover the kinematics of angular motion. We start by formally defining angular velocity and acceleration. Then we'll cover the (tangential) motion of a point on a rotating body. With that, we develop the equations for displacement and velocity vs. time.

3.1 Angular Displacement, Velocity, and Acceleration

Angular displacement has already been defined, and symbolized as θ. The consistent unit for angular displacement is (of course) the radian, which is abbreviated "rad." Just as in linear displacement, we need to define which direction is positive (typically counterclockwise).

> **NOTE**
>
> Sometimes defining CCW angles positive is not convenient. For instance, I'll usually call θ positive for rotation when it would make the car move forward, whether for an axle shaft, driveshaft, or crankshaft. The important thing is to define it clearly.

If we take the change in angular displacement, $\Delta\theta$, over a "short" time Δt, we get the angular velocity ω, which is in the consistent units of rad/sec:

General Equation 8-5   **Angular velocity**

$$\omega = \frac{\Delta\theta}{\Delta t}$$

In this equation, the time Δt needs to be short enough so that the velocity has hardly changed. If we now take the change in angular velocity, $\Delta\omega$, over the same "short" time Δt, we get angular acceleration:

General Equation 8-6   **Angular acceleration**

$$\alpha = \frac{\Delta\omega}{\Delta t}$$

ANGULAR ACCELERATION is how quickly the angular velocity is increasing and is called α (alpha). Its consistent units are rad/sec per second, which is rad/sec².

Sometimes you'll see angular acceleration given in rpm/sec, for example when free-revving an engine at 3,000 rpm/sec, or when measuring an engine's flywheel horsepower on the dyno while increasing its speed at 600 rpm/sec. Converting rpm/sec to rad/sec^2 is just like converting from rpm to rad/sec. For $\alpha = 600$ rpm/sec, you get:

$$\alpha = \frac{\Delta\omega}{\Delta t} = \frac{600 \ \cancel{\text{rev}}/\cancel{\text{min}}}{1 \ \text{sec}} \times \left(\frac{1 \ \cancel{\text{min}}}{60 \ \text{sec}}\right) \times \left(\frac{2\pi \ \text{rad}}{\cancel{\text{rev}}}\right) = 62.8 \frac{(\text{rad/sec})}{\text{sec}}$$

$$= 62.8 \ \text{rad/sec}^2$$

Note that with θ positive in forward motion, engines normally don't turn at negative angular velocities (backwards). But they certainly do experience negative angular *acceleration*, where the angular velocity is decreasing (it's slowing down). When you take your right foot off the pedal on a level road, the engine slows (along with the car) as its torque is reduced.

3.2 Tangential Motion on a Rotating Body

A good example of tangential motion is how fast the tip of a lawn-mower blade is moving. It is the motion of a point along a circular arc about a rotation center.

Let's start with a point at the edge of a flywheel with radius r (**Figure 11**, left). As the flywheel turns, this point moves through a tangential "displacement" s along an arc.[7] The displacement s depends on the flywheel rotation angle θ as:

General Equation 8-7   **Motion along arc at radius r**

$$s = r\theta$$

which is the same amount we saw in equation 8-3 for the belt/pulley system. Tangential velocity is the rate of change in the point's position along this arc, so it is the radius r times ω, the rate of change of the angular displacement θ:

General Equation 8-8   **Tangential velocity**

$$v = r\omega$$

For the 6,000 rpm and radius of 6 inches in **Figure 11**, this comes to 314 ft/sec, or 214 mph!

Do the same for the tip speed of a 21-inch mower blade turning 3,600 rpm, and you should get 330 ft/sec, or 225 mph, so watch your feet.

Note 7: Here, s isn't a proper displacement in the "straight-line" sense that we've studied. But hang in there—we're really after the tangential velocity and acceleration.

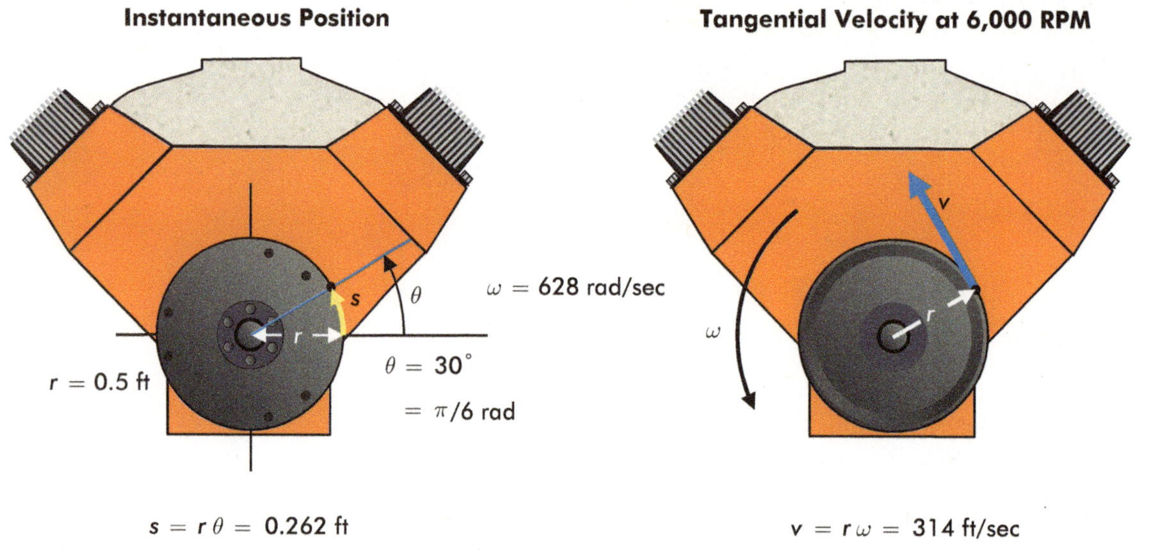

Instantaneous Position

$\omega = 628$ rad/sec

$r = 0.5$ ft

$\theta = 30°$

$= \pi/6$ rad

$s = r\theta = 0.262$ ft

Tangential Velocity at 6,000 RPM

$v = r\omega = 314$ ft/sec

FIGURE 11 While calculating the arc length s traveled by a point at radius r on the flywheel (left), we can also calculate the point's tangential velocity (right).

The relationship between tangential and angular acceleration is very similar to that for angular velocity, by the same reasoning:

General Equation 8-9   **Tangential acceleration**

$$a = r\alpha$$

Using this to calculate the tangential acceleration of the edge of the flywheel in **FIGURE 12**, we get about 31.4 ft/sec², a bit less than 1 g. Note that if the angular velocity were constant, the angular acceleration would be zero, and so would the tangential acceleration at radius r.

Angular Acceleration of 600 RPM/Sec

$\alpha = 62.8$ rad/sec^2

$r = 0.5$ ft

$a = r\alpha = 31.4$ ft/sec^2

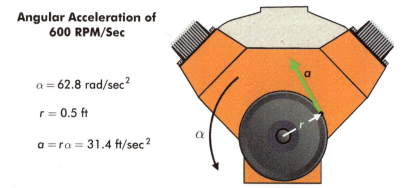

FIGURE 12 A crank/flywheel assembly with angular acceleration α of 600 rpm/sec, or 62.8 rad/sec². The tangential acceleration at the flywheel rim is 31.4 ft/sec².

Remember that the tangential velocity and acceleration of a point are perpendicular to the radius drawn to that point.

> **KEY CONCEPT**
>
> Why do we care about the tangential acceleration of a part of the flywheel? The reason is that there is mass in each part of the flywheel, so when the flywheel is sped up (or slowed down), we are accelerating this mass. We'll use this fact in introducing rotational inertia.

3.3 Angular Kinematic Equations

Now that we have all the terms and tools, let's go through angular kinematics. As with straight-line motion, acceleration changes velocity, and velocity changes displacement. As a result, the kinematic equations for rotation are similar to the equations for translation. We can simply take the equations from Chapter 2 and replace the symbols for translation with the corresponding ones for rotation.

This means that for motion at a *constant* angular velocity, the angular displacement becomes

Equation 8-10   **Constant ω**

$$\theta = \theta_0 + \omega t$$

where θ_0 is the initial displacement. If an object has a constant angular acceleration α, and begins with an angular velocity ω_0 at $t = 0$, the angular velocity increases from there:

General Equation 8-11   **Constant α**

$$\omega = \omega_0 + \alpha t$$

When angular acceleration is constant, angular displacement increases with the square of time. The general equation for this, starting at a displacement of θ_0 and a velocity of ω_0, is

General Equation 8-12   **Constant α**

$$\theta = \theta_0 + \omega_0 t + \frac{1}{2}\alpha t^2$$

Finally, we can find the angular displacement θ when starting at θ_0, while angular velocity increases from ω_0 to ω at a constant angular acceleration α:

General Equation 8-13   **Constant α**

$$\theta = \theta_0 + \frac{\left(\omega^2 - \omega_0^{\,2}\right)}{2\alpha}$$

Compare these to the translational equations in Chapter 2; they have exactly the same form.[8] The same will be true for Newton's laws.

Remember that direction is part of the definition of angular displacement, velocity, and acceleration, just as it is for their translational counterparts. Turning in the right direction is just as important as turning at the right speed.[9] In other words, θ, ω, and α are really vectors.

> **NOTE**
>
> Although angular displacement, velocity, and acceleration are vectors, this book won't use vector notation for them.

Now that we have kinematics under control, let's move on to dynamics.

4   Angular Dynamics

We're ready to relate rotational acceleration to its cause: torque. While doing so, we'll also define rotational inertia.

4.1 Newton's First Law for Rotational Motion

When a body is given a spin at a certain angular velocity, it will keep rotating with that velocity (same speed and direction) unless a torque makes it change. A smooth flywheel on a low-friction bearing comes close to this, and will spin for a very long time if given a good push. An even better example is Earth, which spins about once every 24 hours; it rotates at a constant rate because there is very little resistance to its motion. From this comes Newton's first law for angular motion.

> **FUNDAMENTAL LAW**
>
> A rotating body that is not acted on by a resultant torque has a constant angular velocity.

So just as a large moving mass has a strong tendency to keep moving at a constant straight-line velocity (that is, it has a large amount

---

Note 8: The other straight-line kinematic relationships from Chapter 2 can also be converted to angular motion. That has not been done here, to save space.

Note 9: One of my professors told us about a project for a canning factory, which his design team had been working on for months. Finally, they delivered and installed the machine on the assembly line. When they tested it, everything ran at the right speed, but the shaft driving the conveyor belt (and the conveyor belt itself) ran *backwards*. Right speed, wrong velocity.

of inertia), a large rotational inertia has a strong tendency to keep rotating at a constant angular velocity. This is how a flywheel on an engine's crankshaft smoothes its rotational motion. An engine with a large flywheel tends to run at a constant speed, resisting the fluctuating torque pulses from the cylinders firing that might otherwise twist the driveline and shake the car. While the crank speed variation with a large flywheel isn't zero, it is less than it would be with a smaller flywheel, or with no flywheel. However, that same inertia also makes it harder to accelerate the car when you want to.

4.2 Torque in Motion

In Chapter 7 we discussed the pure torque from a pair of equal and opposite offset forces, which we called a couple. An example is the couple produced by a lug wrench, as in **FIGURE 13**. But here it is also turning counterclockwise once every two seconds (30 rpm). With two 150 lbf forces acting on it, each 1 foot from center, the torque applied is 300 ft-lbf (maintaining this much torque by hand while turning it this fast would be a pretty good trick).

On the right side of **FIGURE 13** we have the business end of a V8 engine, also producing 300 ft-lbf. Because the crankshaft is held on its center by the main bearings, the torque produced at the flywheel is also a pure, rotating couple.

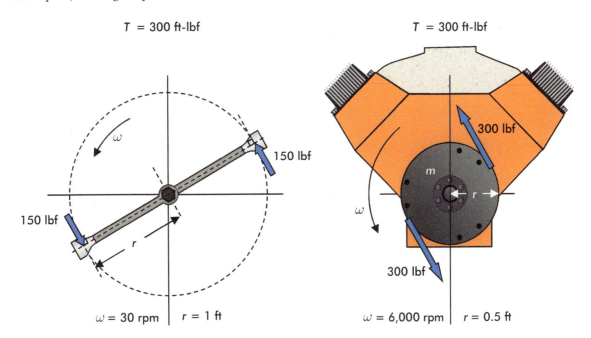

In either case, you can imagine two forces with magnitude F to be "chasing" the lever arms around in a circle with radius r, the forces always staying perpendicular to the lever arms. For the wrench, the two forces F are each 150 lbf, and the lever arm is 1 foot, so the total torque

FIGURE 13 A moving torque produced by a lug wrench or a crankshaft is a rotating couple.

is 300 ft-lbf. For the engine, which has attachment bolts 6 inches out from the crank centerline, we can think of the couple as two 300-lbf tangential forces (as shown), or as 100-lbf tangential forces applied at each of the six attachment points.

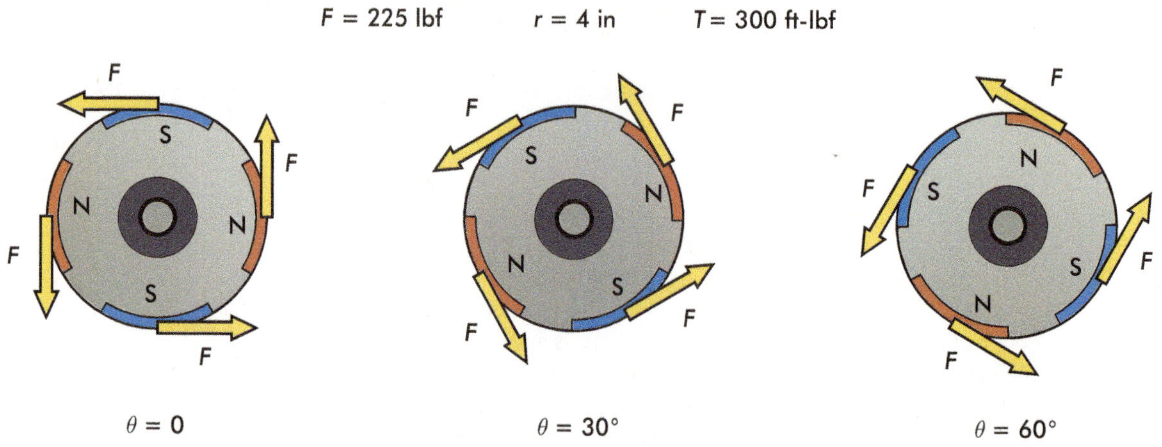

F = 225 lbf r = 4 in T = 300 ft-lbf

$\theta = 0$ $\theta = 30°$ $\theta = 60°$

FIGURE 14 The magnetic forces (in yellow) acting on the rotor of an electric motor follow it as it rotates, producing a near-constant torque.

In an electric motor you don't have to imagine the forces chasing a lever arm around—it's how they work. **FIGURE 14** shows the rotor from an electric motor with four permanent magnets (two north poles outward, and two south), each with a 225-lbf tangential force acting on it. With the rotor's radius of 4 inches (0.333 feet), the four forces create a total of 300 ft-lbf. As the rotor turns, the magnetic field in the stator follows it to maintain a nearly constant torque on the rotor (the forces do vary slightly in size as the rotor turns).

4.3 Newton's Second Law for Rotational Motion

It takes a good "twist" to get a large flywheel to accelerate. You can also imagine that it takes a much larger torque to accelerate a 15-pound flywheel than a 15-pound driveshaft, because of its larger diameter; it has more *rotational inertia*. But how do we quantify it? First we'll go through Newton's second law for rotation; the next section calculates rotational inertia.

Newton's second law for rotation has the same form as for the straight-line acceleration in Chapter 3; torque replaces force, rotational inertia replaces mass, and angular acceleration replaces translational acceleration.

FUNDAMENTAL LAW

The angular acceleration of a body is proportional to the resultant torque acting on it, inversely proportional to the rotational inertia of the body, and is in the same direction as the torque.

The larger the net torque, the larger the angular acceleration of a given rotational inertia. Recall the suggestion to think of physical quantities as measurements with a size. **ROTATIONAL INERTIA** I is the "size" of the resistance to change in angular velocity. As an equation,

---

Fundamental Law Eq. 8-14a   **Newton's second law for rotation**

$$T_{Res} = I\alpha$$

This calculates the resultant torque required to give a given rotational inertia an angular acceleration α. Equation 8-14a can also be rewritten to solve for the angular acceleration:

Equation 8-14b

$$\alpha = \frac{T_{Res}}{I}$$

The rotational inertia allowing a desired angular acceleration α from a given T_{Res} is then

Equation 8-14c

$$I = \frac{T_{Res}}{\alpha}$$

CAUTION

Don't forget that in all these formulas, consistent units must be used: torque in ft-lbf (or N-m), angular acceleration in rad/sec², and rotational inertia in slug-ft² (or kg-m²).

4.4 Rotational Inertia

We've defined Newton's second law, but left the term "rotational inertia" a bit vague. We need to make it more concrete, to know how to create a large inertia where we want, and keep it low where we want. Even when we want a large inertia, we'd typically still want it to be as light as possible (low mass), to keep the car light. How would we do that?

Let's do a thought experiment that begins with translational motion. We'll start with $F = ma$ in a straight line, and then turn it into a rotating system. On the left side of **FIGURE 15**, you see a stationary mass m with a force F acting on it. On the right, you see the same mass with the same force acting on it, but made to go in a circle of radius r by a massless, inextensible (can't be stretched) rod.[10] The rod is perpendicular to the force and the acceleration.

Note 10: Engineering is full of assumptions like this infinitely stiff rod with no mass, which couldn't actually exist, but is used to make a point. We say that parts like this are made of "Un-Obtainium."

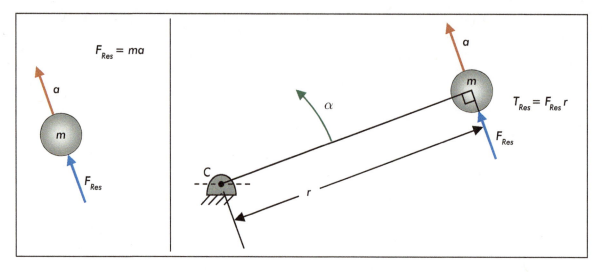

FIGURE 15 A force acting on a mass and accelerating it (left). By constraining the mass to rotate about point C (right) and using $T=I\alpha$, we will find the rotational inertia in terms of the mass and the radius.

Since the direction of the force and the motion of the mass are perpendicular to the rod, the rod doesn't affect the acceleration of the mass. So force still equals mass times acceleration:

$$F_{Res} = ma$$

We can multiply both sides of this equation by r to get:

$$F_{Res}r = mar$$

The perpendicular force F_{Res} acting at the distance r from the center of rotation produces a resultant torque, $T_{Res} = F_{Res}r$, which can be substituted into the left side of the equation:

Equation 8-15
$$T_{Res} = mar$$

Now that we have torque in the equation, we're halfway there. Since the acceleration a equals $r\alpha$, equation 8-15 can be rewritten, using equation 8-9, as:

$$T_{Res} = m(r\alpha)r = mr^2\alpha$$

Compare this to equation 8-14a, which says $T_{Res} = I\alpha$. For both to be correct, it must be true that, for a single mass at radius r:

Physical Law Eq. 8-16   **Single mass at radius r**
$$I = mr^2$$

While **FIGURE 15** doesn't look like a typical flywheel, we can learn a couple of important things from equation 8-16.

KEY CONCEPT

The more mass m, the more inertia, so doubling the mass doubles the rotational inertia. But doubling the radius r increases inertia by *four* times. Placing the mass out as far as possible produces the most rotational inertia with the least mass.

If you've ever seen a "rim flywheel" on the side of an old farm tractor, you can understand its design. These flywheels carried almost all of their mass in the rim at a radius r (see the left side of **FIGURE 16**). They only have a set of thin spokes supporting the outer mass. Neglecting the mass of the spokes and hub (assuming they are made of Un-Obtainium), all the mass can be considered to be at the radius r, and its rotational inertia is effectively $I = mr^2$.

The rotational inertia of any rigid body is the sum from all of its parts, so you can calculate it piece by piece. In the middle part of **FIGURE 16**, we have three masses of $m/3$ each, adding up to the total mass, m. The rotational inertia of each mass is $mr^2/3$, adding up to $I = mr^2$.

Most rotating objects don't have all their mass positioned at one radius, so it takes more work to calculate their rotational inertia. If an object is simple enough, like a flat disc (**FIGURE 16**, right), you can sum the inertias of all its pieces using calculus. The resulting formulas for many shapes have been put into tables. For example, the flat disc has a rotational inertia of $I = \frac{1}{2}mr^2$, only half the inertia of a rim flywheel with the same mass and outside radius.

NOTE

With the disc, not all of the mass is situated at the edge, where it would create the most rotational inertia per unit mass. The mass near the center hardly contributes at all. Therefore, the disc ends up with a rotational inertia much less than mr^2.

If the object is too complicated, you can use a computer to do FEA (Finite Element Analysis) and split the object into thousands or millions of pieces, and add up all their rotational inertias.[11]

Note 11: Computers are very good tools for doing boring tasks like that very fast and very accurately. They haven't replaced human creativity, though. Not yet, anyway.

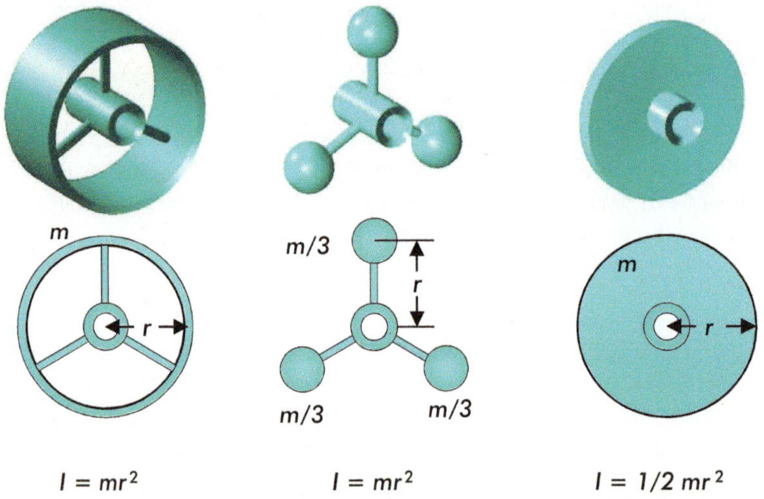

FIGURE 16 A "rim flywheel" (left) has almost all its mass at radius r, so it has $I=mr^2$. The middle object has all three masses at radius r, so $I=mr^2$. A flat disc (right) only has $I=\frac{1}{2}mr^2$, because much of its mass is close to the center.

$I = mr^2$ $I = mr^2$ $I = 1/2\, mr^2$

REMINDER

The dimensions for rotational inertia are mass times length squared (mass × length²). In the SFS system, this is in slug-ft², and in MKS units is kg-m².

4.5 Resultant Torque

Just as combining forces acting on a body produces a *resultant force*, we can combine two or more torque components acting on a body to form the *resultant torque*:

Physical Law Eq. 8-17

$$T_{Res} = T_1 + T_2 + ...$$

Remember that the resultant torque is what accelerates a rotating object. If there are three torques, you add the three, and so on. For example, on a crankshaft you might have one torque from the cylinder pressure, one from bearing friction, and one from friction between the piston and the cylinder. Don't forget to include all the torque components that are big enough to make a difference.

4.6 Free-Rev Acceleration

PROBLEM: Imagine that the engine in **FIGURE 17** is operating at WOT (wide-open throttle) in neutral, and the engine produces a steady resultant torque of 300 ft-lbf on the crankshaft. The flywheel weighs 30 pounds and has a diameter of 14 inches. Calculate the acceleration of the crankshaft and flywheel. Assume the crank's inertia is equal to the flywheel's.

SOLUTION: We want to use equation 8-14b to calculate the acceleration. But first we need to find the rotational inertia of the flywheel. Its mass is 0.93168 slugs, and its radius is 0.58333 feet.

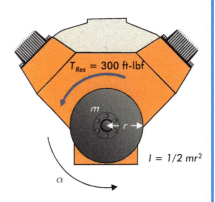

$T_{Res} = 300$ ft-lbf

$I = 1/2\ mr^2$

We'll assume the flywheel is almost the same thickness throughout its cross-section, making it "uniform," and use $I = \frac{1}{2}mr^2$ to calculate the rotational inertia:

$$I = \frac{1}{2}mr^2 = \frac{1}{2} \times 0.93168 \text{ slugs} \times (0.58333 \text{ ft})^2$$

$$= 0.15851 \text{ slug-ft}^2$$

FIGURE 17 An engine free-revving, so the engine torque is only accelerating the rotational inertia of its own crankshaft and flywheel. Here, $T_{Res} = I\alpha$.

Remember we are assuming the crank inertia is equal to the flywheel inertia, so the total inertia is 0.31703 slug-ft². Now we can find the acceleration:

$$\alpha = \frac{T}{I} = \frac{300 \text{ ft-lbf}}{0.31703 \text{ slug-ft}^2} = 946 \text{ rad/sec}^2$$

In each second, the angular velocity increases by 946 radians/second, or about 9,030 rpm. At that rate, it would only take 1/3 of a second to go from 3,000 to 6,000 rpm (see **FIGURE 18**). (This seems like a decent first estimate of the acceleration, but it's a bit on the high side.)

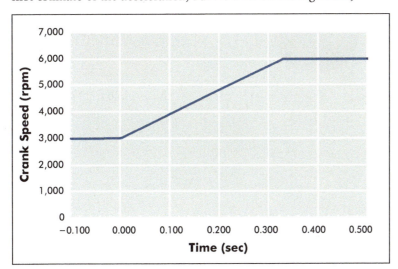

FIGURE 18 The 300 ft-lbf torque accelerates our 0.317 slug-ft² free-revving engine at a rate of 946 rad/sec², or 9,030 rpm/sec.

Other examples of accelerating a rotational inertia are spinning up a turbo after you hit the gas, and the deceleration of a tire/wheel unit when braking hard enough to slide the tire.

4.7 Newton's Third Law for Rotational Motion

Just as a force doesn't appear out of thin air, neither does a torque. When a torque acts on one object, a second torque of equal size acts on another object, in the opposing direction. This is expressed as Newton's third law for rotation.

FUNDAMENTAL LAW

If body A exerts a torque on body B, then body B exerts an equal and opposite torque on body A.

When a wrench puts a CW 100 ft-lbf torque on a nut, the nut exerts a 100 ft-lbf torque CCW on the wrench. When a transmission puts torque into a driveshaft, the driveshaft puts an equal and opposite torque into the transmission (it pushes back). When cylinder pressure exerts a counterclockwise torque on the crankshaft (see **FIGURE 19**), it also exerts an equal clockwise torque on the block. This is why quickly revving the engine in neutral rocks the engine (and whole car) the opposite way.[12] A good way to describe the situation in **FIGURE 19** is that when combustion pressure pushes the angled connecting rod into the crankshaft, it needs something to push against (it's from the piston pushing sideways on the cylinder wall).

FIGURE 19 When cylinder pressure produces a 300 ft-lbf counterclockwise torque on the crankshaft/flywheel, Newton's third law demands that it also produce a 300 ft-lbf clockwise torque on the engine block.

$T_{Flywheel}$ = 300 ft-lbf CCW

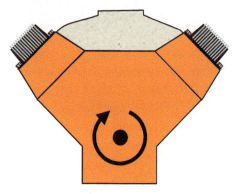

T_{Block} = 300 ft-lbf CW

Note 12: Once I was sitting next to a Dodge Dart with a 340 when they quickly revved the engine in neutral. The car looked like it rocked 2 inches; a combination of light car, big engine, and soft springs.

It's easier to see in an electric motor. **FIGURE 20** diagrams a four-pole motor (four permanent magnets in the rotor). The stator is held stationary, and the rotor turns to drive a shaft (not shown). On the left, there is no current in the stator windings, so it does not exert any torque on the rotor. On the right, current runs through the windings, to produce a magnetic field in a pattern that produces a CCW torque on the rotor by attraction and repulsion. [13]

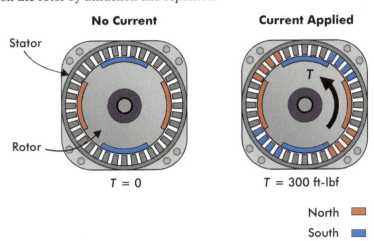

No Current   **Current Applied**

Stator

Rotor

$T = 0$ $T = 300$ ft-lbf

North

South

FIGURE 20 Without current (left), the stator has no magnetic field, so there is no torque. With current applied, a magnetic field pattern develops in the stator, which attracts/repels the rotor to produce a CCW torque. Note that only the strongest north and south poles are shown on the stator, to simplify the diagram—weaker ones are left gray. They are actually all a little bit different, depending on their windings.

FIGURE 21 shows the forces acting in the motor as it rotates through three different angles, from 0 to 60°. There is a 225-lbf tangential force on each rotor magnet, producing 300 ft-lbf CCW total on the rotor. For each force on the rotor, there is an equal and opposite force on the stator. Together they produce an equal and opposite torque on the rotor.

> **KEY CONCEPT**
>
> Equal and opposite reaction *forces* cause equal and opposite reaction *torques*. Newton's third law for torque follows directly from his third law for force.

Note 13: **FIGURE 13** in Chapter 5 has more detail. You can confirm for yourself which direction the torque should be in, by imagining the attractive/repulsive forces between the poles on the stator and on the rotor.

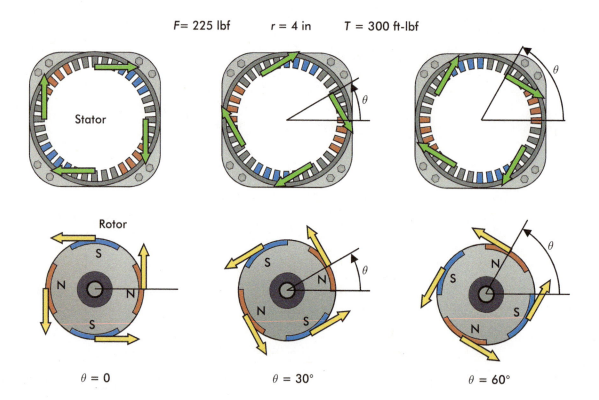

$F = 225$ lbf $r = 4$ in $T = 300$ ft-lbf

Stator

Rotor

$\theta = 0$ $\theta = 30°$ $\theta = 60°$

FIGURE 21 In an electric motor, opposing forces act on the stator (green arrows) and rotor (yellow arrows). In doing so, equal and opposite torques also act on the stator and rotor. As the rotor turns, the forces all move with it.

Notice how the pattern of north and south poles in the stator travels around in a circle, following the rotor to push/pull it around. By changing the current in the windings, the magnetic field in the stator turns to follow the rotor, while the stator itself stays in place.

Besides the tangential forces shown, radial forces are also produced between the stator and rotor. But since they are directed at the center of rotation, they produce no torque (remember?). Therefore we can ignore them here.

5 Summary

This chapter has dealt with angular kinematics, angular dynamics, and relating angular motion to straight-line motion. Major points are:

- As used in this text, the terms "angular" and "rotational" are interchangeable.
- In angular dynamics, an angle is best described as an amount of rotation.
- The most natural way to measure angles in physics is in radians, the ratio of an angle's circular arc length to the arc's radius.
- For a given mass, an object's rotational inertia is greatest when its mass is placed farthest from the center of rotation.
- Angular and straight-line kinematics have similar traits.
- Angular and straight-line dynamics have similar traits.

The last two points are summarized in TABLE 3 and TABLE 4. TABLE 3 summarizes the kinematics of constant-acceleration motion, for both translational and rotational cases.

	Translation	Rotation
Displacement	$x = x_0 + v_0 t + \dfrac{1}{2} a t^2$	$\theta = \theta_0 + \omega_0 t + \dfrac{1}{2} \alpha t^2$
Velocity	$v = v_0 + at$	$\omega = \omega_0 + \alpha t$

TABLE 3 Translational and rotational kinematics with constant acceleration.

TABLE 4 shows the quantities in Newton's second law for both translational and rotational motion. The bottom line shows the equation of motion (Newton's second law) for each case.

	Translational Motion		Rotational Motion	
	Term	**Symbol**	**Term**	**Symbol**
Inertia	Mass	m	Rotational inertia	I
Forcing Action	Resultant force	F_{Res}	Resultant torque	T_{Res}
Acceleration	Linear acceleration	a	Angular acceleration	α
Equation of Motion	$F_{Res} = ma$		$T_{Res} = I\alpha$	

TABLE 4 The terms for translational and angular dynamics, along with their equations of motion (Newton's second law).

We've taken a good step toward understanding the rotating parts in a car, but pure rotation doesn't describe a whole lot of our situations. Rotation is almost always connected to translation, which, after all, is a car's purpose. Chapter 9 will build on our learnings here, and add in a few applications.

Major Formulas

Definition	Equation	Equation Number
Belt Drawn In by Pulley of Radius r	$s = r\theta$	8-3
Converting an Angle in Revolutions to Radians	$\theta_{Radians} = 2\pi\theta_{Revolutions}$	8-4a
Converting an Angle in Degrees to Radians	$\theta_{Radians} \cong \dfrac{\theta_{Degrees}}{57.3^\circ}$	8-4b
Angular Velocity = Change in Angular Displacement ÷ Change in Time	$\omega = \dfrac{\Delta\theta}{\Delta t}$	8-5
Angular Acceleration = Change in Angular Velocity ÷ Change in Time	$\alpha = \dfrac{\Delta\omega}{\Delta t}$	8-6
Tangential Displacement Along Circular Arc	$s = r\theta$	8-7
Tangential Velocity	$v = r\omega$	8-8
Tangential Acceleration	$a = r\alpha$	8-9
Angular Velocity as a Function of Initial Velocity, with Constant Acceleration	$\omega = \omega_0 + \alpha t$	8-11
Angular Displacement as a Function of Initial Displacement, and Initial Velocity, with Constant Acceleration	$\theta = \theta_0 + \omega_0 t + \dfrac{1}{2}\alpha t^2$	8-12
Angular Displacement as a Function of Velocity, with Constant Acceleration	$\theta = \theta_0 + \dfrac{\left(\omega^2 - \omega_0^2\right)}{2\alpha}$	8-13
Resultant Torque = Rotational Inertia × Angular Acceleration	$T_{Res} = I\alpha$	8-14a
Rotational Inertia for a Single Mass m at Radius r from Center of Rotation	$I = mr^2$	8-16
Resultant Torque - Sum of All Torques Applied	$T_{Res} = T_1 + T_2 + \dots$	8-17

9 Angular Dynamics Applications

Driving Mechanisms; Converting Between Torque and Force

This David Kimble drawing of the C6 Corvette powertrain bares the whole propulsion process: producing power in the engine, transmitting it through the driveline, and on to the rear tires. Placing the engine in the front, but transmission in the rear, helps split the vehicle weight equally between front and rear axles. *Illustration by David Kimble with permission of General Motors Co.*

Chapter 9

The focus is on rotational mechanisms, such as belt drives, crank trains, gear trains, and wheels/tires. Accelerating force is calculated from engine and axle torque, and maximum acceleration is calculated for a Corvette Z06. Longitudinal tire slip is related to longitudinal force. Synchronizer torque during a shift is calculated.

David Kimble

Contents

Key Symbols Introduced

Symbol	Quantity	SFS Units	MKS Units
N_{Fric}	Number of Friction Surfaces	dimensionless	dimensionless
k_R	Torsional Stiffness	ft-lbf/rad	N-m/rad
R_{Slip}	Longitudinal Slip Ratio	dimensionless	dimensionless
N_i	Number of Teeth on Gear i	dimensionless	dimensionless

1   Introduction

As mentioned in the last chapter, angular dynamics is very important in automotive applications. Crankshafts, camshafts, gears, driveshafts, and wheels all rotate and carry torque. But by itself, rotation leaves us spinning. Our goal is almost always to cause translation, and Chapter 4 showed that traction *force* is what accelerates the car. So why bother with all the rotation and torque? The answer is that rotating machinery is very effective in transferring power from engine to tires, to produce the traction force. (Power will be covered in Chapter 14.)

You *could* instead have a linkage moving back and forth between a piston and your drive axles, like in an old steam locomotive (see **FIGURE 1**). Although the mechanism would be smaller in a car, it would still have the same problems. It would be heavy, carve out a huge amount of room for its motion, and be a real vibration problem—have you noticed the huge counterweights on a steam locomotive's wheels?

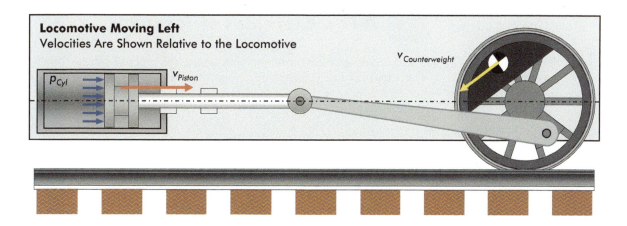

Locomotive Moving Left
Velocities Are Shown Relative to the Locomotive

p_{Cyl} v_{Piston} $v_{Counterweight}$

Spinning in place, the driveshaft shown in **FIGURE 2** transfers the power in less space, with less mass and vibration. Besides the other advantages, a rotating driveline also allows an easy way to select the right engine speed for the conditions, using a transmission (gearbox).

FIGURE 1 A car could use translational motion to drive the wheels, as the linkage in a locomotive does. But it would be large, heavy, and rough.

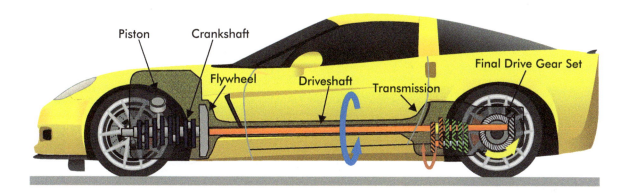

Piston Crankshaft Flywheel Driveshaft Transmission Final Drive Gear Set

But if you looked closely at the powertrain/driveline in **FIGURE 2**, you'd see it converts force to torque, and vice versa, several times:

1. The force on the piston produces a connecting rod force, and crankshaft torque.
2. The torque on each driving gear causes a tooth force between it and its driven gear.
3. The tooth contact force causes a torque on that driven gear.
4. The torque on the drive axle produces a traction force at the tire patch.

Converting between torque and force is therefore key in making a car tick. We'll examine the gears, wheels, and belt drives that do this.

FIGURE 2 Automotive drivelines like in this **Corvette Z06** use rotating parts, taking up much less room than a linkage.

2   A Few Examples of Torque

Since torque comes from applying forces, the types of torque follow the types of forces listed in Chapter 5. We won't cover them all here, because most of them are easily found from the forces. Some others will be brought in as we go.

2.1 Pressure Torque

Here, I'll mainly talk about gas-pressure torque. A gas is a fluid like air, fuel vapor, or exhaust gas. The gas in an engine's cylinder has a certain pressure when brought in (as an air/fuel mixture), and when compressed, burned, and exhausted. The pressure pushes on the piston, which acts on the rod to produce crank torque (see **FIGURE 3**). This is the torque generation process in an internal combustion engine. The term "internal combustion" refers to burning the air/fuel mixture inside the cylinder, rather than outside it, as in a steam engine (an "external combustion" engine).

Note that the torque varies with crank angle, since the cylinder pressure and the effective length of the lever arm change. This is what causes the pulsating torque that can shake the car when you lug the engine.

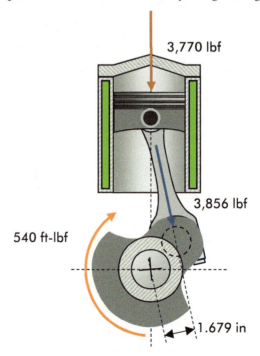

3,770 lbf

3,856 lbf

540 ft-lbf

1.679 in

FIGURE 3 Gas-pressure torque at a given crank angle. The changing pressure and lever arm create a pulsating torque.

In the same way, when you compress the refrigerant within a belt-driven A/C compressor, a gas-pressure torque acts negatively on the compressor's drive pulley.

2.2 Friction Torque

As mentioned in Chapter 5, friction is a huge factor in operating and controlling a vehicle:

- Varying the friction torque in the clutch by modulating the pedal can either launch a vehicle smoothly (when you're alone) or kill the engine (like when you try to impress someone, or you're at the front of a long line of cars).
- Varying the brake friction torque (by brake pedal force) varies the car's deceleration.
- Tightening bolts produces friction torque in the threads that resists loosening.
- Limited-slip differentials typically use friction torque in clutches and gears to keep one axle from turning much faster than the other.

FIGURE 4 is a cutaway of a clutch being driven by the crankshaft. The clutch disc is sandwiched between the flywheel and the pressure plate, and splined to the transmission input shaft. When the clutch is fully engaged, the springs squeeze the disc hard enough that it can't slip. When you push on the clutch pedal, the linkage (not shown) pushes on the release bearing, reducing the clamping force. Varying the force on the pedal controls whether and how much the clutch will slip, by varying the friction torque.

FIGURE 4 A clutch disc (alone, left) is clamped between the flywheel and pressure plate (right). The normal force (clamping force) F_N causes the friction torque to drive the disc.

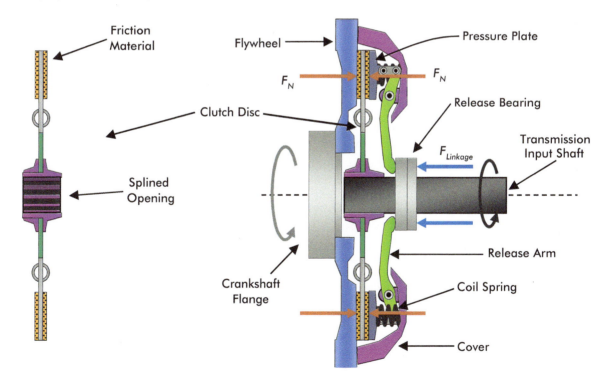

As long as the clutch isn't slipping, its output torque is determined by, and equal to, the torque applied to it. The maximum torque the clutch can produce while locked up is

$$T_{Fric,Max} = \mu_{Static} N_{Fric} F_N r_{Fric}$$

where μ_{Static} is the static friction coefficient, N_{Fric} is the number of friction material surfaces (two here), and r is the radius from the shaft center to the friction material center. When you slip the clutch, it drives through kinetic friction:

$$T_{Fric} = \mu_{Kinetic} N_{Fric} F_N r_{Fric}$$

where $\mu_{Kinetic}$ is the kinetic friction coefficient.

Friction torque can vary greatly with surface finish, lubrication, temperature, etc. The more important the situation, the more closely it needs to be controlled. Automatic transmission fluids, for example, have additives to provide the right balance between high clutch friction (for fast shifts, which could be harsh) and low friction (for a slow shift, which creates more heat in the clutch). To function smoothly, fluids in Posi-Trac® differentials may have to be changed often to replace the additives that get ground up during clutch slippage.

2.3 Deformation Torque

Torsional deformation results from a body being twisted. Torsion bars and axle shafts are the most obvious examples, but crankshafts, vehicle bodies, and steering shafts also twist, affecting how they operate.

When a linear spring is compressed, it pushes back with a resistive force; when a shaft is twisted, it pushes back with a resistive torque. Torsional deformation can be described by a TORSIONAL STIFFNESS:

$$T_{Torsion} = - k_R \theta$$

Here, θ is the rotational deflection of the body from its unstressed position, and k_R is the rotational stiffness. The negative sign means that the torque from the spring pushes in the direction opposite the deflection. In consistent units, θ is in radians, and torsional stiffness is in ft-lbf/rad for SFS units, and N-m/rad for MKS. Most data are given in ft-lbf/deg or N-m/deg, and need to be converted for use in dynamics.

Chrysler used torsion bars in their front suspensions for many years (similar to FIGURE 5), as did VW Bugs and Porsche 911s, and some

Formula 1 cars do today. Many four-wheel-drive pickups also use them, but antiroll bars for suspension tuning are currently the most common application.

When you use a socket wrench with a long ⅜″ extension, you may notice that the socket doesn't turn quite as much as the wrench. The twisting of the extension is often called "lost motion," and can be a pain when you can't get much swing on the wrench in the first place. You'd be better off then with a ½″ drive, which will be three times as stiff, and only twist a third as much.

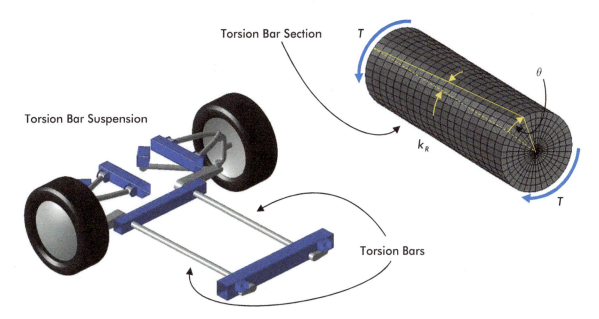

Torsion Bar Section

Torsion Bar Suspension

k_R

Torsion Bars

Torsional spring stiffness doesn't always come from twisting the piece that deforms. Isolators in clutches have *linear* coil springs, mounted at a distance from the center of rotation, that are compressed or stretched when torque is applied to the clutch (see **FIGURE 6**). They help cushion the driveline from the engine torque pulses. The rotational stiffness for the springs as a set is:

FIGURE 5 A torsion bar (right) is a torsional spring, pushing back with a torque proportional to the twist angle θ. In a torsion bar suspension (left), torsion bars resist vertical motion of the spindle and wheel.

Physical Law Eq. 9-3

$$k_R = N k_T r^2$$

Here, N is the number of springs, k_T is the translational stiffness for one spring, and r is the distance from the center of the clutch disc to the spring center.

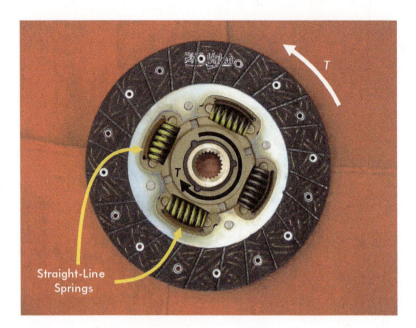

Straight-Line
Springs

FIGURE 6 This clutch disc has four linear springs between the inner hub (bronze color) and the outer friction disc, which together form a torsional isolator.

In vehicle design, a lot of attention is paid to torsional stiffness of car bodies, partly for a solid feel, but also for better handling and simpler suspension tuning. First off, a stiffer body twists and shakes less when a wheel hits a given bump. But further, if the body doesn't move so much at the suspension attachments, vertical wheel motion strokes the shocks (dampers) more, and they damp out the wheel motion better.

3 Belt Drives

Let's look closer at how belt drives work. **FIGURE 7** shows a simple belt-drive system using a six-rib belt (section shown to the left). It has a 6-inch (diameter) crankshaft pulley driving a 4-inch air conditioner (A/C) compressor pulley. The crank pulley turns at 2,100 rpm. When the A/C clutch is not engaged, the compressor pulley just rotates on its bearing, and takes practically no torque to drive—just enough to overcome friction. Because of this, the tensions in the upper and lower belt spans are roughly equal, at say 100 lbf.

We'll start by finding the belt speed and using that to calculate the compressor speed.

EXAMPLE

Crankshaft Pulley   **A/C Compressor Pulley**

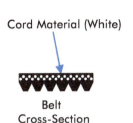

Cord Material (White)

Belt
Cross-Section

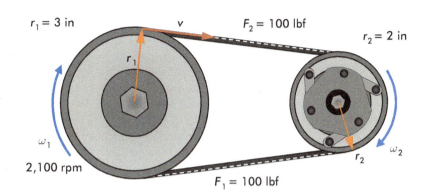

$r_1 = 3$ in v $F_2 = 100$ lbf
$r_2 = 2$ in
r_1
ω_1
2,100 rpm
r_2 ω_2
$F_1 = 100$ lbf

PROBLEM: What is the belt speed v in the figure? What is the angular velocity ω_2 of the A/C pulley?

SOLUTION: To begin, note that the belt speed is the same throughout the whole system as it feeds around each pulley and through each span, and it's equal to the tangential velocity of the belt cord:

FIGURE 7 A simple belt-drive system that only drives an A/C compressor pulley. With no torque load on the A/C compressor, the belt span tensions are equal. Clockwise rotation is here considered positive.

Equation 9-4   **Belt speed**
$$v = r_1 \omega_1 = r_2 \omega_2$$

Next, we convert the crank pulley speed from 2,100 rpm to 219.91 rad/sec, and its radius from 3 inches to 0.250 feet. Then equation 9-4 gives a belt speed of 55.0 ft/sec. (Remember that converting from rpm to rad/sec was covered in Chapter 8. You multiply by $2\pi/60$.)

Using the rightmost parts of equation 9-4, next we solve for the angular velocity of the A/C compressor:

$$\omega_2 = \frac{r_1}{r_2}\omega_1 = \frac{0.25000 \text{ ft}}{0.166667 \text{ ft}} \times 219.91 \text{ rad/sec} = 329.87 \text{ rad/sec}$$

This is 3,150 rpm, 50% faster than the crank pulley. Since the A/C pulley is ⅔ as large, it spins ½ as fast. No surprise, right?

Of course the whole idea of a belt-drive system isn't just to spin another pulley, but to *drive* it with torque. Let's see what it takes to drive this A/C compressor at constant speed.

EXAMPLE

PROBLEM: We engage the clutch to drive the compressor, which takes 5 ft-lbf of torque to turn (this torque depends on how hard the A/C has to work to keep you cool). What are the new tensions in the spans? What is the torque at the crank pulley?

SOLUTION: We're at constant speed, so there is no torque from acceleration. To drive the A/C pulley, the tension F_1 in the lower span must be higher than the tension F_2 in the upper span. In this simple system, the tension *increase* ΔF in the lower span will be equal to the tension *decrease* in the upper span (the tight and loose spans are equal in length, so their stretching amount from driving the compressor is equal and opposite, and so are their tension changes). Their average must still be 100 lbf. **FIGURE 8** shows a free-body diagram for each pulley, with the span tensions acting at the cord radius on each one. To find the new tensions, we balance the torques on Pulley 2, knowing the torque from the compressor is -5 ft-lbf:

$$F_1 r_2 - F_2 r_2 = -T_2 = 5 \text{ ft-lbf}$$

$$\left(F_1 - F_2 \right) r_2 = 5 \text{ ft-lbf}$$

Since the pulley radius r_2 is 0.166667 feet:

$$F_1 - F_2 = \frac{5 \text{ ft-lbf}}{0.166667 \text{ ft}} = 30 \text{ lbf}$$

For F_1 to be 30 lbf higher than F_2, and to have their average be 100 lbf, the tension F_1 in the lower span must be 115 lbf, and the tension F_2 in the upper span is 85 lbf.

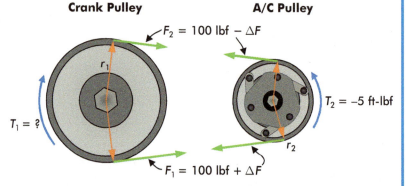

Crank Pulley   **A/C Pulley**

$F_2 = 100 \text{ lbf} - \Delta F$

$T_2 = -5 \text{ ft-lbf}$

$T_1 = ?$

$F_1 = 100 \text{ lbf} + \Delta F$

FIGURE 8 When driving the A/C compressor, the tension must be higher in the lower span than in the upper. These free-body diagrams show how torque is produced on each pulley as a result.

Now for the torque T_1 on the crank pulley. We know the tensions acting on it (F_1 and F_2) and the radius, so we can calculate it in one fell swoop (this could have been done in other ways, but this one took the fewest swoops):

$$T_1 = \left(F_1 - F_2 \right) r_1 = \left(115 \text{ lbf} - 85 \text{ lbf} \right) \times 0.250 \text{ ft}$$

$$= -7.5 \text{ ft-lbf}$$

Notice a couple of things about these results. First, the torque on the crank pulley from the belts is CCW, so it's negative. The crankshaft, rotating CW, has to work *against* the torque from the tensions by putting a positive CW torque into the pulley to drive it. Second, the belt tensions have 50% more leverage on the crank pulley because of its bigger radius, so the torque on it is 50% larger than on the compressor. There is a trade-off here: with the smaller A/C pulley, the belt drive spins the compressor faster than the crank, but more crank torque is required to do so. There is a balance of nature here that will be explained in Chapters 12–15 on energy and power.

KEY CONCEPT

Belt drives are the same as chain drives in calculating torque transfer and speed ratios, except that most belt drives have a little slippage because they drive through friction. The only exceptions are cogged belts.

V-ribbed belts power many automotive accessories, like water pumps, power steering pumps, and generators. Cogged belts have been common in cam drives, and are often used to drive superchargers and oil pumps on race cars.

4 Turning Axle Torque into Driving Force

In a typical go-kart, the torque from the engine is transferred to the rear axle through a chain drive. Given a certain amount of axle torque, how do we find the force pushing the kart?

The rear axle is centered by a bearing (see **FIGURE 9**), and the axle torque T_{Axle} is applied to the wheel. We expect the wheel/tire to produce two forces: $F_{Pavement}$ that pushes rearward on the pavement, and $F_{Bearing}$ that pushes forward on the frame, through the bearing. We'll ignore vertical forces from vehicle weight, because they could be looked at separately.

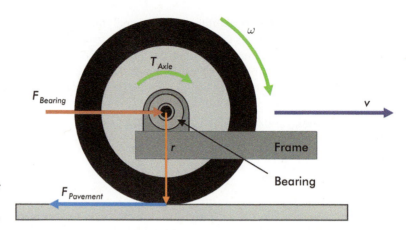

FIGURE 9 A driven wheel/tire "converting" axle torque to thrust. It exerts forces forward on the bearing and rearward on the pavement.

Per Newton, the bearing and pavement forces each have their reaction pairs acting on the wheel/tire (see **FIGURE 10**). How large is the driving force, F_{Thrust}? Neglecting the tire/wheel mass and rotational inertia, the forces on the axle and pavement are equal in magnitude, opposite in direction (you can work out why). The two form a couple, which must offset the axle torque:

Equation 9-5

$$F_{Thrust}\, r = T_{Axle}$$

So if the kart has 60 ft-lbf of axle torque, and the tire radius is 4 inches (0.33333 feet), the driving force from the tires is:

$$F_{Thrust} = \frac{T_{Axle}}{r} = \frac{60 \text{ ft-lbf}}{0.33333 \text{ ft}} = 180 \text{ lbf}$$

The force on the bearing is then also 180 lbf. Since you divide axle torque by the tire radius to get force, you can get a larger forward force by using a smaller radius tire. With a 3-inch radius tire, you would get 240 lbf. On the other hand, a larger tire radius would *reduce* the driving force at the road.

T_{Axle} = 60 ft-lbf

r = 4 in

F_{Thrust} = ?

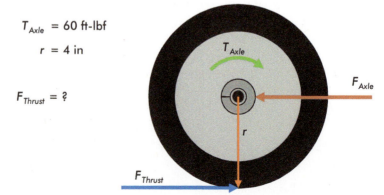

FIGURE 10 A free-body diagram of the torque and the horizontal forces acting on the wheel/tire assembly.

So while tall tires may look cool on your car, you'll need more axle torque for the same acceleration. But tall tires will also slow your engine down for a given vehicle speed, which usually helps fuel economy.

5 | Rolling: Rotation and Translation

Once when I was playing with a toy tractor, I asked my mom how far it would go when the rear tire turned once. She told me it was a little more than three times the diameter of the tire, which was a pretty good answer—she could have just told me to get a life.[1] In any case, this question turns out to be critical in linking tire rotational motion to vehicle motion.

Rolling is a combination of rotational and translational motion; if a tire doesn't slip, it will translate by its circumference as it rotates one revolution, and the vehicle with it. **FIGURE 11** shows a tire with a 1-foot radius doing exactly that.[2] It starts at the left and ends up at the far right, after moving 2π feet (6.28 feet), and rotating by the angle of 2π radians. In between, we have snapshots of the motion at 2 and 4 radians, where it's moved 2 and 4 feet, respectively.

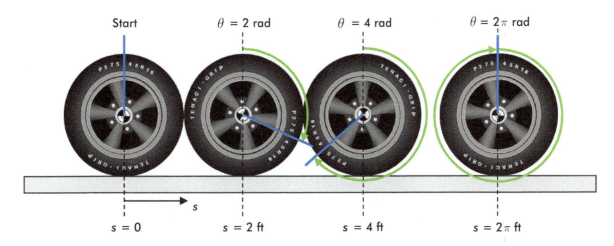

| Start | $\theta = 2$ rad | $\theta = 4$ rad | $\theta = 2\pi$ rad |

| $s = 0$ | $s = 2$ ft | $s = 4$ ft | $s = 2\pi$ ft |

The only difference between a belt drive and a rolling tire is that on the belt drive, the rotation center is stationary; on the rolling tire, the contact patch and road are stationary. So to make a long story short, the tire center translation is tied to the tire rotation by its radius:

FIGURE 11 When a tire rolls by a distance equal to its radius (here 1 foot), it rotates by 1 radian. In one revolution it rolls 2π times its radius (here 6.28 feet) and rotates by 2π radians.

General Equation 9-6   **Rolling displacement (no slip)**

$$s = r\theta$$

Note 1: I don't remember how old I was, but I was either too young to care about this, or too old to be playing with toy tractors.

Note 2: If you look closely, the tire is a P275/45R16, which has a radius of 12.7 inches, not 1 foot; so sue me.

In the same way, angular velocity is related to the translational velocity:

General Equation 9-7   **Rolling velocity (no slip)**

$$v = r\omega$$

This is shown on the left side of **FIGURE 12**. Likewise, the right side of the figure shows that translational acceleration is linked to the angular acceleration of the tire:

General Equation 9-8   **Rolling acceleration (no slip)**

$$a = r\alpha$$

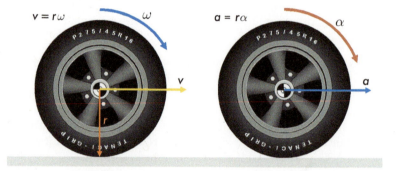

FIGURE 12 The translational velocity v of a tire (and the vehicle) is equal to its radius r times its angular counterpart ω, assuming no slip. Ditto for acceleration.

You can see that things are pretty simple if the tire is perfectly round and it doesn't slip. There are a couple of flies in the ointment, though. When you put a load on the tire, the contact patch at the bottom flattens out, so the wheel center is closer to the ground than it would be with no load. The other is that when braking or accelerating, tire slip makes things a bit more interesting.

6   The Rolling Radius of Real Tires

Now you know how to convert wheel angular velocity ω to vehicle velocity v by multiplying by the tire radius r. But once you put a load on the tire, it flattens out on the bottom. So which radius is it? Is it the radius measured from the center down to the road, where the tire flattens? Or is it to the top edge, where it stays round? Assume you don't have the tire to measure.

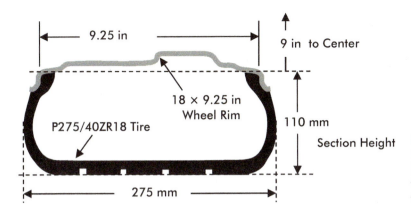

Not having the tire, the easiest way to find its effective radius is to calculate it from tire-chart data. Let's say you have a P275/40ZR18 tire (cross-section in **FIGURE 13**).[3] The tire specs I looked up on the Internet show that it rotates 781 revolutions/mile. Let's use equation 9-6, but use the *effective* radius r_{Eff}, with 781 revolutions for θ and 1 mile for s:

$$r_{Eff} = \frac{s}{\theta} = \frac{1 \text{ mile}}{781 \text{ revolutions}} \times \frac{5{,}280 \text{ feet}}{1 \text{ mile}} \times \frac{1 \text{ revolution}}{2\pi \text{ radians}}$$

$$= 1.076 \text{ feet}$$

Notice that in the conversions, the miles and revolutions cancel out, leaving feet/radian. Radians are dimensionless (unitless), so feet/radian is the same as feet.

The tire acts like it has a radius of 1.076 feet. Let's compare this with the radius calculated from the tire dimensions. We just need to add half of the wheel diameter to the section height, which is 40% of the section width (being a 40-series tire). Remember that taking 40% of something is the same as multiplying by 0.40:

$$r = 9 \text{ in} + \left(275 \text{ mm} \times \frac{1 \text{ in}}{25.4 \text{ mm}} \times 0.40 \right) = 9 \text{ in} + 4.3307 \text{ in}$$

$$= 13.331 \text{ in} = 1.111 \text{ ft}$$

We found out before that the tire rolls like it has a radius of 1.076 feet, while using the tire size data results in 1.111 feet. The difference is 0.03489 feet, or 0.4187 inches. That's about 3% smaller than the dimensions say.

It's roughly true that a tire rolls like it has a radius from the center of the tire to the center of the contact patch, where the weight of the car "squishes" the sidewall down, rather than to the "round" part of the

---

Note 3: The "P" stands for passenger tire, "275" is the section width in millimeters, "40" is the percentage of section height to section width, "Z" is the speed rating, "R" is for radial, and "18" is wheel diameter in inches.

tread. The $\frac{9}{10}$ of an inch difference between the rolling radius and the calculated radius sounds like a believable amount for load deflection. Measuring from the center of the wheel to the road surface gives a pretty good approximation.

Note that this won't work for tires with a very flexible tread, like drag slicks. They are made to "grow" as they spin faster, and calculating vehicle speed from axle speed gets more complicated. Drag racers use the *front*-tire rotation speed to measure vehicle speed.

7 | A Closer Look at Rolling

Does it seem odd that the tire moves at the same velocity as the car, but the tire grips the pavement mainly with static friction, implying no motion? **FIGURE 14** shows the velocities at five different points on a tire. On the left, the tire is locked up, with the car moving 45 mph (66 ft/sec), pure straight-line motion. All the points have the same velocity v as the wheel center. In the middle figure, the tire is stationary, but rotating at 600 rpm. All of the points on the tread have the same instantaneous speed (66 ft/sec with a 1.05-foot radius), from $v = r\omega$. But they have different velocities because of their different directions.

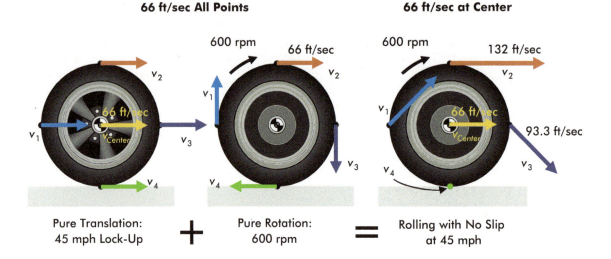

66 ft/sec All Points   **66 ft/sec at Center**

Pure Translation: Pure Rotation: Rolling with No Slip
45 mph Lock-Up   600 rpm   at 45 mph

FIGURE 14 Rolling with no slip (right) combines translation (left) and rotation (middle), so that $v = r\omega$. The velocity at the contact patch is zero. (It's instructive to add each point velocity from the first two figures to get the ones in the third.)

With translation and rotation combined (at right), the bottom of the tire is indeed temporarily stationary on the pavement. We have rolling with no slip, so this really is static friction. To a first approximation, that's right for real tires. But we'll see that it's little more complex.

8 | Longitudinal Slip and Longitudinal Tire Force

Realistic tire forces aren't typically covered in a physics book, but here it can't be avoided. The last section showed that you can only get rolling without slipping if the translation and rotation velocities match just right, so that $v = r\omega$. Otherwise the tread must slip. But it's also the slip that determines the amount of force produced by the tire.

8.1 Positive Longitudinal (Accelerating) Forces

We've seen that applying a torque to a wheel creates a longitudinal force on the tire patch, which is forward when in the same direction as the rotation (as in FIGURE 15). But when a vehicle accelerates, the tire turns faster than if coasting at the same vehicle speed. This is LONGITUDINAL SLIP, which you could call a "lost motion." The longitudinal traction force from a given tire depends on the amount of the slip.

But slip is not so much a matter of sliding tread as it is stretching and compressing of the carcass (cord) and tread. Under torque, the carcass is stretched behind the contact patch and compressed in front of it (FIGURE 15, left). You can then imagine the "bunched up" carcass material being fed into the contact patch (FIGURE 15, right). A section of carcass originally occupying a 20° arc (unstressed) might only occupy 19° of contact patch. This reduces the distance the tire center (and the car) travels for a given wheel rotation. As the tire rolls, the carcass material gets farther back in the contact patch and steadily stretches. This puts a rearward tug on the tread near the rear of contact, which causes some tread blocks to deflect (squirm), and maybe slide.

FIGURE 15 A driven tire experiencing longitudinal "slip," where its forward velocity v is less than you'd calculate from the radius r and the angular velocity ω. Note that the "stress lines" in the sidewall are not visible in most tires (drag slicks not included).

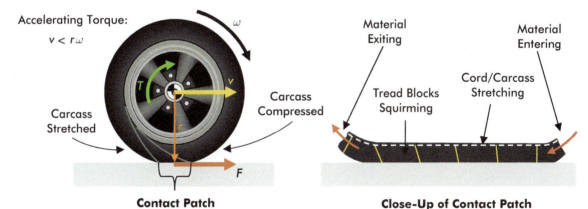

Contact Patch   **Close-Up of Contact Patch**

We measure slip by calculating the SLIP VELOCITY, the difference between the amount of actual forward motion at the wheel center (and vehicle), and the motion you'd expect from the tire's effective rolling radius and angular velocity:

Equation 9-9

$$v_{Slip} = r\omega - v$$

> **NOTE**
>
> The stress lines in the sidewall in **Figure 15** are shown only for illustration, but become visible wrinkles in drag slicks, which operate at low air pressure. There you can easily see the wheel rim pulling on the contact patch through the cords in the sidewall.

EXAMPLE

Problem: If the rotation speed of the axle is 5 rev/sec (300 rpm), the tire's effective rolling radius is 1 foot, and the velocity of the car is 30 feet/sec (about 20.5 mph), what is the slip velocity?

Solution: Using equation 9-9:

$$v_{Slip} = r\omega - v = \left(1\ \text{foot} \times 5\frac{\text{rev}}{\text{sec}} \times \frac{2\pi\ \text{rad}}{1\ \text{rev}}\right) - 30\ \text{ft/sec}$$

$$= 1.4159\ \text{ft/sec}$$

In other words, the angular wheel velocity (from which your speedometer measures speed) and the tire radius say the vehicle should go 31.416 feet in one second, but you actually only go 30 feet.

The amount of slip is usually given as a **SLIP RATIO**, dividing the slip velocity by the velocity if there were no slip:

Equation 9-10

$$R_{Slip} = \frac{v_{Slip}}{r\omega}$$

For our last example, this is:

$$R_{Slip} = \left(\frac{1.4159\ \text{ft/sec}}{31.4159\ \text{ft/sec}}\right) = 0.0451$$

Our slip ratio of 0.0451 is 4.51%. During this acceleration, your speedometer would be reading close to 5% higher than actual. In a single equation, the longitudinal slip ratio is: [4]

Equation 9-11

$$R_{Slip} = \frac{r\omega - v}{r\omega}$$

An extreme example of longitudinal slip is a stationary burnout. Here, the vehicle's velocity v is zero, so the slip ratio calculates to 1.00, or 100% (all of the rotation goes into slip).

Note 4: There are several definitions of longitudinal slip. I chose one that calculates the fraction of a wheel's rotation that goes into slip.

With higher axle torque, the tread in the rear of the contact patch does start to slide, as in the rear quarter of the footprint in **FIGURE 16**. The more torque applied, the farther forward the "sliding point" moves, until it all slides.

This leads to an interesting conflict. On one hand, the traction in the sliding portion of the tread is reduced, because kinetic friction is lower than static friction. But the sliding friction heats the tread, so it's "stickier" when it reenters the front of the contact patch, where there is no sliding. The tread gets warmer and stickier each time around.

NOTE

Most tires generate their maximum traction forces while undergoing 10% slip or so, with a moderate amount of sliding in the rear part of the tread. The whole tread is warmed, increasing traction, especially in the part that isn't sliding.

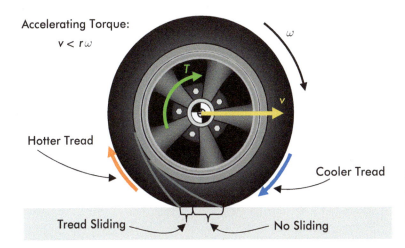

Accelerating Torque:

$v < r\omega$

Hotter Tread

Cooler Tread

Tread Sliding

No Sliding

FIGURE 16 With enough slip, some of the tread slides, which creates frictional heating. This generates better traction as long as the sliding isn't excessive.

Heating the tread made for some especially dramatic drag racing in the mid-1960s, when Top Fuelers found the fastest way down the track was to spin the tires for the entire quarter-mile (**FIGURE 17** is *during* the race, not before). Heating the tires helped traction more than losing the static friction hurt it.

But the days of quarter-mile smoke clouds didn't last long. It was soon realized that spinning the tires *before* the run made a lot more sense. The tires were still heated from friction, but used mostly static friction of the tread on pavement during the entire race. Plus, it would lay a nice clean strip of rubber on the track for the tires to grab on to.

Of course this prerace smoke fest is called a **BURNOUT**; liquid bleach was often used to break traction, by running the tires through a "bleach box" beforehand. Sometimes gasoline was used, to make a fire burnout. They were even more dramatic, but a little dicey. Now only water is allowed as a burnout agent, which seems to be working just fine.

8.2 Negative Longitudinal (Braking) Forces

When a reverse torque is applied to the wheel (see **FIGURE 18**), a rolling tire produces a braking force. Again, it acts through some combination of static friction and kinetic friction, but now each section of carcass fed into the front of the contact patch has been stretched by the torque. The center of the tire travels *faster* than $r\omega$, so the "slip" is negative, and 20° of tread might take up 21° of contact patch. But equation 9-11 still works to calculate the slip ratio.

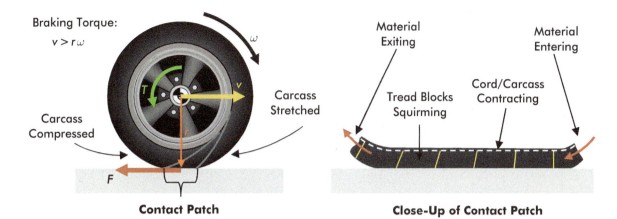

Braking Torque:

$v > r\omega$

Carcass Compressed

Carcass Stretched

F

Contact Patch

Material Exiting

Material Entering

Tread Blocks Squirming

Cord/Carcass Contracting

Close-Up of Contact Patch

All this means your speedometer will read a bit low during a hard stop, but you probably aren't concentrating on the speedo right then anyway, right? Like in acceleration, as the slip ratio increases, more and more of the slip is from sliding tread in the contact patch. When a wheel is locked up, $r\omega$ is zero, making a slip ratio of $-\infty$, per equation 9-11.

FIGURE 18 A braked wheel and tire moving to the right. The tread now enters the contact patch stretched, rather than compressed as in acceleration.

8.3 Traction Force vs. Longitudinal Slip

We've said that producing maximum tire traction force is a balancing act, between maximizing the stationary contact area between tread and road, and letting some of the tread slide to heat the tire. **FIGURE 19** gives a good example of the outcome, showing data from a low-profile sports car tire carrying 1,000 lbf vertical load. This is measured data, hence the lack of perfect smoothness.

As just discussed, when there is no longitudinal slip (0% on the graph), the tire's angular velocity ω equals v/r, and it produces no longitudinal force. When the tire spins faster than this, the slip is positive, and it produces a positive (accelerating) force; if slower, then negative slip and negative (braking) force.

For small slip, the traction force increases 240 lbf for each percent of slip. In this linear range, there isn't much tread sliding, just carcass and tread deflection. If this could continue, we would get our peak traction of 1,200 lbf at 5% slip. But as the slip increases, more sliding occurs. This is where our conflict between losing static contact and gaining tire heat kicks in.

At around 4% slip, the straight steep line becomes less steep. Traction force is still increasing with slip, but not so quickly. Maximum forward traction comes with a longitudinal slip ratio of about 12% (a ratio of 0.12). Note that the maximum traction of 1,200 lbf is 1.2 times the 1,000 lbf load on it; we'd say its friction coefficient is 1.20. With increased slip, the traction force drops off, but slowly. So if you wanted to ride the top of the longitudinal force curve, you could operate with slip somewhere between 11% and 15% without being much off the maximum. You might design a traction control system to do this.

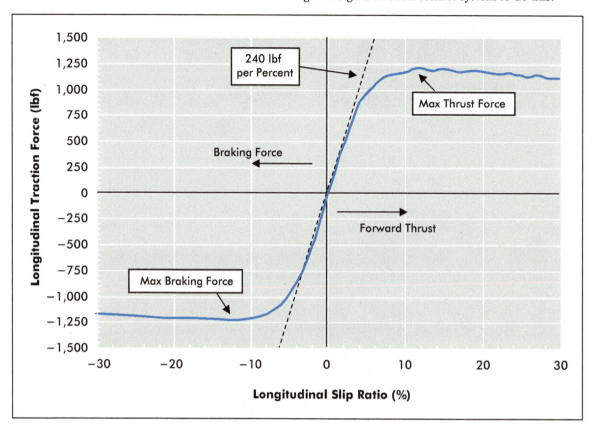

FIGURE 19 Longitudinal traction force for a performance tire with 1,000 pounds vertical load. Maximum traction occurs at around 12% for acceleration, and at −12% for braking.

For braking, we see much the same. It starts off linear, levels out, and produces a maximum (negative) force of −1,200 lbf. Maximum braking force comes at a slip ratio of about −12%, and drops off with greater negative slip, again slowly. So for quickest braking, you'd want to stay somewhere between 11% and 20% slip. In designing an anti-lock brake system (ABS), you'd take advantage of that.

9 | Gear Sets

Often, the torque output from an engine (or motor) isn't right for the job you need to do. Pulling a boat up a ramp requires more axle torque than most engines put out. How do we increase the torque from the engine's crankshaft to what's needed at the axle shaft?

Gear sets are the most common form of changing driving torque and shaft speed. They're named for their effect on output speed; when the driven shaft turns slower, it's called a "reduction" set, and if it turns faster, an "overdrive" set. There are two main types of gears:

external gears, the usual type with teeth on the outside, and internal gears, which (you'd never have guessed) have them on the inside. FIGURE 20 shows two external gears in mesh.

Obviously, when the first gear rotates by one tooth, the second rotates by one tooth in the opposite direction. Their teeth must be traveling the same (tangential) velocity where they mesh, or they'd break each other off. The tangential velocity is the radius times the angular velocity, so $r_1\omega_1 = -r_2\omega_2$ (the negative sign signifies opposite rotation). Rearranging this, their angular velocity ratio is inversely proportional to the ratio of their radii:

Equation 9-12   **External gear ratio**

$$\omega_2 = -\left(\frac{r_1}{r_2}\right)\omega_1$$

But since they have the same tooth spacing at the gear radius, their ratio of tooth count N to radius is the same; that is $N_1/r_1 = N_2/r_2$. In terms of tooth counts, equation 9-12 becomes:

Equation 9-13   **External gear ratio**

$$\omega_2 = -\left(\frac{N_1}{N_2}\right)\omega_1$$

You can use the same ratio for their angular displacements:

Equation 9-14   **External gear ratio**

$$\theta_2 = -\left(\frac{N_1}{N_2}\right)\theta_1$$

This is no surprise: if Gear 2 is *larger* (more teeth), it turns through a *smaller* angle. In FIGURE 20, if $\theta_1 = 360°$ CCW, then $\theta_2 = 240°$ CW. The speed ratio is 0.667, and the reduction is 1.50.

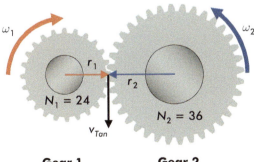

ω_1

r_1

r_2

$N_1 = 24$

$N_2 = 36$

v_{Tan}

Gear 1   **Gear 2**

ω_2

FIGURE 20 An external gear set, with Gear 1 driving Gear 2. Gear 1 has 24 teeth, and Gear 2 has 36 teeth. Their gear reduction is 1.5.

That takes care of the angles and speeds; what happens to torque? When one gear pushes another gear through its teeth, it creates forces at the contact point (see **FIGURE 21**). The forces push on each gear with the same magnitude, but in the opposite direction. Knowing the torque T_1 on Gear 1, we can calculate the tooth force. The torque T_1 is Fr_1, so the force is: [5]

Equation 9-15

$$F = \frac{T_1}{r_1}$$

With this amount of force pushing on Gear 2 a distance r_2 from its center, the torque T_2 is:

Equation 9-16

$$T_2 = -\left(\frac{r_2}{r_1}\right)T_1$$

Again, the negative sign indicates opposite rotation.

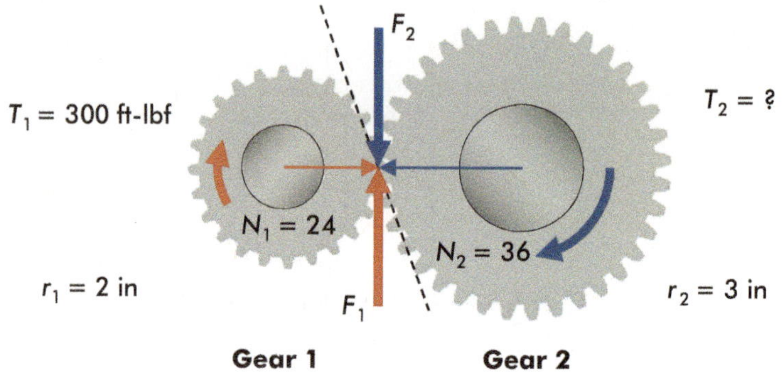

FIGURE 21 Gear 1 has 300 ft-lbf of torque applied to it. Since the tooth forces acting on each gear are equal, what is the torque on Gear 2?

In **FIGURE 21**, $T_1 = 300$ ft-lbf, $r_1 = 2$ inches, and $r_2 = 3$ inches, so the torque on the second shaft is:

$$T_2 = -\left(\frac{r_2}{r_1}\right)T_1 = -\left(\frac{3 \text{ in}}{2 \text{ in}}\right)\times 300 \text{ ft-lbf} = -450 \text{ ft-lbf}$$

Notice that it's okay to use inconsistent units in a ratio; they cancel out when divided through.

Note 5: The tooth force isn't really perpendicular to the radius, but along the dashed line in **FIGURE 21**. This difference in direction is called the **PRESSURE ANGLE**. Ignoring this won't change the torque, but will underestimate the tooth forces, by about 6% here.

The larger gear rotates at a slower rate, but has greater torque. Since the ratio of teeth on the two gears is the same as the ratio of their radii, equation 9-16 also works with tooth counts:

Equation 9-17

$$T_2 = -\left(\frac{N_2}{N_1}\right)T_1$$

Note that in using equation 9-17, we don't need the gear radii or the tooth force, just the tooth counts. But while the tooth force would be 1,800 lbf here, smaller gears with the same torque multiplication would have higher tooth forces. Of course this will affect gear durability.

10   Gear Trains

One gear-reduction set is not always enough to get the torque that's needed. Gears mesh best when they aren't too different in radius, so the gear ratio between them is *usually* 2 or less, meaning a couple of gear sets are often necessary. This is called a gear train, as seen in a transmission.

A simple gear train is shown in FIGURE 22. The input shaft drives Gear 1 with torque T_1, and Gear 1 drives Gear 2. Gear 2 drives Gear 3 through the countershaft, and Gear 3 drives Gear 4 on the output shaft. The tooth counts are as shown in the figure. If T_1 is 240 ft-lbf, what is the torque T_4 at the output shaft? If the input shaft turns 6,000 rpm, how fast does the output shaft turn?

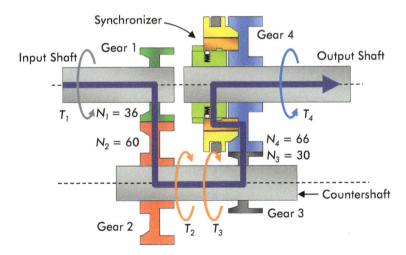

FIGURE 22 A cross-section of a simple gear train that increases torque through gear reductions. The power flow is shown by the large, snaking arrow.

To find the torque T_4, we can step through one mesh at a time. The torque on Gear 2 is:

$$T_2 = -\left(\frac{N_2}{N_1}\right)T_1 = -\left(\frac{60}{36}\right)\times 240 \text{ ft-lbf} = -400 \text{ ft-lbf}$$

Since Gears 2 and 3 are the only two gears on the countershaft, they have the same torque, so $T_3 = T_2$. Then the torque applied by Gear 4 on the output shaft is:

$$T_4 = -\left(\frac{N_4}{N_3}\right)T_3 = -\left(\frac{66}{30}\right)\times(-400 \text{ ft-lbf}) = 880 \text{ ft-lbf}$$

Overall, the increase in torque is equal to the ratio 880/240 = 3.67. How does the speed vary? Applying equation 9-13 twice, we get:

$$\omega_4 = \left(\frac{-N_1}{N_2}\right)\times\left(\frac{-N_3}{N_4}\right)\times\omega_1 = \left(\frac{-36}{60}\right)\times\left(\frac{-30}{66}\right)\times 6,000 \text{ rpm}$$
$$= 1,636.4 \text{ rpm}$$

which divides the input speed by 3.67. Besides increasing the torque in two stages, the gear train has returned the rotation direction to the same as the input. Front-engine rear-wheel-drive transmissions are typically constructed this way.

11   Maximum Acceleration and Maximum Traction

If we know the maximum torque an engine can make, the gear ratios from engine to axle, and the tire radius, we can find the maximum vehicle acceleration, given enough traction. For example, a 2006 Corvette Z06 with an LS7 engine has a peak torque of 480 ft-lbf at 4,800 rpm. Its manual transmission has a first-gear reduction of 2.66 and a final-drive reduction of 3.42 (see **Figure 23**). Using this, the maximum axle torque in first gear is the maximum engine torque times the product of the gear ratios:

$$T_{Axle} = 480 \text{ ft-lbf}\times 2.66\times 3.42 = 4,366.6 \text{ ft-lbf}$$

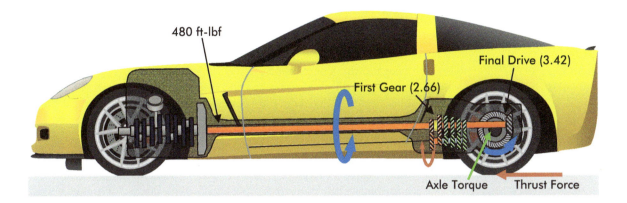

480 ft-lbf

Final Drive (3.42)

First Gear (2.66)

Axle Torque   Thrust Force

That seems like a lot of axle torque. How much force could that make at the rear tire patches? The Z06's P325/30ZR19 tires have a rolling radius of 1.073 feet (calculated from their 783 rev/mile). So, the thrust force would be:

$$F_{Thrust} = \frac{T_{Axle}}{r_{Tire}} = \frac{4,366.6 \text{ ft-lbf}}{1.073 \text{ ft}} = 4,070 \text{ lbf}$$

FIGURE 23 The Corvette Z06 power-train with the first-gear power flow highlighted in orange, including the reduction sets in the transmission.

Hmmm. Can the rear tires really apply that much force? Given a test weight of 3,330 lbf (including the driver), a tire friction coefficient of 1.10, and 50% of the weight on the rear tires, the most force that the rear tires could apply is about:[6]

$$F_{Max} = \mu F_N = 1.1(3,330 \text{ lbf})(0.5) = 1,832 \text{ lbf}$$

The available axle torque could produce 4,070 lbf at the pavement, easily enough to break the rear tires loose. Even considering torque lost to friction in the driveline, the driver will need to feather the throttle pedal to keep the tires planted (or use traction control).

With the traction available, what is the maximum acceleration? Using Newton's second law:

$$a_{Max} = \frac{F_{Max}}{m} = \frac{1,832 \text{ lbf}}{103.416 \text{ slugs}} = 17.71 \text{ ft/sec}^2$$

This is about 0.55 *g*'s, vs. the 1.22 *g*'s you could get by using the max axle torque. This is an example of **TRACTION-LIMITED ACCELERATION**; the tires are the limit to the acceleration you can get, at least in first gear at 4,800 rpm.

Note 6: The load on the rear tires will actually be a little more than 50% of the vehicle weight because of load transfer to the rear. We'll see in Section 4.2 of Chapter 15 that the traction force capability of this Z06 is considerably higher than what we estimated here.

12 | Getting Your Shift Together

Early manual transmissions were brutal to shift. Emile Levassor, who designed the transmission for the Panhard–Levassor in 1891, said as much.[7] To select a gear ratio, one gear was slid along its main shaft into mesh with its mate. If their speeds didn't match, they would clash or grind against each other. You needed to have real skill to avoid "grinding the gears."

Cadillac's 1928 introduction of the Syncromesh transmission made it much easier. As today, the gear pairs are always meshed, but only the pair transferring power is coupled to both its shafts. A **SYNCHRONIZER**, similar to the one in **FIGURE 24**, is used to select the gear. The synchronizer performs two functions: 1) it causes the speeds of the shaft and floating gear to match, then 2) couples them to each other rigidly. The synchronizer hub is splined to the shaft, so the synchronizer assembly rotates at shaft speed.

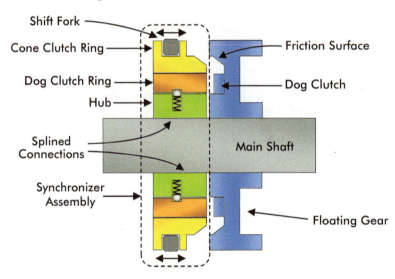

Shift Fork

Cone Clutch Ring

Dog Clutch Ring

Hub

Friction Surface

Dog Clutch

Splined Connections

Main Shaft

Synchronizer Assembly

Floating Gear

FIGURE 24 The cone clutch in the synchronizer is what brings the speed of the floating gear to the speed of the shaft, so the dog clutches can lock the gear to the shaft.

These steps are shown in **FIGURE 25**. On the left, the shift fork (pushed by the shift lever and linkage) presses the cone clutch against its mating surface on the gear. Once the resulting friction torque matches the gear and shaft speeds closely enough, the dog clutch engages (right). A dog clutch doesn't drive through friction, but by dogs (matching notches) that interlock. They are very efficient, but can't engage smoothly unless synchronized in speed. Once the shift is done, the gear drives the shaft through the dogs.

Note 7: His quote in French was "C'est brutale, mais ça marche," meaning "It's crude, but it works."

Cone Clutch Application

Shift Completed

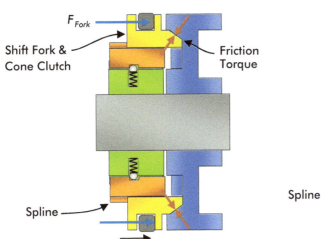

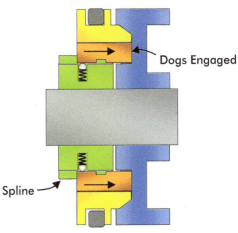

F_{Fork}

Shift Fork & Cone Clutch

Friction Torque

Dogs Engaged

Spline

Spline

Let's try our hand at calculating the friction torque necessary to do a quick shift.

FIGURE 25 During the shift, the friction in the cone clutch brings the rotational velocities of the gear and shaft together (left), so the dog clutches can engage (right). The gear is then driven by the shaft through splines and dogs.

EXAMPLE

PROBLEM: You're just hitting 6,600 engine rpm with a transmission similar to the one in **FIGURE 26**, and shift from first to second. The first gear ratio is 2.78, and second is 1.93. If you engage second gear in 0.100 seconds, what is the angular acceleration of the clutch disc during the shift? What is the friction torque in the second gear synchro if the clutch disc rotational inertia is 0.00799 slug-ft²?

SOLUTION: First we'll need to find the clutch disc angular velocities right before and after the shift. Right before, the clutch's angular velocity, ω_0, is 6,600 rpm, which is 691 radians per second. Right after the shift, the clutch disc has slowed to match up with second gear. The clutch's new angular velocity ω_1 will be the original velocity times the ratio of the gear reductions (the ratio-ratio, if you will):

$$\omega_1 = \omega_0 \times \left(\frac{1.93}{2.78}\right) = (691 \text{ rad/sec}) \times (0.694) = 480 \text{ rad/sec}$$

This is about 4,600 rpm. Now let's assume that the angular acceleration of the disc is constant (i.e., the synchro torque is constant). Then:

$$\alpha = \frac{\omega_1 - \omega_0}{\Delta t} = \frac{480 \text{ rad/sec} - 691 \text{ rad/sec}}{0.100 \text{ sec}} = -2,110 \text{ rad/sec}^2$$

or about −20,000 rpm/sec! A quick shift with that much rpm drop decelerates the clutch disc pretty hard.

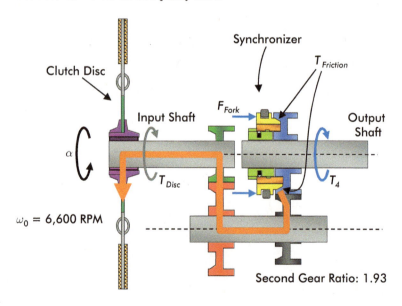

FIGURE 26 During a 1-2 shift, the second-gear synchronizer torque decelerates the clutch disc through the reduction gears. How large is $T_{Friction}$?

Now let's find the synchro torque to do the shift. We'll do it in two steps: 1) find the torque on the clutch disc itself, and then 2) find the torque at the second-gear synchro.

Torque on disc: This is easy. We use Newton's second law, equation 8-14, for angular motion:

$$T_{Disc} = I\alpha = \left(0.00799 \text{ slug-ft}^2\right) \times \left(-2{,}110 \text{ rad/sec}^2\right)$$
$$= -16.86 \text{ ft-lbf}$$

Torque on synchronizer: There is a 1.93 speed-reduction ratio from the clutch disc to the second-gear synchro, meaning the torque at the synchro is *higher* by that ratio:

$$T_{Friction} = T_{Disc} \times 1.93 = \left(-16.86 \text{ ft-lbf}\right) \times 1.93$$
$$= -32.54 \text{ ft-lbf}$$

This doesn't sound like a lot of friction torque, but it has to be exerted by a synchronizer that isn't very big around (not much leverage) and is coated in oil (low friction coefficient). So it may take a lot of shift effort to get enough friction to shift in 0.100 seconds.

> **NOTE**
>
> Think this through and it's clear that reducing clutch disc inertia is a good thing for quicker shifting and putting less wear on the synchros. The clutch in a standard car is over-sized somewhat, to better dissipate the slippage heat from launching the car's maximum possible load. But a sports car is lighter, and doesn't carry or tow a lot. So it may use a lighter clutch to reduce the disc inertia, allowing better shifts.

13 | Summary

The combinations of gears, pulleys, chains, etc., to drive machines are endless, so it's best to remember just a couple of basic concepts that get you through most any of them. One common factor is the use of levers. FIGURE 27 shows a belt drive, a gear set, and a wheel/tire set. In each, a torque is applied to one part of the system, producing a force somewhere else.

In the belt drive (left), the torque on the lower pulley acts through a lever (the pulley radius) to produce tension in the belt span. In the gear set (middle), torque on the upper gear acts through a lever (the gear radius) to produce a tooth contact force. In the tire/wheel (right), the torque on the wheel acts through a lever (the tire radius) to produce a force on the pavement. There is no real difference in these mechanisms, except in the way the forces are then used. The belt and gear drives use the force to produce torque on another pulley or gear. The tire produces a pavement force we use directly.

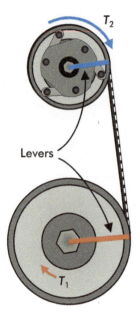

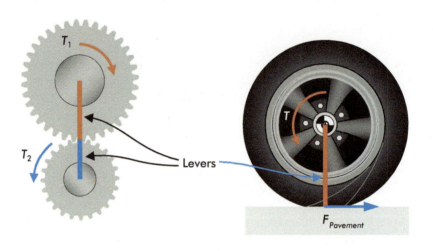

FIGURE 27 All the mechanisms in this chapter boil down to using levers, whether as a pulley radius, a gear radius, or a tire radius.

Viewing these mechanisms as variations of levers allows you to use two recurring relationships:

$$T = Fr$$

for relating torque to force, and:

$$v = r\omega$$

for relating speeds. With this viewpoint and these two equations, you can calculate results for most mechanisms you'll run into.

Major Formulas

Definition	Equation	Equation Number
Maximum Friction Torque in a Locked Clutch	$T_{Fric,Max} = \mu_{Static} N_{Fric} F_N r_{Fric}$	9-1a
Torque in a Slipping Clutch	$T_{Fric} = \mu_{Kinetic} N_{Fric} F_N r_{Fric}$	9-1b
Torsional Spring Torque	$T_{Torsion} = -k_R \theta$	9-2
Torsional Stiffness Using Axial Springs	$k_R = N k_T r^2$	9-3
Belt Speed	$v = r_1 \omega_1 = r_2 \omega_2$	9-4
Axle Torque and Thrust Force	$F_{Thrust} r = T_{Axle}$	9-5
Vehicle Displacement; Rolling Tire with No Slip	$s = r\theta$	9-6
Vehicle Velocity; Rolling Tire with No Slip	$v = r\omega$	9-7
Vehicle Acceleration; Rolling Tire with No Slip	$a = r\alpha$	9-8
Longitudinal Tire Slip Velocity	$v_{Slip} = r\omega - v$	9-9
Longitudinal Slip Ratio	$R_{Slip} = \dfrac{r\omega - v}{r\omega}$	9-11
Velocity Ratio for Meshed Gears	$\omega_2 = -\left(\dfrac{N_1}{N_2}\right)\omega_1$	9-13
Approximate Tooth Force in Meshed Gears	$F = \dfrac{T_1}{r_1}$	9-15
Torque Ratio for Meshed Gears	$T_2 = -\left(\dfrac{N_2}{N_1}\right)T_1$	9-17

10 Dynamics in a Plane Basics

Forces, Moments, and Motion in Two Dimensions

An SRT Viper and two ORECAs avoid the marbles exiting turn 17a onto the very flat front straight at Sebring in 2013, skillfully demonstrating "dynamics in a plane." *Photo by Randy Beikmann.*

Motion in a plane is covered, such as rounding a curve on flat ground, or hills and valleys on a straight road. The centripetal acceleration and the force necessary to follow a curved path are found. Motion is predicted for a motorcycle jump. We introduce rotation of the car body, the forces and moments acting on it during braking, and the effects of its rotational inertia.

Contents

Key Symbols Introduced

Symbol	Quantity	SFS Units	MKS Units
r	Curve Radius	ft	m
$a_{Centripetal}$	Centripetal Acceleration	ft/sec^2	m/sec^2
$F_{Centripetal}$	Centripetal Force	lbf	N
a_{MaxLat}	Maximum Lateral Acceleration	ft/sec^2	m/sec^2
θ_z	Rotation (Displacement) in Yaw	radians	radians
M_z	Moment in the Yaw Direction	ft-lbf	N-m
I_z	Yaw Moment of Inertia	slug-ft^2	kg-m^2
K	Radius of Gyration	ft	m

1 Introduction

You see a curve ahead and it looks very tight. What goes through your mind? You're probably deciding at what speed your car can take the curve, based on its cornering capability (see **FIGURE 1**). The tighter the curve is, the more capability you need from your car (and from yourself), or the more you need to slow down. But what do we mean by *tight*, and what is *cornering capability*? And why do you need to go slower through tighter curves?

To answer all this, we'll examine dynamics in a *horizontal* plane. Have a look at the track in **FIGURE 2**. In Position 1, the car accelerates out of Turn 1, its velocity angled to the northeast, as is its acceleration. In Position 2, it brakes for Turn 3, its velocity due west, its acceleration due east. These are both straight-line motion as we're used to, just following different lines.

But Position 3 in Turn 3 shows something truly new—curved motion—in this case at a constant speed around a constant radius curve. The car's velocity is tangent to the curve, but the acceleration is *perpendicular* to the velocity, pointed at the center of the curve. Since the acceleration is not in line with the motion, it doesn't change your speed—it keeps you on the curve.

This is what car guys call **LATERAL ACCELERATION**, because it is sideways relative to the car. In physics it's called **CENTRIPETAL ACCELERATION**. Centripetal (*sen-TRIP-uh-tul*) means "center-seeking"—the acceleration is toward the center of the curve. Once centripetal acceleration is found, we use Newton's second law to find the **CENTRIPETAL FORCE** (here a *lateral force*) required to push your car into the turn.

FIGURE 1 You don't need to be a professional to have a little fun in the curves. Here, the author hustles a Pontiac Solstice through Canada Corner at Road America with his son, Aric. *Photo by permission of Randy Beikmann.*

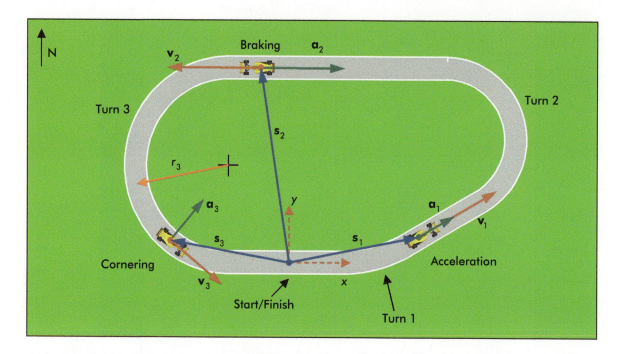

FIGURE 2 Displacement, velocity, and acceleration vectors for three positions around an oval track. Positions 1 and 2 show straight-line acceleration, while Position 3 shows centripetal acceleration.

Motion in a plane also allows rotation. For example, a hockey puck sliding flat in a rink moves in the plane of the ice (down the ice and across it), but it can also rotate. For a car body, the rotation could be its slow rotation as it rounds a curve—as you round a 90° curve, your car also rotates 90°. But if you spin out exiting the curve, your car might rotate an extra 180° or so. This type of rotation is called **YAW**. Bringing in rotation allows us to scrutinize more aspects of the car design, such as how the tire forces must balance in order to prevent a spin. We'll examine this briefly in this chapter, but will go into more depth in Chapters 16 and 17.

We'll also study dynamics in a *vertical* plane, to describe motion going over hills, through valleys, or when leaving the ground. This part will extend the straight-line dynamics covered in Chapters 4 and 6 to planar motion. With that, let's get to it.

2 | Newton's Discovery of Centripetal Force and Acceleration

To me, Newton's most amazing leap of thought came from noticing that when you shoot something like a cannonball through the air horizontally, it falls in a curved path (see **FIGURE 3**, left). If you shoot it faster, it will curve less sharply and go farther before hitting Earth. He then imagined that if you shot the cannonball fast enough, the curve could exactly follow the curve of Earth. It would always be "falling" toward Earth, but in a circular path at a constant height, so it would never drop. He'd figured out that gravity could cause a circular orbit! He knew then that the force that caused an apple to drop toward Earth was the same force that caused the moon to orbit Earth, and Earth to orbit the sun.

This was a huge discovery, and quite a shock to the established thinking of the day, which was that "heavenly bodies" like the sun and the stars followed different rules of motion than the "lowly" bodies on Earth. This was how they had explained that the stars and sun could move in a constant motion seemingly forever, but that a moving object on Earth eventually comes to a stop. Newton realized that the rules were the same for both, but that the heavenly bodies could move "unchanged" because in space there was no friction to slow them down.

Force Causing Orbit

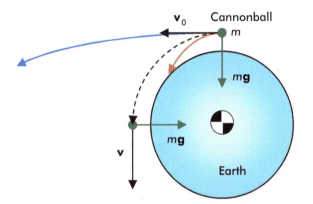

Centripetal Acceleration

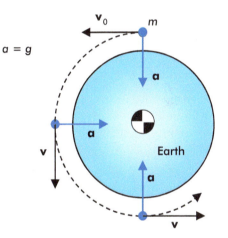

But Newton wasn't done: It was one thing to imagine the cannonball orbiting Earth, but how did he merge this with his second law of motion? After all, he had said that force causes *acceleration*, and when the cannonball follows the circular orbit, its speed never changes.

While it's true that as the cannonball orbits in its circle its speed never changes, its velocity does change—constantly. Remember that velocity is defined by speed *and* direction, and since the direction of travel around a circle is always changing, the cannonball is always accelerating—toward the center of Earth (**Figure 3**, right). Newton defined the acceleration toward the center of the curved motion as centripetal acceleration, and the gravitational force on the orbiting body was just enough to cause it:

Figure 3 Newton's thought process of gravity causing an orbit around Earth. A cannonball shot horizontally (left) could go into a circular orbit (dashed, curved path) *if* the initial velocity $\mathbf{v}_0$ were just right. Gravity causes the centripetal acceleration $\mathbf{a}$, perpendicular to the motion (right). Note that if shot faster, it would go into an elliptical or hyperbolic orbit.

$$\mathbf{F}_{Gravity} = m\mathbf{a}_{Centripetal} \qquad \text{Circular orbit}$$

So *F* still equals *ma* when the mass follows a curved path. Newton used his theories of dynamics and gravity to write equations of motion for the planets, *and* created the math (calculus) to solve them and produce the answers we use.[1] His equations, written in the 1600s, predicted orbits so closely that no improvements were made until the early 1900s. It took Einstein's general theory of relativity to correct the small errors that were noticed when using Newton's laws to predict the orbit of the planet Mercury.

Of course, gravity isn't the only force that can cause an object to follow a curved path. A rock swung on the end of a string is held to a circular path by the tension force in the string. A ball circling a roulette wheel is pushed toward the center by the rail at the edge (a normal force), much like a car is when sliding along a curved guard rail. And of course in a turn, your motorcycle is held on the curve by its lateral tire forces.

3 Dynamics of Rigid-Body Motion in a Plane

The rest of this chapter will spell out the dynamics for three cases of rigid-body motion in a plane:

1. Translation in two independent, perpendicular directions, with constant acceleration in each direction.
2. Translation following a circle centered about a point, as in taking a curve at constant speed.
3. Rotation about an axis perpendicular to the plane (such as yaw).

The first example above can produce either curved or straight-line motion, as we'll see. And note that while a rigid body is undergoing either of the first two types, it will often yaw at the same time.

Using the first two, we can compare the two distinct types of translational acceleration in a plane: straight-line and centripetal. On the left side of FIGURE 4 is rectilinear (straight-line) acceleration, the type we've studied for several chapters; the force and acceleration are in line with the velocity. But here the velocity is at an angle between north and east. The acceleration is at the same angle, and changes the velocity vector only by increasing its length, i.e., vehicle speed (v_2 being greater than v_1).

On the right is centripetal acceleration, where the car follows a circular path at a constant speed. The acceleration, pointed radially (at the curve center), changes the velocity vector by changing its direction.

Note 1: The universities were closed due to the plague, or Black Death, so Newton used his time off to develop his gravitational theory and invent calculus. I usually worked in a gas station on my breaks.

Straight-Line Acceleration   **Centripetal Acceleration**

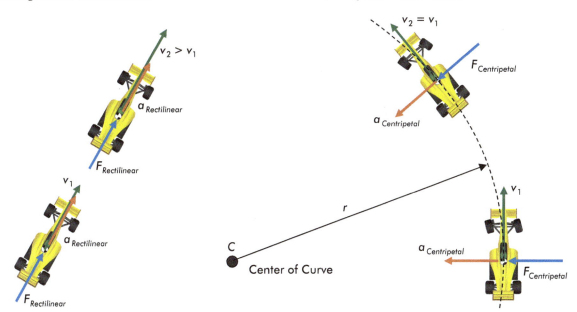

While straight-line and centripetal accelerations are shown separately in **FIGURE 4**, they can certainly happen simultaneously. For example, race drivers are usually still braking lightly while they steer into a corner, and bringing on some throttle before exiting it. Then the "straight-line" part of acceleration (which increases speed) becomes *tangential* acceleration, being in line with the velocity and tangent to the car's path. The centripetal acceleration part is then called *radial* acceleration, being directed radially toward the center of the curve.

FIGURE 4 Two ways to change a car's velocity. Rectilinear acceleration (left) changes its speed. Centripetal acceleration (right) changes its direction. In either case, $F_{Res} = ma$. Note that both can happen at once, as when speeding up while exiting a curve.

> **NOTE**
>
> The vector concepts we've covered are the tools to keep this straight. It may help to review Chapter 7 if something's not clear.

The next section will cover translation in a plane with constant acceleration, Section 6 will cover centripetal acceleration, and Section 8 will bring in rotation of the vehicle itself.

4 # Translation in a Plane with Constant Acceleration

With constant acceleration, the physics of translation in a plane is handled very similar to straight-line motion. It's just that the forces acting on the body, now in two perpendicular directions, can act with different strengths. For example, for the motion of a motorcycle in a jump, gravity acts vertically, but not horizontally. This section will take us through the kinematics, then the dynamics.

4.1 Kinematics of Translation in a Plane

To illustrate translation, let's set up the motion of a motorcycle during a jump from Ramp A to Ramp B (see **Figure 5**). Its horizontal and vertical displacement components are called s_x and s_y; velocity components are represented by v_x and v_y; and acceleration components by a_x and a_y. All component vectors are shown as dashed lines, while each resultant vector is shown as solid.

We say that during the jump, the bike is moving in a **VERTICAL PLANE**, because one of its directions of motion is vertical. In contrast, racing on a flat oval track is motion in a **HORIZONTAL PLANE**.

Figure 5 The kinematics of a motorcycle jump. This is motion in a vertical plane, and the displacement, velocity, and acceleration are vectors in the plane.

The displacement **s** goes from the origin to the mass. The velocity **v** is always tangent to the path of motion. The acceleration **a** could generally have both an x- and a y-component. But here the only acceleration is downward (from gravity), so the x-acceleration is zero.

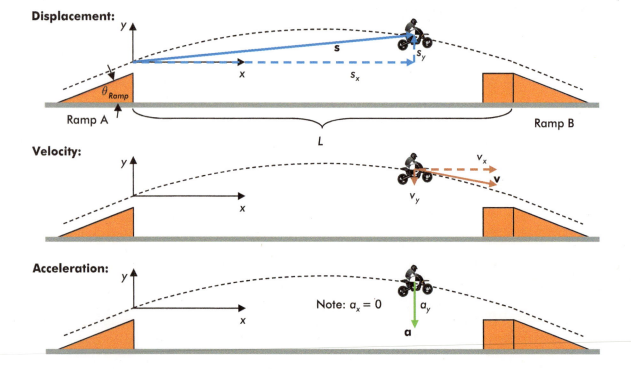

With these assumptions, we just adapt the straight-line kinematic equations to the two directions in the plane. For the velocity components, with constant acceleration,

General Equation 10-1a

$$v_x = v_{x0} + a_x t$$

and

General Equation 10-1b

$$v_y = v_{y0} + a_y t$$

where v_{x0} and v_{y0} are the initial velocity components in the x- and y-directions. These velocities produce displacements of:

General Equation 10-2a

$$s_x = s_{x0} + v_{x0}t + \frac{1}{2}a_x t^2$$

and:

General Equation 10-2b

$$s_y = s_{y0} + v_{y0}t + \frac{1}{2}a_y t^2$$

where s_{x0} and s_{y0} are the initial x- and y-displacements. Equations 10-1 and 10-2 are written in *component form*, dealing separately with the x- and y-components. Separate equations for x and y, each dependent on time, are an example of "parametric equations" as covered in math and engineering texts. These equations can also be put in *vector form*. For velocity, equations 10-1a and 10-1b are then written together as:

General Equation 10-3

$$\mathbf{v} = \mathbf{v}_0 + \mathbf{a}t$$

For displacement, equations 10-2a and 10-2b are combined as:

General Equation 10-4

$$\mathbf{s} = \mathbf{s}_0 + \mathbf{v}_0 t + \frac{1}{2}\mathbf{a}t^2$$

Compare these equations to the ones in Chapter 2, and you'll see very little difference.

CAUTION

These equations only apply if the acceleration is constant. They *will not* apply to centripetal acceleration, which constantly changes in direction.

So Step 1 is accomplished; we can track translation in a plane. Now on to the dynamics.

4.2 Dynamics of Translation in a Plane

Remember that the acceleration of a body is proportional to the force applied, and is in the same direction. This means the acceleration vector is always parallel to the force vector.

KEY CONCEPT

At any instant in time, the x-acceleration is (only) caused by the resultant x-force, and the y-acceleration is (only) caused by the resultant y-force. In other words, x-forces don't affect y-motion, and y-forces don't influence x-motion. In this section, accelerations in both directions are constant, so the motions can be treated separately.

In component form, Newton's second law for translation in a plane is in two equations:

Fundamental Law Eq. 10-5a

$$F_x = ma_x$$

Fundamental Law Eq. 10-5b

$$F_y = ma_y$$

and in vector form, simply:

Fundamental Law Eq. 10-6

$$\mathbf{F} = m\mathbf{a}$$

Mathematically, the vector form makes it more obvious that the acceleration is parallel to the resultant force and along the same line of action (as in **FIGURE 4**). While the force and acceleration are vectors, the mass is a scalar. For equation 10-6 to be true, $\mathbf{a}$ must be parallel to $\mathbf{F}$; the ratio of a_x/a_y is the same as F_x/F_y.

5 Specializing to the Motorcycle Jump

Now let's jump this bike. We'll use our equations to track its motion, neglecting air resistance. As **FIGURE 6** shows, it will leave Ramp A with a speed v_0 at the ramp angle θ_{Ramp}.

5.1 Dynamics and Kinematics

First let's deal with the forces on the bike. Once clear of Ramp A, the only force acting on it is gravity, which acts straight downward, producing acceleration only in the negative y-direction.

> **KEY CONCEPT**
>
> Think about this before we use a single equation. There is no force acting in the x-direction, so there is no x-acceleration; the horizontal velocity will stay constant. On the other hand, v_y starts off upward (positive), but downward acceleration will reduce it, and then make it negative, bringing the bike back down.

Let's write the equations for the x- and y-accelerations, from equations 10-5a and 10-5b:

$$a_x = \frac{F_x}{m} = \frac{0}{m} = 0 \qquad \textbf{\textit{x}-acceleration}$$

$$a_y = \frac{F_y}{m} = \frac{-mg}{m} = -g \qquad \textbf{\textit{y}-acceleration}$$

As expected, the motorcycle experiences no acceleration in the x-direction, and a negative acceleration equal to $-1\,g$ in the y-direction (again, see **FIGURE 6**).

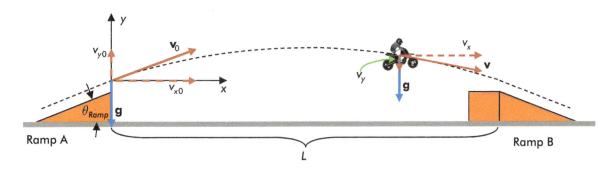

Now all we need to do is find the initial velocity components, so that we can use equations 10-1. We said that the motorcycle leaves the ramp at speed v_0 at the angle of the ramp, θ_{Ramp}. Adapting equation 7-6:

FIGURE 6 The bike has initial speed v_0 leaving Ramp A. After leaving the ramp, it's at the mercy of gravity, which causes the downward acceleration of 1 g.

Equation 10-7
$$v_{x0} = v_0 \cos\theta_{Ramp}$$

Equation 10-8
$$v_{y0} = v_0 \sin\theta_{Ramp}$$

Plugging these values into equation 10-1, the equations of motion in the x- and y-directions become:

Equation 10-9   **x-direction**

$$v_x = v_0 \cos \theta_{Ramp}$$

Equation 10-10   **y-direction**

$$v_y = v_0 \sin \theta_{Ramp} - gt$$

As expected, the x-velocity stays at its initial value, but the y-velocity starts off at its initial velocity, and then decreases steadily. Then, using equations 10-2, the displacements starting from $(s_{x0}, s_{y0}) = (0, 0)$ are:

Equation 10-11   **x-direction**

$$s_x = \left(v_0 \cos \theta_{Ramp} \right) t$$

Equation 10-12   **y-direction**

$$s_y = \left(v_0 \sin \theta_{Ramp} \right) t - \frac{1}{2} gt^2$$

Too abstract? These equations were used to plot the curved motion of the bike in **Figure 6**. Plus, we'll plug in some numbers and find out where it lands.

5.2 Making a Successful Jump

We want the bike to come back down to $y = 0$ exactly when the motorcycle has reached Ramp B (where $s_x = L$).[2] A little shorter and you bounce off the catch ramp, much longer and you land hard on flat ground. So we have two conditions: $s_x = L$ and $s_y = 0$ when $t = t_F$.

With that, we have everything necessary to use equations 10-11 and 10-12 and find the right combination of speed and length for the jump. There are three methods:

1. Set the initial speed v_0, find t_F when the bike lands ($s_y = 0$), then find s_x and note $L = s_x$.
2. Set L and find v_0 by trial and error, so that $s_y = 0$ at $s_x = L$.[3]
3. Combine the equations for s_x, s_y, and v_0, then solve for the right v_0 that makes $s_x = L$ when $s_y = 0$.

Which one to use depends. If you can only reach a certain speed, and want to know how long a jump you can make, the first method works. To find the jump speed for a set distance, you either use the second (trial and error), or the third (direct). We'll use the first method here, and save the third for Chapter 11.

Note 2: A little too long is better than too short, so in real life I'd plan for a few extra feet as insurance.

Note 3: Not the slickest way, but better than making the jump by trial and error.

So, how far could you jump if your initial velocity v_0 = 100 ft/sec (68.2 mph) and the ramp angle θ_{Ramp} is 21°? Putting the initial speed and ramp angle into equation 10-12 for vertical motion, we can find the "time of flight"; when the bike lands, it's back down to s_y = 0:

$$s_y = 0 = \left(v_0 \sin\theta_{Ramp}\right)t - \frac{1}{2}gt^2 = \left(100 \text{ ft/sec}\right)\times 0.35837 \times t - \frac{1}{2}\left(32.2 \text{ ft/sec}^2\right)\times t^2$$

or:

$$\left(35.837 \text{ ft/sec}\right)\times t - \left(16.1 \text{ ft/sec}^2\right)\times t^2 = 0$$

This has *two* distinct answers (as equations in t^2 usually will). The easy one here is t = 0, when the bike is at the top of Ramp A, where we already said s_y = 0. With that out of the way, we can solve for t_F at landing by dividing the equation by t (now assuming t is not zero, or we couldn't divide by it):

$$\left(35.837 \text{ ft/sec}\right) - \left(16.1 \text{ ft/sec}^2\right)\times t = 0$$

Then the end time t_F is:

$$t = 2.2259 \text{ sec} = t_F$$

The length L of the jump is equal to how far it's gone (s_x) at 2.2259 sec. Using equation 10-11:

$$L = \left(v_0 \cos\theta_{Ramp}\right)t_F = \left(100 \text{ ft/sec}\right)\times\left(0.93358\right)\times\left(2.2259 \text{ sec}\right) = 207.81 \text{ ft}$$

So the bike will go about 208 feet with this launch speed and angle. Air resistance would shorten the jump, and any bounce from the suspension during launch would also change it, by tweaking the launch angle.

KEY CONCEPT

Did you notice that we didn't need to know the mass of the bike? Just like a falling object, a projectile like this is only acted on by gravity, so a larger mass is offset by a larger weight force. The acceleration is −1 g regardless. But with a heavier bike, it would take more power to reach the 100 ft/sec launch speed in the same run-up distance.

6   Centripetal Acceleration

The previous section showed that you could get curved motion from having constant acceleration in both directions in a plane. But that won't take you around most curves on a road.

Let's say you want to follow a constant-radius curve at a constant speed. When you do, your velocity vector will stay the same size, but continuously change in direction. This motion requires an acceleration that also changes constantly, being directed toward the center of the curve: *centripetal acceleration*. We will soon find the "canned" formula for centripetal acceleration, but first let's work an example to illustrate it.

6.1 Step by Step Through Centripetal Acceleration

You enter the 300-foot radius curve in FIGURE 7, planning to hold a constant 45 mph (66 ft/sec) speed. Is that reasonable, or are you in over your head?

As you enter the curve (at point A), you head due east. Your velocity is 66 ft/sec east and 0.0 ft/sec north. But once turning, your direction of travel changes, changing your velocity.

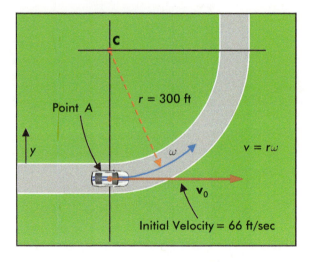

FIGURE 7 You are traveling 45 mph (66 ft/sec) as you enter a 300-foot radius curve. What will your centripetal acceleration be? Is that okay?

At this point we don't know much about centripetal acceleration, but we do know that acceleration is the rate of change of velocity. So let's find the velocity change as we round part of the curve (see FIGURE 8), and then divide that by the change in time it takes to do it.

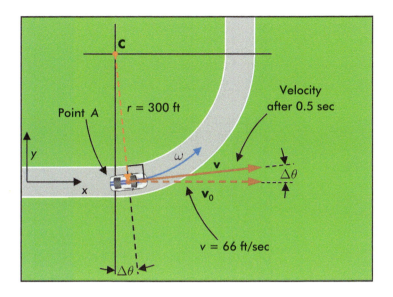

FIGURE 8 After ½ second, the speed is still 66 ft/sec, but the velocity vector is rotated slightly northward; the northward velocity component has increased.

First let's calculate the velocity ½ second after the beginning of the turn. Since the car is following a curve of radius r with velocity v, it has an angular velocity $\omega = v/r$ around the curve center C. In our example:

$$\omega = \frac{v}{r} = \frac{66 \text{ ft/sec}}{300 \text{ ft}} = 0.220 \text{ rad/sec}$$

As the vehicle rotates about the curve's center, its velocity vector turns at the same rate. Since a radian is about 57.3°, that makes ω about 12.61 degrees/second, so it would take about 7.14 seconds to round the entire 90° curve. At $t = 0.50$ seconds, the velocity has changed direction by an angle of $\Delta\theta = \omega\Delta t$:

$$\Delta\theta = \omega\Delta t = \left(0.220 \text{ rad/sec}\right)\times 0.5 \text{ sec}$$
$$= 0.110 \text{ rad} = 6.3025 \text{ degrees}$$

What is our new velocity? The speed hasn't changed, so the velocity vector has the same length, but is turned 0.110 radians north of east (see **FIGURE 9**). The eastward velocity is now:

$$v_x = v\cos\Delta\theta = 66 \ \frac{\text{ft}}{\text{sec}} \times \cos\left(0.110 \text{ rad}\right) = 65.601 \text{ ft/sec}$$

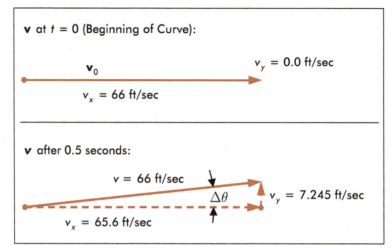

v at $t = 0$ (Beginning of Curve):

$\mathbf{v}_0$

$v_y = 0.0$ ft/sec

$v_x = 66$ ft/sec

v after 0.5 seconds:

$v = 66$ ft/sec

$\Delta\theta$

$v_y = 7.245$ ft/sec

$v_x = 65.6$ ft/sec

FIGURE 9 After 0.5 seconds rounding the curve, the velocity has changed direction by 0.11 radians (about 6.3°) CCW. The y-velocity has increased by 7.2454 ft/sec.

So the velocity in the eastward direction hardly changed. What happened to the northward velocity? It started off at zero, and after rotating through the angle $\Delta\theta$ of 0.11 radians, it is:

$$v_y = v \sin\Delta\theta = 66 \; \frac{\text{ft}}{\text{sec}} \times \sin(0.11 \text{ rad}) = 7.2454 \text{ ft/sec}$$

In 0.5 seconds, the northward velocity increased by 7.2454 ft/sec. Changing at that rate, the northward (y-) acceleration is:

$$a_y = \frac{\Delta v_y}{\Delta t} = \frac{7.2454 \text{ ft/sec}}{0.5 \text{ sec}} = 14.49 \text{ ft/sec}^2$$

Without any change in speed or sleight of hand, you can see a definite acceleration in the y-direction. This is our centripetal acceleration, being perpendicular to the direction of motion, aimed at the curve center. If it's right, our brute force method says you're at less than half a g. But let's go through the "official" calculation of centripetal acceleration and see how it compares.

6.2 The Formula for Centripetal Acceleration

We've shown that we can calculate the centripetal acceleration necessary to follow a curve, but the way we did it was slow. We need an answer that is more accurate, quicker, and more insightful.

In **FIGURE 10**, we again have the vehicle rounding a curve at speed v, starting due east. Again we'll let a short time Δt go by while the vehicle goes through the small angle $\Delta\theta$. But we won't plug numbers in right away, so that the change in time could become *very* small, and we can use any speed or radius.

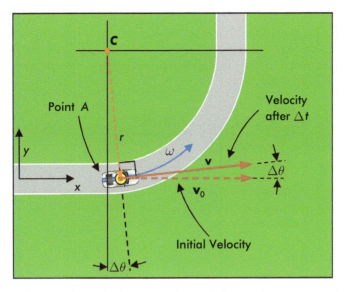

FIGURE 10 Now we'll find the vehicle's velocity change after a very short time, after it turns a very small angle (smaller than pictured).

Again the angular velocity $\omega = v/r$. And the angle covered, after a short time Δt, is $\Delta\theta = \omega \Delta t$. Taking that, and substituting v/r in for ω,

Equation 10-13

$$\Delta\theta = \left(\frac{v}{r}\right)\Delta t$$

After the vehicle travels through the angle $\Delta\theta$, the velocity is rotated by the same angle. From **FIGURE 11**, you see that the change in the y-velocity is:

$$\Delta v_y = v\sin\Delta\theta$$

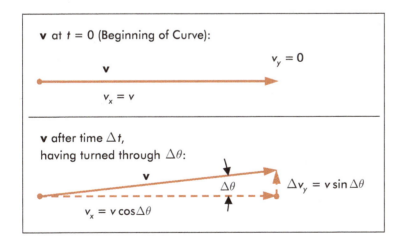

FIGURE 11 The velocity vector at the beginning of the curve (top) and after Δt seconds (bottom). The y-velocity has increased by $v\sin\Delta\theta$.

To simplify the calculation of Δv_y, we can use the fact that when an angle θ is small, the hypotenuse of the right triangle (as in **FIGURE 12**) is almost the same length as the lower "adjacent" side (both shown here as length L), and the "opposite" side is almost equal to $L\theta$.

This useful manipulation is called the **SMALL ANGLE APPROXIMATION**: The sine of a small angle is almost exactly equal to the angle, in radians.

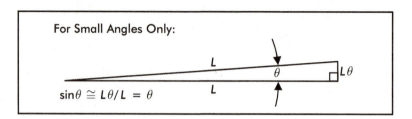

FIGURE 12 The small angle approximation for its sine.

Now let's look at the change in velocity after a very short time, such as $\frac{1}{1000}$ of a second. Under this condition, the angle $\Delta\theta$ will be small enough to say that $\sin\Delta\theta = \Delta\theta$. The equation for Δv_y is now:

$$\Delta v_y = v\sin\Delta\theta = v\Delta\theta$$

In words, the change in the y-velocity is equal to the speed multiplied by the change in direction angle. If we substitute our expression for $\Delta\theta$ from equation 10-13, this becomes:

$$\Delta v_y = v\times\frac{v}{r}\Delta t = \frac{v^2}{r}\Delta t$$

We're almost there. Now, we divide both sides by the change in time, and note that the result is equal to the y-acceleration:

$$\frac{\Delta v_y}{\Delta t} = \frac{v^2}{r} = a_y$$

The way this problem was set up, the y-acceleration is pointed toward the center of the turn, so it is the centripetal acceleration:

General Equation 10-14

$$a_{Centripetal} = \frac{v^2}{r}$$

And, since force still equals mass times acceleration, the centripetal force pushing the mass at the center of the curve is:

Physical Law Eq. 10-15

$$F_{Centripetal} = m\frac{v^2}{r}$$

This is the force that must be produced by your tires in a curve, the string swinging a ball in a circle, etc.

Sometimes it is convenient to write the centripetal acceleration in terms of the curve radius and the angular velocity around the curve. Knowing that $v = r\omega$, we can square that and substitute it into equation 10-14, producing

General Equation 10-16

$$a_{Centripetal} = r\omega^2$$

EXAMPLE

Now let's redo our calculation for that 45 mph, 300-foot curve in Section 6.1. Using equation 10-14, the exact answer is

$$a_{Centripetal} = \frac{v^2}{r} = \frac{(66 \text{ ft/sec})^2}{300 \text{ ft}} = 14.52 \text{ ft/sec}^2$$

or about 0.451 g's. Our brute force answer from before of 14.49 ft/sec² is only off by about 0.2%! That's pretty darned good. You could verify that equation 10-16 gives the same answer.

Now that we know that our 0.45-g centripetal acceleration is a good answer, can the car handle that? A modern car would do fine, but some passengers think you are overdoing it above 0.4 g's (not just parents-in-law). But add rain or gravel, and you may need to slow down.

6.3 Shouldn't Doubling the Speed Double the Centripetal Acceleration?

Equation 10-14 says that centripetal acceleration goes up with the *square* of the speed. In the same curve, two factors increase the centripetal acceleration in proportion to speed:

1. The velocity vector is larger, so rotating its direction by a certain angle produces a larger change in velocity.
2. The velocity's direction angle changes more quickly.

So in calculating centripetal acceleration, you multiply by speed v for two reasons, not one; it grows with v^2, not v. Doubling the speed does quadruple the centripetal acceleration, so think twice before taking a curve at twice the posted speed.

6.4 With 10% More Cornering Traction, How Much Faster Can I Take a Curve?

Centripetal acceleration is toward the center point of the curve you're taking, making it roughly lateral to your car. So we usually call it lateral acceleration, and the cornering capability a_{MaxLat} depends on how much lateral force the tires can produce. Since centripetal acceleration grows with the square of speed, improving traction 10% doesn't increase your cornering speed by 10%. From equation 10-14, your maximum cornering speed v_{Max} for a given curve is

Equation 10-17

$$v_{Max} = \sqrt{r}\sqrt{a_{MaxLat}}$$

where a_{MaxLat} is the maximum centripetal acceleration the car can produce. If the traction force is 10% greater, the maximum possible acceleration will be 10% greater. Using equation 10-17 to find the ratio of velocity $v_{Max,2}$ (with 1.1 times the baseline cornering traction) to $v_{Max,1}$ (the baseline), we find

$$\frac{v_{Max,2}}{v_{Max,1}} = \frac{\sqrt{r}\sqrt{1.1}}{\sqrt{r}\sqrt{1}} = \sqrt{1.1} = 1.0488$$

The maximum cornering speed with 10% better traction is only about 5% higher. If you increase your traction by a certain percentage, your maximum cornering speed goes up by about half that percentage. That hardly seems fair.

But don't despair, because increased cornering speed doesn't just save time in the corners. Cornering sets your "minimum speeds" around an oval track, road course or autocross course. Faster cornering lets you brake later and start down the next straight at a higher speed.

6.5 What About a Curve's "Tightness"?

Equation 10-15 says that three things affect the traction force required for a vehicle to make a curve: 1) the vehicle's mass m, 2) the vehicle's speed v, and 3) the radius r of the curve. The only part of this that pertains to the road is the radius r. The smaller the radius, the higher the centripetal acceleration in the turn. The centripetal acceleration and force are proportional to $(1/r)$, so we could call this the "tightness." But it has a better name.

Bear with me for a minute. The radius r of the curve is measured in feet (or meters), but it can also be thought of as the number of feet you must travel to turn through an angle of one radian; that is, feet/radian (or meters/radian). This works out because radians have the dimensions of feet/feet, or 1, which is usually ignored. So $(1/r)$ can be thought of as radians/foot, or the angle the road curves for every foot traveled along it. Another name for $1/r$ is the **CURVATURE**. Curvature is sometimes called ρ, but we're using that for density. Since we won't use curvature much in equations, we'll stick with $1/r$.

> **NOTE**
>
> Curvature and tightness of the curve are really the same thing.

Figure 13 shows the effect of curvature on centripetal acceleration. Each car entered its curve at the same time and at the same constant speed. The curve on the right has half as large a radius, meaning it has twice the curvature.

While the car on the left has made it through 30° of its 400-foot radius curve, the car on the right has gone through 60° of its 200-foot radius curve. The velocity has been changing twice as quickly on the tighter curve, so the centripetal acceleration a is twice as high, and the car requires twice the centripetal force F to stay on the curve.

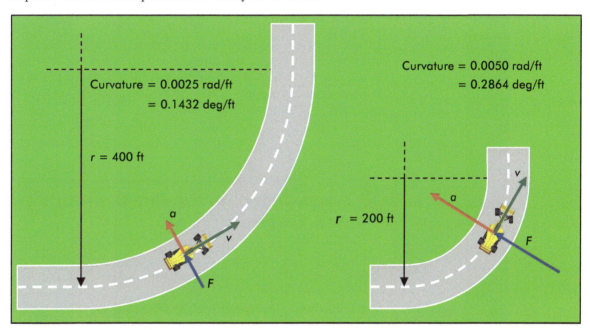

The larger a curve's radius, the less curvature it has, and the road becomes "straighter." What if you're driving in a straight line? The straight line can be thought of as the edge of a circle that has a really large (infinite) radius. Then the curvature is $1/\infty$, which is zero. This makes good physical sense, since it says that a straight road doesn't curve.

Figure 13 Two cars enter two curves at the same speed. The curve on the right has twice the curvature (half the radius), so it requires twice the centripetal acceleration.

KEY CONCEPT

Saying that the curvature for a straight line is zero may not seem too profound. There is a larger point to it, though, than showing off a flair for the obvious. It pays to do a sense-check on a formula by using extreme values (like zero or infinity), or calculating the answer a second way, before using it.

7   Torques and Moments

In Chapter 7 we called a torque that results from a force acting on a body a *moment*. In that sense, "moment" means some quantity (like force or mass) multiplied by some power of distance (like distance, distance^2, distance^3, etc.).

A torque is then a force multiplied by distance to the *first* power, and called the *first* moment of a force. In Section 4.4 of Chapter 8 we showed how rotational inertia is an inertia (mass) multiplied by the *second* power of distance, so it is the *second* moment of an inertia.

In the next section, we will look at the dynamics of a rigid body with several forces applied to it, at separate points. There we will need to consider the forces and moments (of force) applied to it. In this chapter we'll begin referring to a moment of a force by the symbol M, as in the upcoming "split-μ braking" problem.

8   Fixed Coordinates, Car Coordinates, and Rotation

In a horizontal plane, a rigid body can have three independent motions: two perpendicular translations and one rotation. Each is called a **DEGREE OF FREEDOM** (or DOF). The motions and forces can be given relative to a *stationary* coordinate system or a *moving* coordinate system.

In fixed coordinates (like a flat track), an example would be:

1. East/west translation, which we might call x;
2. North/south translation, which we might call y; and
3. Yaw, which we will call θ_z.[4]

But a moving coordinate system, such as one relative to the car, is often more convenient when describing the forces required for motion:

1. Longitudinal (fore-aft), typically called x;
2. Lateral (side-to-side), typically called y; and
3. Yaw, which is again called θ_z.

We saw in Chapter 4 that a stationary coordinate system is an inertial reference frame, so accelerations relative to it can be used directly in Newton's laws. A "car-centered" system is not. But we won't measure the car's motion relative to the car-centered system—the car's motion is always zero relative to itself! Instead we will take the car's motion vectors, as measured in the stationary system, and split them into longitudinal and lateral components relative to the car. We'll first go through fixed and then car-centered coordinates.

Note 4: The vertical direction would be called z, so rotation about the z-axis is called θ_z.

8.1 Motion in Fixed Coordinates

The most straightforward way to describe motion in a plane is to do it in a fixed coordinate system. It's easier to visualize, and you're already in an inertial frame for using Newton's laws. In **Figure 14** you can see the velocity and acceleration vectors of a vehicle going through a curve at a constant speed v. The left side of the figure shows these as resultant vectors. On the right they are shown as component vectors directed relative to the fixed coordinate system. Rotation as seen from overhead is called yaw, which changes slowly as the vehicle follows the curve, or quickly if it spins out. We'll normally call yaw positive CCW looking down, as when following a left-hand curve. (Note that the traditional SAE coordinate system has CW rotation as positive when viewed from overhead.)

CAUTION

Note that here ω is the yaw rotation of the car about its own center of gravity, not motion of its CG around the curve. Where this could cause confusion, we'd have to use subscripts, like ω_{Curve} for curved motion, and $\omega_{Vehicle}$ or ω_Z for yaw.

Figure 14 A car's translational and rotational motions during centripetal acceleration. The left side shows translation in resultant vectors, while the right side shows it in component vectors relative to the fixed coordinate system axes.

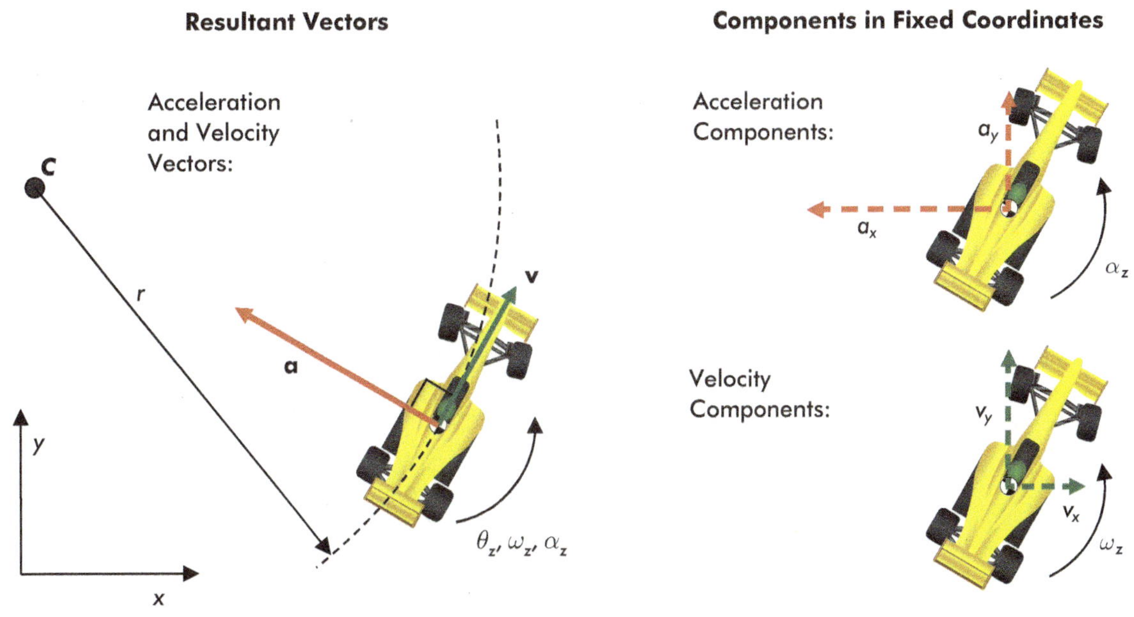

Resultant Vectors

Components in Fixed Coordinates

So in addition to the translational equations of motion in equations 10-5 and 10-6, we now have an equation of motion for vehicle yaw.

The yaw moment on the car is equal to the car's (yaw) rotational inertia times its angular acceleration in yaw:

Fundamental Law Eq. 10-18

$$M_z = I_z \alpha_z$$

Here we must be careful to take the moment about the CG. The forces causing the moments about the car's CG could be from tire traction, crosswinds, etc. This equation of motion is really Newton's second law for rotation (equation 8-14 in Chapter 8), but written here for a body that isn't on a fixed rotation axis.

8.2 Motion in Car-Centered Coordinates

As a driver, it's more natural to visualize your motion in terms of a coordinate system that moves with you and your car.[5] And since we think of tire forces relative to the vehicle, it makes sense to think of its acceleration and velocity in the same terms. When we brake, we have fore-aft acceleration and forces. Cornering forces and crosswinds act left/right. Most onboard data channels would be measured relative to the vehicle.

The CG is the obvious choice for the origin here, since for dynamics, we'll be calculating forces and moments relative to it. Fore-aft, the x-direction, is called longitudinal. Left/right, the y-direction, is called lateral. Longitudinal velocity and acceleration are v_x and a_x, and for lateral they are v_y and a_y.

But though we are stating the force and motion components in car coordinates, they are still the same vectors as in fixed coordinates, just restated. As an example, you can see that the left side of **Figure 15** is the same as in **Figure 14**. It's the same dynamics. But now the right side splits the motion vectors into car-centered components, rather than ground. The lateral (y-) acceleration is the centripetal acceleration vector.[6] If the car were speeding up or braking, it would also have an x-acceleration.

Yaw displacement is still symbolized as θ_z, and still measured relative to a fixed coordinate system, positive CCW looking down.

Note 5: Anyone with a high opinion of themselves should be able to relate to a coordinate system that moves with, and revolves around, them.

Note 6: This is only approximate. We can split the vectors into components *tangent* (parallel to) and *radial* (perpendicular to) to the path, and the centripetal acceleration will equal the radial acceleration. But if the car is in a "drift" with the tail out, even a bit, lateral to the car won't exactly be radial to the path.

Resultant Vectors

Acceleration
and Velocity
Vectors:

$\theta_z, \omega_z, \alpha_z$

Components in Car-Based Coordinates

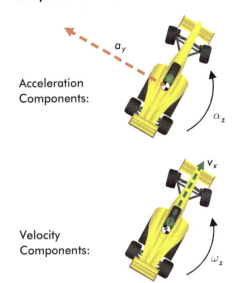

Acceleration
Components:

a_y

α_z

Velocity
Components:

v_x

ω_z

KEY CONCEPT

Car-centered coordinates are handy because we can refer to accelerations as longitudinal and lateral, instead of east/west and north/south. Just keep in mind it's not an inertial frame, so be careful using Newton's laws.

FIGURE 15 The same motion as in **FIGURE 14**. The left side shows translation in resultant vectors, while the right side shows it in component vectors relative to the car-coordinate axes, *x* being longitudinal and *y* being lateral.

In car-based coordinates, we still use equation 10-18 for yaw dynamics. Speaking of yaw, let's take a look at how this equation applies when braking.

8.3 An Example of Transient Motion (Split-μ Braking)

A transient motion is one that changes with time, as opposed to, say, constant acceleration. We can't use the equations developed here to predict transient motion for very long, but at least we can get an idea of how it starts off. Split-μ braking is one example.[7] It's one of the more challenging driving maneuvers, because the tires on each side of the car see a different friction coefficient μ. Let's imagine the driver is doing it wrong and has locked the tires (see **FIGURE 16**).[8] Since the tires are sliding, their friction forces are set by their kinetic coefficients of friction, $\mu_{Kinetic}$, and the friction force on each acts directly opposite its velocity.

Note 7: Remember that μ is the Greek letter mu, so this is pronounced "split-mew."

Note 8: It may not be good braking technique, but it's an easy way to illustrate transient motion. It can also be a good way to make the car slide in a straight line—more or less.

Here, the left tires are on good, clean pavement ($\mu_{Kinetic} = 1.2$), while the right tires are on some gravel ($\mu_{Kinetic} = 0.4$). Because of the difference in traction, the braking forces on the left side (F_{LF} and F_{LR}) are triple those on the right side (F_{RF} and F_{RR}). This creates unbalanced forces that cause a CCW moment on the car, accelerating it into a CCW spin. What are the translational and angular accelerations of the car just after the tires lock up?

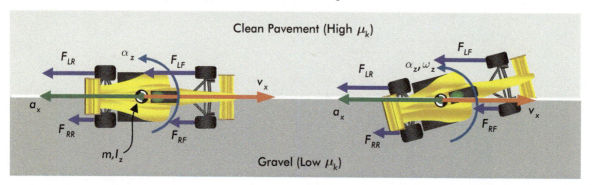

Clean Pavement (High μ_k)

Gravel (Low μ_k)

FIGURE 16 Braking on surfaces with two friction coefficients (split-μ braking).

Let's assume the car weighs 1,932 lbf (mass = 60 slugs), each tire carries one fourth of this load during braking, the car's rotational inertia is 900 slug-ft², and its track width is 5 feet. (Note that the term "track width" is used for the distance between tire patch *centers* on the same axle.)

The car's x-acceleration is in the direction opposite its velocity, and is calculated as usual. Based on kinetic friction coefficients, the left-side braking forces add up to 120% of 966 lbf, or 1,159.2 lbf; on the right, 40% of 966 lbf, or 386.4 lbf. The x-acceleration is then

$$a_x = \frac{F_{LF} + F_{LR} + F_{RF} + F_{RR}}{m} = \frac{-1{,}159.2 \text{ lbf} - 386.4 \text{ lbf}}{60 \text{ slugs}}$$

$$= -25.76 \text{ ft/sec}^2 = -0.8 \text{ g's}$$

So the car decelerates at 0.8 g's, just as if the friction coefficient were 0.8 on both sides. Now for yaw. The left side forces are offset from the CG by half the 5-foot track width, creating a moment that tries to turn the car in the positive (CCW) yaw direction. The same is true for the right, except it tries to turn the car in the negative (CW) direction.[9] We add the moments and divide by the yaw inertia to find the yaw acceleration:

$$\alpha_z = \frac{\left(F_{LF} + F_{LR}\right)^{W_{Track}}\!\big/\!2 - \left(F_{RF} + F_{RR}\right)^{W_{Track}}\!\big/\!2}{I_z} = \frac{1{,}159.2 \text{ lbf} \times 2.5 \text{ ft} - 386.4 \text{ lbf} \times 2.5 \text{ ft}}{900 \text{ slug-ft}^2}$$

$$= 2.15 \text{ rad/sec}^2$$

Note 9: In normal braking, the left and right braking forces are about the same. If they are equal, the CW and CCW moments cancel, and the car brakes smooth and straight with little steering.

After braking like this for 1 second, the car would be rotating at a little over 2 radians per second, or about 120 degrees/second. The point is that the car doesn't instantly change from pointing straight to spinning—it builds up over time.

> **KEY CONCEPT**
>
> Because of inertia, a car can't turn into a curve or go into a spin instantly, although it may happen very quickly. The larger the rotational inertia, the longer it takes to respond, all else being equal.

This is transient motion because the moments spinning the car don't stay constant. Soon after the car begins yawing, some tires move closer to the center of the car's path and some move farther off (**Figure 16**, right). Also, they'll move from the low-friction surface to high friction, or vice versa. Our simple answer isn't exact for long, but you get the idea that the car would be spinning really well in a couple of seconds.

9   Radius of Gyration

Rotational inertia was covered in Chapters 8 and 9, but it deserves another look. Rotational inertia is often given in terms of the object's mass, and its **RADIUS OF GYRATION**, which is the distance from the CG to where the mass acts like it's situated. It's a good way to quantify and visualize the effectiveness of a flywheel design, or the effects of the mass locations of the heavy components (like the engine, transmission, etc.) in a vehicle.

9.1 Flywheels

Have a look at **Figure 17**. On the left is a solid disc with constant thickness, whose rotational inertia is $I = \frac{1}{2}mr^2$. On the right is a rim flywheel (mass concentrated at the rim), with the same mass m, and a (smaller) radius K chosen to produce the same inertia I, so that $I = mK^2$. Setting both their inertias equal:

$$\frac{1}{2}mr^2 = mK^2$$

Dividing through both sides by m, and taking the square root of both sides, we find that for a solid, uniform disc:

$$K = \sqrt{\frac{1}{2}} \times r \cong 0.707r$$

We can say that the solid uniform disc acts like all its mass is distributed at a distance K of about 70.7% of its radius r. We call the distance K the *radius of gyration* of the disc.

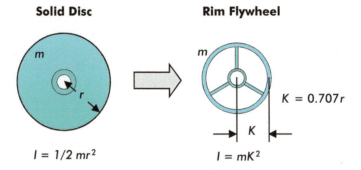

FIGURE 17 A uniform disc with mass m (left) and a "rim flywheel" with the same mass and inertia I, but with a radius just 70.7% as large (right). We say the radius of gyration K of the solid disc is 70.7% of its actual radius.

The radius of gyration can be used to describe the rotational inertia of *any* object, in terms of its mass:

Equation 10-19a

$$I = mK^2$$

Here,

Equation 10-19b

$$K = \sqrt{\frac{I}{m}}$$

Equation 10-19a clearly spells out that any rotational inertia is the second moment of an inertia, with the inertia m being multiplied by the second power of the distance K.

Do you have to think of rotational inertias in terms of their radius of gyration? No, but it makes it easy to see that, compared to a solid disc, the rim flywheel can produce the same inertia in a smaller space with the same mass, or with less mass in the same space.

9.2 A Car's Polar Moment of Inertia

Some cars have a strong tendency to keep pointing in whatever direction they're moving. Other cars are made to react to a driver's steering very quickly, and may be almost "twitchy." Many factors affect this, one being the vehicle's **POLAR MOMENT OF INERTIA** I_z. We've called it "yaw inertia," but car guys usually call it the "polar moment."

FIGURE 18 shows a street rod (left) and a Formula car (right), each weighing 1,288 lbf (mass is 40 slugs). Even so, the rotational inertias of the two cars are much different. The street rod has the engine toward the front, and the transmission in the rear. Its rotational inertia I_z is 541.7 slug-ft². By contrast, the Formula car is rear-mid-engined, putting the engine and transmission almost directly on the car's CG. Tucking the major masses in close like this produces much less rotational inertia, at 336.4 slug-ft².

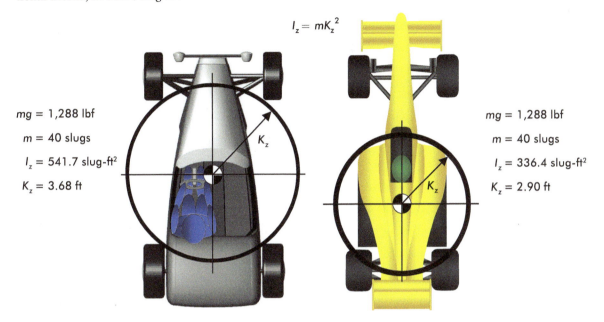

$$I_z = mK_z^2$$

mg = 1,288 lbf

m = 40 slugs

I_z = 541.7 slug-ft²

K_z = 3.68 ft

K_z

mg = 1,288 lbf

m = 40 slugs

I_z = 336.4 slug-ft²

K_z = 2.90 ft

K_z

No matter how the masses are spread out in the car, we can visualize it by its radius of gyration (again, **FIGURE 18**). Knowing the car's mass and its polar moment of inertia (through testing or calculations), we can imagine its mass spread out in a thin ring centered about the CG, with a radius K_z so that it produces the same rotational inertia as the car itself:

FIGURE 18 The street rod and Formula car have the same mass, but because of mass placements, the street rod has 60% higher rotational inertia, and a 27% larger radius of gyration. Each inertia is shown as if the car's mass were contained in the thin ring at its radius of gyration K_Z.

Equation 10-20a
$$I_z = mK_z^2$$

where K_z is the car's radius of gyration for yaw. In this way, you can summarize a car's mass distribution as a sort of "average distance" from the center of gravity:[10]

Equation 10-20b
$$K_z = \sqrt{\frac{I_z}{m}}$$

Using this, the radius of gyration for the street rod is 3.68 feet, about 9 inches farther out than the 2.90 feet for the Formula car.

Note 10: For purists, it's really the square root of the average *squared distance*. Just sayin'.

10 | Summary

This chapter certainly finds "new horizons" with rigid-body dynamics in a plane, but most of it simply expands on previous concepts. We found that it was easy to handle situations with constant x- and y-accelerations by separately working the dynamics in those two directions, using vector components. We added rotation of the rigid body itself by simply using angular dynamics, now taking moments about its CG. These three degrees of freedom (two translations and one rotation) enabled us to handle any constant-acceleration motion in a plane.

The truly new concept here is *centripetal acceleration*, which is perpendicular to the velocity. Rather than changing speed, centripetal acceleration changes the direction of the velocity, causing curved motion. On a constant radius curve at constant speed, the acceleration is purely centripetal, and is equal to v^2/r. We referred to the quantity $1/r$ as the curvature, or "tightness" of the curve.

The centripetal acceleration is caused by centripetal force. In vehicle dynamics, this is from the lateral tire forces, and centripetal acceleration is usually called lateral acceleration. Just remember that's not exact; for dirt-trackers in a four-wheel drift, a good amount of the centripetal acceleration is directed forward in the car, though always at the center of the curve.

Tire imbalance is another place centripetal acceleration is important. **Figure 19** shows the tire from Chapter 7, having a 0.9 oz. "heavy spot" 8 inches off-center. Imagine it's rolling at 60 mph and has a 1-foot radius, given it an angular velocity ω of 88 rad/sec. Using $F = mr\omega^2$ to calculate the (orbiting) force needed at the wheel bearing to keep the wheel on center, I get 9.02 lbf. That may not sound like much, but you know the result. At certain (resonant) speeds, the imbalance sets up a resonant vibration of the suspension and steering systems. Not fun.

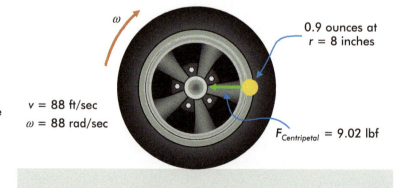

Figure 19 Having a tire imbalance is like having a "heavy spot" (orange) at the wheel rim. Here, we have 0.9 ounces at 8 inches out. At 60 mph and a 1-foot tire radius, this small imbalance would require a centripetal force of 9.02 lbf to keep the wheel centered.

ω

0.9 ounces at $r = 8$ inches

$v = 88$ ft/sec
$\omega = 88$ rad/sec

$F_{Centripetal} = 9.02$ lbf

Major Formulas

Definition	Equation	Equation Number
Velocities in a Plane from Constant Acceleration	$v_x = v_{x0} + a_x t$	10-1a
	$v_y = v_{y0} + a_y t$	10-1b
Displacements in a Plane from Constant Acceleration	$s_x = s_{x0} + v_{x0}t + \dfrac{1}{2}a_x t^2$	10-2a
	$s_y = s_{y0} + v_{y0}t + \dfrac{1}{2}a_y t^2$	10-2b
Velocities in a Plane in Vector Form from Constant Acceleration	$\mathbf{v} = \mathbf{v}_0 + \mathbf{a}t$	10-3
Displacements in a Plane in Vector Form from Constant Acceleration	$\mathbf{s} = \mathbf{s}_0 + \mathbf{v}_0 t + \dfrac{1}{2}\mathbf{a}t^2$	10-4
Newton's Second Law in Component Form	$F_x = ma_x$	10-5a
	$F_y = ma_y$	10-5b
Newton's Second Law in Vector Form	$\mathbf{F} = m\mathbf{a}$	10-6
Centripetal Acceleration from Speed and Curve Radius	$a_{Centripetal} = \dfrac{v^2}{r}$	**10-14**
Centripetal Force	$F_{Centripetal} = m\dfrac{v^2}{r}$	10-15
Centripetal Acceleration from Angular Velocity Around the Curve	$a_{Centripetal} = r\omega^2$	**10-16**
Maximum Speed Possible in a Curve	$v_{Max} = \sqrt{r}\sqrt{a_{MaxLat}}$	10-17
Yaw Moment = Rotational Inertia $\times$ Angular Acceleration in Yaw	$M_z = I_z \alpha_z$	10-18
Radius of Gyration	$K = \sqrt{\dfrac{I}{m}}$	10-19b
Polar Moment of Inertia	$I_z = mK_z^2$	10-20a

Equation numbers shown in bold above have an alternative calculus-based derivation demonstrated in Appendix 5.

11 Dynamics in a Plane Applications

Dealing with Changes in Direction

Jim Clark's Lotus (5) tails John Surtees's Honda (7) in a tight left-hander at the 1967 Dutch Grand Prix. Clark's eventual win was the first for the Lotus 49 and its Ford-Cosworth DFV engine. *Photo by permission of Ford Motor Company.*

The advantage of following the "racing line" in a curve is found, as well as the relative benefits of improving cornering capability vs. adding power on an oval track. We examine how tires produce cornering forces. "Weightlessness" when topping a hill is examined.

Contents

Key Symbols Introduced

Symbol	Quantity	SFS Units	MKS Units
W_{Eff}	Effective Track Width	feet (ft)	meters (m)
r_{Line}	Radius of the Racing Line	feet (ft)	meters (m)
λ	Tire Slip Angle (Lateral Slip)	radians	radians
C_λ	Tire Cornering Stiffness from Lateral Slip	lbf/rad	N/rad
γ	Tire Camber Angle	radians	radians
C_γ	Tire Cornering Stiffness from Camber Angle	lbf/rad	N/rad
θ_x	Vehicle Roll Angle	radians	radians

1   Introduction

In the early 1970s, Indycars began sprouting wings to produce down-force and increase their maximum cornering forces. In the mid-1970s, I remember a race announcer describing that because they ran so much wing, they could go around the whole track without lifting off the throttle. The wings added lots of drag, greatly reducing their top speed on the straights. But their downforce increased their cornering speeds enough that the net result was a faster lap!

This chapter will use the tools developed so far to investigate a few aspects of lapping race courses: why you want to follow the racing line and hit a curve's apex, how improved cornering on a road course or oval track does more than just speed you up in the turns, and how tires develop their cornering forces in the first place (it's more than just a matter of the rubber's friction).

We'll also throw in a few comments about "centrifugal force," a concept that needs to be used carefully (note that we made it through every curve in the previous chapter without using the term!). We'll also cover the effects of going over a rise in the road; besides giving you a feeling of weightlessness, it also reduces the load on the tires and therefore their available traction.

Remember that for a car going through a curve, lateral acceleration and centripetal acceleration are essentially interchangeable terms.

2   Skid Pad Testing

A car magazine reports from their skid pad tests that a Shelby Mustang can achieve a maximum lateral (centripetal) acceleration a_{MaxLat} of 1.01 g's. If they use a 200-foot-diameter skid pad (see **Figure 1**), what is the car's speed v_{Max} during the test? How much time does a lap take?

With the 200-foot diameter the radius is 100 feet, which we'll assume is out to the car's CG. Using equation 10-17 for lateral acceleration, and knowing that 1.01 g's is 32.522 ft/sec²:

$$v_{Max} = \sqrt{r a_{MaxLat}} = \sqrt{(100 \text{ ft}) \times 32.522 \frac{\text{ft}}{\text{sec}^2}} = 57.03 \text{ ft/sec}$$

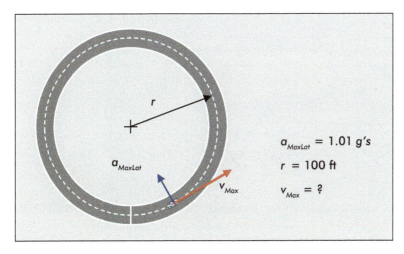

Figure 1 Car magazines routinely test cars for their maximum lateral acceleration on a skid pad. If they report a lateral acceleration a_{MaxLat} of 1.01 g's, and r was 100 feet, what was the maximum speed v_{Max}?

This is about 38.9 mph. You don't need to go fast to corner hard, looping a 100-foot radius curve. As for timing a lap around the pad, we first find the "curved displacement" s of a lap:

$$s = r\theta_{Curve} = (100 \text{ ft}) \times 2\pi = 628.32 \text{ ft}$$

Dividing this distance by the speed, we find that the time around the pad is 10.827 seconds. You could easily write a formula for calculating lateral acceleration from the pad radius and the time around it.

3 | Why Am I Thrown Out in a Curve?

When you go through a curve fast you need to hang onto something, or sit in a seat with some good lateral support. Otherwise, you'll go "flying across the car." Or will you?

When you corner, the tires' lateral forces push the car at the center of the turn. The tires act on the car, but not on you. To go along with the car, something needs to push *you* into the curve, or you'll keep traveling in a straight line (see **Figure 2**).[1] For moderate cornering, the friction between you and the seat will do. In hard cornering it takes more, like a "real" bucket seat that has side bolsters. Racing seats have side bolsters that are more like walls.

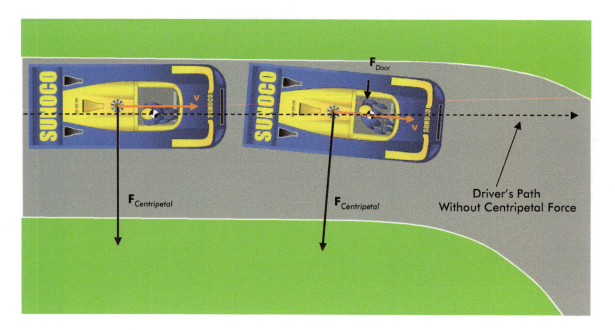

Figure 2 You are not "thrown out of" a curve. Entering the turn (left), the car begins to accelerate laterally, but the CG of the (unrestrained) driver is unaffected, and continues in a straight line (right).

In F1 and Indycars, cornering acceleration might be up to 4 g's, so a driver who weighs 160 lbf needs 640 lbf pushing him into the turn. To do this without bruising, the entire cockpit is like a seat wrapped around the driver. Drivers also hook straps to the helmet to support the head because in a 4-g turn, it takes an 80-lbf sideways force to accelerate 20 pounds of head and helmet. The neck muscles might get a bit sore otherwise.

So to stay put in the car, you do need to be held by a force. But even if you're not, you don't fly across the car—the car flies across you!

Note 1: The car in the figure is a Porsche 917/30, with which Mark Donohue dominated the Can-Am series in 1973. Its twin-turbocharged flat-12 engine produced somewhere around 1,500 horsepower in qualifying trim, and somewhere north of 1,000 during the race!

3.1 Doesn't "Centrifugal Force" Push Me Against the Door?

No, nothing pushes you into the door; the door is pushed *into you* as the car follows the curve. Once the door gets to you, you feel a force, as it pushes on you to accelerate you into the curve.

Now, Newton's third law does say that when the door pushes you into the curve, you exert an outward reaction force on the door. This outward force *could* be called a "centrifugal force," but remember that it doesn't exist until you are accelerated by the centripetal force from the door. It's an *effect* of your motion, not the cause of it.

So using the concept of centrifugal force can make working some problems easier, but it doesn't help you understand the situation—in fact, it can make you think things that are not true. Using the term "centrifugal force" is like using a pipe to extend a wrench for more torque. You shouldn't do it unless you know the risks and use caution.

4   When Cutting Corners Is a Good Thing

You've either heard or read that the fastest way through a curve is to start wide, clip the apex (the inner edge) partway through, and then angle back to the outer edge. It's called "taking the racing line," as shown in **FIGURE 3**. Instead of following the inside or outside edges of the track, we follow the (black dashed) curve that has radius r_{Line}. How much difference does taking the line make, and when does it matter most?

The racing line simply increases the radius of your path (it's less tight), so you can go through the curve at a higher speed v using your maximum lateral acceleration a_{MaxLat}. While it's true that cars running a lot of aerodynamic downforce can corner at more g's at higher speeds, we'll assume here that a_{MaxLat} doesn't change with speed.

This curve has a width w of 30 feet, a 103-foot radius from center (point C) to mid-road, and turns 90°. Assuming a 6-foot-wide car, that leaves 24 feet of track width to maneuver in without putting a tire on the grass; so the track has an effective width w_{Eff} of 24 feet. If the car hugged the inside of the curve, its CG would then follow a radius r_{Inner} of 91 feet (103 ft − 12 ft), and at the outer edge the path would have a radius r_{Outer} of 115 feet (103 ft + 12 ft).

Now let's look at the racing line. We start the turn early, follow a longer arc, clip the apex halfway through, and arc back out wide, exiting the curve late.

As the bold black arrows show, the curve center of the racing line (point C′) is a distance r_{Line} from the *outer* edge of the track at our new curve entry point, and the same distance from the *inner* edge of the track at the apex, halfway through the curve. Using this, we'll solve the problem.

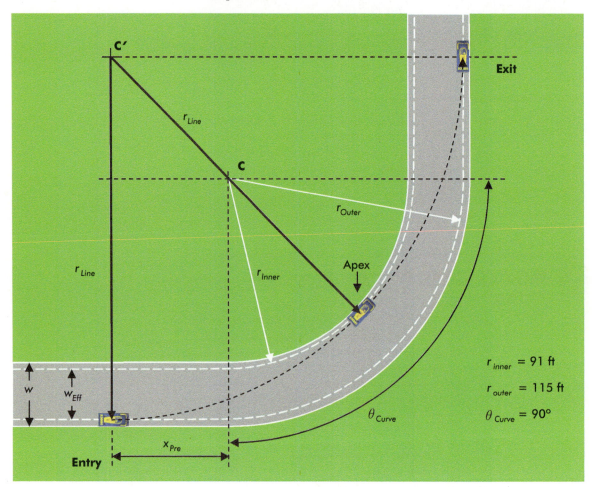

FIGURE 3 "Straightening out" a curve. The curve radius can be made much greater than for either the inside or outside lanes (white dashed arcs), by following the racing line: turning in wide and early, clipping the apex, and exiting wide and late (black dashed arc).

For reference, let's calculate the speed the car could carry through the innermost curve if it can corner at 1.2 *g*'s (38.64 ft/sec²). Using equation 10-17 from Chapter 10:

$$v = \sqrt{r a_{MaxLat}} = \sqrt{(91 \text{ ft}) \times (38.64 \text{ ft/sec}^2)} = 59.298 \text{ ft/sec}$$

or about 40.4 mph. The time it takes to go through the curve would be

$$t_{Curve} = \frac{s_{Curve}}{v} = \frac{r\theta_{Curve}}{v} = \frac{91 \text{ ft} \times \dfrac{\pi}{2}\text{rad}}{59.298 \text{ ft/sec}} = 2.41 \text{ sec}$$

You could instead round the outer edge at 66.66 ft/sec, in a time of 2.71 sec. The upside is entering and exiting the corner at a 12.5% faster speed, and traveling faster on the straights. The downside is spending 0.30 sec more in the curve, by following a longer path.

To do better, we need to follow the racing line. But when do we turn in, when do we exit, and what radius do we follow? Using the geometry in **Figure 4**, we start by turning early by some distance x_{Pre}. Remember that the distance from the new curve center C' to the apex is equal to the distance from C' to the outer edge of the track at curve entry, and both of these equal the new radius r_{Line}. Mathematically, we're saying that:

Equation 11-1

$$r_{Line} = r_{Inner} + \frac{x_{Pre}}{\sin\left(\dfrac{\theta_{Curve}}{2}\right)} = r_{Outer} + \frac{x_{Pre}}{\tan\left(\dfrac{\theta_{Curve}}{2}\right)}$$

These relationships are straight from the figure. Solving for x_{Pre} (the pre-turn) and doing some simplifying gives:

Equation 11-2   **Early turn-in**

$$x_{Pre} = \frac{w_{Eff} \sin\left(\dfrac{\theta_{Curve}}{2}\right)}{1 - \cos\left(\dfrac{\theta_{Curve}}{2}\right)}$$

We take this answer for x_{Pre} and plug it into the first part of equation 11-1, to find the radius of the curve on the racing line:[2]

Equation 11-3   **Racing line radius**

$$r_{Line} = r_{Inner} + \frac{w_{Eff}}{1 - \cos\left(\dfrac{\theta_{Curve}}{2}\right)}$$

Note that although we pictured a 90° curve here, we never assumed that in deriving the formulas. So they are good for angles from 0° to 180°.

Note 2: Aren't geometry and trig awesome?

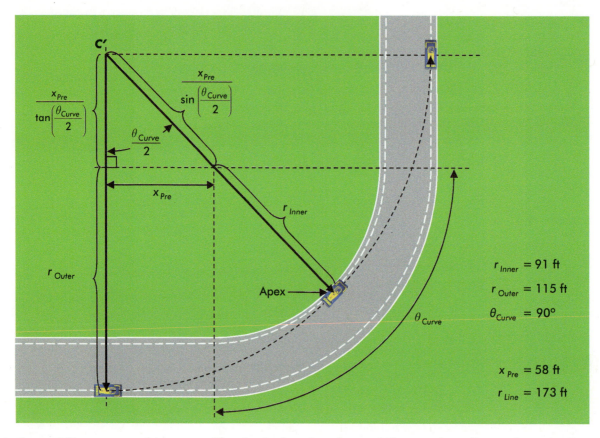

$$\frac{x_{Pre}}{\tan\left(\frac{\theta_{Curve}}{2}\right)}$$

$$\frac{x_{Pre}}{\sin\left(\frac{\theta_{Curve}}{2}\right)}$$

$$\frac{\theta_{Curve}}{2}$$

x_{Pre}

r_{Inner}

r_{Outer}

Apex

θ_{Curve}

$r_{Inner} = 91$ ft

$r_{Outer} = 115$ ft

$\theta_{Curve} = 90°$

$x_{Pre} = 58$ ft

$r_{Line} = 173$ ft

FIGURE 4 The geometry of the problem. We solve it by setting the two values of r_{Line} equal (from C' to the curve entry, and from C' to the apex).

Plugging in the values for our 90° curve, the radius on the racing line has increased to 173 feet (cornering values are compared in **TABLE 1**)! Now you can corner way faster, at 81.75 ft/sec. You do spend more time turning, but since you start turning 58 feet before the curve and finish 58 feet after, you cover more track while doing so (you shorten the straights). Plus, taking the curve faster, you can brake less entering the curve, and exit into the next straight faster. We'll look at the overall effect in an oval track example soon.

TABLE 1 With a cornering capability of 1.2 g's, the racing line increases our curve speed, from the 66.7 ft/sec at the outside radius, to 81.8 ft/sec, 23% higher. We do spend more time in the lengthened curve, but cover more track during that time.

Curve	Radius (ft)	v at 1.2 g's (ft/sec)	Time in curve (sec)
Inner	91	59.3	2.41
Outer	115	66.7	2.71
Racing Line	173	81.8	3.32

On what type of curves does taking the racing line matter most? The smaller the curve's radius, the wider the track, and the smaller angle it turns through, the more cutting the curve helps—as you can gather by studying equation 11-3. In fact, many shallow chicanes can be taken almost straight, as long as you're not restricted by traffic. It also says that being alongside another car (which restricts your "line") will slow you the most in tight curves. So try not to go through chicanes or hairpins side-by-side.

The racing line is generally the fastest, except in extreme conditions. Some "amusement ride" go-karts are such dogs that you can hug the inside of the curves full out. Then you minimize lap time by minimizing lap distance (follow the inside of a curve, then drive straight-line to the inside of the next one). Of course, if it's that easy, you might as well go home.

4.1 Motorcycle Cornering

If you took a motorcycle through the same curve at the same speed (81.75 ft/sec), how many g's would the bike have to pull? Note: this is not a trick question.

Remember, the racing line depends on your maneuvering room on the track: the effective width w_{Eff}. With our 6-foot-wide car, the 30-foot-wide track had an effective width of 24 feet. But the "vehicle width" of the bike on the track is the tread width, maybe 6 inches. Let's give the bike a little extra room and call the effective width of the track 29 feet. Then using equation 11-3,

$$r_{Line} = 91 \text{ ft} + \frac{29 \text{ ft}}{1 - \cos(45°)} = 190.01 \text{ ft}$$

With this curve radius, and the speed v of 81.75 ft/sec:

$$a_{Centripetal} = \frac{v^2}{r} = \frac{(81.75 \text{ ft/sec})^2}{190.01 \text{ ft}} = 35.172 \text{ ft/sec}^2$$

So the bike only needs to corner at about 1.09 g's to stay with a car cornering at 1.2 g's. Although motorcycles typically can't corner quite as hard as cars, at least as long as the cars keep all four tires on the pavement, they can get some of this back by using more of the track.

5 | Oval-Track Lap Times

The racing line makes you faster in the turns, but you spend more time turning. What's the effect on lap times? For that matter, what's the advantage of greater lateral acceleration? Is it more important, or less so, for a slower car (one with less straight-line acceleration)?

To find out, let's take four 90° turns identical to our previous problem (103-foot radius), join them with four straights (two 500-foot and two 300-foot), and make the oval track in **Figure 5**. To calculate lap times, we'll combine the cornering calculations we just developed with the acceleration/braking equations from Chapter 3. We'll compare lap times for several situations:

1. Inside path, outside path, and racing line for a car with 0.6-*g* acceleration, −1.1-*g* braking, and 1.0-*g* cornering.
2. Lateral acceleration increased to 1.1 g.
3. Longitudinal acceleration reduced by 50% to 0.3 g, with 1.0-*g* braking and 1.1-*g* cornering (this part produced a surprise).

Besides comparing lap times, we'll also check where time was saved: corners vs. straights.

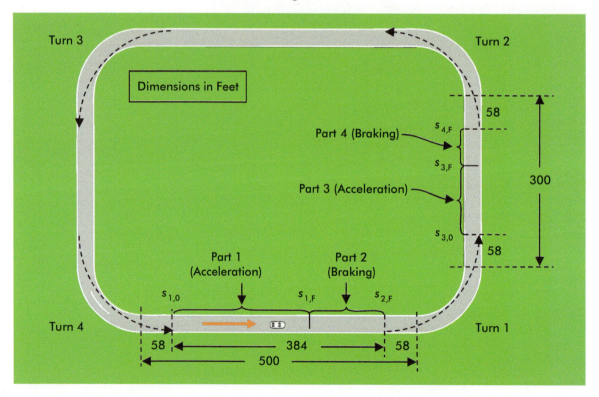

Figure 5 An oval track with four turns identical to the one in the previous problem, and straights of 500 feet and 300 feet.

So that we can fairly compare average lap speeds, let's find the total distance of one lap, going down the middle of the track. Each curve has a mid-track radius of 103 feet, and turns 90° (which is $\pi/2$ radians). The arc length traveled along the middle of each curve is:

$$s = r\theta = 103 \text{ ft} \times \frac{\pi}{2} \text{ rad} = 161.8 \text{ ft}$$

Four of these curves make a total of 647.2 feet. Add that to 1,600 feet of straights, and we have a 2,247.2-foot lap. This is a fairly tight track.

5.1 The Lanes vs. "The Line" for the 0.6 / −1.1 / 1.0 g Car

Now we calculate the speed through the curves from our previous racing line radius of 172.94 feet, and a cornering capability of 1.0 g's (32.2 ft/sec²). Using equation 10-17, we find that

$$v_{Max} = \sqrt{ra_{MaxLat}} = \sqrt{(172.94 \text{ ft}) \times (32.2 \text{ ft/sec}^2)} = 74.62 \text{ ft/sec}$$

So all four curves can be taken at 74.62 ft/sec, making that the beginning and ending speed on each straight. Referring to **FIGURE 5** then,

$$v_{1,0} = v_{2,F} = v_{3,0} = v_{4,F} = 74.62 \frac{\text{ft}}{\text{sec}}$$

Speaking of the straights, how long were they? 500 and 300 feet, right? On the racing line, we start each curve 58 feet (actually 57.94 feet) early, and end it 58 feet late. That lops almost 116 feet off each straight, so now they are 384.12 feet and 184.12 feet, respectively.

As in Chapter 3 ("Going Faster by Stopping Faster"), we use equation 3-7 to find $s_{1,F}$, the end of the acceleration zone/beginning of braking on the front straight:

End of acceleration

$$s_{1,F} = \frac{-2a_2 s_{2,F} + v_{2,F}^2 - v_{1,0}^2}{2a_1 - 2a_2}$$

Using our beginning and ending straight speeds, the acceleration rate a_1 of 19.32 ft/sec² (0.6 g's), and the braking acceleration a_2 of −35.42 ft/sec² (−1.1 g's), this means:

$$s_{1,F} = \frac{-2 \times \left(-35.42 \frac{\text{ft}}{\text{sec}^2}\right) \times (384.12 \text{ ft}) + \left(74.62 \frac{\text{ft}}{\text{sec}}\right)^2 - \left(74.62 \frac{\text{ft}}{\text{sec}}\right)^2}{2 \times 19.32 \frac{\text{ft}}{\text{sec}^2} - 2 \times \left(-35.42 \frac{\text{ft}}{\text{sec}^2}\right)} = 248.55 \text{ ft}$$

The car accelerates for 248.55 feet on the front straight, so it decelerates for 135.61 feet. On the second ("300-foot") straight, we'll use the same equation, but call the acceleration "Part 3" and deceleration "Part 4". Here, $v_{3,0}$ = 74.62 ft/sec, a_3 = 19.32 ft/sec^2, $v_{4,F}$ = 74.62 ft/sec, a_4 = $-$35.42 ft/sec^2 ($-$1.1 g's), and $s_{4,F}$ = 184.12 feet. Using these values gives:

$$s_{3,F} = \frac{-2a_4 s_{4,F} + v_{4,F}^2 - v_{3,0}^2}{2a_3 - 2a_4} = 119.14 \text{ ft}$$

So the car accelerates for 119.14 feet, and decelerates for 64.98 feet of the 184.12 feet. Still following Chapter 3, we find the time spent in acceleration and braking (t_1 and t_2) for the "500-foot" straight. First, find the speed at the end of the acceleration, using equation 3-8:

$$v_{1,F} = \sqrt{2a_1 s_{1,F} + v_{1,0}^2}$$

$$= \sqrt{2 \times \left(19.32 \frac{\text{ft}}{\text{sec}^2}\right) \times 248.55 \text{ ft} + \left(74.62 \frac{\text{ft}}{\text{sec}}\right)^2} = 123.18 \text{ ft/sec}$$

The time t_1 spent accelerating is the change in velocity divided by the acceleration, so:

$$t_1 = \frac{v_{1,F} - v_{1,0}}{a_1} = \frac{123.18 \text{ ft/sec} - 74.62 \text{ ft/sec}}{19.32 \text{ ft/sec}^2} = 2.513 \text{ sec}$$

Doing the same for the braking section produces t_2 = 1.371 sec, for a total time on the front straight of 3.884 sec. Using the same procedure for the second straight, we find a top speed $v_{3,F}$ = 100.86 ft/sec, time t_3 = 1.358 sec, t_4 = 0.741 sec, and a total time of 2.099 sec for the 300-foot straight.

Now let's add up all the times in the corners and compare it to the amount of time in the straights. The time in the four curves is (4 × 172.94 ft × π/2)/(74.62 ft/sec) = 14.56 sec. The time in the straights is 2 × 3.884 sec + 2 × 2.099 sec = 11.965 sec. The total lap time is 26.525 sec, about 55% of which is spent turning. Calculating and adding in lap times, for staying in the inside and outside lanes, produces the data in TABLE 2.

	v in Curves	Time in Curves	v_{Max} Straight 1	v_{Max} Straight 2	Time in Straights	Total Time	Lap Speed
	(ft/sec)	(sec)	(ft/sec)	(ft/sec)	(sec)	(sec)	(ft/sec)
Inside	54.13	10.56	124.22	102.13	18.89	29.46	76.280
Outside	60.85	11.87	127.30	105.85	17.83	29.70	75.663
"The Line"	74.62	14.56	123.18	100.86	11.97	26.53	84.704

Look closely. Compared to staying in either lane, following the racing line produces the slowest top speed on each straight; but because the straights are shorter, almost 7 seconds less is spent going straight! Because the curves are longer, about 4 seconds more was spent turning, but at a much higher speed. Overall, it cuts 3 seconds off the lap time. It may seem ironic that spending more time in the curves (the "slow part") saves time, but the curves aren't nearly as slow as before, and they cover 464 more feet of track.

TABLE 2 Lap time comparison for the oval track in **FIGURE 5**, accelerating at 0.6 g's, braking at −1.1 g's, and cornering at 1.0 g's. The "racing line" is over 2.9 sec faster than staying in your lane around the track.

5.2 Increased Cornering to 1.1 g's

Next, let's compare times with increased lateral acceleration, from 1.0 g's to 1.1 g's, taking the racing line. Rather than slogging through more calculations, the results are shown in **TABLE 3**.[3]

	v in Curves	Time in Curves	v_{Max} Straight 1	v_{Max} Straight 2	Time in Straights	Total Time	Lap Speed
	(ft/sec)	(sec)	(ft/sec)	(ft/sec)	(sec)	(sec)	(ft/sec)
1.0 g's Cornering	74.62	14.56	123.18	100.86	11.97	26.53	84.704
1.1 g's Cornering	78.27	13.88	125.42	103.58	11.59	25.48	88.195
Change	+3.65	−0.68	+2.24	+2.72	−0.38	−1.05	+3.49

As mentioned before, increasing the cornering capability by 10% only increases the cornering speed by about 5%, so the time in the curves is only reduced by 5% (from 14.56 seconds to 13.88). But the time in the straights is also reduced significantly, by about 3.1% (from 11.97 seconds to 11.59).

TABLE 3 Comparing lap times on the racing line with 0.6 g's acceleration and −1.1 g's braking, but then changing 1.0 g's cornering to 1.1 g's. The time is cut by over a second. Most of the improvement comes in the turns, but over ⅓ of the total reduction (36%) is in the straights.

Note 3: Problems like these are what make spreadsheets worth using.

5.3 Increased Cornering of a Slow-Accelerating Car

Let's repeat the cornering comparison (1.1 g's vs. 1.0 g's) for a car that only has half the straight-line acceleration, at 0.3 g's. The car brakes at −1.1 g's, the same as before.

	v in Curves	Time in Curves	v_{Max} Straight 1	v_{Max} Straight 2	Time in Straights	Total Time	Lap Speed
	(ft/sec)	(sec)	(ft/sec)	(ft/sec)	(sec)	(sec)	(ft/sec)
0.3/−1.1/1.0 g Car	74.62	14.56	106.77	91.45	12.91	27.47	81.806
0.3/−1.1/1.1 g Car	78.27	13.88	109.35	94.45	12.45	26.34	85.315
Change	+3.65	−0.68	+2.58	+3.00	−0.46	−1.13	+3.509

TABLE 4 Comparing lap times for a car with only 0.3 g's acceleration and −1.1 g's braking, but then changing 1.0 g's cornering to 1.1 g's. This slower car sees a larger reduction in lap time (1.13 seconds) from better cornering, with 0.68 cut in the turns. But a higher proportion of the cut (now 41% vs. 36%) is in the straights.

TABLE 4 shows that the lap time reduction from better cornering is slightly greater for the slower car than for the faster one (a 4.11% reduction vs. 3.96%). Why? Because the slower car can't recover as easily from slowing down for curves. It needs to hold every bit of speed it can in each curve, in order to have any speed down the next straight. Losing speed in the corners isn't quite as damaging for the faster car.

5.4 The Unexpected

An interesting thing to notice is that the 0.3-g accelerating car with 1.1-g cornering is 0.2 seconds *faster* than the 0.6-g accelerating car with 1.0-g cornering! This is directly compared in **TABLE 5**. By saving 0.68 seconds in the curves, it overcomes losing 0.48 seconds on the straights.

	v in Curves	Time in Curves	v_{Max} Straight 1	v_{Max} Straight 2	Time in Straights	Total Time
	(ft/sec)	(sec)	(ft/sec)	(ft/sec)	(sec)	(sec)
0.6/−1.1/1.0 g Car	74.62	14.56	123.18	100.86	11.97	26.53
0.3/−1.1/1.1 g Car	78.27	13.88	109.35	94.45	12.45	26.34
Change	+3.65	−0.68	−13.83	−6.41	+0.48	−0.19

TABLE 5 Comparing the lap times for 0.6 g's acceleration vs. 0.3 g's, both with −1.1-g braking. With a 10% advantage in cornering acceleration (1.1 g's vs. 1.0 g's), the "slower" car wins out.

If the race began from a standing start, the 0.6-g car would take the lead, but the 0.3-g car would eventually track it down and start an interesting duel. The 0.3-g car would catch up in the curves, and the 0.6-g car would pull away on the straights. This would probably result in a blocking match, with the 0.3-*g* car trying to pass the 0.6-*g* car coming out of a corner, then trying to hold the inside line going into the next curve. The 0.6-*g* car, once passed, would lose contact in a few laps. Likely it would actually be decided by who made the fewest mistakes.

Of course this is a tight track, but it still seems stunning that increasing the lateral (cornering) acceleration by 10% (from 1.0 to 1.1 *g*'s) will outdo increasing the straight-line acceleration by 100% (from 0.3 to 0.6 *g*'s). I guess it says that for this track, you can spend money on your engine, but first spend money on the tires and suspension.

5.5 The Moral

To see how a 3.65 ft/sec advantage in the curves can overcome 13.83 ft/sec and 6.41 ft/sec lower top speeds on the straights, see **FIGURE 6.** It plots the cars' speeds as height above the track. The only places the faster car really takes advantage of its better straight-line acceleration are on long straights. Its advantage builds up on acceleration, but quickly disappears during braking. Because of its better cornering, the 0.3-*g* car is 3.65 ft/sec faster in every curve, but it can also brake later, making it faster in all the braking zones.

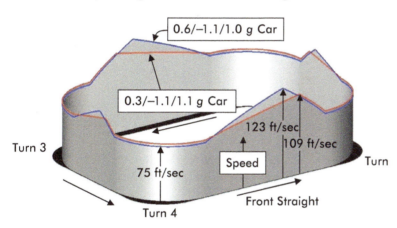

FIGURE 6 The 0.3/−1.1/1.1 *g* car has a small advantage in speed over most of the track, overcoming the large advantage the 0.6/−1.1/1.0 *g* car has mid-straight.

This exercise turned out to be a good example of looking for the unexpected. The original intent here was to show that good cornering was more important to a car with less straight-line acceleration. Accidentally, it also showed the relative importance of straight-line acceleration vs. cornering *g*'s on an oval track or road course. I knew that poor cornering was hard to make up for with power, but had never tried to put numbers to it before.[4]

The moral is the same as in Chapter 3: "Not going slow is more important than going fast."

Note 4: Straight-line acceleration still comes in handy to pass, or to avoid being passed, on the straights.

6 | A Closer Look at the Motorcycle Jump

In Chapter 10 we calculated the length of a motorcycle jump, given its speed and the ramp angle. What if we knew the angle and jump length, but wanted to find the right speed, or to look at all the effects at once? Let's say the ramp angle θ_{Ramp} is 21° and the jump length L is 200 feet, and we need the jump speed v_0.

We could do it by trial and error, but another way gives more insight. We already know that the x- and y-motions don't affect each other (they're independent), so the time t_F for the bike to travel between ramps is the length L divided by the initial x-velocity, which is equal to $v_0 \cos \theta_{Ramp}$. This means the time for the jump is:

Equation 11-4   **Time for jump**

$$t_F = \frac{L}{v_0 \cos \theta_{Ramp}}$$

We also know that at the same time (landing), the y-displacement has returned to zero (the bike has dropped to the same level as the ramp):

$$s_y = v_{y0} t_F - \frac{1}{2} g t_F^{\,2} = 0$$

Knowing the initial y-velocity v_{y0} equals $v_0 \sin \theta_{Ramp}$, this becomes:

$$\left(v_0 \sin \theta_{Ramp} \right) t_F - \frac{1}{2} g t_F^{\,2} = 0$$

We can substitute the expression for the landing time t_F from equation 11-4 into this, to find the necessary initial speed:

Equation 11-5   **Necessary launch speed**

$$v_0 = \sqrt{\frac{gL}{2 \cos \theta_{Ramp} \sin \theta_{Ramp}}}$$

Solving this for v_0 was more involved than I want to show here, but the answer makes good sense.[5] It says that the longer the jump (larger L), the faster you need to leave the ramp to make it. On the other hand (see **Figure 7**), if the ramp had a larger angle θ_{Ramp}, it would have a larger sine ($\sin \theta_{Ramp}$), and the jump speed could be reduced, to a point. A 45° ramp allows the lowest jump speed v_0.

Note 5: The steps in getting to this equation are in Appendix 5, if you'd like to follow them.

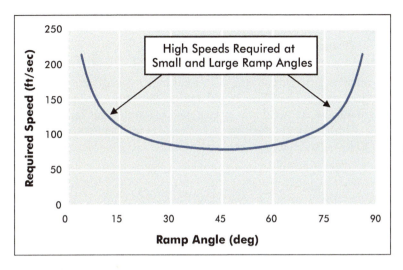

Figure 7 The required ramp speed vs. ramp angle, to make a 200-foot jump. The required speed is lowest for a 45° ramp, but really it "ramps up" below 20° (where you don't stay airborne for long) and above 70° (where you mainly go up, rather than across).

Before making too much of the result, let's check the formula. For our 200-foot jump from a 21° ramp:

$$v_0 = \sqrt{\frac{gL}{2\cos\theta\sin\theta}} = \sqrt{\frac{32.2\dfrac{\text{ft}}{\text{sec}^2} \times 200 \text{ ft}}{2 \times \cos(21°) \times \sin(21°)}} = 98.105 \text{ ft/sec}$$

Then, using $t = L/v_0\cos\theta$:

$$t_F = \frac{200 \text{ ft}}{98.105 \dfrac{\text{ft}}{\text{sec}} \times 0.93358} = 2.1837 \text{ sec}$$

To be correct, the y-displacement must be zero at the same point in time, since it is landing:

$$s_y = (v_0\sin\theta)t_F - \frac{1}{2}gt_F^2$$

$$= 98.105\frac{\text{ft}}{\text{sec}} \times \sin(21°) \times 2.1837 \text{ sec} - \frac{32.2\dfrac{\text{ft}}{\text{sec}^2} \times (2.1837 \text{ sec})^2}{2} = 0$$

So the motorcycle lands (reaches $s_y = 0$) right when it's supposed to.

Why don't most stunt drivers use a 45° ramp to jump farther for the same speed? If you think it through, launching yourself at 45 degrees would also mean you land at 45 degrees, with a pretty good vertical velocity. I'd rather go a little faster and land a little flatter, thank you.[6]

Note 6: At high speed, running onto a 45° ramp from flat ground might also be a little harsh, even with a curved ramp.

7   Lateral Forces from Car and Truck Tires

No one thing determines the way a car handles and feels more than the way the tires produce lateral forces. Not only do they limit the maximum lateral acceleration, but also the way their lateral forces build with slip angle changes the responsiveness and stability of the car. Here we'll go through how lateral forces are generated, and some of their handling effects.

7.1   Basic Tire Mechanics

If you steadily push sideways on a stationary car at the front axle, you'll notice that it moves laterally, up to a point. The harder you push, the more it moves. The front tires deflect the same all the way across the contact patch, since it is well locked to the ground (unless you can push harder than I can!).

In the same way, a tire does not produce a lateral force while rolling down the road, unless it deflects sideways (see **FIGURE 8**). However, the tire deflection is no longer the same along the whole contact patch, as you'll soon see. The more lateral force required to make a vehicle follow a curve, the more a given tire has to deflect sideways. In this way, the tire carcass (the inner structure of the tire) acts like a lateral spring.

FIGURE 8 This 1980 McLaren M30 (Alain Prost's first F1 ride) shows how tires develop cornering forces. Note how the right front is practically ripping itself off the rim. *Photo by Randy Beikmann at Road America with permission of owner Sean Allen.*

A tire produces a controlled lateral force only when it's 1) rolling, and 2) moving in a different direction than where it's pointed, deflecting the tire sideways.[7] If you want to start a right turn, you need to

---

Note 7: A locked tire produces a lateral force when you slide sidewise, but you can't control it by steering (see Section 7.3).

produce a lateral force to the right. But the tire must be turned to the right more than its velocity is. This difference in direction is called the **SLIP ANGLE**. **FIGURE 9** shows an undriven tire with a 9° slip angle. In this text, the slip angle is λ (spelled "lambda" and pronounced *LAM-duh*). Note that slip angle is called α in most textbooks, but we use α for angular acceleration. To avoid confusion, we use λ for slip angle. Since λ is the Greek letter for L, let's say it stands for *lateral* slip.

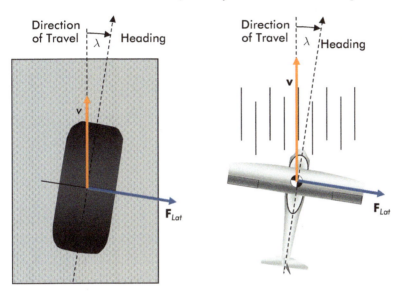

FIGURE 9 On the left is an undriven tire experiencing lateral slip, its direction of travel being different than the heading by the slip angle λ. This produces a lateral traction force F_{Lat} perpendicular to the wheel. The term "slip" originates from an airplane's slipping sideways through the air (right). The airplane is shown doing a maneuver called a "forward slip," where it is intentionally flown sideways, to increase drag and slow it down—good for landing a plane with ineffective flaps. Not shown is the fact that the plane is banked to the left, producing a lateral force canceling the one shown.

Now let's see what a tire does while producing a lateral force. **FIGURE 10** shows two tires from overhead, both traveling toward the top of the page. Each has its top half cut away so we can see its lower carcass. On the left is a straight-rolling tire, its velocity in line with its heading (moving where the wheel points). As you'd expect, the tire's carcass is symmetric, and it doesn't produce a lateral force. The contact patch is centered.

CAUTION

A tire's slip angle is the difference in direction between its heading and its motion (velocity *v*). Don't be misled by the term "slip." The first few degrees of slip are mostly from carcass and tread deflection, not tread sliding on pavement.

On the other hand, the tire on the right of **FIGURE 10** has its wheel turned 4° to the right, while still traveling straight up the page (i.e., a 4° slip angle). The tread fed into the front of the contact patch stays in line with the velocity, making it run at an angle to the wheel. This pulls the contact patch and carcass to the right, and this deflection builds up to cause the lateral force F_{Lat}, perpendicular to the wheel. Notice how the tire's cross-section is nearly undeformed before being fed into the contact patch, but noticeably warped after exiting.

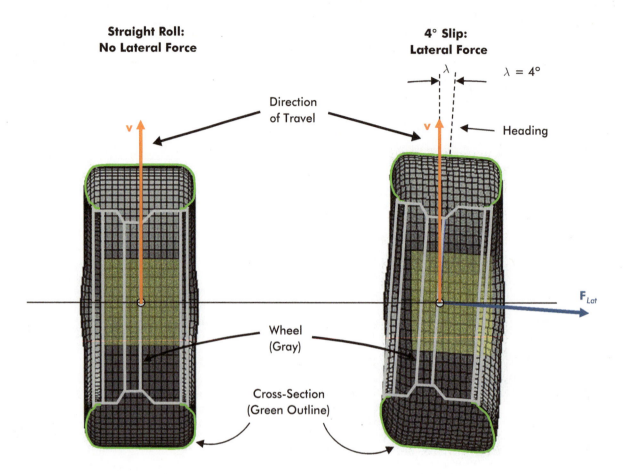

**Straight Roll:
No Lateral Force**

**4° Slip:
Lateral Force**

$\lambda = 4°$

Direction
of Travel

Heading

Wheel
(Gray)

F_{Lat}

Cross-Section
(Green Outline)

FIGURE 10 The lower carcass of a tire, seen from overhead through the wheel. The contact patch is shaded yellow. The tire on the left rolls straight, while at right the wheel is turned 4°, creating a 4° slip angle λ and a lateral force.

In a tire's normal working range, the larger the slip angle, the greater the lateral force. So to corner harder, you angle the wheel more. Notice that this doesn't necessarily mean the *wheel* is steered. Rear wheels, over 99% of which are not steered, must also be pointed in a different direction than they're traveling, in order for the rear tire to produce a lateral force. To do so, the car body must angle into the curve, acting to "steer" the wheel into it.

The relationship between the slip angle and the cornering force of a performance tire is shown in **FIGURE 11**. It looks much like the longitudinal slip characteristics we saw in Chapter 9 (especially in **FIGURE 19**), because similar things are going on. From 0° to about 1.2° slip, the lateral force increases linearly with slip angle, because most of the tread is in stationary contact with the road, meaning the slip is primarily from carcass and tread defection.

With more slip angle, the deflection forces on the rear part of the tread are great enough to start it sliding, reducing its grip. So from 1.2° to 5° slip, it takes more deflection to increase the lateral force; it continues to increase, but more slowly than before. Past 5°, the cornering force decreases, because so much of the tread is sliding on the pavement instead of being "hooked" to it. Frictional heating of the sliding tread makes up for some of this loss.

Sliding of the rear portion of the tread near the traction limit also produces a shorter static part of the contact patch (fore-aft in vehicle), giving the tire less resistance to steering. This is what produces the drop in steering effort you feel in the wheel; your tires are telling you they're "saturated." Note that some tires give you more warning than others.

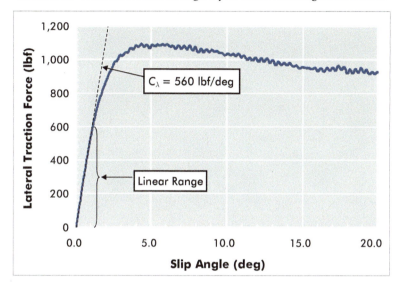

FIGURE 11 Cornering force for a performance tire with 1,000 lbf of load. Up to about 1.2° slip, cornering force is directly proportional to slip angle. This slope (560 lbf/degree here) is called the cornering stiffness C_λ.

In normal driving, the most important part of this is the linear range from zero to about 1.2°, as shown in FIGURE 11. Here, a given percentage increase in slip produces the same percentage increase in lateral force. The slope of this straight-line portion is called the tire's CORNERING STIFFNESS C_λ. For these small slip angles, the lateral force can be calculated as:

Equation 11-6   **Linear range**

$$F_{Lat} = C_\lambda \lambda$$

For small slip angles, you can just calculate the cornering force from equation 11-6. For larger angles, you'd need to look up the force in a chart, or on a graph. FIGURE 11 shows that with 1,000 lbf load, the maximum cornering force this tire produces is 1,080 lbf. So you would quote a "coefficient of friction" of 1.08 for this tire in cornering. Performance driving uses the full range of the traction curve, while everyday street driving uses the linear range.[8]

The force data that go into these charts are measured by a tire dynamometer that loads a rolling tire against a traction surface like a flat belt (see FIGURE 12). The tire is held by an arm at a certain slip angle and CAMBER (tilt) angle. The tire forces are then measured by gauges built into the arm. Note that this same type of machine is what is used to find the longitudinal tire forces discussed in Chapter 9. In fact, the machines can also measure forces from combined lateral and longitudinal slip (as when you bring on throttle exiting a corner, while still unwinding the steering).

Note 8: If you spend much time on the street in the nonlinear range, you won't have your license very long.

FIGURE 12 A tire force dynamometer, as used at Calspan's tire research facility. It can measure longitudinal and lateral tire forces at a wide range of normal loads, applied torques, slip angles, and camber (lean) angles. *Photo by permission of Calspan Corp.*

The cornering force is usually called a "lateral" force, but it isn't entirely lateral relative to the car's motion (review **FIGURE 10**). Because the wheel is mounted on a bearing, the force of an undriven tire must be parallel to the wheel's rotation axis, which is not perpendicular to the direction of travel.[9] This will produce a small rearward force component when cornering, which will tend to decelerate the car.

We can't do justice to tire forces here. Whole books can be (and have been) written about them. To sum it all up, a given tire's lateral force capability changes with:

1. Normal force (load on the tire)
2. Road surface (material type, moisture, evenness, texture, temperature, loose stuff)
3. Inflation pressure
4. Tire temperature
5. Tread depth
6. Speed of travel
7. Camber angle
8. Tire aging and degradation (temperature cycling, sun and ozone exposure, flat spots)

This isn't even a complete list, but the point here is to hit the major factors and to emphasize that predicting a tire's performance is involved, but manageable. This is what tire companies, car manufacturers, and race teams do to prepare their vehicles. Of course, road conditions are the most variable factor and also the hardest to predict. That's up to you, the driver.

Note 9: If any tread force *were* produced in the plane of an undriven wheel, it would produce a moment about the wheel's spin axis, which could only be balanced by bearing friction. That would be a lot of friction.

7.2 Load Effects on a Real Tire

Increasing load on a tire doesn't always increase traction as much as we would expect. For one thing, friction between rubber and pavement doesn't keep growing with load, given a certain contact patch area. For another, if the tread blocks (if any) and tire structure aren't stiff enough to hold the contact patch straight, the tread will squirm. In cornering, this comes into play when the load transfers from the inside tires to the outside. The lateral force capability lost by the inside tires is greater than what is gained by the outside ones.

7.3 Locked-Up Tires

To produce a lateral force, a tire must deflect laterally. To do this in a way that controls the car, the tire must be rolling. When it's locked, it's just a sliding piece of rubber on pavement, producing a force in the direction opposite its velocity. If all tires are locked, the car slides in a straight line regardless of steering. FIGURE 13 shows the same slide with two steer angles, which doesn't affect the slide one bit.

In other words, steering a locked-up tire doesn't do much of anything! But don't interpret this as meaning it's better to lock up the rear tires than the front ones, because then "at least you can still steer." It's not that simple.

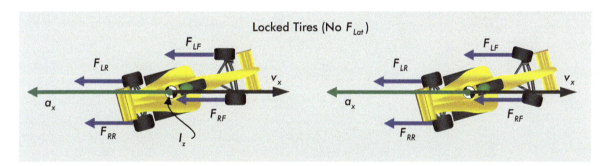

Locked Tires (No F_{Lat})

FIGURE 14 shows the car braking with the rear wheels locked. The front tires are rolling, and have a slight steer angle, which produces a lateral force. The lateral force puts a moment on the car that accelerates it in clockwise rotation (yaw rotation), so without correction, things quickly get out of hand.[10] The more the car rotates, the larger the slip angles of the front tires, and the larger their lateral force. This accelerates the angular motion even more, feeding the spin. But once the car has swapped ends, the (rolling) front tires become the rear tires, and it happily slides in a straight line.[11] Why the difference?

FIGURE 13 When the wheels are locked, tire forces don't depend on the steer angle.

Note 10: In fact, the major advantage of antilock brakes (ABS) isn't so much short stopping distances, but that keeping the tires rolling makes steering control more effective.

Note 11: Anyone who's done a J-turn by locking the rear tires with the parking brake is familiar with this.

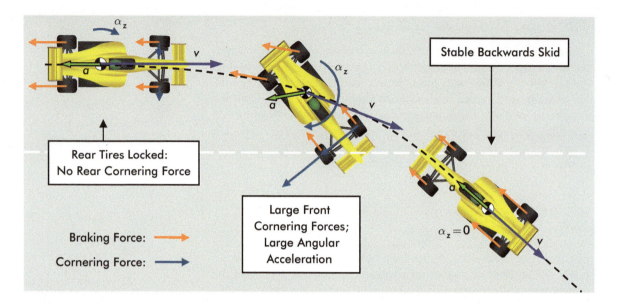

Rear Tires Locked:
No Rear Cornering Force

Braking Force: $\longrightarrow$

Cornering Force: $\longrightarrow$

Large Front
Cornering Forces;
Large Angular
Acceleration

$\alpha_z = 0$

FIGURE 14 With locked rear tires, having rolling front tires will produce yaw acceleration that, left to its own, will take you around.

If you duplicated **FIGURE 14**, but with locked front tires and rolling rears, you'd see that when the car rotates a little bit clockwise, the lateral forces from the rear tires produce a counterclockwise moment that straightens the car out. Rolling rear tires do the same thing for a car that the rudder does for an airplane and the fletch does for an arrow; they stabilize the car's yaw and keep it on a relatively straight line. Even in normal cruising, having more cornering stiffness at the rear axle than the front provides a similar stabilizing effect, called *understeer*.

7.4 Combined Longitudinal and Lateral Traction Forces

We've looked separately at longitudinal (Chapter 9) and lateral tire forces. But we may well accelerate longitudinally (speeding up or braking) while accelerating laterally (cornering). Mark Donohue described this in his "American Method" of combined braking and cornering, which differed at the time from the more typical style of braking first, then cornering. His logic was that if he wasn't using the tire's entire traction for braking, he should use what was left for cornering (on braking, his technique is now called *trail braking*). Let's briefly discuss what this means to performance.

A tire's contact patch can only produce so much traction, which must be divided up between lateral and longitudinal forces. If it's near its limit in one, it can't produce much of the other. This is often described by a tire's "friction circle" (though it's more of an oval). This limits how much you can accelerate exiting a curve, or brake while turning into it. Since this interesting topic is described in many automotive engineering and performance driving books, we will leave it at this.

8 Roll

In car and truck tires, cornering forces are produced when a rolling tire is "yawed" so that its velocity is in a different direction than its heading. Motorcycle tires work completely differently—they work by leaning. When cornering, a bike *rolls* into the turn (see **Figure 15**).

Roll, which we'll call θ_x, is rotation of a vehicle about its longitudinal (*x*-, or fore-aft) axis. Cars roll outward in a curve, while motorcycles must roll inward.

Cars roll at most a few degrees, even in hard cornering (it may feel like more). Rolling motion is a side effect of cornering in cars and trucks, one which I could do without.

For motorcycles, it's a different story. Their roll is key. It's the only way they can provide the lateral force to make the curve while balancing the forces and moments on the bike. The tire's camber angle (lean), not slip, produces most of the cornering force. As **Figure 15** shows, the camber angle of the tires is roughly equal to the bike's roll angle, and there is very little steer angle.

Figure 15 When motorcycles roll, they really roll. The resulting camber of the tires is what produces the lateral tire forces. Notice there is hardly any steer angle, even in this hard turn. *Photo by permission of owner Doug Cornett.*

9   Tire Force Generation in Motorcycles

Motorcycles are driven very differently from cars. Instead of steering into a curve, you initially steer *out* of the curve, which leans you *into* it. Apparently, a lot of riders don't even realize this, because they drive by instinct. But tugging on the *right* handlebar banks you into a *left* turn, and vice versa. Good to know in a tight spot.

Once into the curve, the handlebars are almost straight. With little slip angle to work with, the tire's camber angle must create the main cornering force. This is shown in **FIGURE 16**.

With a given load F_{Load} and a given camber angle γ (spelled "gamma" and pronounced *GA-muh*), the tire will produce a certain lateral force F_{Lat}. Without getting into the details, as the tire leans, its contact patch material becomes warped from a banana shape into a straight line. This distortion of the carcass results in the tire pushing sideways, producing a lateral force F_{Lat}, called "camber thrust."

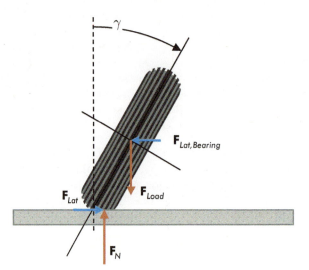

FIGURE 16 A motorcycle tire leaning at a camber angle γ of 30°. With a given load F_{Load}, the contact patch produces a lateral force F_{Lat} to the right.

As with car tires, motorcycle tire forces are routinely measured on dynamometers. A graph of lateral force vs. camber angle is shown in **FIGURE 17**. Here, a tire with a 350-lbf load on it is shown to produce an increasing lateral force with increasing camber.

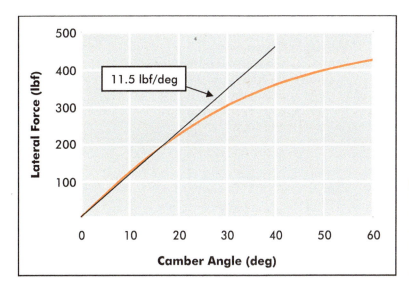

FIGURE 17 An example of cornering force data for a motorcycle tire under 350 lbf load. Up to about 20°, the cornering force is roughly proportional to camber angle. This slope (11.5 lbf/degree) is called the camber stiffness C_γ.

From 0° to 20°, the graph is roughly linear. Just as we related lateral force in car tires to slip angle λ, by a cornering stiffness C_λ in the linear range, we can relate motorcycle tire lateral forces to camber angle γ. So in its linear range, we'll approximate the cornering force of a motorcycle in terms of a camber stiffness C_γ:

Equation 11-7   **Linear range**

$$F_{Lat} = C_\gamma \gamma$$

While camber thrust produces the majority of the cornering force in a motorcycle, the rider does use slip angle to adjust the exact cornering force needed at a certain lean angle. You might say that camber puts it in the ballpark, while lateral slip fine-tunes; but the rider is probably unaware of doing it.

NOTE

Although car tires mainly work through slip angle, some camber does help cornering. Many racers and auto-crossers lean the tops of the tires in a degree or two (negative camber) to increase the grip of the outside tires in the curve.

10 A Vertical Curve, and "Weightlessness"

You've probably driven over a few hills having a sharp crest that make you feel like your stomach is up in your throat. What gives you that roller-coaster feeling?

The full answer is that it depends on your suspension, but let's leave that part out and assume the car's body follows the road exactly.[12] Just as following a horizontal curve requires centripetal acceleration, so does a vertical curve like the crest of a hill. In **FIGURE 18** (left) is a sideview of a hill that has a certain curvature $1/r$ at its crest. At a speed v, the acceleration toward the center of the curve is:

$$a_{Centripetal} = \frac{v^2}{r}$$

At the crest, this acceleration is straight downward. So what's causing it? There are only two vertical forces acting on the car: 1) its weight mg, which acts straight down, and 2) the normal force F_N of the road, acting straight up (**FIGURE 18**, right).

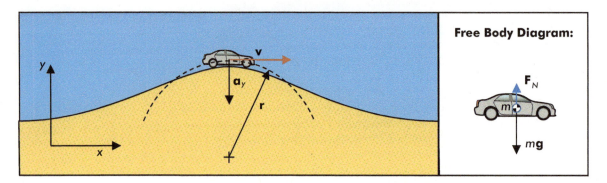

FIGURE 18 Following the crest of a hill requires downward centripetal acceleration. Some of the car's weight causes that, instead of pressing the tires against the road. This reduces traction and hurts braking and cornering. You may also experience "weightlessness."

If you were traveling on flat, level ground, there would be no vertical acceleration, requiring the normal force to equal the car's weight. Following the curve of the hill, things are obviously different, but what changes? The car's weight is the same as before. Yet the car is accelerating downward in following the curve, and there must be a downward resultant force causing this.

Note 12: My kids' favorite road had swells that felt like a carnival ride in most vehicles. But we took a Cadillac CTS over it, and the suspension just "soaked it up." It spoiled the fun, but I'll take its road manners.

The only way this can happen is if the upward acting normal force F_N is *reduced* from, um, normal. We know that the centripetal acceleration a_y is $-v^2/r$, negative because it is downward. Using $F_y = ma_y$, and putting in the forces:

$$F_N - mg = m\left(\frac{-v^2}{r}\right)$$

Let's rework that to solve for the normal force supporting the car, when at the crest of the hill:

Equation 11-8   **For $v^2 \leq gr$**

$$F_N = m\left(g - \frac{v^2}{r}\right)$$

This says that the normal force between tires and road starts off at mg when the vehicle is stopped ($v = 0$) or when the road is flat ($r = \infty$). So far, so good. With increasing speed or a sharper crest (smaller r), the normal force gets smaller. When $(v^2/r) = g$, the normal force becomes zero, meaning the tires are *just* losing contact with the ground. This is significant in itself, because unloaded tires produce no traction.

But your normal force doesn't have to go to zero to cause a problem. If it drops to half the vehicle weight, your traction drops by about half. So there are two good reasons to watch your speed topping a sharp hill: 1) you can't see anything in the road on the other side, and 2) you can't stop or turn as fast as usual because of lower traction. Bad combination.

What else happens when $(v^2/r) = g$? The vertical support force on you from your car seat also goes to zero; you are "free-floating" relative to the car. Your weight is still pulling you down; you just aren't feeling the usual resistance to downward acceleration. Your body is so accustomed to a support force (at your feet or at your seat), that without it, you misinterpret this as "weightlessness."

The same would be true if you were orbiting Earth in a space station 200 miles up. Your weight would be reduced very slightly, only because you're farther away from Earth's center, but you and the ship are both using weight to accelerate toward the center of Earth in the orbit. No force results between you and the ship, so you feel weightless.

11 | Summary

This chapter has shown how useful it is to combine physics with math. In finding the "racing line" through a curve, the hardest part was drawing the figure showing the radius r_{Line} at different points in the curve. The rest of the problem was basic trigonometry and algebra (although some patience and stubbornness also helped).

We've also shown how powerful it is to combine different applications of physics. Combining the straight-line acceleration and braking from Chapters 2–6 with the lateral acceleration of Chapter 10, we could suddenly calculate lap times on an oval track. There's no reason we couldn't do the same for a road course, the main difference being the added complexity of different curve radii and lengths of straights.

Please note that none of the examples done here are perfectly accurate. For example:

1. Real curves don't have constant radii.
2. Cars don't speed up with constant acceleration. They accelerate more slowly at higher speeds, due to aerodynamic drag and power limitations.
3. Cars can't transition from maximum braking to maximum cornering instantly, or from maximum cornering to maximum acceleration.

Even so, the examples worked have pointed out some relevant trends for performance and strategy, given the track dimensions and your car's acceleration capabilities. From this you can at least build some intuition on where to look for improvements, and how to choose between trade-offs.

Knowing these basic factors will also help you absorb the lessons and retain the skills taught in performance driving books and schools. You'll find that many of the "counter-intuitive" tips they give are clear, once you know the physics behind them.

Race teams have very sophisticated simulation programs for lap times, including engine torque curves, aerodynamic drag and downforce, more complete tire data, banking, elevation changes, etc. The only difference between their work and ours is the level of detail and accuracy. The fundamentals are the same — the conclusions may be somewhat different. They will often find a better racing line than the one we assumed in this chapter.

Looking it all over, the conclusion is that although the basic laws of physics are few, and "boringly" reliable, they can help you choose the best trade-offs in setup, and point you to unexpected areas of improvement.

Major Formulas

Definition	Equation	Equation Number
Pre-Turn to Take a Curve on the "Racing Line"	$$x_{Pre} = \frac{w_{Eff}\,\sin\left(\dfrac{\theta_{Curve}}{2}\right)}{1-\cos\left(\dfrac{\theta_{Curve}}{2}\right)}$$	11-2
Radius of the Racing Line Curve	$$r_{Line} = r_{Inner} + \frac{w_{Eff}}{1-\cos\left(\dfrac{\theta_{Curve}}{2}\right)}$$	11-3
Launch Speed Required for a Motorcycle Jump	$$v_0 = \sqrt{\frac{gL}{2\cos\theta_{Ramp}\,\sin\theta_{Ramp}}}$$	11-5
Cornering Force in Terms of Slip Angle and Cornering Stiffness, Linear Range	$$F_{Lat} = C_\lambda \lambda$$	11-6
Cornering Force in Terms of Camber Angle and Camber Stiffness, Linear Range	$$F_{Lat} = C_\gamma \gamma$$	11-7
Normal Force on Tires When Cresting a Hill with Vertical Radius r	$$F_N = m\left(g - \frac{v^2}{r}\right)$$	11-8

12 Energy Basics

Work, Motion, and Heat,
and the First Law of Thermodynamics

The Corvette of Gavin, Milner, and Westbrook converts kinetic energy
to heat as it brakes hard for a turn, at the 12 Hours of Sebring, 2012.
Photo by Richard Prince (rprincephoto.com).

Work is defined, as well as other types of energy: kinetic, potential, chemical, heat, etc. Each is categorized as either mechanical or thermal energy. The law of conservation of energy is described, and the issues in converting thermal energy into work are introduced.

Contents

Key Symbols Introduced

Symbol	Quantity	SFS Units	MKS Units
W	Work Performed	ft-lbf	N-m (joules)
E_K	Kinetic Energy	ft-lbf	N-m (joules)
E_G	Gravitational Potential Energy	ft-lbf	N-m (joules)
E_{Strain}	Elastic Strain Energy	ft-lbf	N-m (joules)
U	Internal Energy	ft-lbf	N-m (joules)
Q	Heat Added	ft-lbf	N-m (joules)
Q_H	Heat Added at High Temperature	ft-lbf	N-m (joules)
Q_L	Heat Rejected at Low Temperature	ft-lbf	N-m (joules)

1   Introduction

You're headed down a straight at 110 mph, and need to haul it down for a tight turn. You have a lot of speed to scrub off before you get there. This heats up your brakes lap after lap, so you've got air ducts routed to cool them. But when you are braking, what exactly are you "scrubbing off," and how much of it? And why is there heat going into the brakes, when there is no flame to produce it? You can only explain this if you understand *energy*.

While Newton's laws are intuitive and easy to visualize, energy can be less obvious. It usually can't be seen, many of its effects can't be seen, and some of its forms don't even seem related. The *work* done by a force in pushing a car seems very different from *heat*, but they both are important types of energy, and are measured in the same units.

Although it works behind the scenes, the science behind energy is as strong as for Newton's laws. **THERMODYNAMICS** is the science of heat and motion (*thermo* for heat and *dynamics* for motion), and how each can be converted to the other. It is a very powerful and wide-ranging part of physics. Thermodynamics takes situations that would be very complicated using Newton's laws, and turns them into child's play. We can, for example:

1. Handle situations that combine different types of energy (from forces that do work, moving masses, and electrical current, for example).
2. Examine the "big picture" of a situation, rather than bogging down in the details.
3. Predict how much work we can "extract" from a given amount of heat.

So back to braking. The faster your car flies down the track, the more *kinetic energy* it has, carried in its moving mass. As you brake and reduce speed, its kinetic energy decreases. But the kinetic energy that is "lost" by slowing down isn't lost; it's converted by friction into heat that goes into the brakes. To maintain their temperature, your brakes need to give off the same amount of heat to the air that flows across them.

This chapter will introduce the different types of energy, and then cover the *first law of thermodynamics*, which states that energy can never be created or destroyed; it can only be transferred between types, by a *thermodynamic process*.

But it would be wrong to leave it at that, because not all forms of energy are created equal. Besides applying energy concepts, Chapter 13 will drop the other shoe: the *second law of thermodynamics*, which restricts which way processes can occur, and how completely. For example, it says that on its own, heat always moves from hot areas to cold. It also places severe restrictions on the efficiency of heat engines, like the one in your car or at the electrical power plant.

Anticipating this, we introduce each type of energy in this chapter according to its thermodynamic capability, as either *mechanical* or *thermal*. While this grouping may seem arbitrary now, it will make working with thermodynamics much clearer.

2   Background and History

Largely on his own, Newton was able to take the groundwork laid by Galileo and develop his three laws of dynamics that revolutionized science. It doesn't take anything away from this to acknowledge that dynamics is fairly simple and neat, once you make the right assumptions. In fact, Newton's genius was in seeing the simplicity behind what looked complex.

By contrast, forming the concept of energy and the laws controlling it was more subtle and messy, involved a lot of squabbling, and took many more investigators. For one thing, it's more difficult to describe energy than force or mass. It's more abstract. In fact, the thermodynamics books I own deftly avoid defining energy, instead giving examples of it. (In one of them, the term "energy" by itself doesn't even appear in the index.) Fortunately, dictionaries can't avoid defining things. I'd say *Merriam-Webster's Collegiate Dictionary* (Eleventh Edition, 2003) gives it the best shot (my italics):

> **ENERGY**: A fundamental entity of nature that is *transferred* between parts of a system in the production of *physical change* within the system and usually regarded as the *capacity for doing work*.

While not specific enough to tell you "what energy really is," it does hit a few key points. The first is that energy can be *transferred* between parts of a system (such as heat transfer from hot gases in a cylinder to the cylinder walls, or from hot brakes to the air).

The second, and major point in my opinion, is that transferring energy is what drives physical change; nothing happens without it. The forces propelling or stopping a car either cause, or are caused by, energy being transferred from one area to another, and usually from one form to another. Heat (thermal energy) produced from the chemical energy of fuel burned in an engine creates *work* that propels the vehicle through the driveline and tires. This *changes* the vehicle's speed, or *changes* its position while working against aerodynamic drag, rolling resistance, or gravity (as when climbing a hill).

As for energy being the *capacity for doing work*, this is partially correct, but far too optimistic. Work can be completely converted to other forms of energy quite easily, but most forms of energy cannot be completely converted to work. For now, just take this part with a grain of salt.

2.1 Scientific Discoveries

Because it takes so many forms, it wasn't initially clear that heat and work were different faces of the same thing: energy. So they were studied separately, each measured in their own units. Work was measured as force multiplied by distance, while heat was measured by how much it would increase the temperature of a known mass of water. Their main connection was the fact that it took heat to drive machines that produce work, such as steam engines.

In the 1700s, scientists thought that heat was an indestructible massless fluid contained within materials, which they called *caloric*.[1] Hotter objects would have more caloric than colder ones. When a hot object touched a cold object, caloric was thought to flow from the hot object to the cold one until they were at the same temperature.

This theory was successful in making many predictions about heat, so most scientists had warmed to it. For one thing, it had shown that each substance has a predictable relationship between the heat transferred to it and its temperature change. For example, the amount of heat to

---

Note 1: "Caloric" comes from the Latin *color* for heat, from which we also get "calorie."

raise the temperature of one pound of water by one degree Fahrenheit was called a BTU (British Thermal Unit).[2]

There were problems with caloric theory, however. For instance, it didn't explain the frictional heating produced in grinding or machining (like drilling and boring). In this case, no hot object supplied caloric to heat up the work piece. To get around this, some theorized that the machining chips somehow released caloric when they were removed from the work piece.

In the late 1700s, Benjamin Thompson (aka Count Rumford) was directing the making of cannon for the Bavarian Army. He was curious about the source of the heat when cannon were bored (see **Figure 1**). He suspected that it wasn't from the machined chips, so he used a dull cutting tool to show that without removing *any* chips, machining still produced enough heat to boil water poured on the cannon.

So the heat didn't come from removed material, but from the scraping motion of the tool *working* against the cannon surface (kinetic friction). Rumford correctly concluded that there was no such thing as caloric, but that heat was itself motion within the material. Although he was correct, his idea took a while to catch on. He fought caloric theory into the mid-1800s.

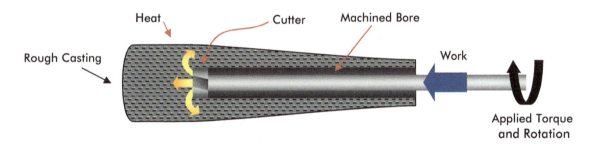

Also, in the early 1800s, James Prescott Joule produced a firm connection between the work done by lowering a weight and the heat it could produce. His most cited experiment was to drive a paddle wheel to stir water, and measure its increase in temperature (see **Figure 2**). He found that driving the paddle with a weight of 800 pounds, dropping by one foot, would increase the temperature of one pound of water by about one degree Fahrenheit.[3] **Figure 2** uses today's more accurate value of 778 ft-lbf of work to create the 1° F rise (Joule was pretty close).

Figure 1 In a boring experiment on cannon, Count Rumford discredited the "caloric theory" of heat. When a dull cutter was used and removed no chips (the supposed source of the machining heat), it still produced a tremendous amount of heat.

Note 2: The metric system later used a similar definition for the calorie, which is the amount of heat required to raise the temperature of one gram of water by one degree Celsius.

Note 3: You could drop a 200-lbf weight by 4 feet, a 100-lbf weight by 8 feet, etc., to get the same result.

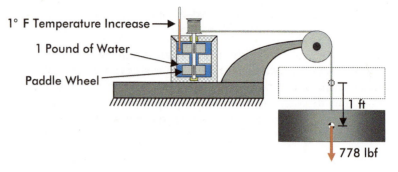

FIGURE 2 A setup similar to what Joule used to measure the temperature increase of water from stirring it with a paddle driven by a dropping weight. The energy of a 778-lbf weight, dropping by 1 foot, produces a 1° F increase of 1 lbm of pure water.

Joule also performed experiments to find how much resistive heat is produced by electrical current flowing through a wire (**FIGURE 3**). He found that the heat created per second was proportional to the current multiplied by the voltage across the resistor. [4]

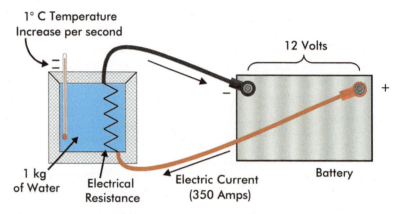

FIGURE 3 A modern metric version of Joule's experiment, converting electrical energy from a battery into heat. The temperature increase depends on the voltage applied to the resistor, the current that results, and the amount of time it flows.

So a certain amount of work could be converted to the same amount of heat, but what about converting heat to work? During the beginning of the Industrial Revolution, increasing the efficiency of steam engines was a major deal, to save money on coal. Joule's work showed that the first ones were less than 1% efficient, but many improvements were made by inventors like James Watt, who increased it to more than 2%. [5]

About the same time, the French engineer Sadi Carnot (pronounced *car-NOH*) was very interested in whether there was a maximum efficiency for a heat engine, and what it was. After brilliant study, he concluded in 1824 that even an ideal heat engine could never convert 100% of its fuel energy into mechanical work. He also stated that the maximum efficiency would be achieved when the heat was supplied to the engine at a constant, high temperature, and then rejected at a constant, low temperature. His ideal engine operating process is called the Carnot Cycle, and is the theoretical "gold standard" for engine efficiency. Unfortunately, it's nearly impossible to operate a

---

Note 4: Since we haven't defined the units of electricity, we'll speak in proportionalities rather than in formulas/equations. For our discussion, electric current flow is analogous to the amount of fluid flow in a pipe, voltage is analogous to pressure, and resistance is analogous to a flow restriction.

Note 5: Today's power plants, using steam turbines, are about 30% efficient at converting heat to work at the turbine shaft.

real Carnot engine, mainly because its output is so small it can barely overcome friction.

Joule found that the work of stirring a fluid produced an equal amount of heat in the fluid, raising its temperature; energy was conserved. This was the *first* law of thermodynamics. But Carnot's work made it clear that converting heat to work is never easy or complete. His discoveries led directly to the *second* law of thermodynamics, which we'll investigate in Chapter 13.

In one of the great ironies of science, Carnot used the now-discredited caloric theory to get to his findings. His father, Lazare Carnot, had shown that a water wheel is its most efficient when the stream smoothly enters the buckets, and after dropping to the lower level, exits almost motionless with no splashing. Carnot reasoned that if it worked for a fluid like water, it should work for a "fluid" like heat, and concluded that transferring heat into a cylinder at a constant temperature, then out at a constant temperature, was the "smoothest" way.

The irony is that if he'd realized that heat was *not* a fluid, he might never have made the connection with his father's work that lead to his great discoveries.

2.2 Thermodynamic Processes

In automotive use, our most basic energy uses are to propel and to brake the car. In either case, we change its speed by causing an exchange of energy — a **THERMODYNAMIC PROCESS** — to occur.

In propelling a typical car, we partially convert its fuel energy to work, in an internal combustion engine. The resulting torque from the turning crankshaft is transmitted by the driveline to drive the wheels (as in Chapter 9). Mechanical energy "flows" down the driveline.

Braking is a simpler process, using friction to convert kinetic energy to heat. But it's not the only process creating heat while the car moves. Friction and "windage" in the driveline sap some of the mechanical energy, converting it to heat. The work done pushing against aerodynamic drag and rolling resistance also turns to heat. In fact, at a steady speed on a level road, *all* of the engine's work soon becomes heat!

KEY CONCEPT

Everything we do to propel and control a car, cool the engine, signal a turn, etc., involves exchanges of energy through thermodynamic processes.

In the next few chapters, we will discuss many processes that change the state, or condition, of things. We can use some of these to our advantage, while many work against us. We can't cover them all, but we will show the general way of handling this very broad field.

3 The Basics of Energy

Despite the difficulty in defining energy, we all have some sense of it. We know we need fuels like gasoline or diesel to power most cars. Natural gas, fuel oil, and wood heat our houses. Electricity powers our lights, air conditioners, fans, computers, and some cars.

Every physical event involves physical change. Energy is the cause behind each, such as:

1. The change in height of a lifted weight.
2. The change in speed of an accelerating vehicle.
3. The change in temperature of a heated object.
4. The change in pressure of a compressed gas.
5. The change in length of a stretched spring.
6. The change in voltage of electrons as they flow through a (resistive) copper wire.

There are many types of energy to consider when designing or driving a vehicle. But from a practical (thermodynamic) standpoint, they can be grouped into two main categories.

1. **MECHANICAL ENERGY**, which has to do with the motion of whole objects, like a car moving down the highway or a crankshaft spinning. It can also be in other forms, each of which could ideally be converted 100% to such mechanical motion. These include work, and the energy stored in a compressed spring, an electric battery, or a raised weight.

2. **THERMAL ENERGY**, often called heat, which has to do with the random motion of atoms or molecules within a material, such as the vibration of atoms in an aluminum sheet, or the fast-moving gas molecules in the "dead air" in an air tank. Thermal energy raises the temperature of your engine, your tires, or your brakes. It can also be stored as chemical energy, such as in fuels like gasoline, diesel fuel, and ethanol.

> **NOTE**
>
> This grouping of energy types may seem artificial at first, but we'll see that there is a fundamental difference in their usefulness. Mechanical energy is much more versatile and valuable than thermal energy.

Let's compare the two types in **FIGURE 4**. Both involve the motion of molecules in a solid material (say, a hood), but they have a clear distinction. At left is an example of **KINETIC ENERGY**. All the molecules move to the right with the same speed, in rigid-body translation. Kinetic energy is a form of mechanical energy.

On the other hand, the right side of **FIGURE 4** shows the invisible, random vibration of the molecules that make up the object. This is a type of thermal energy called *internal energy*. The motions may be too small to see, but you'll feel their effect when you touch a hot object.

Kinetic Energy

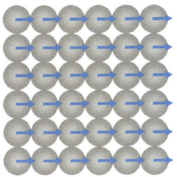

Mechanical Energy in a
Solid Material

(Coordinated Particle Motions)

Internal Energy

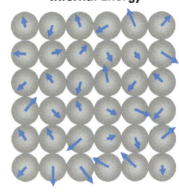

Thermal Energy in a
Solid Material

(Random Particle Motions)

Molecules:

Velocity: ➡

KEY CONCEPT

Mechanical energy is associated with coordinated motions like the motion of a whole vehicle. Thermal energy is associated with random motions, such as within the vehicle's materials. A moving object has both types of energy, but the thermal energy is there even when the object is stationary.

FIGURE 4 Rigid-body motion (left) is a form of mechanical energy, all the molecules moving in a coordinated way. Internal energy (right) is a form of thermal energy, the molecules vibrating back and forth randomly, on average not getting anywhere.

Within mechanical and thermal energy, each has three categories:

1. Energy appearing as actual motion in a body (as those in **FIGURE 4**)
2. Potential energy (which has the potential to cause these motions)
3. Transfer of energy, through a thermodynamic process

For example, mechanical energy can be stored as potential energy in a chemical battery (such as lead-acid or lithium-ion), and thermal energy can be stored in a chemical fuel (such as gasoline or diesel fuel). The transfer of mechanical energy is called work, and the transfer of thermal energy is called heat. [6]

Note 6: The professor in my heat transfer course said that for us the term "heat transfer" was redundant, since heat *meant* transfer of thermal energy. It could have just been called "heat."

> **NOTE**
>
> "Heat" has two meanings in physics. Many use it broadly, as we have so far, for any form of thermal energy. But now we'll limit it to the *transfer* of thermal energy from one area to another, such as when a radiator transfers heat to the air.

The next two sections will sort out the types of mechanical and thermal energy. We'll begin with mechanical, using work to define it. Kinetic energy and the other forms of mechanical energy follow. After that will be thermal energy, starting off with internal energy.

4   Types of Mechanical Energy

Since mechanical energy involves coordinated motion of mass, we'll introduce it by pushing on a mass to get it moving. This push performs work, our most basic type of mechanical energy. Then we'll see what changes it causes, such as increasing the speed of a mass and its kinetic energy of motion, or the height of a mass and its gravitational potential energy.

> **NOTE**
>
> Unless it's specifically mentioned, we will ignore friction and air resistance, electrical resistance, battery inefficiencies, etc., in this section.

4.1 Work

The concept of work is illustrated by the difference between a jack stand and a jack. A jack stand will hold up a car for days (or years) without any effort on your part or any power source. But it will never *lift* a car—that takes *work*—and a jack. With a floor jack, you apply a force to move its handle and pump hydraulic fluid against a piston that lifts the car. A jack stand pushes upward to *hold* against a force (the car's weight), but never produces motion, or any work.

If you've ever had to push a car down the road, you know that the farther you push it, the more tired you get. Even if it's easy to push (it takes a small force), pushing it for a long distance will wear you down. If it's hard to push (requiring a large force), you'll tire out in much less distance. Somehow, the force and distance combine to define the total amount of "effort" it takes:

WORK is the effort required when a force acts through a distance. The work W done on a moving object, by a force F, is equal to that force multiplied by the displacement s the object moves in the same direction as the force.

As a formula, we'd say:

Physical Law Eq. 12-1   **Constant force**

$$W = Fs$$

Loosely, work equals force times distance, as with the car being pushed in **FIGURE 5**. It takes both. If you apply a force to the car and no motion results (parking brake on?), you have done no work. Or, if there is motion, but it takes no force (say it's coasting with no friction), there is also no work done.

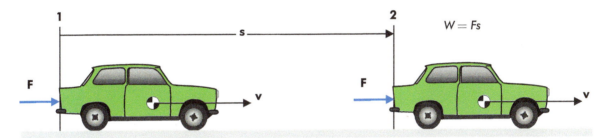

It's very important in calculating work to only include the part of the force acting in the same direction as the motion. With the horizontal motion of **FIGURE 6**, only the horizontal component of the angled cable force (F_x, or $F \cos\theta$ here) does work. So here, the work done is just $F_x s$, or $Fs \cos\theta$. Note that while accelerating the car's mass, more work is required to tow it, and the cable pulls more level. Once at a steady speed, the cable just pulls against drag on the car, and it becomes more vertical.

FIGURE 5 A disabled 1991 Trabant being pushed down the road through a displacement s by a force with magnitude F. The work W done by the force equals Fs.

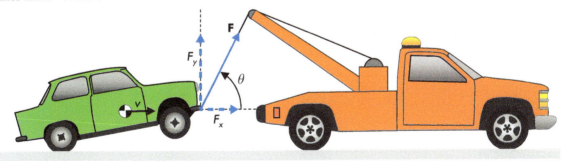

Since work is defined as force × displacement, its units are force units times length units:

The standard SFS unit for work is the **FOOT-POUND** (ft-lbf). The standard MKS unit is the **NEWTON-METER** (N-m), but when referring to

FIGURE 6 The towed Trabant. Note that the vertical cable force F_y does no work while towing, but the horizontal part F_x does, since it acts parallel to the motion.

energy, a newton-meter is usually written as a **JOULE** (pronounced like "jewel"), so 1 N-m = 1 joule.

Notice that the dimensions of energy and torque are both force × length, which may cause some confusion. But of course torque and energy are two different things. In torque, the force is multiplied by the length of a lever arm *perpendicular* to it. In work, the force is multiplied by the length of the motion *parallel* to it.

A critical aspect of work is that it's a process performed on something; work itself isn't stored as work. So where does the energy go? One place it might end up is in kinetic energy.

4.2 Kinetic Energy

Let's see what happens if you take a car that starts at rest, and push on it with a constant tire thrust *F*, on a level road, with no friction or aerodynamic drag (see **FIGURE 7**).

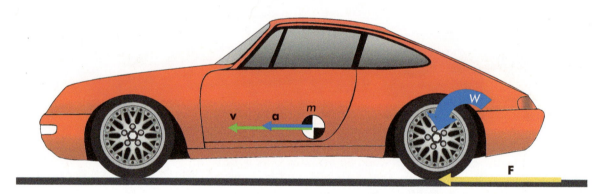

FIGURE 7 A 1993 Porsche 911 accelerated from rest by the constant force *F* produced by the work *W* from the (rear-mounted) engine.

With no other forces, the resultant force on the car is the thrust *F*. This force is then equal to the mass times the acceleration of the car: *F* = *ma*. Starting from rest, after a time *t* the car moves forward by the displacement $s = \frac{1}{2}at^2$. Starting with "work = force × displacement" and substituting our known force and displacement:

$$W = Fs = (ma)\left(\frac{1}{2}at^2\right) = \frac{1}{2}m(at)^2$$

With constant acceleration from rest, *v* = *at*. Substituting *v* for *at* in the above equation gives:

$$W = \frac{1}{2}mv^2$$

This is the work done on the car during the acceleration, but where did that energy go? The only thing changed is the car's speed, so the work must have gone into the car's motion as kinetic energy E_K, so that $E_K = W$. Since the work done here is ½ mv^2, the kinetic energy is:

Physical Law Eq. 12-2   **Translational kinetic energy**

$$E_K = \frac{1}{2}mv^2$$

So the kinetic energy is directly proportional to the car's mass, and to the *square* of its speed. Let's calculate the kinetic energy for the Porsche in **FIGURE 7** after 1 second and 2 seconds of acceleration at 20 ft/sec². Using a weight (with driver) of 3,630 lbf, its mass is 112.73 slugs. The kinetic energy after 1 second, when the car has accelerated to 20 ft/sec, is:

$$E_K = \frac{1}{2}mv^2 = \frac{1}{2}\times 112.73 \text{ slugs}\times\left(20 \text{ ft/sec}\right)^2 = 22{,}550 \text{ ft-lbf}$$

After 2 seconds, at 40 ft/sec, its kinetic energy is four times as high, because it has gone four times as far (40 feet vs. 10 feet) while being acted on by a constant force (2,250 lbf):[7]

$$E_K = \frac{1}{2}mv^2 = \frac{1}{2}\times 112.73 \text{ slugs}\times\left(40 \text{ ft/sec}\right)^2 = 90{,}200 \text{ ft-lbf}$$

What if your acceleration isn't constant in getting to speed? The final kinetic energy would be the same. What if there were friction or air resistance? This would increase the work required to get to speed, because some of the work is transformed into heat, rather than kinetic energy. But the kinetic energy is still ½ mv^2. So no matter how you get there, kinetic energy increases (rapidly) with increasing speed. This is important not only in acceleration, but also in braking: doubling your speed quadruples the amount of energy you need to "scrub off" with brakes or "bleed off" with a parachute.

KEY CONCEPT

Increasing the speed of a mass requires that a net force do work on that mass, increasing its kinetic energy. So kinetic energy results from acceleration.

Note that vibration is a type of coordinated motion, so it's also a form of kinetic energy.

Note 7: You can double-check the force and distance numbers from $F = ma$ and $s = \frac{1}{2}at^2$.

4.3 Gravitational Potential Energy

How much work is done in lifting a car with a hoist (see **Figure 8**) at a steady speed? With the speed constant, no work (or force) goes into increasing its kinetic energy. So the upward force *F* is equal to the car's weight *mg*, and the work done lifting a car by the height *h* is:

$$W = mgh$$

Of course the hoist has to lift itself, too, but we're only considering the work *mgh* the hoist does on the car. Since all that work goes into lifting the car, it is converted to *gravitational potential energy*:

Physical Law Eq. 12-3

$$E_G = mgh$$

Here *h* is measured relative to some origin (in our case, the floor). Something's missing, you might think. If I measure height *h* from a different point, is there a different amount of energy stored? Yes and no. More energy is stored relative to the origin, but potential energy only matters when something drops or rises a certain distance—so you only care about the *change* in height, and the *change* in E_G.

Figure 8 The work done lifting a weight *mg* a distance *h* is *mgh*. This much energy is then stored as gravitational potential energy.

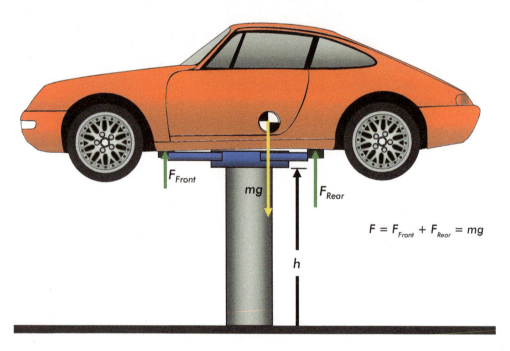

$$F = F_{Front} + F_{Rear} = mg$$

It's called **GRAVITATIONAL POTENTIAL ENERGY** because it could *potentially* be released and cause motion. It's considered mechanical energy because, ideally, it could be *fully converted* into other forms of mechanical energy.

4.3.1 Exchanging Kinetic and Potential Energy

The simplest example of storing and releasing gravitational potential energy is what happens when a ball is thrown straight up (dropping a Porsche off a hoist didn't seem like an appropriate example). The 1-pound ball (mass = 0.031 slugs) in **FIGURE 9** starts off at $h_0 = 0$ with an upward velocity v_0 of 32.2 ft/sec. From equation 12-2, the starting kinetic energy E_K is 16.1 ft-lbf. The ball stops at the height h_{Max}, when all its initial kinetic energy has been converted to potential energy:

$$\frac{1}{2}mv_0^2 = mgh_{Max}$$

Canceling the mass m, you can solve for the maximum height reached:

Equation 12-4   **Object thrown straight up from $h_0 = 0$**

$$h_{Max} = \frac{v_0^2}{2g}$$

In **FIGURE 9**, h_{Max} is 16.1 feet, reached at $t = 1$ second. Doing the same for a falling object ($t = 1$ to $t = 2$ seconds), the increase in kinetic energy is equal to the reduction in potential energy ($\Delta E_K = -\Delta E_G$).

FIGURE 9 A ball thrown straight up gradually slows as its kinetic energy is converted to gravitational potential energy. At the top ($h = h_{Max}$), it has zero kinetic energy and maximum potential energy. On the way down, the energy conversion is reversed.

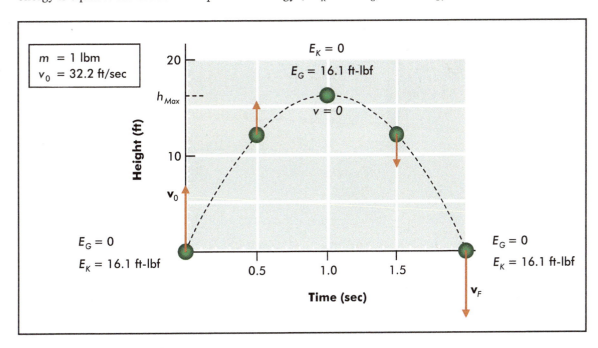

If an object starts at rest, and then falls from height h_0 to h_F, it reaches a final downward velocity v_F of

Equation 12-5   **Falling object**

$$v_F = \sqrt{2g\left(h_0 - h_F\right)}$$

Using this, you'd find the final speed in **FIGURE 9** is 32.2 ft/sec, same as when the ball was thrown.

4.4 Elastic Strain Energy in Translational Springs

When you pull back on the rock in a slingshot, you are doing work on the rubber bands in the sling. That energy is stored in the rubber bands.[8] When you let go, the energy is transferred to the rock as kinetic energy, and off it flies. The energy stored in the stretched rubber bands is called **ELASTIC STRAIN ENERGY**. *Strain* is the stretching or twisting that happens to an object being deformed, such as in the leaf springs in a truck, or the coil springs in a car. Materials that are good at storing and returning this energy are called *elastic* materials.[9]

For example, the spring in **FIGURE 10** pushes back with a force $-kx$, so as you compress it, you are doing work on it. As you push farther (increasing x), the required force increases, so more work is done for each extra inch of deflection.

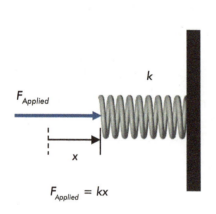

$$F_{Applied} = kx$$

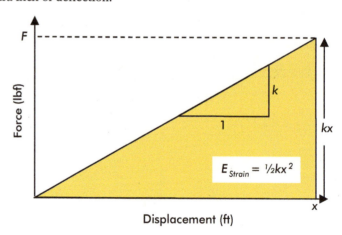

$$E_{Strain} = \tfrac{1}{2}kx^2$$

FIGURE 10 The applied force on a linear spring increases with deflection x. As it is compressed, the spring stores elastic strain energy equal to the shaded area under the force curve.

The work put into a given deflection x is the area under the force-displacement curve (**FIGURE 10**, right):

$$W = \frac{1}{2}x\left(kx\right) = \frac{1}{2}kx^2$$

Note 8: All right, a bit gets turned into thermal energy. Just humor me for now.

Note 9: Some of the better materials for elasticity might be surprising. For instance, steel is more elastic than rubber—not softer, but more complete in returning elastic strain energy.

If the material is perfectly elastic, all this work is stored as elastic strain energy:

Physical Law Eq. 12-6   **Translational spring**

$$E_{Strain} = \frac{1}{2}kx^2$$

As usual, the units for this must be consistent. The stiffness must be in lbf/ft (or N/m), the displacement in feet (or meters), and the energy in ft-lbf (or joules).

> **NOTE**
>
> It's worth noting that equation 12-6 correctly calculates the energy stored whether the spring is stretched or compressed from its free (unstressed) length. Stretching a spring by a distance x produces a displacement of $-x$. Since $(-x)^2 = x^2$, the answer is the same.

Coil springs in the typical car are very elastic, so most of the energy stored during deflection in one direction is returned to kinetic energy when they rebound. All the proof you need is to see a car with worn-out shock absorbers (suspension dampers) bouncing down the road. In doing this, it converts kinetic energy (from vertical vehicle motion) to potential energy (in the springs and against gravity), and vice versa. With little energy dissipation in the shocks, this can go on for a while.[10]

Some texts call this sort of energy "internal potential energy." We'll avoid it here because it's close to the term "internal energy," which we use for the thermal energy within an object.

While it's not elastic strain energy, a trapped pressurized gas can also store energy, as in an air spring. For instance, when you use a bicycle pump, the gas pushes back a couple of inches as you let up on the handle (in fact, it has a "springy" feel). Compressed air is also a convenient way to transmit energy, such as with air tools. Gases don't compress as linearly as metals, however. You'll see this in Chapters 13 and 15.

4.5 Work in Rotation

Let's think a little more about that accelerating Porsche in the earlier example. The car is accelerated by a forward traction force doing work to increase the car's kinetic energy. The funny thing is, the work comes from a crankshaft, through the transmission and axle shaft. So obviously, rotating machines can also do work, but how is it calculated?

An easy way to visualize work in rotation is to imagine starting a lawnmower. **Figure 11** shows a simple pull-start mechanism. The

---

Note 10: I remember sitting next to a car that bounced for a good 30 seconds *at a stoplight*. I don't know how the driver could even keep the thing on the road.

tension force F in the rope pulls on the edge of the pulley of radius r. Let's say that F is constant. As the pulley turns by an angle θ, the rope unwraps a distance s equal to $r\theta$. Since the force F and the motion of the rope are parallel, the work W done is:

$$W = Fs = Fr\theta$$

Since the force is always a perpendicular distance r from the center, the torque T on the pulley equals Fr. Substituting T for Fr gives:

Physical Law Eq. 12-7   **Constant torque**

$$W = T\theta$$

Remember that this formula is so simple only because it assumes consistent units (ft-lbf or joules for work, ft-lbf or N-m for torque, and radians for angle). It's not so pretty with rotation in degrees, for example.

Note that the two equations for the work, $W = Fs$ and $W = T\theta$, describe the same situation from two viewpoints: 1) pulling the rope with a force, and 2) turning the pulley with a torque.

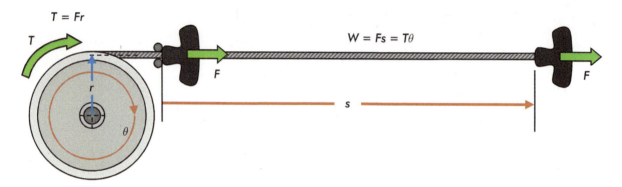

FIGURE 11 A pull-starter producing torque T from force F. The pulley turns an angle θ while the handle moves a displacement s. The work done is force × displacement, or torque × angle. Two points of view, two equations, same result.

As mentioned, your body knows the definition of work. As long as there is no motion, a force or torque can be produced with little trouble. And the first tug on a starter cord might not seem like much effort, but it accumulates. Each pull requires the same work, so after 10 or 12 pulls on a stubborn mower, you know you've done some real (physical) work.[11]

Note 11: Especially on a hot, humid day. Actually, the work done isn't affected much by the temperature or humidity, but your body's ability to shed heat and recover from the exertion are.

EXAMPLE

PROBLEM: Suppose the rope in **FIGURE 11** is pulled a distance so that the pulley turns three revolutions (6π radians). How much work is done if the radius is 2 inches (0.166667 feet) and you pull on the rope with a force of 60 lbf?

SOLUTION 1: Using force × distance, the rope is pulled a distance $s = r\theta = 0.166667$ ft × 6π rad = 3.14159 ft, so $W = Fs = 60$ lbf × 3.14159 ft = 188.5 ft-lbf.

SOLUTION 2: Using torque × angle, the pulley turns by the angle $\theta = 6\pi$ radians. The force in the rope produces a torque of $Fr = 60$ lbf × 0.16667 ft = 10.0 ft-lbf. The work is then $W = T\theta = 10.0$ ft-lbf × 18.8496 rad = 188.5 ft-lbf.

This same thinking is used in calculating the work done by axle torque in propelling a car. You just multiply the axle torque by the rotation angle in radians, per equation 12-7.

4.6 Kinetic Energy in Rotation

We can find rotational kinetic energy in the same way that we derived equation 12-2 for straight-line motion. Assuming the work from a constant torque goes into the kinetic energy of a rotational inertia I with final angular speed ω:

Physical Law Eq. 12-8

$$E_K = \frac{1}{2} I \omega^2$$

A good example is a flywheel. When it spins, it carries kinetic energy. Notice that doubling the rotational inertia I of the flywheel doubles the kinetic energy. But doubling the *speed* from, say, 3,000 rpm to 6,000 rpm *quadruples* the kinetic energy. So dumping the clutch at 6,000 rpm instead of 3,000 gives the clutch and driveline four times the energy to contend with, causing a lot more stress and wear, and risking more damage to each.

4.7 Elastic Strain Energy in Rotational Springs

Just as deflected straight-line springs store potential energy, torsional springs also do. The resulting formula for the elastic strain energy is similar to equation 12-6:

Physical Law Eq. 12-9   **Torsional spring**

$$E_{Strain} = \frac{1}{2} k_R \theta^2$$

The torsional deflection could be positive or negative, and still store the same energy.

4.8 Electrical Charge and Electrical Current

You might be surprised to find electrical energy listed as a form of mechanical energy. But from our thermodynamic standpoint, it is one. The ideal electric motor would completely convert electrical energy into work, and the ideal electrical generator would completely convert work into electrical energy (see **Figure 12**).

Electricity won't be explained in detail here, but you need to know that there are two types of electrical energy, both involving electrons: 1) stationary (static) electrons stored as a charge, such as in a capacitor, and 2) electric current, in which electrons flow from one point to another (as through a cable or a motor).[12]

Nearly all the electricity we consume is the result of a steam turbine driving a generator at a power plant. Energy-wise, driving an electric drill motor from the wall plug in your garage is like driving the drill bit directly by the turbine shaft at the power plant. Most cars, including most hybrids, make their own electrical energy by driving a generator with the engine.

4.9 Chemical Potential Energy in Electrical Batteries

Certain chemical combinations are capable of producing an electrical voltage when they are brought into contact. This can be used to drive an electrical current when hooked to a circuit to conduct it. This concept is used in chemical batteries and in fuel cells.

Our main interest is in storage batteries (see **Figure 12**) capable of storing the energy produced by a car's belt-driven generator, a hybrid

---

Note 12: Individual electrons are not static, but the charge they produce together can be.

powertrain's motor/generator, or an electrical power plant.[13] They can then return that energy in the form of an electric current later.

Charging from Belt-Driven Generator   **Discharging by Driving Starter**

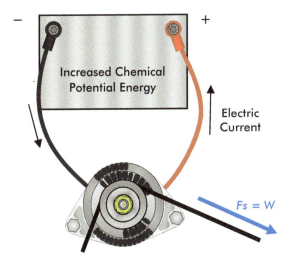
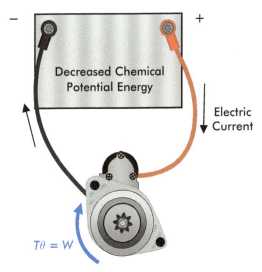

Most automotive batteries are of the lead-acid type, first developed in the mid-1800s. They can release a lot of energy in a hurry, making them perfect for powering starters. But they are not ideal for storing the large quantities of energy needed to assist hybrid vehicles, or to power electric cars. Because of their lower mass, nickel-metal hydride or (increasingly) lithium-ion are used in most hybrids and electrics. In Chapter 15 we'll compare the energy storage capability of batteries to conventional fuels.

FIGURE 12 A generator receives work and then drives a reverse current into the battery, causing a chemical reaction to store energy (left). The opposite reaction releases electrical energy to drive a starter that produces work (right).

Fuel cells are similar to batteries, except that instead of charging them, you feed them fuel. The fuel is typically hydrogen, which reacts with oxygen brought in from the air. Most fuel cells use a platinum catalyst and a proton exchange membrane (PEM) to break every hydrogen molecule into two hydrogen atoms (each of which has one proton and one electron) and then separate each atom into a proton and an electron. The proton migrates through the PEM and the electron travels around the electric circuit to mate up with a different proton, to re-create a hydrogen atom.[14] The hydrogen then reacts with oxygen. So besides the electric current, fuel cells also produce water.

Note 13: There are also the "single-use" types, such as the alkaline batteries you can buy for cameras, toys, etc. They are little good for automotive use, so we'll skip them here.

Note 14: Protons and electrons are apparently not monogamous.

4.10 Magnetic Field Energy

Energy is also stored in the field of a magnet or an electromagnet. This is the source of the spark in a gasoline engine. An electric current is run through an inductor (the circular primary windings in the "coil") that creates a magnetic field. By opening this circuit suddenly, a much larger voltage is created across it, which induces a voltage across a second inductor (the secondary windings in the coil) large enough to jump the spark plug gap. The energy in the resulting spark is from the energy previously stored in the coil's magnetic field.

5   Types of Thermal Energy

Thermal energy has several forms. Each form is random particle motion, could cause this motion, or transfers it. Thermal energy is often called "heat," but we'll typically reserve the term for *transfer* of thermal energy from one place to another.

5.1 Internal Energy

The random motion of atoms or molecules is a form of thermal energy. Being within a material, it is called internal energy.

INTERNAL ENERGY is the thermal energy contained within a material, resulting from random particle motion, and is given the symbol U. It consists of the kinetic energy of the particles, and the potential energy in the chemical bonds between atoms and in the molecular bonds between molecules (all of which we'll simply call "bonds").

It may not be obvious that there is particle motion in a solid material. It certainly doesn't look that way. The molecules are held closely together and in place by the bonds between them, but only *on average* (FIGURE 13, left). Despite being held together more or less in place, the molecules vibrate randomly against each other, stretching and compressing the bonds. An analogy is springs holding pairs of billiard balls together.

If you add heat to the material, its internal vibrations get stronger, the molecules moving farther and faster. In other words, a hotter material has more molecular motion, and more internal energy. When you touch a hot object, some of this motion is transferred to you, and your increased internal energy is what you feel as "heat."

Add enough heat to the material, and the average particle speed increases to where the bonds can't hold the molecules in position, so the molecules slide by each other and "swim around" (FIGURE 13, right). The solid becomes a liquid. The molecules are not organized and are free to move, but they remain close. They still have a net attraction to each other.

**Thermal Motion in a
Solid Material**

**Thermal Motion in a
Liquid Material**

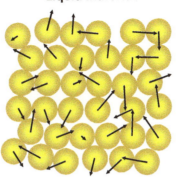

FIGURE 13 Internal energy in a solid (left) and a liquid (right). In the solid, bonds hold the molecules in position, though the molecules still "bounce" against each other randomly. In the liquid, the bonds cannot tightly restrict the molecules' more energetic motion, so they move more freely.

Keep adding heat and eventually the atoms or molecules speed up enough to completely overcome the attractive bonds between them, and the material becomes a gas (**FIGURE 14**). The molecules are much more spread out, which is why gases are so "light" (low density). Gases, unlike liquids or solids, expand to fill the container they are in. The internal energy of a gas is contained in the kinetic energy of the individual particles.

**Thermal Motion in a
Gaseous Material**

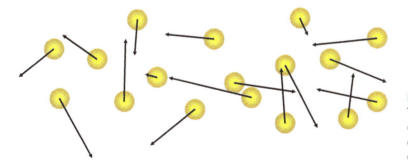

FIGURE 14 Internal energy in a gas. The high average speed of the molecules allows them to overcome the attractive forces (molecular bonds) and spread far apart.

Remember that whether solid, liquid, or gas, internal energy is the disorganized motion of its particles, as opposed to kinetic energy, the coordinated motion of a body. Here, every molecule moves with a different velocity, with no pattern.

KEY CONCEPT

From this discussion, you see that for a given material (say, H_2O), the gas (steam) has more internal energy than the liquid form (water), and the liquid has more internal energy than the solid form (ice).

5.2 Heat

By **HEAT** we specifically mean the *transfer* of thermal energy from one place to another. Heat is given the symbol Q. There are three types of heat transfer:

1. **CONDUCTION**—transfer through contacting material (a solid, or a stationary fluid).
2. **CONVECTION**—transfer through material contact involving moving fluids.
3. **RADIATION**—emission or absorption of electromagnetic radiation by a material (sometimes called radiant heat).

Braking (as shown in **FIGURE 15**) happens to give good examples of all three types:

1. The hot brake rotor and caliper transfer heat to the wheel hub and caliper bracket by *conduction*. The energy spreads by contact, from hotter parts to cooler ones.
2. Cool air blowing over the brake rotor picks up heat through *convection*. This is similar to conduction in that heat is transferred by contact, except that new fluid (the cool air) is constantly moving in to receive energy from the hot brakes. Again, heat transfers from hotter material to cooler. Convection can be natural, where heat drives the fluid motion; or forced, where it is driven by a fan, vehicle motion, etc.
3. The hot surface of the rotor *radiates* heat outward as electromagnetic radiation. Radiation can transfer energy without contacting another material. For example, you can measure the rotor temperature with an infrared "gun," which directly measures the infrared radiation emitted by the rotor (that's why you can do it from a distance). These are pretty cool.

Direct Contact of Solids

Conduction

Contact w/Moving Fluid

Convection

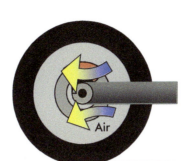

No Contact

Radiation

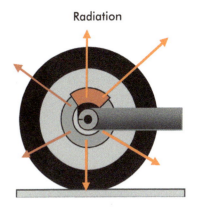

Electromagnetic radiation describes a wide range of waves, including radio waves, microwaves, infrared waves, light waves, ultraviolet waves, X-rays, and gamma rays. These all are the same basic phenomenon at different energy levels (and listed in order of increasing energy). The energy level depends on the radiating surface's temperature. For brake surfaces, usually a few hundred degrees, this is mostly infrared rays but sometimes also visible light. The surface of the sun is about 10,000° F, so it emits much more of its energy as visible light.

FIGURE 15 Hot brakes shed their heat by conduction (left), convection (middle), and radiation (right). Conduction is shown between solids, but it can also happen between solids and fluid. But the fluid usually starts moving from temperature and density differences, and we then have convection.

5.3 Chemical Potential Energy in Fuels

When a typical fuel burns, its molecules break up into their atoms or smaller molecules, combine with oxygen in a chemical reaction, and release heat. The heat is from the **CHEMICAL ENERGY** of the air/fuel mixture. The burning process is called **COMBUSTION**.

Let's look at the relatively simple example of burning methane, the main component of natural gas.[15] On the left side of **FIGURE 16** is one molecule of methane, called CH_4 because it's made of one carbon atom (C) and four hydrogen atoms (H). To burn completely, one methane molecule needs to combine with four oxygen atoms, contained in the two oxygen molecules shown. The methane and oxygen are called the **REACTANTS** because once a spark is supplied, they will react (burn, in this case) and rearrange the atoms into the combustion **PRODUCTS** (**FIGURE 16**, right). The total numbers of hydrogen, carbon, and oxygen atoms are the same before and after the reaction:

Burning methane

$$CH_4 + 2O_2 \rightarrow CO_2 + 2H_2O + heat$$

Note 15: This is shown so you can see what goes into combustion, not to make you a chemistry expert. Note that we are ignoring other gases in the air, which do not enter into the main combustion process.

Besides producing combustion products, combustion produces a lot of heat, which is the whole point. The energy is released from the chemical bonds that held together the atoms in the reactants, which stored more energy than the bonds in the combustion products do. We needn't go through the details here.[16]

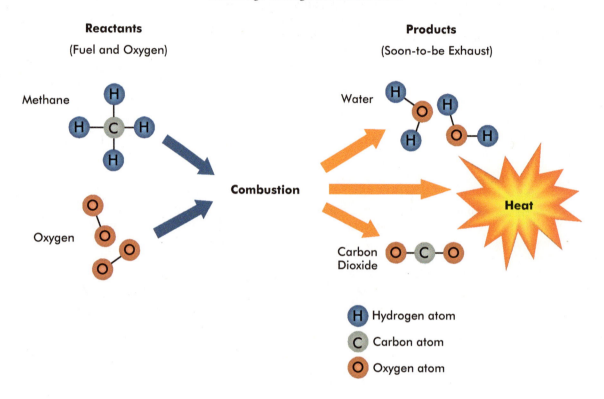

Reactants

(Fuel and Oxygen)

Products

(Soon-to-be Exhaust)

Methane

Water

Combustion

Heat

Oxygen

Carbon Dioxide

H   Hydrogen atom

C   Carbon atom

O   Oxygen atom

FIGURE 16 Combustion of methane with pure oxygen. When methane is burned, it produces water, carbon dioxide, and a lot of heat.

In fossil fuels like petroleum, coal, and natural gas, the chemical energy was stored in plant material millions of years ago. The energy had been radiated by the sun and collected by the plants' leaves. In ethanol and methanol, newly grown plants (like sugar cane, corn, or wood) are used to produce the fuel.

A fuel's heat is what drives an engine, so the amount it releases will largely determine how far it can push our cars down the road. Each fuel produces a certain amount of heat per pound, and per gallon. Let's look at the amount of heat produced by a few typical fuels. The heat-release numbers we'll use are called the "lower heating value," or LHV. This assumes that the energy is transferred from the combustion products until they are at their original temperature, as gases.

Per pound, gasoline releases about 16 million ft-lbf of heat, diesel fuel about 15 million ft-lbf, and ethanol about 9 million ft-lbf. For reference, a pound of coal has about 11 million ft-lbf. Per gallon, gasoline releases 90.5 million ft-lbf, diesel fuel about 100 million ft-lbf, and

Note 16: Partly because we don't need to know the details, just the results, and partly because it would show my lack of expertise in chemistry.

ethanol 59 million ft-lbf.[17] Just as for methane, the combustion products of all these are water and carbon dioxide.

> **NOTE**
>
> Note that we named two forms of chemical potential energy: mechanical, as in batteries, and thermal, as in fuels. In reality they overlap. For example, hydrogen can be burned in an engine or it can power a fuel cell.

5.4 Nuclear Energy

Nuclear fuel isn't likely to be carried in a vehicle you'll buy or drive (unless you work for the Department of Defense). However, it is an energy source for plug-in hybrids and electric vehicles, since some power plants convert nuclear energy into electrical energy.

Nuclear energy is released by converting some of the mass of a fuel (usually uranium-235) into energy, by fission (splitting the atom). We think of mass and energy as being different, but Albert Einstein showed that mass is actually a very concentrated form of energy.[18] With the right fuel and conditions, you can convert some mass to energy. Einstein's famous equation calculates the energy from the converted mass:

Fundamental Law Eq. 12-10   **Energy from converting mass**

$$E = mc^2$$

where c is the speed of light, about 984,000,000 ft/sec, or 300,000,000 m/sec. This says that just a little bit of mass transforms into an absolutely huge amount of energy. Converting *one ounce* of mass creates 1,900,000,000,000,000 ft-lbf of heat, equal to over 20 million gallons of gasoline! No other source packs energy so tightly. Fully reacting one pound of uranium-235 transforms less than a gram of its mass into as much heat as 730,000 gallons of gasoline.

Note 17: Note that ethanol has 35% less energy content per gallon than gasoline. In amounts lower than 10%, ethanol hardly hurts fuel economy, especially if the engine is tuned for its increased octane rating (resistance to knock). Above 10% or so, ethanol's reduced energy takes its toll on mpg.

Note 18: This is the only spot in the book where mass and energy will be treated like the same quantity. Everywhere else they will be treated distinctly, as in everyday life.

6 The Conservation of Energy (The First Law of Thermodynamics)

Now that we've covered the different energy types, we can fully appreciate the first law of thermodynamics, which states the conservation of energy.

FUNDAMENTAL LAW

The First Law of Thermodynamics: Energy cannot be created or destroyed, only converted from one form to another. It doesn't appear out of nowhere or disappear without a trace.

"Losing" energy is like losing your keys: they didn't cease to exist, they just aren't where you can use them. "Lost" energy is energy that has "slipped away" from where you want it, to where it is useless (gone to the surroundings) or harmful (possibly overheating something).

The conservation of energy is one of the "bedrocks" physics is built on. Several times it has been questioned, such as in the heat from radioactive materials, and in quantum mechanics. But it has always held true.

Knowing that energy cannot be created or destroyed gives you a very clear and powerful way to look at physics. Since energy can't be created, you can quickly decide that some things are impossible, like propelling a car without an energy source (the work to push the car down the road against drag has to be produced from something—try running on an empty tank). You also know that the more force it takes to push the car down the road, the more energy is required, and the more fuel is consumed. Since energy also has to *go* somewhere, it allows you to easily calculate how much heat is produced when you slip the clutch, and how much the clutch material's temperature would go up. This all leads to a systems approach.

7 A Systems Approach to Energy

When financial analysts examine a company, they don't count every salary, purchase, and sale the business makes. They look at the bottom line. Regardless of the details, they know that profit equals income minus outgo. Using a thermodynamic point of view—a SYSTEMS APPROACH—is very similar.

Imagine you're sizing a radiator for a car, so you need to know how much heat the engine will reject to it. You don't need to know how much heat transfers into the coolant from each cylinder's combustion, or how much is produced by each friction force (in the cam followers, piston rings, and bearings). While nature accounts for each of these, you don't have to.

Instead, you can calculate the heat the coolant "picks up" as it passes through the engine, by measuring 1) the amount of coolant flow per second, and 2) how much higher its temperature is, outlet vs. inlet. This is a "system view" of the engine's heat rejection, rather than a detailed one. The same amount of heat must then be "dumped" per second to the air flowing through the radiator, to keep the overall coolant temperature constant.

We work with two types of thermodynamic systems, closed and open, referring to whether mass can cross the system boundary.

A **CLOSED SYSTEM**, often called a *control mass*, is defined by a mass and a boundary containing it. The contained material, usually a working fluid, cannot cross the boundary, although the boundary itself might move. Energy, such as heat or work, can cross. A closed system where energy cannot cross the boundary is called an **ISOLATED SYSTEM**.

The air trapped by the cylinder and piston in **FIGURE 17** (left) is a closed system, being confined within the dashed boundary.[19] But heat flows into it, heat flows out of it, and it puts out work across the moving boundary (the piston face). Air is the working fluid, because it is used to perform work on the piston. Note that here the closed system approach assumes the fuel's heat flows in through the walls, ignoring the fact that the air and fuel are usually burned *within* the cylinder instead.[20]

Closed System: Mass is "Trapped"   **Open System: Mass Crosses Boundary**

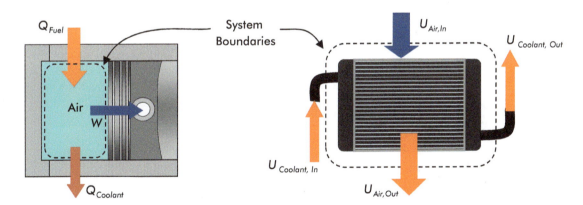

An **OPEN SYSTEM**, often called a *control volume*, is defined by a boundary containing some volume of space. The working fluid(s) can cross the boundary, which may be able to move. Heat and work can also cross.

FIGURE 17 On the left is a closed system, the air trapped in the boundary formed by an engine's cylinder. Heat and work can cross the boundary. To the right is an open system, where coolant and air flow through a radiator, both crossing the boundary. Energy is carried in and out with the fluid masses.

Note 19: Note that when the piston moves, so does the boundary for the trapped air.

Note 20: The Stirling engine is an external combustion engine, which does receive and reject heat through the cylinder wall.

The radiator in FIGURE 17 (right), is an example of an open system. Mass and energy can both cross the system boundary. Coolant flows through the radiator, entering "hot" and leaving "warm." Air enters "cold" and exits "warm." Within the radiator, heat is transferred from coolant to air, though we don't "see" it at this level. We see the increase in the internal energy U_{Air} being equal to the decrease in $U_{Coolant}$.

NOTE

What we say here is good for any system, which might be as simple as your brakes, or as complicated as a gasoline-electric hybrid powertrain. It can simplify your view of just about anything.

Regardless of the type of system, the first law of thermodynamics holds:

Fundamental Law Eq. 12-11   **Thermodynamic process**

$$E_{In} = E_{Out} + \Delta E$$

In other words, any energy entering the system either comes out of the system as some form of energy, or it changes the energy of the system itself. It has to end up somewhere. This plays out in several ways, as we'll see next.

7.1 Closed Systems (or Control Masses)

Most systems are really open systems, but we often examine them as closed systems to gain insight in a simpler way. For example, the operating cycle of an IC engine (internal combustion engine) is shown in FIGURE 18. At left, we start with cool air trapped in the cylinder. Then we 1) compress it by performing work on it, 2) add heat from fuel, 3) let it expand to perform work on the piston (the power stroke), and 4) reject heat, to return to the beginning state. This simplified operation is called the Air Standard Otto Cycle. It is called "air standard" because it keeps the original "air" trapped in the cylinder, adding combustion heat through the cylinder walls, and "Otto" for Nikolaus Otto, who first successfully produced gasoline engines.

FIGURE 18 Production of work in a heat engine, viewed as a closed system. It goes through four thermodynamic processes: compression, heat addition, expansion, and heat rejection. Occurring over two strokes, they make up a simplified Otto Cycle.

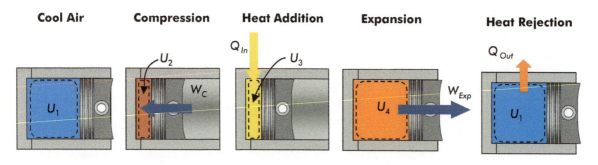

Cool Air   **Compression**   **Heat Addition**   **Expansion**   **Heat Rejection**

Each of these four steps is a thermodynamic process. For example, the compression process does work W_C on the trapped air and increases its internal energy from U_1 to U_2. We can use the first law of thermodynamics to balance the energy flow in each process, one by one. The general formula for a process sets the net energy into the system equal to the net change in the energy of the system:

Physical Law Eq. 12-12a   **Thermodynamic process**

$$Q_{In} + W_{In} = Q_{Out} + W_{Out} + \Delta U + \Delta E_K + \Delta E_G$$

where E_K is kinetic energy and E_G is gravitational energy. Usually we don't need all these terms; we don't have significant changes in the altitude or kinetic energy of the air in the engine above, for example. You'll often see this same equation written as:

Physical Law Eq. 12-12b   **Thermodynamic process**

$$Q + U_1 + E_{K,1} + E_{G,1} = W + U_2 + E_{K,2} + E_{G,2}$$

Here, Q is the net heat added, and W is the net work out. Either of equations 12-12 can have other types of energy added if necessary.

> **NOTE**
>
> Don't get too hung up with naming the terms exactly like equations 12-12. You can take considerable artistic license with them, as long as you follow the first law in equation 12-11. You may also need to include other energy terms, such as elastic strain or magnetic field energy.

The whole point of operating an engine is to repeatedly produce work, stringing thermodynamic processes into a thermodynamic cycle. The air temperature at the end of the cycle in **FIGURE 18** had to be equal to that in the beginning, for it to repeat. If we add up the energy balances of each process, for a steady operating cycle, using equation 12-12a:

Physical Law Eq. 12-13a   **Thermodynamic cycle**

$$Q_{In} + W_{In} = Q_{Out} + W_{Out}$$

or, starting with equation 12-12b:

Physical Law Eq. 12-13b   **Thermodynamic cycle**

$$Q = W$$

Note how the system energy change dropped out of equations 12-12 when we applied it to a cycle in equation 12-13, since each cycle returns us to the same temperature, pressure, etc.

7.2 Fuel Economy at a Steady Cruise

If the Porsche in FIGURE 19 has 150 lbf of drag, how much work is done propelling it for 1 mile (5,280 feet)? What is its fuel economy in mpg, with the engine operating at 25% efficiency?

The amount of work done in 1 mile is 150 lbf × 5,280 feet = 792,000 ft-lbf. As for fuel economy, recall that a gallon of gasoline has 90.5 million ft-lbf of energy. At 25% thermal efficiency, this would produce 22.6 million ft-lbf of work (we'll ignore driveline losses). Since we need 792,000 ft-lbf of work per mile, the distance the car could go on 1 gallon of gasoline would be:

$$\frac{s}{V_{Fuel}} = \frac{s}{W} \times \frac{W}{V_{Fuel}} = \left(\frac{1 \text{ mile}}{792,000 \text{ ft-lbf}}\right) \times \left(\frac{22,600,000 \text{ ft-lbf}}{1 \text{ gallon}}\right) = 28.5 \text{ mi/gal}$$

where V_{Fuel} is the volume of fuel burned, in gallons.

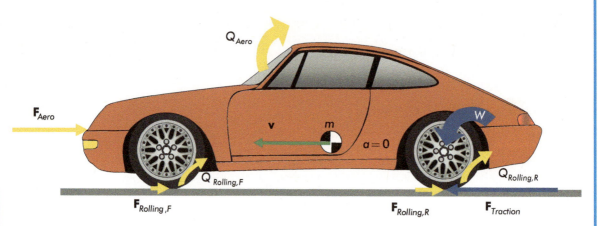

FIGURE 19 A car traveling on a level road with 150 lbf total aerodynamic drag and rolling resistance. The traction force does 792,000 ft-lbf of propulsion work per mile.

FIGURE **20** (left) is an energy diagram for the car's engine, showing the high temperature heat Q_H put in by the fuel, low temperature heat Q_L lost to the surroundings, and the work W put out to the car. On the right, the car receives the work, and rejects heat Q_{Drag} to the surroundings. Obviously, higher drag would require more work, and require more fuel.

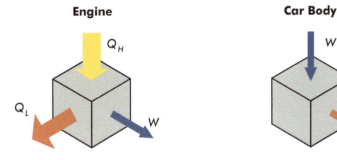

FIGURE **20** Energy diagrams for the engine and the car body in FIGURE **19**. Note how the propulsion work W supplied to the car is completely offset by the heat Q_{Drag} being dissipated through drag.

As in FIGURE **20**, the heat added, heat rejected, and net work for the engine are often symbolized as Q_H, Q_L, and W. So you'll typically see the energy balance for an engine as:

Physical Law Eq. 12-14   **Heat engine (steady state)**

$$Q_H = Q_L + W$$

While we considered the engine here a closed system, this also applies to when we're more realistic, treating it as an open system.

7.3 Open Systems (or Control Volumes)

Open systems are much the same as closed, except you need to track the energy entering and leaving the system with the mass flow. This is especially important when gases are crossing the boundary, because they can do work by expanding while they are in the system.[21]

This is a good way to examine an engine as a whole, as shown in FIGURE **21**. Here we have mass flow across the boundary, with an air/fuel mixture coming in, exhaust gases going out, and coolant flow both in and out. The chemical energy in the air/fuel mixture carries the heat input, and the heat rejected is the exhaust heat plus the net heat to the coolant. As before, work crosses the boundary, but we only see the net work output (not compression and expansion).

We can use the same equation 12-14 for the energy balance of this engine. This works for an engine at a steady operating condition, but what if it's changing? When a cold engine has just started (see FIGURE **21**), not only is there energy coming in and going out, some

---

Note 21: The "expansion energy," of the exhaust gases say, involves adding a term for *enthalpy*, which we don't have the background to develop here.

goes to heating the engine's mass, increasing its temperature and internal energy U. This condition is changing, and so is called *transient*.

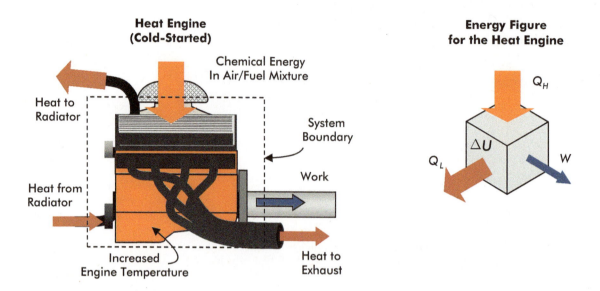

**Heat Engine
(Cold-Started)**

Chemical Energy
In Air/Fuel Mixture

Heat to
Radiator

System
Boundary

Heat from
Radiator

Work

Increased
Engine Temperature

Heat to
Exhaust

**Energy Figure
for the Heat Engine**

Q_H

ΔU

Q_L

W

FIGURE 21 During warmup of a cold-started engine, we have an additional energy term. Besides the heat rejected and work put out, some of the fuel's energy goes into increasing the engine's internal energy U.

In the transient case, we can just add a term for the change in internal energy:

Physical Law Eq. 12-15   **Heat engine (transient conditions)**

$$Q_H = Q_L + W + \Delta U$$

Since some of the heat goes into warming the engine, this suggests less will go into producing work, reducing the engine's efficiency until it warms up.

Other important open systems are turbines and centrifugal compressors (such as in a turbocharger), A/C compressors, power steering pumps, and water pumps. Each of these either transforms energy in the working fluid to work (the turbine), or uses work to put energy into the working fluid.

8   Summary

We've seen that while energy is a little abstract, it isn't mysterious. It follows very well-defined rules that we can take advantage of in designing, setting up, or driving a car.

This is fortunate, because energy is involved in everything we do. In practice, changing the motion of a mass involves some form of energy transfer, whether putting work into kinetic energy, converting kinetic energy into heat through friction, etc. In every case, the total amount of energy afterward equals the total energy beforehand; *energy is always conserved*. It can't be created or destroyed. In fact, the conservation of energy (the first law of thermodynamics) is probably the most general fundamental law of physics.

What makes thermodynamics such a powerful tool here is that it applies to any dynamics situation you can imagine, regardless of what energy source causes the forces and torques. It also tackles problems that are very complicated, allowing you to skip to the bottom line. You can see through the problem and ignore the "fluff."

For example, you don't need to understand exactly how electric motors work, to know they can't produce more work than the electrical energy they're fed; you can make quick work of outlandish claims. Likewise, if someone says they have an engine that takes in 1 million foot-pounds of fuel energy and produces 1.2 million ft-lbf of work, you know right off not to believe them. You don't even need to waste time looking at the engine's construction, materials, etc. They are either wrong or lying, and you don't want to be involved either way. (Chapter 13 will show you should be just as skeptical if someone claims an engine is even 75% efficient.)

The most important thing to remember in using thermodynamics is to define the problem clearly. Because it is so general, *you* must decide where to draw the line. Determine what is inside the system being examined, and what is outside it. *Then forget everything that doesn't affect the system*. Here are examples of what might be left:

1. For an engine, you'd have heat in, then work and heat out.
2. For an electric motor, you'd have mechanical energy in (in the form of electrical energy), and work and heat out.
3. For a transmission or a belt drive system, you'd have work in, then work and heat out.
4. For a moving car body, you have work going in, gravitational potential energy and kinetic energy changing, and heat (from drag and rolling resistance) going out.

> **NOTE**
>
> Thermodynamics is meant to make your job easier, not harder. It's meant to simplify, not complicate.

Before this chapter, we discussed how to take the engine's torque to produce a thrust force and propel a moving car. But until now we couldn't discuss how to *produce* the torque from the engine, or for that matter, an electric motor. Having energy concepts under our belt will open things up quite a bit for Chapter 13, and even more in Chapters 14–15 (power).

And finally, after griping that other books haven't really defined energy clearly, I suppose I should do more than complain. So here's my best shot, drawing on Count Rumford's description of heat, keeping in mind that motion can be coordinated or random:

> Energy is a conserved physical quantity that is transferred in all physical processes, and either is or could be transformed into motion of mass.

Major Formulas

Definition	Equation	Equation Number
Work During Translation	$W = Fs$	12-1
Kinetic Energy in Translation	$E_K = \dfrac{1}{2}mv^2$	**12-2**
Gravitational Potential Energy	$E_G = mgh$	12-3
Elastic Strain Energy, Straight-Line Spring	$E_{Strain} = \dfrac{1}{2}kx^2$	**12-6**
Work During Rotation	$W = T\theta$	12-7
Kinetic Energy in Rotation	$E_K = \dfrac{1}{2}I\omega^2$	12-8
Elastic Strain Energy, Torsional Spring	$E_{Strain} = \dfrac{1}{2}k_R\theta^2$	12-9
Energy by Conversion from Mass	$E = mc^2$	12-10
First Law of Thermodynamics, General	$E_{In} = E_{Out} + \Delta E$	12-11
First Law for a Process, Optional Forms	$Q_{In} + W_{In} = Q_{Out} + W_{Out} + \Delta U + \Delta E_K + \Delta E_G$	12-12a
	$Q + U_1 + E_{K,1} + E_{G,1} = W + U_2 + E_{K,2} + E_{G,2}$	12-12b
First Law for a Cycle, Optional Forms	$Q_{In} + W_{In} = Q_{Out} + W_{Out}$	12-13a
	$Q = W$	12-13b
Heat Engine (Steady State)	$Q_H = Q_L + W$	12-14
Heat Engine (Transient)	$Q_H = Q_L + W + \Delta U$	12-15

Equation numbers shown in bold above have an alternative calculus-based derivation demonstrated in Appendix 5.

13 Energy Applications

Energy Balance and Conversion,
and the Second Law of Thermodynamics

Two Audi R18 e-tron quattros tail each other at Sebring in 2013.
They used a regenerative system that built up kinetic energy in a
flywheel during braking, which was returned as added power during
acceleration. *Photo by Randy Beikmann.*

Thermodynamics is applied to pendulums, levers, gears, braking, and "perpetual motion" machines. Conversion from work to thermal energy, and back, is detailed. The second law of thermodynamics, some of its implications, and entropy are explained.

Contents

Key Symbols Introduced

Symbol	Quantity	SFS Units	MKS Units
R_{Comp}	Compression Ratio	dimensionless	dimensionless
T	Absolute Temperature	°Rankine (°R)	°Kelvin (°K)
S	Entropy	ft-lbf/°R	joules/°K

1   Introduction

This chapter is a bit different than the other "applications" chapters. Besides applying concepts from Chapter 12 to convert between energy forms, we'll also introduce the *second law of thermodynamics*. But this is a natural time to do it, because of the severe limitations the second law puts on how you use energy.

The first part of this chapter takes advantage of the conservation of energy, the first law of thermodynamics. Some problems here would be very tedious if we used Newton's laws, but nature's natural accounting of energy makes them quick, using thermodynamics. Most of them are fairly basic, but illustrate very useful points.

But then we'll begin to look at the *mechanical* reason for the first law to hold, in an idealized engine.[1] We'll examine a "simplified gas" made of a single particle bouncing in a cylinder, and show 1) why the pressure and temperature increase during compression, 2) why they decrease during expansion, and 3) how the drop in the gas's internal energy equals the work done on the piston. That section may seem fairly deep, but the *ideas* behind it are straightforward. In Chapter 15 we'll add heat to it and make it an engine.

The next section moves on to the second law of thermodynamics. Only the second law explains why, though we are swimming in energy, we always worry about having enough to use. There is no "energy shortage," only a shortage of energy that "cooperates." We'll go through the second law in an unconventional way, because 1) I think this gets at the real meaning better (if not the detailed calculations), and 2) we haven't covered the background for the conventional explanation

---

Note 1: Before working this, the mechanism of converting heat to work always bugged me. Now it seems solid.

anyway.[2] In my opinion, many standard texts haven't explained the real meaning of the second law as well as they have the math.[3]

In short, the *first* law is certainly restrictive, requiring energy to be conserved. But it *would* allow heat to flow from cold to hot on its own, run engines at 100% efficiency, and even use the energy in the surrounding air to propel our cars. The *second* law is much more restrictive, and brings us back to reality.

2   Professor Williams's Pendulum

One of the most effective demonstrations of the conservation of energy was given by my physics instructor, Professor Williams. He made a pendulum of a bowling ball hanging from a 25-foot rope, tied to the lecture hall ceiling. He'd start by putting his head right up against the concrete wall, holding the bowling ball stationary against his nose (**FIGURE 1**, Position 1), and letting it go.

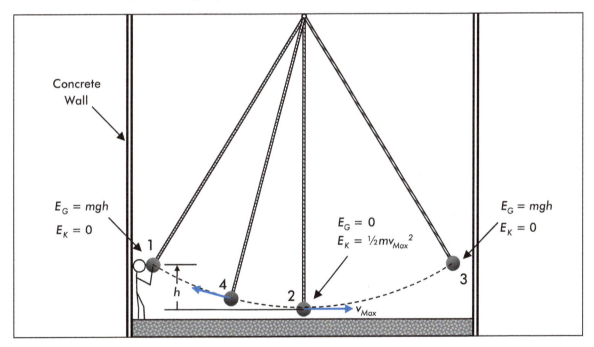

While it moved away, he'd explain how the bowling ball's total energy is constant, so when the ball drops through its swing, it speeds up, and

FIGURE 1 The bowling-ball pendulum swings away from the professor's head, then comes back toward it. The professor's skull is not at any risk, because the ball cannot swing higher than the release point without energy somehow being added to it.

Note 2: Usually ideal gas laws and heat transfer are covered first, as well as plenty of math. But the second law is crucial to understanding energy usage, and not including it would be irresponsible on my part.

Note 3: One that does a very good job is *Thermodynamics: An Engineering Approach*, by Çengel and Boles, 2007, on pages 311–313 and 352–356.

when it climbs it slows down. It trades potential energy for kinetic energy, and vice versa, as it passes through positions 2, 3, and 4.

By the time the ball returned to Position 1, it had again traded all of its kinetic energy for potential energy. Having no kinetic energy left, it stopped moving just when it seemed it would hit his nose (I sat too far back to see if he blinked).

I thought this was a great way of "practicing what he preached," because if the conservation of energy weren't reliable, the bowling ball might have smacked his face. But it always holds, so he could have done it a million times and not worried.

3   Looping the Loop

A good example of simplifying with energy principles is a loop-the-loop. Suppose a Hot Wheels car is released at point 1 in FIGURE 2. If the loop has a radius of 6 inches, what is the minimum starting height for the car to stay in contact with the track throughout the loop? Let's assume no friction, no flex in the track, and ignore the kinetic energy in the car's rotation.

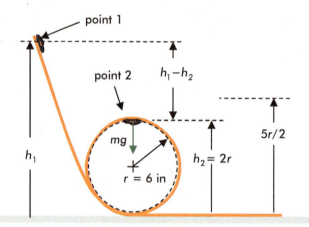

FIGURE 2 There is a minimum height the Hot Wheels car must drop to make the loop without falling off the track. Using energy equations makes it easy to find.

Since the loop is round (constant radius), the car would lose contact most easily at the top (point 2). This is the slowest part of the loop (the lowest speed), and gravity is trying to pull the car directly away from the track. At point 1, the total energy of the car, from potential and kinetic energy, is the same as at point 2. Using equation 12-12b, and filling in the values:

$$E_{G,2} + E_{K,2} = E_{G,1} + E_{K,1}$$

$$mgh_2 + \frac{1}{2}mv_2{}^2 = mgh_1 + 0$$

This can be solved for the speed at point 2:

Equation 13-1   **Speed after frictionless descent**

$$v_2 = \sqrt{2g\left(h_1 - h_2\right)}$$

Note that this speed is the same as for a free-falling object, as in equation 12-5, even though the free-falling object is going straight down, and our car is moving horizontally.

Next we find the speed v_2 it takes to *just* keep the car touching the track, using Newton's second law. This is like cresting a hill (Chapter 11), except we are on the *inside* of the vertical curve. So the normal force from the track at point 2 pushes straight *down* on the car (higher speeds increase the normal force). Together, the normal force and gravity cause the centripetal acceleration to follow the loop. Since up is positive:

$$-F_N - mg = -\frac{mv_2^{\,2}}{r}$$

We want the speed v_2 to be barely enough to keep contact with the track, so the normal force F_N is zero. Setting it to zero, cancelling the mass terms, and rearranging:

$$\frac{v_2^{\,2}}{r} = g$$

Solving this gives us our minimum speed at point 2:

Equation 13-2   **Minimum speed allowable at point 2**

$$v_2 = \sqrt{gr}$$

Along with the results from the energy balance in equation 13-1, we have two answers for the speed at point 2, so they had better be equal:

$$\sqrt{gr} = \sqrt{2g\left(h_1 - h_2\right)}$$

This means the minimum drop from point 1 to point 2 must be:

Equation 13-3   **Minimum drop to make the loop**

$$\left(h_1 - h_2\right) = \frac{r}{2}$$

With a track radius r of 6 inches, the drop is 3 inches. Add that to the 12-inch diameter of the loop, and the minimum starting height is 15 inches from the bottom of the loop.

In general, this says that you could start from a height of ¾ of the diameter of the loop (½ of the radius, as in the figure). From experience, this looks too low, so I checked the answer several times. But we neglected the fact that some energy is diverted to friction, track flex, and the car's rotational motion. In reality, you would have to start off higher to make up for these.

4

Levers: Getting Something for Nothing?

We move things with levers all the time, whether in using a jack, a gear shift, a clutch linkage, or a pry bar. The idea of most levers is to multiply the force you apply, to make it easier to move another object. But if you push on one side of a lever with 100 lbf, and the other side lifts a 200-lbf weight, are you getting more work out than you put in? Wouldn't that violate the first law of thermodynamics?

If you push down with force F_1 on the left arm of the lever in **FIGURE 3**, the right side pushes up on the box with a force F_2 that depends on the ratio of the lever-arm lengths, L_1 and L_2. Assuming that the lever itself is balanced, or light enough not to matter, we then get:

Physical Law Eq. 13-4   **Rigid, frictionless lever**

$$F_2 = F_1 \left(\frac{L_1}{L_2} \right)$$

FIGURE 3 A diagram of a lever, set up to lift a weight *mg* by applying a force F_1. The applied force is smaller than the lifted weight, by the lever ratio.

This comes from the fact that without any acceleration of the lever, the CCW torque F_1L_1 on point C equals the CW torque F_2L_2. The leverage ratio L_1/L_2 is also called the **MECHANICAL ADVANTAGE**. You can see that lifting a weight of 200 lbf with a 100-lbf input force takes a leverage ratio of 2, so L_1 must be $2L_2$.

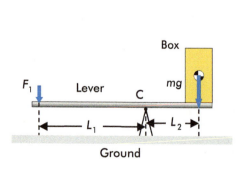

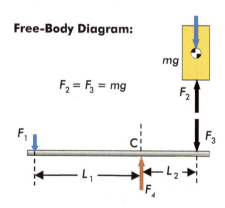

What about work in vs. out? The output work is easy. Assuming the weight is lifted 6 inches (0.5 feet), the 200-lbf force does 100 ft-lbf of work. On the input, we know the force, but need to find the distance moved. From **FIGURE 3** and **FIGURE 4**, you can show that the distance moved on each side is proportional to the lever length, so:

$$\frac{d_1}{d_2} = \frac{L_1}{L_2}$$

Since the input arm length is twice the output arm length, the input force moves twice as far as the output (that was probably obvious). The input distance (**FIGURE 4**) is then 1 foot.

Lever Motion and Force Ratios

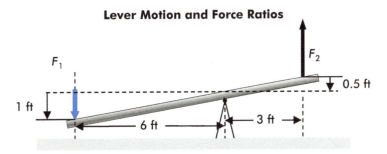

FIGURE 4 The end with the smaller force moves farther than the side with the larger force. Barring friction, this makes the work in equal the work out. So it all works out.

That makes the input work 100 lbf × 1 foot = 100 ft-lbf, exactly the same as the output work. This must always be true for any frictionless, rigid lever:

$$F_1 d_1 = F_2 d_2$$

so:

$$W_1 = W_2$$

The moral of the story is that (big surprise) you don't get something for nothing. A lever that doubles the lifting force cuts the lifting distance by half; you trade distance for force. That's a trade you're quite willing to make if it means popping that rusted ball joint free from the steering knuckle. If there is any friction in the pivot, you do lose some mechanical advantage because part of the input work is done against friction.

5   Work Input and Output in Gear Reduction Sets

In Chapter 9 we looked at a gear train that multiplied the torque input, but in doing so reduced the output speed. We called this a gear reduction set. Based on the lever problem, can you guess the relationship between the input and output work (**FIGURE 5**), neglecting friction?

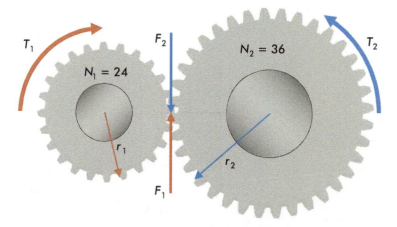

FIGURE 5 An external gear set, with a 1.5 gear ratio. The torque is multiplied by 1.5, but the output rotation angle is only ⅔ of the input. Work in equals work out.

We know from Chapter 9 that the output torque is the input torque multiplied by the ratio of the gears' tooth counts:

Equation 9-17

$$T_2 = -\frac{N_2}{N_1} T_1$$

Also, the output angle is the input angle times the inverse of the tooth count ratio:

Equation 9-14

$$\theta_2 = -\frac{N_1}{N_2} \theta_1$$

We can use these to calculate the output work vs. input work:

$$W_2 = T_2 \theta_2 = -\frac{N_2}{N_1} T_1 \times \left(-\frac{N_1}{N_2} \theta_1 \right) = T_1 \theta_1 = W_1$$

The larger gear carries a larger torque, but turns a smaller angle, so that the work in equals the work out. The first law of thermodynamics says the same for any pure mechanism.

KEY CONCEPT

Pure mechanisms never create energy, they only transfer it.

6 | Work Done by Centripetal Force

A question on my exam went something like "How much work is done by the centripetal force on a car that weighs 3,220 lbf, as it circles a 200-foot skid pad once, at 40 mph?"[4] We say that work is force times distance. You can calculate the centripetal force (see **FIGURE 6**), and the distance traveled is the circumference of the track. But the work should be zero, because the car could coast around the skid pad and make it back to the beginning point at the original speed (assuming no drag). Common sense said the work *had* to be zero, but why?

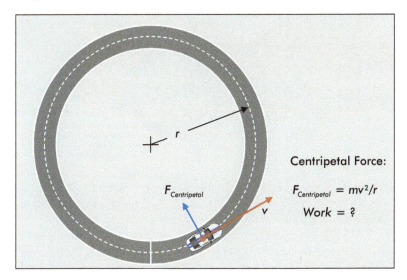

Centripetal Force:

$$F_{Centripetal} = mv^2/r$$

$$Work = ?$$

FIGURE 6 How much work does the centripetal force do in one lap around a skid pad? Remember, work depends on the force applied in the direction of motion.

The detail I forgot was that work is equal to distance times the component of the force that is in the same direction as the motion. Here, there is *no force in the direction of motion*. It is always perpendicular to the velocity, as the figure shows. This means you could (theoretically) corner without losing mechanical energy. In reality some is lost to heat in the curves, especially in hard cornering with large slip angles.

7 | Braking Heat

Let's work an example, first using Newton's laws, and then using only thermodynamics. Let's say we have a 3,630 lbf Porsche 911 with sport tires, giving it a 60–0 braking distance of 100 feet. How much heat will its brakes produce in braking from 60 to 30 mph (**FIGURE 7**)?

Note 4: This question sticks in my head because I missed it on the exam. It was especially galling to me because I knew the answer I put down was going to be wrong, since the result didn't make sense.

7.1 Solution Using Newton's Laws

In detail, this solution took about a page and a half to cover. Rather than drag it out, here's a summary. We assume the braking deceleration is constant, and find it using equation 2-18b. Knowing that $v_0 = 60$ mph $= 88$ ft/sec, and $s_F = 100$ feet, we find that $a = -38.72$ ft/sec^2 (or -1.2 g's). The car's weight of 3,630 lbf makes its mass 112.73 slugs. The braking force is then $-4,364.9$ lbf.

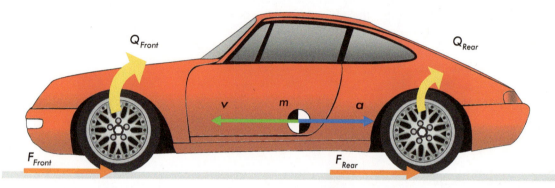

FIGURE 7 A Porsche 911 braking from 60 to 30 mph. As the brakes create friction, they convert kinetic energy to heat Q_{Front} at the front wheels and Q_{Rear} at the rear, for a total heat generation Q.

The kinetic energy converted to heat is equal to the negative work done on the car by the brakes. To find this we need to know how far it went while braking from 60 mph to 30 mph. With the acceleration of -38.72 ft/sec each second, the time to brake from 88 ft/sec to 44 ft/sec is 1.13636 seconds. Knowing the time, we can use equation 2-14 to find that the distance traveled is 75 feet. The work done is then simply force ($-4,364.9$) times the distance (75 feet), giving us the work done on the car, $-327,400$ ft-lbf.

All the work done by the brake friction goes to heat Q, so $Q = W = -327,400$ ft-lbf of heat produced. Note that the *heat added* is negative, the heat being *rejected* by the brakes.

7.2 Solution Using Only Thermodynamics

Taking the entire car as the thermodynamic system, the only thing crossing the system boundary is the heat Q shed by the brakes. We use equation 12-12b for the energy balance:

$$Q + E_{K,1} = E_{K,2}$$

$$Q + \frac{1}{2}mv_1^2 = \frac{1}{2}mv_2^2$$

Then the heat must be the difference in kinetic energy while braking:

Equation 13-5   **Heat from braking**

$$Q = \frac{1}{2}m\left(v_2^2 - v_1^2\right)$$

Again, Q is the heat fed *into* the car. With the initial speed of 88 ft/sec and the final speed of 44 ft/sec, this gives us the same answer as before:

$$Q = \frac{1}{2} \times 112.73 \text{ slugs} \times \left(\left(44 \text{ ft/sec}\right)^2 - \left(88 \text{ ft/sec}\right)^2 \right) = -327,400 \text{ ft-lbf}$$

Notice that we only needed the car's mass, and the beginning and ending speeds. We didn't need to know how hard it braked, or whether the braking force was steady or not. All that matters is that the braking is done quickly enough so that aerodynamic drag and rolling resistance don't do much of the negative work on the car.

If you use equation 13-5 to calculate the heat given off in braking to a full stop, you'll find it to be 436,500 ft-lbf. Braking to half speed dissipates ¾ of that. What would this mean in road racing? Most of the braking heat comes from the first part of braking, because that's when your kinetic energy changes fastest vs. speed. A little more power, producing a little more top speed, means a lot more heat to dissipate from the brakes.

NOTE

F1 cars currently use an energy recovery system (ERS), which includes an electric motor/generator that charges the battery with the braking energy, and returns much of it on acceleration. Production hybrid-electric vehicles do the same, which is most effective in the frequent braking of stop-and-go city driving.

8 | Perpetual Motion Machines (Or Not)

At Kansas State, I took a couple of thermodynamics classes by Professor Crank.[5] Once he told us about a letter he received from an inventor who had an idea for an electric car that used no energy, at least once the car was rolling. An electric motor would drive the rear axle, and the front axle would drive an electrical generator that then powered the electric motor at the rear (see **FIGURE 8**).

Note 5: Is that the perfect name for a professor in mechanical engineering or what?

Cables

Motor

Generator

T_{Rear} ω

W

T_{Front} ω

FIGURE 8 The inventor's electric car. The rear axle is powered by the electric motor, which is in turn powered by the generator on the front axle. This system cannot propel itself, because it violates the first law of thermodynamics.

My professor answered him, explaining that since energy can neither be created nor destroyed, the current from the front generator (even ideally) could only produce enough work at the rear motor to offset the work to drive the generator. There would be no energy left to push against vehicle drag, so you couldn't even hold a constant speed. Worse, since there *will* be friction and electrical resistance in the motor/generator system, the system would actually slow the car down quicker than if it were coasting!

Prof. Crank soon received another letter from the inventor, who had a "new" idea for his electric car. It still had an electric motor at the rear, but a "small but powerful" generator at the front. "Small" evidently meant that the generator took little energy to drive, while "powerful" meant it put out a lot. Crank answered that a generator can't produce more energy than the energy put into it, etc. After receiving a third letter, he gave up (referring the inventor to the department head).

Our professor knew the invention couldn't work before seeing any details because nature, by the first law of thermodynamics, doesn't allow systems to produce more energy than they consume. This would be called a *perpetual motion machine*, since it could produce energy forever without energy input. It sounds good, but no one has ever done it, and I would bet our house that no one ever will. The U.S. Patent Office won't even accept patent applications for any invention claiming perpetual motion, without a working demonstration.

We shouldn't be too harsh on the would-be inventor, though.[6] Most people don't fully understand electricity or know thermodynamics. There is a tendency to recognize that an electric motor can push a car down the road (providing mechanical energy), but to forget that it took that much mechanical energy (actually more) to create it in the first place. Again, nature doesn't allow you to "get something for nothing" when it comes to energy.

Note 6: Many good, creative inventors have done their work solely by intuition. But a good background in physics would help them avoid many dead ends. It helps to know the rules if you're going to play the game.

9   Kinetic Energy Is Not Momentum!

Although we aren't doing much with momentum in this book, we will need it in the next section, and defining it here will head off a potential point of confusion. Besides, momentum fits in well with our systems viewpoint of looking at the "overall situation" rather than minute details.

Kinetic energy increases with mass and speed, so the heavier the car and the faster it travels, the more kinetic energy it has. When I first learned this, I figured it was the same thing as momentum. But they are two different things.

MOMENTUM is defined as an object's mass multiplied by its velocity, or $m\mathbf{v}$. Note that since velocity is a vector, momentum is a vector along the same line.

So while kinetic energy ($E_K = \frac{1}{2}mv^2$) grows with the square of speed, momentum grows linearly. A 3,220-pound car (mass of 100 slugs) going 60 mph (88 ft/sec) has kinetic energy of 387,200 ft-lbf, and momentum of 8,800 slug-ft/sec. Doubling its speed to 120 mph *quadruples* its kinetic energy to 1,548,800 ft-lbf, but only *doubles* its momentum, to 17,600 slug-ft/sec.

The second difference is that momentum depends on the *direction* of the velocity. Assuming east is positive, our 3,220-pound car going east at 60 mph has momentum of +8,800 slug-ft/sec; going west it would be −8,800 slug-ft/sec. But its kinetic energy is 387,200 ft-lbf either way.

Using momentum gives us another way to summarize the effect of a force accelerating a mass. If we apply a constant *resultant* force F to a body with mass m, it has a constant acceleration $a = F/m$. After a certain time Δt, its change in velocity is:

Physical Law Eq. 13-6a   **Constant resultant force**

$$\Delta v = \frac{F}{m}\Delta t$$

or:

Physical Law Eq. 13-6b   **Constant resultant force**

$$m\Delta v = F\Delta t$$

We call $F\Delta t$, a force applied over a period of time, an "impulse." The word comes from the Latin *impulsus*, meaning a pushing against. Even though we assumed a constant known force and acceleration in deriving this, the resulting impulse doesn't actually require that. You could double the force and cut the time in half, and have the same impulse.

Or you could vary the force while it's applied, as long as *its average* multiplied by the impulse time is the same:

Fundamental Law Eq. 13-7   **Variable resultant force**

$$m\Delta v = F_{Ave}\Delta t$$

What good is this? It gives us the ability to work with incomplete information. For example, when an object hits a solid wall and stops, we know the mass m and the velocity change Δv. If we measure the duration Δt of the collision, we can find the average impact force F_{Ave} using equation 13-7. Sometimes we don't even need to know the average force or the impulse duration to get good out of this, as we'll soon see.

> **KEY CONCEPT**
>
> Let's assume a constant force F. Then an impulse $F\Delta t$, the force applied over a certain *time*, causes a change in momentum $m\Delta v$. But the work $F\Delta s$, the force applied over a certain *distance*, causes a change in kinetic energy of $\frac{1}{2}m(v^2 - v_0^2)$.

Combining momentum and energy concepts is very useful, especially when objects collide. We'll make use of that next.

10 Converting Work to Thermal Energy, and Vice Versa

We can easily use the energy balance equations from Chapter 12 to show that when a gas is compressed in a cylinder, the work done on the gas by the piston must equal the increase in its internal (thermal) energy—it heats up. But the *mechanism* of this process is probably less clear. *How* is the work converted to internal energy? When hot gases expand and push the piston, *how* does the internal energy of the gas produce work?

Here we'll examine both compression and expansion, assuming that the piston, cylinder wall, and cylinder head are all perfectly insulated, to make sure no thermal energy in the gas is lost to them as heat. The air will be our closed system.

This may seem quite a leap for us, since we haven't even covered gases in any depth. But we can do this with mechanics alone. I think it's important to show that there is more than one way to solve a problem, and that thermodynamics and Newton's laws agree.

10.1 A Full Cylinder of Air at Constant Volume

We'll start with a piston in a round cylinder, at BDC (bottom dead center), at the end of the intake stroke. We'll assume that the cylinder has been filled with air through a wide-open throttle, so the air is at atmospheric pressure (14.7 psi or 101.3 kPa) at say, 100° F (37.8° C). The cylinder (see FIGURE 9) has a flat head at one end (valves not shown), and a flat piston at the other. We'll "hold" the piston at BDC for now, so we can look closer at how a gas produces pressure.

When air is trapped in a cylinder, its particles (molecules of nitrogen, oxygen, argon, etc.) randomly bounce around, hitting the cylinder walls, the piston, and each other (see FIGURE 9). Each of the billions of collisions per second between these molecules and the piston create a small impulsive force on its surface, for a short time; that is, an impulse $F\Delta t$. Add their effects together and you would get the average pressure force on the piston ($F_{Press} = pA$).

Remember also that the kinetic energy of the gas particles as they randomly fly around forms the internal energy of the gas, each particle having a kinetic energy E_K of $\frac{1}{2}mv^2$. This normally includes the rotational energy of the molecules spinning. To simplify, FIGURE 9 shows no rotational motion, which is really only valid for a noble gas like argon.

The absolute temperature of the gas is proportional to this internal energy, meaning it increases with the square of the particle speeds (more specifically, average of the squared particle speeds). Absolute temperature is measured from where particle motion theoretically stops, which is about −460° F, or −273° C. So 77° F is an absolute temperature of 537 degrees (called degrees Rankine), and 25° C is an absolute temperature of 298 degrees (called degrees Kelvin).

KEY CONCEPT

The internal energy U of any ideal gas is the total kinetic energy of its particles (generally including translational and rotational motion). The pressure depends on the number of collisions the particles have with the container per second, and the impulse of each collision. So internal energy and pressure both change with particle speed. This will get us to our solution.

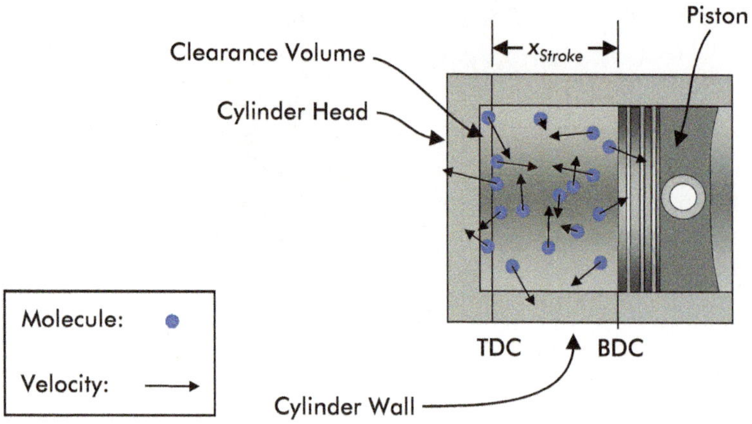

FIGURE 9 In an actual gas, the molecules in the cylinder travel in all three dimensions, banging against the cylinder walls, the head, and the piston. These impacts produce the pressure on each of these surfaces.

Because of the randomness in the particles' directions and speeds, calculating the pressure force on the piston from the impacts in **FIGURE 9** is more complicated than necessary to understand the cause of the pressure. Instead, we'll use a "fictional" gas that is much simpler.

10.2 A Full Cylinder of 1-D Gas, at Constant Volume

To simplify our problem, we'll ignore the particles' rotations. Further, we'll only allow them to travel along lines parallel to the cylinder centerline (hitting the piston and head square-on), making the motion one-dimensional. Then, instead of randomly varying particle speeds, we'll give them all the same speed v_1. With that, we may as well join all the molecules into *one* particle with mass m.

The resulting single-particle gas, moving in one dimension, is shown in **FIGURE 10**. Its motion is like when you bounce a Ping-Pong ball off a table, and then hold a paddle over it, forcing the ball to bounce back and forth between the table and paddle.

Despite the extreme simplification, this 1-D gas keeps the main features of a 3-D ideal gas in a sealed and perfectly insulated enclosure:

1. The collisions are elastic, so the particle's departure speed after rebound is the same as its approach speed.
2. The average pressure force is caused by its impacts with the piston and head.
3. The internal energy of the gas is contained in the kinetic energy of the particle.

An *elastic collision* is one where no mechanical energy is lost. This makes the relative speed between the colliding objects after the collision equal to that before the collision.

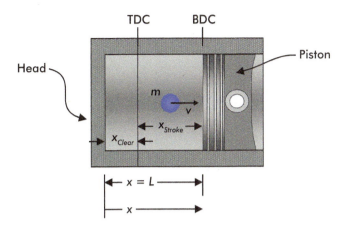

FIGURE 10 With our simplified "gas," the mass of the trapped air is concentrated in a single particle. It "rattles" back and forth between the piston and head, so that its impacts produce pressure, and the resulting average force, on the piston. The piston position is called x, which equals L at the start of compression (BDC). At the end (TDC), the piston has moved left by x_{Stroke}, and $x = x_{Clear}$.

Let's get specific by assuming a 4-inch cylinder bore (0.333 feet), and a 2.75-inch stroke (0.22917 feet). Then each cylinder has a swept volume (volumetric displacement) of 34.558 in³ (or 0.020 ft³). Note that x_{Clear} is the "clearance distance" between the piston and head at TDC.

The pressure in this 1-D gas rises faster during compression than real 3-D air does, so instead of a typical 10:1 compression ratio, we'll set it to 3:1 (the volume at BDC is 3 times that at TDC), to get a realistic pressure after compression.

The known values at BDC and TDC are shown in **TABLE 1**. Note that to get the 3:1 compression ratio, the total volume at BDC is 0.030 ft³, and the volume at TDC is 0.010 ft³. The density of air at BDC (at 100° F and 14.7 psi) is about 0.0709 lbm per cubic foot, giving a mass of 0.0021258 lbm (about 0.964 grams). This is about ⅓ the mass of a Ping-Pong ball, which "weighs" 2.7 grams. This mass stays trapped in the cylinder.

Piston Position	Cylinder Volume (in³)	Cylinder Volume (ft³)	Air Mass (lbm)	Air Mass (slugs)	Pressure (lbf/in²)	Temperature (°F)
BDC	51.837	0.0300	0.002126	0.000066	14.7	100
TDC	17.279	0.0100	0.002126	0.000066	?	?

TABLE 1 Gas properties before compression (at BDC) and after (at TDC).

With a 4-inch bore, the piston face area is 12.566 in². Multiplying by the initial pressure of 14.7 lbf/in² produces a piston force of 184.7 lbf. We want to match this force with our 1-D gas.

So let's "watch" our 1-D gas as it impacts the *stationary* piston, and see what happens (see **FIGURE 11**). It approaches the piston with velocity v_1 and impacts it. With an elastic collision, the particle bounces back with the same speed v_1 as before the impact, but its velocity is now $-v_1$. Its velocity has changed by $-2v_1$.

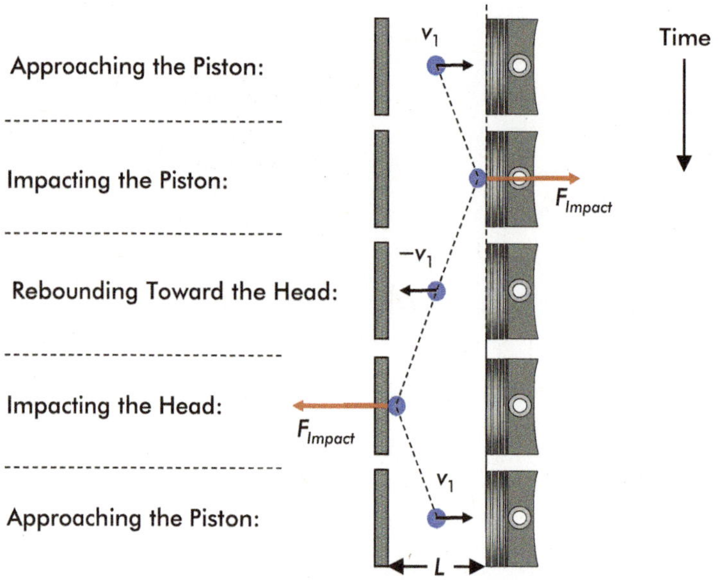

FIGURE 11 At BDC, the gas particle bounces back and forth a distance *L* between the piston and cylinder head, creating an impulse during each impact.

From Section 9, we know that the change in momentum ($m\Delta v$) is equal to the impulse (here $F_{Impact}\Delta t_{Impact}$) during the contact. We don't know F_{Impact} or Δt_{Impact}, but we know $F_{Impact}\Delta t_{Impact}$ because we know $m\Delta v$. These impulses are plotted vs. time in **FIGURE 12**.

Knowing that the particle velocity changes by $-2v_1$ and that there is a time of $2L/v_1$ between impacts (you can verify this), the "pressure force" F_{Press} on the piston at BDC is

Equation 13-8   **Pressure force at BDC**

$$F_{Press,1} = \frac{mv_1^2}{L}$$

It doesn't get much simpler than that. Now, since we know the pressure force, the particle mass, and the distance *L* from the piston to head, we can find the necessary particle speed v_1 to make it all fit together.

NOTE

The physical explanation is the main thing to take away from this discussion. The gas we'll use shows the right trends, but is too simplified to predict performance of a real engine. The equations, derived in Appendix 5, will be written in red, meaning *they should not be used elsewhere.*

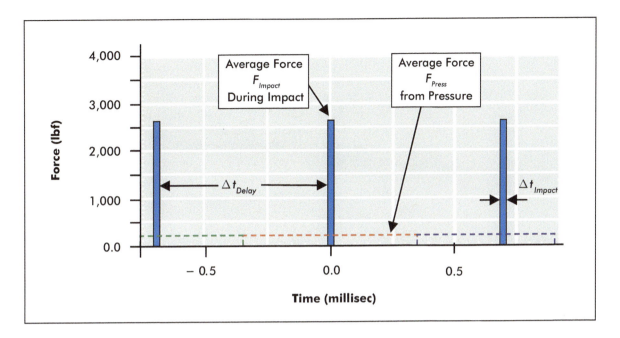

Solving equation 13-8 for the particle speed, and with a piston force of 184.7 lbf, the particle mass of 0.000066 slugs, and cylinder length L at BDC of 0.34375 feet (4.125 inches):

$$v_1 = \sqrt{\frac{F_{Press,1}L}{m}} = \sqrt{\frac{(184.7 \text{ lbf}) \times (0.34375 \text{ ft})}{.000066 \text{ slugs}}} = 980 \text{ ft/sec}$$

FIGURE 12 Force on the piston. Each of the repetitive impacts of the particle on the piston produces a large force F_{Impact} for a short time Δt_{Impact} (in blue); an impulse of $F_{Impact}\Delta t_{Impact}$. We can find the resulting average pressure force F_{Press} on the piston by requiring its impulse to be the same: $F_{Press} \Delta t_{Delay} = F_{Impact}\Delta t_{Impact}$.

Not surprisingly, this very light particle has to travel pretty fast to hit hard enough and often enough to average 184.7 lbf of force on the piston. At that speed, there are only 0.00068 seconds between impacts on the piston, so it receives about 1,470 impacts per second.

KEY CONCEPT

The pressure is from the average force on the piston caused by the repetitive impacts of the gas on it. For a given cylinder length, as the particle speed goes up, the impacts become harder and more frequent. So the pressure is proportional to the *square* of the particle speed.

For the record, the gas exerts the same force on the cylinder head.

10.3 Compression of the 1-D Gas

Now that we know how a gas produces pressure on a piston, let's compress it by crowding it into a smaller space. We move the piston to the left and see what happens to the gas particle's speed. We'll start with v_1 at 980 ft/sec, and assume a *piston* velocity to the left at 40 ft/sec. As shown in **Figure 13**, when the particle impacts the moving piston, it bounces back to the left at the same speed, *relative to the piston*. Since the relative approach speed is 1,020 ft/sec, the departure speed is also 1,020 ft/sec. But this means the actual particle speed after the first impact is 1,060 ft/sec; it sped up by 80 ft/sec, *twice* the piston speed.[7]

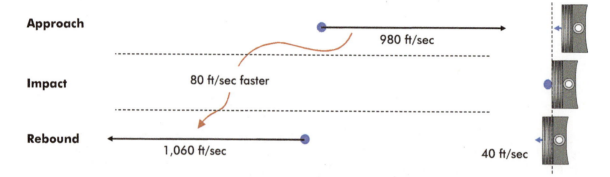

Figure 13 With a piston velocity of −40 ft/sec and a gas particle velocity of 980 ft/sec (top), the particle's approach speed is 1,020 ft/sec. After impact its velocity is −1,060 ft/sec, giving it a departure speed of 1,020 ft/sec (bottom). The diagram is expanded so you can see some piston motion.

Next, the particle bounces off the cylinder head. The next time it hits the piston, it again increases its speed by 80 ft/sec. So after two impacts, the velocity is 1,140 ft/sec, then 1,220 ft/sec, etc. Because of the higher impact speed, each impulse on the piston becomes larger (see **Figure 14**). But you can see that the impulses also become more frequent, both because the particle speed has increased *and* because the distance traveled between impacts has decreased (the piston getting closer to the cylinder head).

KEY CONCEPT

During compression, the force (from pressure) on the piston keeps increasing. Since the gas particle here is speeding up, its kinetic energy (the internal energy for our 1-D gas) also increases, and its temperature increases. Meanwhile the piston, being in motion, does work on the particle during each impact.

Note 7: We are assuming a force is acting on the back side of the piston to keep it at a constant velocity, despite the impacts.

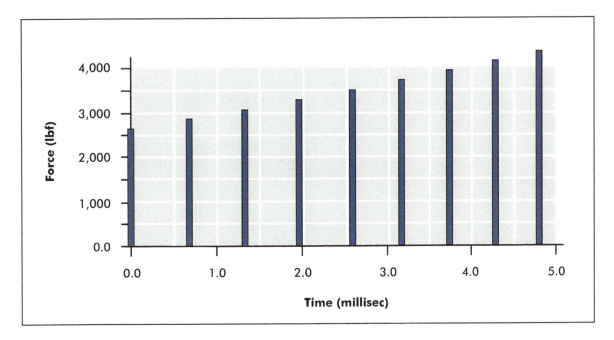

These individual impacts are good to illustrate the pressure mechanism, but we want a smooth relationship between the piston position and the pressure force. By now assuming the piston moves very slowly compared to the gas particle, the force from pressure at any piston position x can be found as

FIGURE 14 During compression, the particle's impulses on the piston get larger and closer together, both factors increasing the pressure and force on the piston. Note that Δt_{Impact} is assumed constant.

Equation 13-9   **1-D gas pressure**

$$F_{Press} = \frac{mv_1^2}{L}\left(\frac{L}{x}\right)^3 = F_{Press,1}\left(\frac{L}{x}\right)^3$$

Remember that x is the distance from the cylinder head to the piston at any time. So at BDC, $x = L$, and at TDC, $x = x_{Clear}$ (refer back to **FIGURE 10**). So equation 13-9 says that as the piston moves closer to the head (smaller x), the pressure force increases, quickly.

FIGURE 15 uses this to plot the force during the compression stroke, starting at point 1 (BDC). Note how the force increases more and more steeply as the piston nears TDC, because each fraction of an inch of motion reduces the volume by a higher percentage than the previous one.

At TDC, the piston position is x_{Clear}. Putting this into equation 13-9, and comparing the result to equation 13-8, the force after compression is just our initial force multiplied by the cube of the compression ratio. Our pressure force after compression is then 4,990 lbf, 27 times greater than before! The corresponding pressure would also be 27 times higher, at 397 psi.

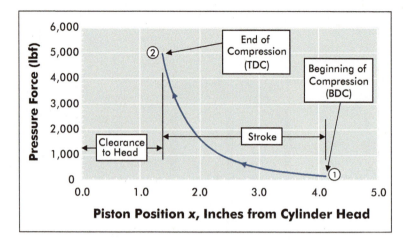

During compression (from Position 1 to 2), the cylinder volume decreases by a factor of 3, and the force from this 1-D gas pressure increases by a ratio of 27 (a much higher ratio than for real gases).

Now that we know how the piston force varies during compression, we can calculate the compression work done on the gas:

Equation 13-10   **1-D Gas**

$$W_{Comp} = \frac{1}{2}mv_1{}^2\left(\left(\frac{L}{x_{Clear}}\right)^2 - 1\right) = U_1\left(R_{Comp}{}^2 - 1\right)$$

So for the compression work, we only need to know the initial internal energy U_1 and the compression ratio R_{Comp}. Remember that U_1 is the particle's kinetic $\frac{1}{2}mv_1{}^2$ at BDC, which comes to 31.7 ft-lbf.

For our engine, this says compression takes 253.6 ft-lbf of work. If we calculate the particle speed after compression (adapting 13-8, we can find v_2 as the square root of $F_{Press,2}x_{Clear}/m$, or about 2,940 ft/sec) we can get the internal energy U_2 at TDC, $\frac{1}{2}mv_2{}^2$, or 285.3 ft-lbf. The increase $U_2 - U_1$ of 253.6 ft-lbf is the same as the work done on the gas, just as the first law requires.

KEY CONCEPT

When a trapped gas is compressed without heat loss, the increase in its internal energy is exactly the same as the work done on it. And note that by compressing it, we increased the gas's internal energy (and temperature) without transferring heat to it, only work.

Since the absolute temperature is proportional to the square of the particle speed, which has tripled, we would find it to be 9 times higher than before compression (again, a larger increase than for a real gas at the same compression ratio).

10.4 Summary: An Air Spring and Dead Cylinders

If we started at Position 2 and let the piston move to the right, the gas would expand while pushing on the piston. The force on the piston would drop, being the same at each position as it was during compression (just reverse the blue arrows in **FIGURE 15**).

What we have with this gas-filled cylinder is an air spring. We increased the (internal) energy of the gas when the piston did the compression work. The gas returned the favor by reducing its internal energy and doing work on the piston during expansion. Because the cylinder was insulated and no gas leaked out, we recovered the compression work completely: the expansion work $W_{Exp} = W_{Comp}$. Plus, we ended up with the same piston force as when we started. We say the compression process was *reversible*.

> **CAUTION**
>
> Again, while the qualities of this 1-D gas are the same as in a real 3-D gas, the details are different, and the equations used here would need to be modified.

Why is this all significant? First, many engines today deactivate some cylinders when their power is not needed (at cruise, for example).[8] The valves in the "dead" cylinders are kept closed, making each cylinder an air spring.[9] And second, we'll work these same ideas into a running engine in Chapter 15.

11   The Second Law of Thermodynamics

The second law of thermodynamics is "the fine print" of the energy accounting rules. Nature not only says energy can't be created or destroyed; it has a lot to say about which direction energy can be transferred, and how efficiently. It's easy to convert mechanical energy 100% into heat (as your brakes do in slowing your car), but

---

Note 8: Some names for this feature are AFM (Active Fuel Management), DOD (Displacement on Demand), VCM (Variable Cylinder Management), MDS (Multi-Displacement System), and ACC (Active Cylinder Control), depending on the manufacturer.

Note 9: You may rightly wonder why shutting down some cylinders would increase engine efficiency, since using fewer cylinders makes the remaining ones work harder. It's because engines using throttles (like gasoline engines) become more efficient as they run closer to wide-open throttle, as less work is done in "drawing" air in past the restrictive throttle; the remaining cylinders run closer to WOT.

engines must always be less than 100% efficient in converting heat into mechanical energy.

Unlike the first law, there are several ways to put the second law, and it's possible that none will strike you as obvious or meaningful on the first take. Let's check out two statements of the second law, and then discuss what it all means. These statements are not different laws, just two different ways of saying the same thing.

FUNDAMENTAL LAW

The Second Law of Thermodynamics (Clausius Statement):

No process is possible whose *only* result is the transfer of heat from a body of lower temperature to a body of higher temperature.

In layman's terms, the Clausius statement says that, on its own, heat always goes from a hot area to a cold one. This shouldn't come as a shock, and probably seems trivial. If you put a 180° F cup of coffee and an 80° F cup in a cooler, then came back and they were 200° and 60°, that *would* be a shock! Even so, the *first* law would allow it, because the total energy would still be the same. The second law is what drives processes in the right *direction*.

Notice that although air conditioners do ultimately move heat from cold to hot, they require the input of work to drive the A/C compressor—the heat doesn't move without being forced to. That violates the "only result" part; somewhere else (such as an electric plant or a car's engine), heat was used to create that work, and some heat went from a hot area to cold in that process. Now let's look at another way to put the second law.

FUNDAMENTAL LAW

The Second Law of Thermodynamics (Kelvin-Planck Statement):

No process is possible in which the *only* results are the absorption of heat from a single reservoir and its complete conversion into work.

The Kelvin-Planck statement is less obvious, but comes straight from the Clausius statement.[10] It just says that we cannot take heat (such as that released by a fuel) and convert it completely to work. Again, the first law would be okay with it, but any effort to get 100% efficiency is doomed from the start.[11] This roadblock applies to all heat engines,

---

Note 10: Clausius, Kelvin, and Planck were scientists who made major contributions in forming the second law.

Note 11: For converting heat to work, the first law has been paraphrased as "You can never get ahead, you can only break even," and the second law as "You can never break even."

whether the internal combustion engine in your car, the boiler and steam turbine at the power plant, or a jet engine.

Carnot found that to maximize engine efficiency, it must use "reversible processes." The compression of the gas in our insulated cylinder was one example of a reversible process. It's only because of reversibility (no heat loss, no leakage) that the work of compression was equal to the work of expansion, and the temperature and pressure ended up where they started.[12]

Like the first law, the second law affects everything you do. It is what says that when coasting on a level road, you will eventually stop, as your kinetic energy bleeds off to heat. Whether from rolling resistance, aerodynamic drag, or braking friction, you will stop. But the second law isn't always bad. When sky diving, you are counting on the second law to convert your potential energy from gravity mgh into heat via your parachute, instead of into kinetic energy $\frac{1}{2}mv^2$ of (literally) terminal velocity before you hit the ground.

11.1 What Causes the Second Law?

You might think that nature has it in for us by having such a frustrating law, but it's nothing personal. It's due to probability. Statistics says that energy will naturally "spread out" into different locations, given the chance. It won't concentrate itself, on the whole, and won't stay concentrated. It tends to even out.

FIGURE 16 shows several examples of this. The first shows two cups of coffee in an insulated cooler that start off 120°F apart in temperature, eventually becoming equal. In the middle example, two containers of oil start off with one level much higher than the other, but open the valve between them and they eventually become the same height. The flow stops once viscous drag dissipates the flow's kinetic energy as heat.

The third example is an important one to us. A car is initially sitting at the top of a hill and is tapped forward just a bit; it coasts down the hill. After going back and forth a few times, it eventually stops at the bottom. At first all the energy in the system is *concentrated* in the car's mechanical energy, as potential energy. As the car coasts, rolling resistance and aerodynamic drag convert its mechanical energy to heat. In the end, all the energy has *spread out* as thermal energy of the surrounding air. We say the energy has *dissipated* (not disappeared).

Note 12: Carnot also found that to maximize the efficiency of a heat engine, the heat should be 1) brought in at a very high temperature, 2) rejected at a very low temperature, and 3) transferred in and out without a temperature difference(!). So although the Carnot engine is a great way to illustrate how the second law affects engine efficiency, no one uses one. Not only are the processes very difficult to achieve, it produces so little work that it might not overcome its own friction.

Coffee in Insulated Box

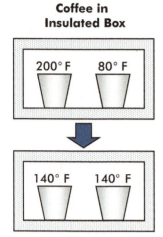

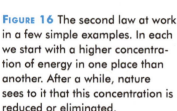

Oil in Connected Containers

Car Coasting Into a Valley

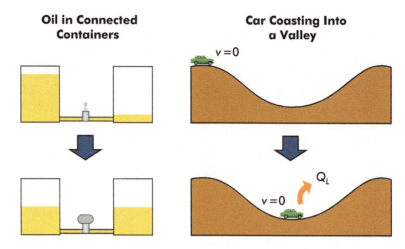

FIGURE 16 The second law at work in a few simple examples. In each we start with a higher concentration of energy in one place than another. After a while, nature sees to it that this concentration is reduced or eliminated.

Where does probability come into all this? Using Newton's laws and statistics, you can't *completely* rule out thermal energy spontaneously doing work; the molecules in the air *could* all start to push the same direction on a car and push it down the road. Statistically, heat *could* become concentrated in one small place on its own.[13]

But (and this is a big but) you can also show that out of the many possibilities, the odds against increasing concentration are so astronomical that they have effectively zero chance of happening. It's less likely for heat to transfer from cold to hot than you winning every Lotto in the world, every week, for the rest of your life (feeling lucky?).

12 | Mechanical Energy, Thermal Energy, and "Energy Quality"

The second law of thermodynamics creates a steep road between mechanical energy and thermal energy. Mechanical energy is easily converted to thermal energy, but thermal energy can only be partially converted to mechanical energy, at best.

These traits are shown in FIGURE 17. On this chart, going "downhill" is easy. For example, converting mechanical energy to heat is like falling off a log.[14] FIGURE 17 reflects this by showing that mechanical energy can be 100% converted to either high- or low-temperature heat (note we're using the term "heat" here for any kind of thermal energy).

Note 13: Similarly, politicians *could* keep all their campaign promises, a chimpanzee with a computer *could* write Shakespeare, and pigs *could* fly.

Note 14: In fact, falling off a log shows how easy it is to completely convert mechanical energy (your gravitational potential energy) into heat, assuming you don't bounce.

It also shows that high-temperature heat can be converted to work, but never at 100% efficiency. While an ideal engine could convert 70%, many practical issues keep it to maybe 30–40%. On the other hand, high-temperature heat can be 100% converted into low-temperature heat, as when burning fuel heats your pistons.

Low-temperature heat gives us the fewest options for use (again, **FIGURE 17**). An example is using the energy in a car's exhaust to drive a turbine and help propel your car. Because the exhaust gas is at a much lower temperature and pressure than when in the cylinder, converting its residual heat to work is relatively inefficient, and the turbine would have to be large. You have to seriously question whether the extra cost and mass is worth utilizing the energy, and the extra mass will work against you by requiring more energy to accelerate the car.

Based on their usefulness, the left side of **FIGURE 17** grades the "quality" of the energy types. Mechanical energy has the highest, high-temperature heat is the next highest, and low-temperature heat the lowest.[15]

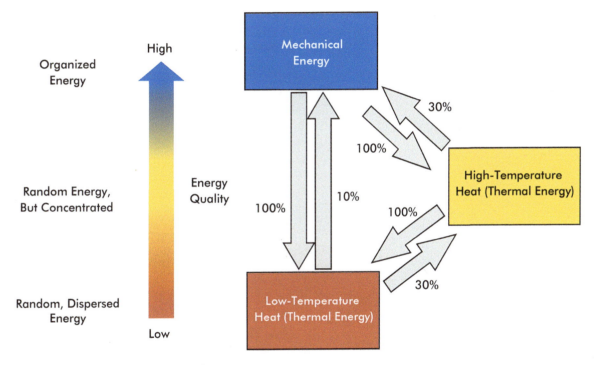

While the "quality" is an important concept in understanding the usefulness of energy, it is not precise. To track this more precisely, we need the concept of entropy.

FIGURE 17 The "quality" of different types of energy. Mechanical is the most useful and highest quality, followed by high-temperature heat, which can at least be partially converted to work. Low-temperature heat is low quality, and has limited use.

Note 15: One of my favorite descriptions of the uselessness of low-temperature heat is from the book *Thinking Physics*, by Lewis C. Epstein, where he calls it "the graveyard of energy." That makes the second law of thermodynamics the "grim reaper."

13   Entropy: A Lack of Concentration

The second law is often explained in terms of **ENTROPY**, which is another word for disorganization or disorder. Unlike energy, entropy is not a household word, so most people have no feel for it. Entropy is a physical measure of the disorganization of a pocket of energy, increasing as energy becomes less concentrated. So it changes in the opposite direction as "energy quality." Mechanical energy has zero entropy, and low-temperature heat has high entropy. Let's look at a dictionary definition:[16]

> Entropy: A measure of the unavailable energy in a closed thermodynamic system so related to the state of the system that a change in the measure varies with change in the ratio of the increment of heat taken in to the absolute temperature at which it is absorbed.

What more can I say? Well, the first part says that entropy is a physical measure of how *unavailable* the energy *in an isolated system* is—for doing work (I know it said "closed," but it meant "isolated"). If one half of a system is hot (a high-temperature *reservoir*) and the other is cold (a low-temperature reservoir), the temperature difference could be used to drive an engine to do work. If that system then has some heat transferred from the hot to the cold reservoir, the total system energy is unchanged. But it has a smaller temperature difference within it, and less ability to drive an engine. Less energy is available to do work, so entropy has increased. If the heat transfer continues, eventually the temperature will be the same throughout the system. It cannot drive an engine at all — no work can be done. Now, with its energy evenly spread out, the system has reached its *maximum entropy*.

The second part tells us how to measure it. If you have a reservoir and it absorbs a little bit of heat, its entropy increase is equal to that amount of heat divided by its absolute temperature (more about that later). So during the heat transfer our cold reservoir receives heat, and its entropy goes up. Meanwhile, the hot reservoir loses heat, and its entropy is reduced, but by a smaller amount. The net entropy increases, just as described above.

Note that a **HEAT RESERVOIR** is any mass of material at a certain temperature, such as a hot oven baking a powder-coated brake caliper. An *infinite reservoir* is a mass of material large enough that transferring heat to it or from it hardly changes its temperature. The air surrounding you outdoors is a good example for most purposes. Low-temperature reservoirs are often called *heat sinks*.

Entropy puts numbers to the "spreading out" of energy, whether from heat moving from a high temperature to a lower one, or mechanical energy being converted to heat.

Note 16: *Webster's Ninth New Collegiate Dictionary*, s.v. "Entropy."

> **KEY CONCEPT**
>
> All processes are *driven by differences*: heat transfer by a difference in temperature; work on a piston by a difference in pressure, electric current by a difference in voltage, etc. Each process is driven in the direction that increases total entropy of an *isolated system*.

By isolated system we mean having a system boundary that separates everything involved in our process from the "outside world." No energy can enter or leave. **FIGURE 16** showed several examples. The coffee cups in the cooler were separated from everything else by the insulated cooler. The oil in the containers could be separated by insulating the room containing them. The car coasting down the hill could be "separated" by considering a large enough box of air around it, so the air would capture any heat from the car.

So let's look back at these processes. Heat was transferred from the hot coffee to the cold coffee, until they were the same temperature; the temperature difference was reduced (actually eliminated); entropy was increased. The oil flowed from the container at the higher level to the other, until they were at the same level; the potential energy turned to heat, and entropy was increased. When the car coasted down the hill and came to a stop, its potential energy was reduced and lost to heat, and entropy increased.

Let's find the entropy increase from that last example, just to show we can (see **FIGURE 18**). Imagine our Trabbi is having starting problems (not too difficult, is it?) and we unsuccessfully try to bump-start it, coming to a stop at the bottom of the 25-foot hill. The entropy definition above says to divide the heat received by a reservoir, by the absolute temperature of the reservoir. Here, the heat from drag and friction is all soon absorbed by the surrounding air, so we just find that heat and divide it by the air temperature. [17]

The change in potential energy from the top of the hill to the bottom is equal to the weight of the car (say, 2,000 lbf) times the change in height:

$$\Delta E_G = mg\Delta h = (2,000 \text{ lbf}) \times (-25 \text{ ft}) = -50,000 \text{ ft-lbf}$$

This change in potential energy would show up as kinetic energy if the car coasted freely, but instead 50,000 ft-lbf of heat was transferred to the air. Let's assume the air has a constant temperature of 75° F, an absolute temperature of 535° Rankine. The entropy change ΔS_{Air} is:

$$\Delta S_{Air} = \frac{Q}{T} = \frac{50,000 \text{ ft-lbf}}{535° \text{ R}} = 93.46 \text{ ft-lbf/°R}$$

Note 17: When the temperature of the reservoir changes during the heat transfer (like either of the two coffee cups), it's more complicated to calculate the change in entropy. Same concept, different details.

The entropy of the car didn't change, because it ended up at the same temperature as it started. So entropy increases by a net 93.5 ft-lbf/°R. The next question is what that means.

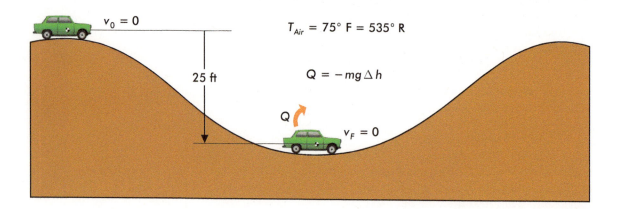

$$v_0 = 0$$

$$T_{Air} = 75°\ F = 535°\ R$$

25 ft

$$Q = -mg\,\Delta h$$

$$Q$$

$$v_F = 0$$

FIGURE 18 Our 2,000-pound Trabant coasts from the top of a 25-foot hill to the bottom and comes to a stop at the bottom, shedding frictional heat in the process. How much is entropy increased?

Remember that an increase in entropy means that we've either lost mechanical energy to heat (as in this example), or lost some capability to produce work. For example, burning a gallon of gasoline in open air heats the surroundings, but does no work. All the heat increases the entropy of the air. Burning it in an engine would have "diverted" much of this heat into producing work, released less heat to the air, and increased entropy less.

Note that although the total entropy in a system must increase, the entropy of *part* of a system is often reduced.[18] After the 200°-F coffee transfers a bit of heat to the 80°-F coffee, the temperatures have hardly changed. The hot coffee has lost one unit of heat Q at 660° R, and the colder coffee received that same amount at 540° R. Since entropy is heat received by a reservoir divided by its temperature, the hot coffee's entropy went *down* by a small amount, but the entropy of the cold coffee went *up* more (dividing Q by 660 gives a smaller number than dividing by 540). Heat transfer from hot to cold produces a net entropy increase, and the larger the temperature difference, the larger the increase in entropy.

On the other hand, when we compressed our ideal 1-D gas and let it re-expand, we got all our work back; neither the compression nor the expansion increased entropy. This is only because it was a thermodynamically reversible process — it left no trace that it happened. The same is true when we compress an ideal (elastic) spring. But every time we actually compress a gas, we have heat transfer out the cylinder and leakage. And every real spring has some internal friction. That makes real processes irreversible, so they all increase entropy.

Note 18: This fact causes a lot of inaccurate statements about the implications of the second law, from people who are only slightly familiar with it. So be forewarned about some claims you'll read.

KEY CONCEPT

Only reversible processes, those that could be undone without leaving a trace, create no entropy increase. All real processes are irreversible, so every one increases entropy.

So even though entropy itself is hard to visualize, it provides the book-keeping system to track the second law of thermodynamics. This is just as important as the conservation of energy. But the entropy of an isolated system is not conserved. It always increases:

Fundamental Law Eq. 13-11   Second law of thermodynamics

$$\Delta S_{Isolated} > 0$$

The only exception is if the system is already at its maximum entropy; it's at uniform temperature, pressure, etc. Then it can't increase.

14 Summary

It's common to get energy wrong. Misstatements by the media confuse and mislead their readers. People spend time and money pursuing inventions that have no hope of being successful, or in buying equipment that won't live up to the promised savings or performance.

But thermodynamics shows that energy is pretty logical, and provides you a check on physical processes without knowing their details. So it can slice through otherwise complicated situations like a knife. You can use it to decide for yourself whether a new type of engine is an ingenious use of physics, or just another attempt to cheat it. Be open-minded, but skeptical.

It's also clear that thermodynamics is an essential tool for bringing different types of systems together, like machinery, electricity, and heat; and that it lets you ignore many intermediate processes in a problem, to concentrate on the big picture and the end result.

No matter what, everything that happens must follow the two basic rules we have laid out, the first and second laws of thermodynamics. Overall, the first law can be summarized as the total amount of energy in an *isolated* system being constant:

Fundamental Law Eq. 13-12   First law of thermodynamics

$$\Delta E_{Isolated} = 0$$

With the definition of entropy, we can also summarize the second law as the entropy of an *isolated* system always increasing:

Fundamental Law Eq. 13-11   **Second law of thermodynamics**

$$\Delta S_{Isolated} > 0$$

Everything else follows from these. These formulas are the most general and compact ways to state these laws, but they don't mean anything if you don't understand what's behind them. The second law especially needs background, and maybe some extra thought, to really have it soak in. But it is what makes mechanical energy more versatile than thermal, because nature prefers to turn mechanical energy into heat, rather than the other way around.

On the practical side, the second law frustrates your attempts to increase efficiency in several ways. First off, it limits the efficiency of producing power from heat. It then shows up in mechanical losses in transmitting power, like friction, drag, and electrical resistance. For example, regenerative braking (in F1's ERS, or in hybrid vehicles) might only return 50%–70% of the original kinetic energy, instead of 100%. It also reduces the fuel economy benefit of extra gear ratios in a transmission, by introducing more frictional losses. These things may still be worth doing, but the extra mass and cost need to be weighed carefully. There are few "no brainers" here.

KEY CONCEPT

With every energy conversion or transfer process, you "lose control" of some of the energy.

On the philosophical side, the second law is the only "one way" sign for the passage of time. We never see processes that decrease entropy. Cars don't coast uphill. Running an engine backwards doesn't produce cold air and fill a gas tank. An exploded blower doesn't assemble and install itself on the top of a Funny Car engine. If we watch a video of any of these, we could tell it's being played backwards. And the second law agrees, because backwards, each would decrease entropy, which isn't allowed.

And, since the universe is an isolated system (as far as we know), its entropy is constantly increasing as a whole. Its energy is becoming more spread out and uniform; less and less organized. But organized pockets, like we organisms, can still manage to exist. We are quite exceptional—the more you know about thermodynamics, the more amazing life seems. All of us are bucking the trend of entropy…at least for a time.

Major Formulas

Definition	Equation	Equation Number
Force Ratio for a Lever (The Mechanical Advantage)	$F_2 = F_1\left(\dfrac{L_1}{L_2}\right)$	13-4
Heat from Braking from Speed v_1 to v_2 (Q Positive into Car, so Q Is Negative)	$Q = \dfrac{1}{2}m\left(v_2{}^2 - v_1{}^2\right)$	13-5
Change in Momentum ($m\Delta v$) Equals Impulse ($F_{Ave}\Delta t$)	$m\Delta v = F_{Ave}\Delta t$	13-7
First Law of Thermodynamics	$\Delta E_{Isolated} = 0$	13-12
Second Law of Thermodynamics	$\Delta S_{Isolated} > 0$	13-11

14 Power Basics

Getting Work Done

As Graham Rahal tops a rise at Mid-Ohio in his Dallara/Honda IndyCar, you can actually see how some of its power is being sapped by aerodynamic drag. The turbulent air over the car is clearly blurring the trees in the background. (And check out the negative camber on those inside wheels!) *Photo by Randy Beikmann.*

Power is defined for straight-line motion, rotational motion, and combinations of both. A car's motion is presented as a "kinetic energy bucket," showing that power is required to sustain motion as mechanical energy is being lost to heat. The power to drive the Thrust SSC, and a Top Fueler, are calculated. The difference between engines and motors is explained.

Contents

Key Symbols Introduced

Symbol	Quantity	SFS Units	MKS Units
P	Power	ft-lbf/sec	N-m/sec
$\eta_{Thermal}$	Thermal Efficiency	dimensionless	dimensionless
$\eta_{Mechanical}$	Mechanical Efficiency	dimensionless	dimensionless

1   Introduction

Power! It's about time—in more ways than one. First off, how did we get to Chapter 14 of a book on automotive physics without talking about it? Until now, we haven't even been able to argue about which is more important: torque or horsepower? You've got to be thinking it's high time we did.

But power itself is about time. It's how fast *work* is done—how quickly mechanical energy is produced. In fact, its definition is work divided by time. So power couldn't really be explained until work was discussed in Chapters 12 and 13. You've paid your dues, so let's get to it.

With the automobile, there's always a job to be done, and only so much time to do it. It takes a certain amount of work to accelerate your car from zero to 60 mph, so the faster you want to get to 60, the faster your engine (and/or electric motor) needs to supply that work. Providing the work in less time means more power.

The power your car has on tap affects everything about the way it performs: its acceleration, its top speed, and its zero to 60 and quarter-mile times. But your engine's *peak* power is only part of the picture, since it doesn't produce it over a wide range of speeds. For most vehicles, performance depends on the horsepower and torque curves (vs. engine rpm).

In some ways introducing power brings nothing new, since it is merely the rate at which work is done. But you'll see that speaking in terms of power can simplify a lot of decisions about tuning or designing a car. Besides, it gives us an excuse to revisit some areas of using energy.

Chapter 15 will apply the power concepts we cover here, and show how power is produced from fuel. Generating power takes a deliberate, coordinated process, with nature working against you every step of the way. This makes power a precious commodity. Once you've created it, you really hate to lose any before it drives the wheels. So we will also cover mechanical losses in the driveline.

We'll see that there is more than one way to produce the right amount of power at the tire patch to make your car perform. A fast-spinning, low-torque engine can produce the same power as a slow-spinning, high-torque engine. Which is better? Let the debate begin.

2   The Basics of Power

In the late 1700s, steam engines were replacing horses in many industrial uses, such as water pumps for mines, and in factories and grain mills. Factory designers were used to figuring how many horses were needed to pull the drawbar powering the machinery. So to sell their steam engines, manufacturers had to accurately predict the horses they could replace. After all the trouble and expense of installing one, it would be a disaster if it couldn't pull the load.

With some rounding here and there, James Watt came up with a quick way to rate his engines' power in "horses."[1] From experience, he estimated that a dray horse could pull with a force of 180 lbf while walking 180 feet/minute in a circular path (see **FIGURE 1**). This would come out to 32,400 ft-lbf of work per minute, which he rounded up to 33,000 ft-lbf/minute. This became the standard horsepower we use today:

$$1 \text{ horsepower} = 33,000 \text{ ft-lbf/minute}$$
$$= 550 \text{ ft-lbf/second}$$

Since most of our calculations use time in seconds, the 550 ft-lbf/sec definition will be the more important one to remember here.

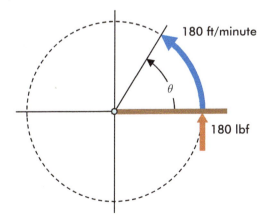

FIGURE 1 James Watt estimated that a horse could pull with 180 lbf at 180 ft/minute, doing 32,400 ft-lbf of work per minute. He rounded this up to a value of 33,000 ft-lbf/minute, calling it the power of "one horse."

Note 1: It's still fairly common to refer to a small engine as a "1-horse" or "5-horse" engine.

Watt kept his calculations on the safe side; the average workhorse did not really pull that hard or that fast; most horses would have rated at about 0.6 horses by his calculations.[2] So a "2-horse" engine *could have* replaced 3 horses, but the last thing he needed was to overpromise.

2.1 Visualizing Power Flow

Just like Watt, we rate engines by what "horsepower" they can produce, but usually at their peak. For several reasons, an engine's power depends on its speed, the most obvious one being that power is how *quickly* work gets done. But power doesn't just have to do with rotating machinery. Watt's horse pulled in a circular path, but if it had pulled the same load in a straight line, at the same speed, it would have produced the same power. If we know the drag forces on a car, we can calculate the drive wheel power needed to maintain a given speed. Neglecting driveline friction, this is the required engine power.

For example, the 1969 Cougar in **FIGURE 2** has a certain amount of kinetic energy E_K at the constant speed it's traveling. To maintain speed, the kinetic energy must be kept constant; it must be replaced as quickly as it's converted to heat, requiring a certain amount of power.

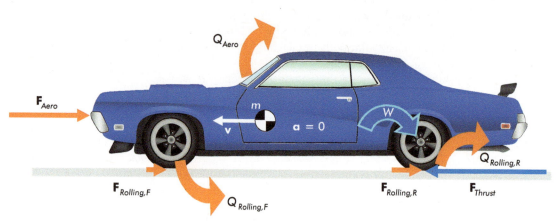

FIGURE 2 The power required to propel a car at constant speed against drag is the amount of work W done per second by the thrust force. Here, blue arrows indicate mechanical energy, and the orange indicate thermal energy.

To increase speed, there must also be work left over after overcoming drag, to increase the kinetic energy by ΔE_K. In terms of energy going into the motion vs. that going out, this is:

$$W = Q_{Aero} + Q_{Rolling} + \Delta E_K$$

Note that the heat created by drag is equal to the negative work the drag does on the car. So here we can use $Q_{Rolling}$ for $W_{Rolling}$, and Q_{Aero} for W_{Aero}. Power being work divided by time, we can then divide this

Note 2: Before criticizing the horses for only producing six-tenths of a horsepower, remember that the rating was for what they could do for eight hours, not just during a "wide-open throttle" acceleration. It's more appropriate to compare them to slower, industrial-type engines.

equation by a short change in time Δt to produce the total required power:

$$P = P_{Aero} + P_{Rolling} + P_{Accel}$$

The term P_{Accel} is the available power to accelerate the car. **FIGURE 3** shows this equation as a kinetic energy "bucket" for the car traveling on a level road. Here, the water level represents the speed of the car, and the amount of water contained (its volume) represents the car's kinetic energy. So just as we've seen, increasing the velocity would increase the kinetic energy.[3]

Now, the power P is how fast you "pour" work W into propelling the car (upper arrow in the figure).[4] At the same time, some kinetic energy is converted to heat (Q_{Aero} and $Q_{Rolling}$) "pouring out" the holes and representing lost power (P_{Aero} and $P_{Rolling}$). To keep a constant speed (constant E_K), the rate at which work pours in (power) must equal the rate at which heat pours out (dissipative power).

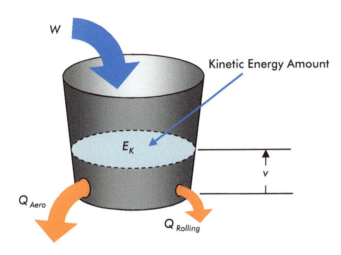

FIGURE 3 Your car as a "kinetic energy bucket." The kinetic energy (amount of water) increases with vehicle speed (the water depth). To stay at a constant speed, the rate at which work "pours into" the car (to push it) must equal the rate at which mechanical energy is converted to heat that "pours out" the holes. To increase speed, more power must be supplied than is being dissipated as heat.

To accelerate the vehicle, increasing its kinetic energy, you must add work faster than it's lost to heat. That is, P_{Accel} is what's left after P_{Aero} and $P_{Rolling}$ are subtracted from P. Then the "kinetic energy" level in the bucket can rise. But then the energy will "flow out the holes" faster, too, because of increased aerodynamic drag (just like water runs out a full bucket faster than a nearly empty one). We'll see that aero drag hits you especially hard at higher speeds when it comes to power.

In the same way, if you quit pouring the work in by shutting off the engine (zero power), the drag would "drain" the kinetic energy out, quickly at first, then more slowly, until the energy bucket is empty (the kinetic energy is zero and the vehicle has stopped).

Note 3: It isn't a perfect analogy. For one thing, for E_K to correctly change with the "height" v, the bucket should be a parabolic funnel. It gets the main idea across, though.

Note 4: Note how I avoided saying that we pour P in the bucket.

3 | The General Definition of Power

We've done just fine up to now finding a vehicle's acceleration from engine torque, gear ratios, and tire diameters. In fact, if we knew the engine's entire torque curve vs. engine speed, we could calculate the vehicle's performance at any vehicle speed. But power is directly linked to how fast mechanical energy is delivered. That's why cars are listed with their power-to-weight ratios, and not their torque-to-weight ratios.

> **KEY CONCEPT**
>
> There is no limit to how much the torque from an engine (and/or electric motor) can be increased by the time it gets to the drive axles, by changing gear ratios. But the power at the axles can never be higher than the total power supplied by the engine (and/or electric motor).

An engine's torque, horsepower, and speed are intertwined: it can't make power without making torque, and it can't make torque without power.[5] The real question is the bottom line: can your engine produce the tire force you want, given the appropriate gearing, at the speed you're traveling? There are many ways to size and tune for the necessary power, but one may be more efficient or drivable than the rest.

Power can show up in many different situations, so we need to have a very general definition:

POWER is the rate at which work is done. In other words, power is the work done divided by the time it takes to do it.

This holds in any situation, whether for power delivered by a spinning crankshaft, power required to accelerate a car in a straight line, in electric current flowing from a generator, or in lifting something with your hands.[6]

Note 5: An "engine" could make torque without horsepower if it's not turning, but it wouldn't have to be much of an engine. I could produce torque without motion just by standing on the end of a breaker bar.

Note 6: We won't cover the electromagnetic theory of an electric motor here, but the power supplied to it is proportional to the voltage across it multiplied by the current going through it.

4   Power in Straight-Line Motion

Even though we most often talk about power from a rotating crank-shaft, its effect is seen where the rubber meets the road—a traction force pushing the car. Straight-line motion is also the easiest place to calculate power.

Remembering that the definition of power is the rate at which work is done, let's calculate it for a car that is traveling at constant speed v, propelled by a thrust force F in the same direction (**Figure 4**). First we find the work done in traveling a displacement Δs, as $F\Delta s$, from equation 12-1. The power P is the work divided by the change in time:

$$P = \frac{W}{\Delta t} = \frac{F\Delta s}{\Delta t}$$

We said that v is constant, so it is equal to $\Delta s/\Delta t$. This means power simplifies to:

Physical Law Eq. 14-1   **Power in translation, force and velocity parallel**

$$P = Fv$$

That is, power = force $\times$ velocity, assuming they are pointed in the same direction. Note that although we produced the equation by as-suming a constant speed, it still works when it's not constant. When the velocity or force (or both) are changing, the answer you get is the instantaneous power. If the force and velocity have an angle θ between them, the power is force multiplied by the component of the velocity in the same direction, that is, $Fv \cos\theta$.

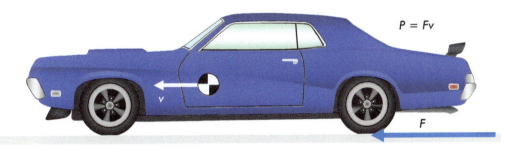

Now we need to define the units of power. Since power is work/time, the units for power are work units divided by time units:

In consistent SFS units, power is in ft-lbf/sec. But automotive engines are typically rated in horsepower (HP), where 1 HP equals 550 ft-lbf/second.

In the MKS system, power is in N-m/sec, or joules/sec. But 1 joule/sec is given the name 1 watt (1 W), in honor of James Watt. Following the

Figure 4 The power of propulsion is equal to the force multiplied by the component of the velocity that is in the same direction as the force.

metric system, 1,000 watts is 1 kilowatt (1 kW). Engines are often rated in kW when using metric units.

To convert between SFS and MKS units, 1 horsepower equals 0.74570 kilowatts, so 1 kilowatt equals 1.34102 horsepower, or 737.56 ft-lbf/sec.

Let's work an example for the Cougar in FIGURE 4, traveling at a constant 60 mph (88 ft/sec) and propelled by a 150-lbf traction force. Using equation 14-1, the required power is

$$P = Fv = 150 \text{ lbf} \times 88 \text{ ft/sec} = 13,200 \text{ ft-lbf/sec}$$

That's the answer in consistent SFS units, but how many horsepower is it? Remember that there is one horsepower per 550 ft-lbf/sec:

$$P = 13,200 \text{ ft-lbf/sec} \times \frac{1 \text{ HP}}{550 \text{ ft-lbf/sec}} = 24.0 \text{ HP}$$

So cruising at 60 mph would only take 24 HP, an amount a large lawnmower engine could put out. Of course, *accelerating* to 60 mph or climbing a hill is a different story.

5   Power in Climbing a Hill

How about an example of power that you can feel? Suppose I keep up an 8-mph speed (11.73 ft/sec) on a 10% hill on the bicycle in FIGURE 5. If the bike and I weigh 240 lbf, how much power does that take, ignoring air resistance?

We can do this problem in two ways. One is to calculate the force along the incline and multiply by the velocity, which is in the same direction. The other is to recognize that the power supplied is going into increasing the bike's height, and therefore its gravitational potential energy $E_G = mgh$; we would then calculate power from the rate of change of E_G, as the bike moves upward at velocity v_y.

The second option seems like less work, and almost like cheating, so I'm picking that.

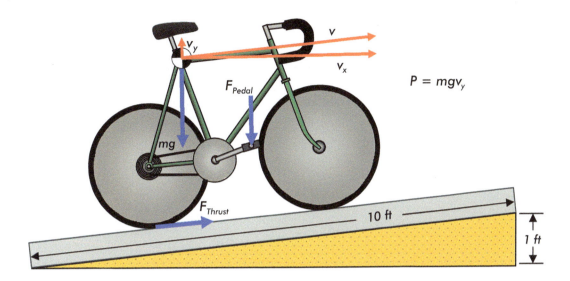

$$P = mgv_y$$

First, let's calculate the upward part of the velocity. Since it's a 10% hill, the upward velocity will be 10% of the forward velocity:

$$v_y = v\frac{s_y}{s} = 11.73 \text{ ft/sec} \times 0.1 = 1.173 \text{ ft/sec}$$

So the bike and I climb upward, our height h increasing at the rate of 1.173 feet every second. The power to lift our 240 lbf of road-hugging weight is the rate of our increase in potential energy E_G:

$$P = mg\frac{\Delta h}{\Delta t} = mgv_y = 240 \text{ lbf} \times 1.173 \text{ ft/sec} = 281.6 \text{ ft-lbf/sec}$$

In horsepower, that is:

$$P = 281.6 \text{ ft-lbf/sec} \times \frac{1 \text{ HP}}{550 \text{ ft-lbf/sec}} = 0.512 \text{ HP}$$

About a half a horsepower for a human that weighs one fifth as much as a horse sounds like too much, but it really isn't unreasonable. Some athletes can produce almost 1 horsepower for a short period of time.

Note that I didn't even mention which gear I'm in, because it doesn't matter. Let's think about the different ways I could produce the required power to climb the hill. If I were in first gear, I'd be furiously pedaling at a high speed (many times per second), but I'd need little pedal force (high speed, low force). But in top gear, I'd move my feet much more slowly, while pushing the pedals pretty hard (low speed, high force). Either way, the power is the same.

FIGURE 5 The power in pedaling a bike up a 10% hill. Neglecting rolling resistance and aerodynamic drag, all the work is done against gravity.

6 | Power in Rotation

Just as we calculated power in translation, we can do it for rotation. In FIGURE 6, the force F creates a torque $T = Fr$ about the center of rotation of the lever. Equation 12-7 writes the work done by the lever as $W = T\Delta\theta$, where $\Delta\theta$ is the change in the angle.

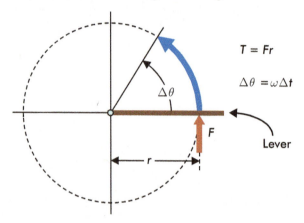

$T = Fr$

$\Delta\theta = \omega\Delta t$

F

Lever

r

$\Delta\theta$

FIGURE 6 In rotation, power can be visualized as a torque turning at a certain angular velocity, or as a force "chasing" a lever arm around in a circle at a certain speed along the dashed arc.

Assuming the lever is turning at a constant speed, we can divide each side of equation 12-7 by the time Δt it took to turn through the angle $\Delta\theta$:

$$\frac{W}{\Delta t} = T\frac{\Delta\theta}{\Delta t}$$

But $W/\Delta t$ is just power, and $\Delta\theta/\Delta t$ is the angular velocity ω, so:

Physical Law Eq. 14-2   **Power in rotation**

$$P = T\omega$$

Of course this requires consistent units, so angular velocity is in rad/sec, torque is in ft-lbf or N-m, and power is in ft-lbf/sec or watts. Just as for straight-line power, equation 14-2 is still correct if torque and angular speed are changing, giving the instantaneous power.

MAJOR POINT

Note how similar the power calculations are for rotational motion and straight-line motion. In straight-line motion, power is how hard you push times how fast you push. In rotation, power is how hard you turn times how fast you turn.

One look at equation 14-2 says that you can increase power by increasing the torque or the crank speed. If you want to make big power with a small engine having low torque, you need to spin it fast; if you don't want to spin it fast, you'll need lots of torque (bigger engine? supercharging?).

This trade-off is like in the bicycle example using first and top gears. In fact, we could calculate the power put into the bike's crank from the torque produced by the pedal forces and its rotation speed. This torque will be in pulses since the angle between the force and lever arm varies. Likewise, the torque in an engine is delivered in pulses. But when we refer to torque, we almost always mean the *average* torque.[7]

Let's calculate the power of an engine at its torque peak, which we'll say is 300 ft-lbf at 4,000 rpm. To find the power there, we first find the angular velocity ω in rad/sec:

$$\omega = 4{,}000 \ \frac{\text{rev}}{\text{minute}} \times \frac{1 \ \text{minute}}{60 \ \text{seconds}} \times \frac{2\pi \ \text{rad}}{\text{rev}} = 418.88 \ \text{rad/sec}$$

Then the power is:

$$P = T\omega = 300 \ \text{ft-lbf} \times 418.88 \ \text{rad/sec}$$
$$= 125{,}700 \ \text{ft-lbf/sec} = 228 \ \text{HP}$$

Note that the *peak* power will be at a higher speed. For instance, if this engine could make 250 ft-lbf of torque at 6,000 rpm, the power at that speed would be:

$$P = T\omega = 250 \ \text{ft-lbf} \times 628.32 \ \text{rad/sec}$$
$$= 157{,}100 \ \text{ft-lbf/sec} = 286 \ \text{HP}$$

Even though the torque is 50 ft-lbf lower at 6,000 rpm, the 50% increase in engine speed means the engine is doing work (much) faster. **Figure 7** compares the work done in 1/100 of a second at each speed. At 4,000 rpm, the engine rotates 240° ($4\pi/3$ radians), creating 1,257 ft-lbf of work. At 6,000 rpm, it rotates 360° (2π radians), creating 1,571 ft-lbf of work.

Figure 7 Comparing the work done in 0.01 seconds by an engine producing 300 ft-lbf torque at 4,000 rpm, vs. 250 ft-lbf at 6,000 rpm. At the faster speed it makes more power by turning through a larger angle (360° vs. 240°), despite the reduced torque.

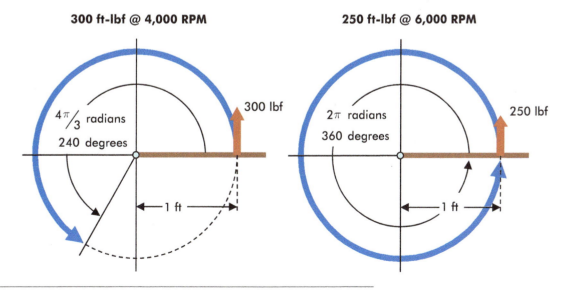

300 ft-lbf @ 4,000 RPM | **250 ft-lbf @ 6,000 RPM**

$4\pi/3$ radians
240 degrees
300 lbf
1 ft

2π radians
360 degrees
250 lbf
1 ft

Note 7: For the most part, the only people who care about the pulsations are noise and vibration geeks like me.

This is why engine torque and power curves (see **FIGURE 8**) look the way they do. The torque curve is relatively flat, and mainly depends on how much air/fuel mixture its cylinders can take in (the volumetric displacement in cubic inches or liters). Simplistically, the torque peak occurs when the engine can fill its cylinders best; street engines are tuned for peak torque in the mid-range of engine speed, racing engines at higher speed. The torque drops at higher speed because the engine can't flow air quickly enough to feed the cylinders. (Supercharging is one way to cram more air in.)

Power always peaks at a higher speed than torque, at a point where torque is dropping so fast that increasing speed can't make up for it, so the power ($P = T\omega$) drops.

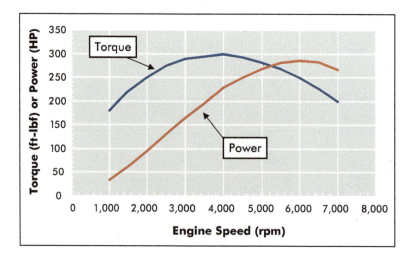

FIGURE 8 Typical torque and power curves for a gasoline engine at wide-open throttle, with peaks of 300 ft-lbf @ 4,000 rpm and 286 HP @ 6,000 RPM. Torque is relatively flat vs. speed, while power increases steadily through most of the operating range.

Torque and power from IC engines usually haven't been rated below 1,000 rpm or so, since they don't run well there.[8] This has been changing though, as manufacturers squeeze more and more fuel economy out of their engines.

7   Power Dissipation

We've mentioned energy losses to friction, calling them mechanical losses. Power dissipation is the rate at which mechanical energy is converted to heat. It is sometimes intentional, but usually a necessary evil.

Note 8: Actually, the driveline "doesn't like it" when the engine runs that slow, often causing excessive vibration and noise.

7.1 Intentional Power Dissipation

When you want to slow a car, you need to dissipate its kinetic energy. For instance, if your brake pads exert a friction force of 1,000 lbf on a rotor, whose surface is sliding past at 50 ft/sec, the power dissipated is:

$$P_{Friction} = F_{Friction}v = -1,000 \text{ lbf} \times 50 \text{ ft/sec}$$
$$= -50,000 \text{ ft-lbf/sec} = -90.9 \text{ HP}$$

The power is negative because the force acts opposite the velocity, so the brakes are dissipating power as heat. You could just as easily look at this as a friction torque on the rotor:

$$P_{Friction} = T_{Friction}\omega_{Rotor}$$

For the same braking dissipation, a larger brake rotor would use less friction force, but with faster sliding.

7.2 Undesired Power Dissipation (Parasitic Losses)

Once an engine or electric motor produces power, some is dissipated on the way to the tires, by mechanical losses:

1. Friction (in bearings, and from sliding contact between gear teeth).
2. Fluid drag (gears immersed in oil, flung oil, etc.).
3. Electrical resistance in a motor and/or generator.

These are called **PARASITIC LOSSES**, because they sap the life's blood of the vehicle—the propulsion power. They are part of the cost of propelling the car, but we are always trying to reduce them. A good example is power lost by a two-piece propshaft to bearing friction (**FIGURE 9**). The friction reduces the propulsion torque. The heat produced is of no use, and may be harmful to the bearing if it overheats. But we must use bearings to position rotating shafts, and all bearings have friction.

FIGURE 9 The support bearing for this propshaft has 1 ft-lbf of friction. The power and torque output are less than at the input, with the difference showing up as heat in the bearing. This is a parasitic loss.

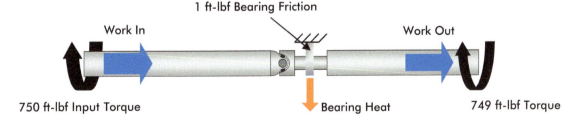

1 ft-lbf Bearing Friction

Work In

Work Out

750 ft-lbf Input Torque

Bearing Heat

749 ft-lbf Torque

The picture is similar for an electric-powered vehicle, except that (typically) heat is converted to mechanical energy at a power plant, converted to electrical energy, and transmitted to your car through the power lines. Here, the goal is to minimize mechanical losses (in a thermodynamic sense) in the power lines, the battery, and the motor.

Not all parasitic losses are due to the driveline. The engine drives accessories that are absolutely necessary (like the cam drive and water pump), and some that are nice to have (like the A/C compressor). Each consumes power to drive it, increasing energy consumption.

8   Power Through a Driveline

Transmissions have been trending toward more ratios, as of this writing. But besides mechanical loss, each gear mesh adds mass to the vehicle, so you might ask whether the advantage of each additional transmission gear ratio outweighs its downside, and you'd be right to ask. In fact, you could ask why we need transmissions at all! You certainly don't want to add unnecessary mass or losses. Drivetrains are used for several reasons:

- To allow changing gear ratios (shifting gears), so you can match the engine's capability to the conditions: more reduction for more axle torque and acceleration, less reduction for a lower engine speed and more fuel economy.
- To enable the engine to stay within its operating speed range (a gear ratio that would be right for a parking lot won't do on a highway).
- To let the crankshaft rotate on a different axis than the drive axle, or in a different direction. For example, the drive axles are transverse (cross-car) and the crank may be longitudinal. Also, you can go in reverse without reversing the engine rotation.

Some transmissions have four gear ratios, some seven, and some an infinite number within a range (as in a continuously variable transmission, or CVT). Large trucks might have 18 ratios. On the other hand, many drag racers run Powerglides (two ratios), and Top Fuelers have *no* transmission (effectively one ratio). What is best depends on how the vehicle is used, what it weighs, and what it costs, so one size doesn't fit all. Here we'll cover power transfer, while Chapter 15 covers gear selection.

First let's examine a lossless driveline, and then add in the losses (if that makes any sense). In Chapter 13 we found that the work coming out of a lossless gear train equals the work in ($W_{Out} = W_{In}$). The increased torque output (assuming a gear reduction) was exactly offset by a reduction in the output rotation angle ($T_{Out}\theta_{Out} = T_{In}\theta_{In}$). Since power is simply how fast work is done, this also means $T_{Out}\omega_{Out} = T_{In}\omega_{In}$, and the power out is equal to the power in:

Lossless gear train

$$P_{Out} = P_{In}$$

You won't find a lossless gear train outside a textbook, so some power will be lost to heat, as in **Figure 10**. To adjust for this, we often use a mechanical efficiency factor:

Equation 14-3   **Real gear train**

$$P_{Out} = \eta_{Mechanical} P_{In}$$

The Greek letter η (spelled eta and pronounced *AY-tuh*) stands for efficiency. The mechanical efficiency is always between 0 and 1 (with 1 being 100% efficiency). Note how using power here lets us ignore gear ratios and torque. It also shows that a driveline can never increase the power supplied by the engine, it can only reduce it.[9]

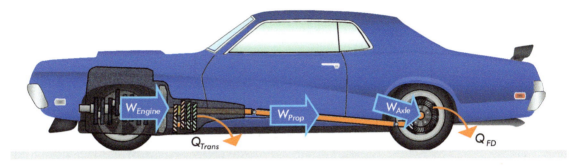

In automatic transmissions and CVT's, we also need to consider the losses from driving the transmission pump. And of course automatics have torque converters, which increase the torque into the transmission, at the expense of transmission input shaft speed. Torque converter efficiencies vary widely, depending on your driving conditions, and are too complicated to detail here.[10]

Figure 10 Mechanical losses in a typical drivetrain. The engine power is reduced by the rate of heat generated in the transmission and final drive gears, before reaching the drive axle. Overall mechanical efficiencies vary from 85% to 95%.

8.1 Torque and Power at Each Point in a Lossless Driveline

Power out of a lossless gear train is equal to the power in, but what about points in between? Let's take a look at the horsepower, torque and speed at the engine, propshaft, and axles.

Figure 11 shows the Cougar with its engine, driveline, and wheels/tires. The figure includes mechanical efficiencies for the transmission and final drive gears, but in this section we assume no losses, so they both are equal to 1. Its engine runs at speed ω_{Engine}, and puts out torque

---

Note 9: That's why this section is named "Power Through a Driveline," not "Power From a Driveline." A gearbox doesn't produce power, it just *transmits* it. Hmmm...maybe that's why they're called *transmissions*.

Note 10: The comedy writer Dave Barry has suggested that the intricate designs of torque converters and automatic transmissions are proof that aliens have been feeding us technology. But I've worked with a few transmission engineers who can explain them, and most don't seem like aliens.

T_{Engine}. The transmission input shaft runs at engine speed, while its output shaft turns the propshaft at the engine speed divided by the gear ratio:

$$\omega_{Prop} = \frac{\omega_{Engine}}{R_{Trans}}$$

The final drive input (the pinion of the ring and pinion) turns at prop-shaft speed, so the ring turns the drive axles at prop speed divided by the final-drive ratio:

$$\omega_{Axle} = \frac{\omega_{Prop}}{R_{FD}}$$

Finally, the rolling tires push the car forward at velocity v. Assuming no tire slip:

$$v = r_{Tire}\,\omega_{Axle}$$

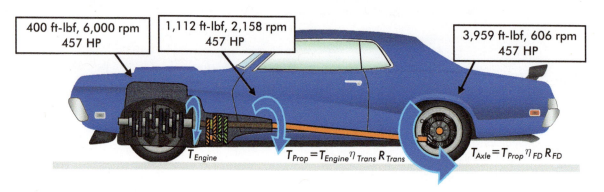

| 400 ft-lbf, 6,000 rpm 457 HP | 1,112 ft-lbf, 2,158 rpm 457 HP | 3,959 ft-lbf, 606 rpm 457 HP |

T_{Engine} $T_{Prop} = T_{Engine}\,\eta_{Trans}\,R_{Trans}$ $T_{Axle} = T_{Prop}\,\eta_{FD}\,R_{FD}$

FIGURE 11 An engine and driveline having gear ratios R and efficiencies η for the transmission and final drive. With no mechanical losses (each η equal to 1), the axle is driven with a torque that is higher than the engine torque by the overall gear ratio. This trades rotation speed for torque, but leaves horsepower unchanged.

All in all, we combine these three formulas to get the vehicle speed from engine speed, in terms of the gear ratios and tire radius:

Equation 14-4   **Manual trans. or locked TCC**

$$v = \frac{\omega_{Engine}\,r_{Tire}}{R_{Trans}\,R_{FD}}$$

This is valid when there is no slip in the driveline, i.e., a manual transmission or an automatic with a locked torque converter clutch (TCC). We are also ignoring tire slip. This is a handy equation to use for calculating vehicle speed from your tach, knowing the tire radius. But remember to use consistent units!

How about the thrust at the tire patches? Following a similar set of calculations to multiply the engine torque by the gear ratios, etc., the thrust force is (assuming we have enough traction),

Equation 14-5   **Manual trans. or locked TCC**

$$F_{Thrust} = \frac{T_{Engine} R_{Trans} R_{FD}}{r_{Tire}}$$

Using this, let's compare the power at the pavement to the engine power. If you multiply the force F_{Thrust} from equation 14-5 by the velocity from equation 14-4 and simplify, the power is:

$$P_{Thrust} = F_{Thrust} v = T_{Engine} \omega_{Engine}$$

But $T_{Engine} \omega_{Engine}$ is just power from the engine, so:

Lossless gear train

$$P_{Thrust} = P_{Engine}$$

as expected. Knowing the first law of thermodynamics, we already knew that it *had* to come out like this, regardless of gear ratios or tire diameter.[11]

CAUTION

Please remember that you don't always use peak engine power or torque in these calculations. It depends on engine speed, and how far your right foot is down.

8.2 Torque and Power at Each Point in a Real Driveline

What if we do have mechanical losses in the gear train? The results are the same, except for a drop in torque at the prop shaft and at the drive axles, depending on the driveline mechanical efficiencies as shown in FIGURE 11. Inserting these into equation 14-5, we get

Equation 14-6   **Real gear train**

$$F_{Thrust} = \frac{T_{Engine} \eta_{Trans} R_{Trans} \eta_{FD} R_{FD}}{r_{Tire}}$$

Note 11: If I had gotten an answer where power in was different than power out (remember, no mechanical losses yet), I would have been checking through a few equations for the error.

This produces a power at the tire patch of:

Equation 14-7   **Real gear train**

$$P_{Thrust} = P_{Engine} \eta_{Trans} \eta_{FD}$$

ignoring tire slip. This is the relationship to use; for a lossless gear train you just set the mechanical efficiencies to 1. How much power might be lost in a typical drivetrain? Let's calculate the power at the tire with 457 engine HP (as in **FIGURE 11**), a transmission efficiency of 96%, and a 94%-efficient final drive. Using equation 14-7,

$$P_{Thrust} = P_{Engine} \eta_{Trans} \eta_{FD} = 457 \text{ HP} \times 0.96 \times 0.94 = 412.4 \text{ HP}$$

FIGURE 12 As power is transmitted from the engine, through the propshaft, to the drive axles, the horsepower is decreased due to mechanical losses.

In this example, 44.6 HP (about 10% of the total) is dissipated as heat, 18.3 HP in the transmission and 27.4 HP in the final drive.[12] **FIGURE 12** plots the reduced power levels resulting from the transmission and final drive losses. Speed is traded for torque, but power comes through relatively unscathed.

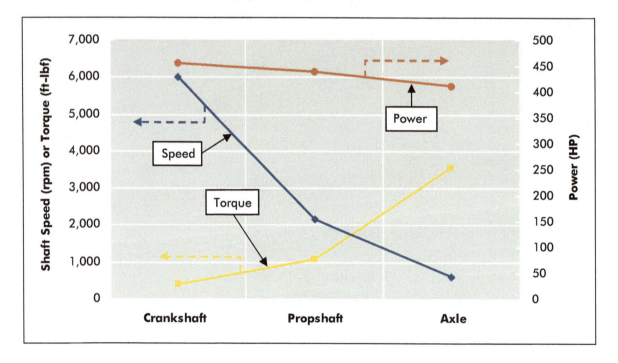

NOTE

The concept of a single mechanical efficiency value for a driveline component is an approximate one. The friction between meshing gear teeth does indeed depend on the power going through, but windage does not. It is due to internal aerodynamic drag and oil impacting moving parts, and depends mainly on operating speed.

8.3 Slipping Clutches

What about when you slip the clutch? How much power is lost? How much torque?

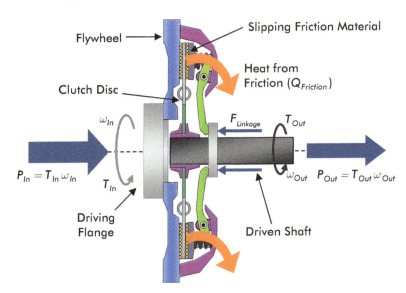

FIGURE 13 A slipping clutch, with output speed less than input speed. Power out is less than power in by the amount of heat dissipated per second by friction.

With bearing friction, we lost torque, but not speed. In driving a slipping clutch, we lose speed, but not torque. In **FIGURE 13**, the flywheel turns at engine speed ω_{In}, while the clutch disc turns at the transmission input shaft speed ω_{Out}. The power dissipated by friction ($P_{In} - P_{Out}$) depends on the *difference* in their angular velocities:

Equation 14-8   **Slipping clutch**

$$P_{Friction} = T_{Friction}\left(\omega_{In} - \omega_{Out}\right)$$

Why could we assume no torque loss across the clutch assembly? In reality, there will be a difference from aerodynamic drag, and from its inertia (if accelerating). But those don't actually affect the power loss. All that matters are the torques acting *at the friction surface*, one on the clutch disc and the other on the flywheel assembly (per Newton's third law). The power dissipation is due to the "speed loss," not a torque loss.

One interesting result of this is that by *increasing* the sliding friction torque in a clutch, you often *reduce* the total amount of heat generated by friction during engagement, by reducing the amount and time of slip. (When the clutch is engaged [no slip], then $\omega_{Out} = \omega_{In}$, and the power loss is zero.) So ironically, engaging the clutch rapidly with high-friction torque is one way to keep it from heating up.

In an extreme example of power dissipation, Top Fuelers control the crank speed/propshaft speed relationship by slipping a pack of clutches. Torque isn't reduced, but the clutch pack dissipates several thousand horsepower and creates lots of heat. They grind off around a tenth of an inch of friction material in a single run. In fact, the clutch housings have their own "exhaust pipe" for the worn material and hot gases to escape!

9   Acceleration and Power

Everyone wants to maximize their car's acceleration performance. The first step is to find out what factors are limiting it, and in what conditions. Does it need more power? More traction? Less drag? To check out the conditions where the first two are important, we'll look at the acceleration resulting from constant power, then the power required for constant acceleration.

9.1 Acceleration at Constant Power

How fast would a car accelerate from speed v_1 to speed v_2, if it could get all its power to the ground (no losses)? To find out, we'll use the first law of thermodynamics. The work W done to accelerate the car is equal to its increase in kinetic energy:

$$W = \frac{1}{2} m \left(v_2^{\,2} - v_1^{\,2} \right)$$

With constant power, the work is also equal to power multiplied by time ($W = P\Delta t$). Putting this into the previous equation gives:

$$P \left(t_2 - t_1 \right) = \frac{1}{2} m \left(v_2^{\,2} - v_1^{\,2} \right)$$

where t_1 is the time at the start of acceleration. Rearranging this gives the ending time, t_2:

Equation 14-9a   **Acceleration at constant power**

$$t_2 = t_1 + \frac{m \left(v_2^{\,2} - v_1^{\,2} \right)}{2P}$$

This can also be solved for the final speed v_2, given the time interval and the initial speed:

Equation 14-9b   **Acceleration at constant power**

$$v_2 = \sqrt{\frac{2P\,(t_2 - t_1)}{m} + v_1^2}$$

So let's check out the theoretical 0 to 60 mph time of a 3,220-pound car with 300 horsepower. First, we convert 300 HP to 165,000 ft-lbf/sec. The 3,220-lbm car has a mass of 100 slugs. Finally, 60 mph is 88 ft/sec. Setting the initial speed and time to zero, the 0 to 60 time is:

$$t_2 = \frac{mv_2^2}{2P} = \frac{(100 \text{ slugs}) \times (88 \text{ ft/sec})^2}{2 \times (165,000 \text{ ft-lbf/sec})} = 2.347 \text{ seconds}$$

Well, that's optimistic. In fact, it's ridiculous! Nobody would expect a 300-HP, 3,220-lbm car to accelerate from 0 to 60 in 2.5 seconds. There's obviously more to it, and it doesn't have anything to do with exaggerated horsepower claims. It has to do with traction limits, which make low speed acceleration at constant power unrealistic. We'll take this up in the next chapter.

9.2 Power in Constant Acceleration

In previous chapters, we've often used constant acceleration, the velocity increasing the same amount each second. How much power is required to do this, neglecting drag? We know the force is constant, using $F = ma$. We also know that the velocity $v = at$. We should be able to find power in terms of acceleration and time. Starting with $P = Fv$, and substituting ma for F, the power at the tire patch is:

Physical Law Eq. 14-10   **Power to accelerate, no drag**

$$P_{Accel} = mav$$

This is always true for the work done to accelerate the mass, with no drag. With constant acceleration, and assuming we start at rest, we can substitute at for v, which gives us $P_{Accel} = ma \times at$, or:

Equation 14-11   **Constant acceleration, from rest**

$$P_{Accel} = ma^2 t$$

This is interesting, because increasing the acceleration by 20% doesn't increase the required power *at a certain time* by 20%, but by 44% (1.2 × 1.2 = 1.44). It not only takes a 20% greater force, but you are also at a 20% higher speed at that point (multiplying the higher force times a higher speed to get power). So ironically, the more power you can use, the sooner you need more.[13] Aerodynamic drag isn't in the equation, yet, but it will only increase this speed effect.

FIGURE 14 shows the power required for the first ten seconds of constant acceleration, for a 3,220-pound car from a standing start. Notice that hardly any power is required at first, because velocity is almost zero. But after a few seconds, the horsepower required becomes huge.

At 0.6-g acceleration (19.32 ft/sec²), it takes 408 HP at the six-second mark, while at 0.72 g's (23.18 ft/sec²), it takes 586 HP. Of the few road cars that have the traction to accelerate at 0.72 g's, even fewer have that much power! And remember, this is only the power going into acceleration, not accounting for drag. It's obvious that with most cars, you can only accelerate at a high constant rate for a short time, before a lack of power limits you.

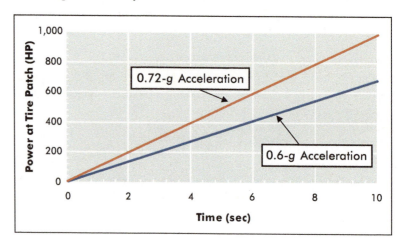

FIGURE 14 Accelerating hard at a constant rate requires a lot of power after only a few seconds. At 0.6 g's, you'd need a net 408 HP at the 6-second mark with a 3,220-pound car. For 0.72 g's, it would take 586 HP.

Equation 14-11 is power vs. time, but what about power vs. displacement down the road? At constant acceleration, from rest, the displacement $s = \frac{1}{2}at^2$, so:

$$t = \sqrt{\frac{2s}{a}}$$

Substituting this in for t in equation 14-11 gives:

Equation 14-12   **Constant acceleration**

$$P_{Accel} = \left(ma^{\frac{3}{2}} \right) \sqrt{2s}$$

If you plug in $1.2 \times a$ into this equation, you find that increasing the acceleration by 20% increases the required power *at a certain distance* by 31%. This effect is plotted in FIGURE 15 for our 0.6-g and 0.72-g acceleration rates, over a quarter-mile. The 0.6-g run needs about 794 HP by the end of the quarter, and the 0.72-g run needs over a

Note 13: Yes, there are obvious parallels with life and politics there.

thousand! The 0.6-*g* finishes the quarter-mile in 11.7 seconds at 154 mph, and the 0.72-*g* run finishes in 10.7 at 169 mph.

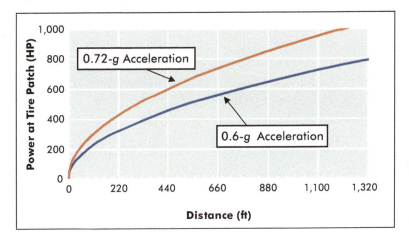

Figure 15 Power vs. distance down the track at constant acceleration. You'd need 794 HP by the quarter-mile mark at 0.6 g's. For 0.72 g's, it would take 1,040 HP.

MAJOR POINT

If these combinations of top speed and ET (elapsed time) for the quarter-mile don't sound typical, it's because they're not. The top speeds are too high for the ETs. The main reason is that very few cars can run the quarter-mile at constant acceleration. Limited horsepower slows the acceleration past a certain point, keeping the top speed down. Chapter 15 will examine this closely.

10 Available Power for Acceleration

You'd like to have all your power available for acceleration, but as we've seen, Nature has other ideas. The second law of thermodynamics says that some power will be dissipated as heat. Let's find what's left for acceleration.

10.1 Power to Overcome Rolling Resistance

As your tires roll, mechanical energy is lost in the wheel bearings and in the flexing of the tires themselves. As the curved tread meets the flat road, it has to straighten out, and must bend into a curve again when it leaves the contact patch. The repetitive bending back and forth converts mechanical energy to heat in the tread and carcass through internal friction, just as bending a piece of wire back and forth makes it hot.

A first approximation of rolling resistance is a constant force. Using that, the power to overcome it is simply force times velocity:

Equation 14-13   **Power lost to rolling resistance**

$$P_{Rolling} = F_{Rolling} v$$

A good first guess for the total rolling resistance is 1% of the car's weight, 32 lbf for the 3,220-pound car we just looked at.

10.2 Power to Overcome Aerodynamic Drag

Aerodynamic forces grow with the square of velocity, so they are not very important at lower speeds, but climb quickly. After about 40–50 mph, aerodynamic forces are the main source of drag. Using equation 5-5 for aerodynamic drag F_D, the power to overcome it is:

$$P_{Aero} = F_D v = \left(\frac{1}{2}\rho v^2 C_D A_{Frontal}\right) \times v$$

or:

Equation 14-14   **Power lost to aero drag**

$$P_{Aero} = \frac{1}{2}\rho v^3 C_D A_{Frontal}$$

Drag power comes from multiplying speed v by a force growing with v^2, so power grows with the *cube* of speed (v^3). This becomes huge at higher speeds. Doubling the velocity requires eight times the power! If you can do 100 mph with 100 HP, going 200 mph takes 800 HP. You're hitting a wall of air.

Let's look at it the other way. Suppose you double the horsepower: how much extra speed does that buy you if drag is your main power sink? Solving equation 14-14 for velocity, and setting velocity and power at the contact patch to their maximum values, gives:

Equation 14-15   **Max drag-limited speed**

$$v_{Max} = \sqrt[3]{\frac{2P_{Max}}{\rho C_D A_{Frontal}}}$$

Note that this is a cube root. Doubling the power P_{Max} multiplies the top speed by the cube root of 2 (about 1.26), for a 26% increase in speed. If your top speed is 100 mph with 100 HP, it will be about 126 mph with 200 HP, or 159 mph with 400 HP (impressive, but not overwhelming). It is easy to see why aerodynamics was ignored when cars drove 40 mph, but got much more attention as speeds increased, and when fuel economy became more important.

10.3 Total Power Required

Now let's summarize these factors in a single formula. The power to the wheels has to equal the power required to a) maintain speed against drag sources, and b) provide the desired acceleration. Putting together all that we know:

Equation 14-16

$$P_{Thrust} = P_{Rolling} + P_{Aero} + P_{Accel}$$

Fleshing this out with the values for each piece, from equations 14-7, 14-10, 14-13, and 14-14, this is:

Equation 14-17

$$P_{Engine}\eta_{Trans}\eta_{FD} = F_{Rolling}v + \frac{1}{2}\rho v^3 C_D A_{Frontal} + mav$$

Here we are again ignoring tire slip. Note that all three drag factors increase with speed. This can be looked at several ways. One is to ask what engine power would be needed to accelerate at a certain rate, given the losses:

Equation 14-18   **Required engine power**

$$P_{Engine} = \frac{F_{Rolling}v + \frac{1}{2}\rho v^3 C_D A_{Frontal} + mav}{\eta_{Trans}\eta_{FD}}$$

Another way would be to ask how much acceleration is produced at a certain speed from a given engine power:

Equation 14-19   **Max acceleration**

$$a = \frac{P_{Engine}\eta_{Trans}\eta_{FD} - F_{Rolling}v - \frac{1}{2}\rho v^3 C_D A_{Frontal}}{mv}$$

In other words, take the engine's maximum power, reduce it by the losses, and what's left over can be used for acceleration. When you're going so fast that $a = 0$, you've reached top speed.[14]

Note 14: You can use this to get a better estimate of top speed than equation 14-15, by trying different speeds until $a = 0$. I could also have explicitly solved equation 14-19 for $a = 0$, but I didn't feel like solving a cubic equation.

11 | Two High-Powered Examples

It's always good to see a couple of outrageous examples of a concept to drive home the point. Now that we have the basics of power down, let's use it to find how much it takes to compete in Top Fuel, and what it took to set the first supersonic land-speed record.

11.1 Top-Fuel Power

In Chapter 2 we found the acceleration of Cory McClenathan's dragster to be about 4 *g*'s (about 132 ft/sec²) for the first 3 seconds in a run at Phoenix in 2010 (see **FIGURE 17** in Chapter 2). In Chapter 6 we calculated the traction force to do this to be about 9,660 lbf. How much power to the ground did it take, including drag?

Taking the speed trace and fitting a curve to it, we can get the acceleration for the entire 3.813-second run (see **FIGURE 16**). Note that the acceleration does tend to tail off after 2 seconds, but is still around 2 *g*'s at the end, when the car is going about 330 mph.

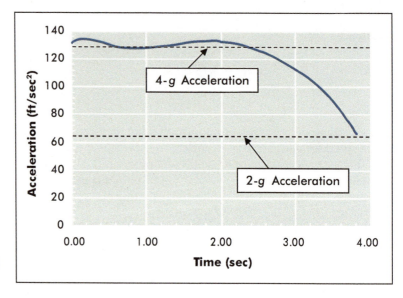

FIGURE 16 Acceleration of McClenathan's dragster. Note that at the end of the 1,000-ft run, it is still accelerating at about 2 g's!

To find the power at the pavement, we calculate the power needed to accelerate the car and add the power to overcome aerodynamic drag, by adapting formula 14-18. But that equation calculates engine power, so it includes mechanical efficiency. We can ignore that since we are looking directly at the power acting at the track. We'll also ignore the rolling resistance of those little front tires, producing

Equation 14-20

$$P_{Thrust} = \frac{1}{2}\rho v^3 C_D A_{Frontal} + mav$$

With acceleration and speed traces, it's easy to calculate power in a spreadsheet. With air density of 0.0709 lbm/ft³ and C_DA of 11.3 ft², we get the power traces in **Figure 17**; impressive numbers, as you'd expect. The highest net power accelerating the car is 6,200 HP, at 2.96 seconds. At the end of the run, it takes 2,500 HP just to overcome drag. The peak of 7,800 HP agrees with recent engine power estimates (8,000–10,000 HP), since we're estimating HP to the ground. Assuming 80% driveline/tire efficiency, 7,800 HP means an engine output of about 10,000 HP.

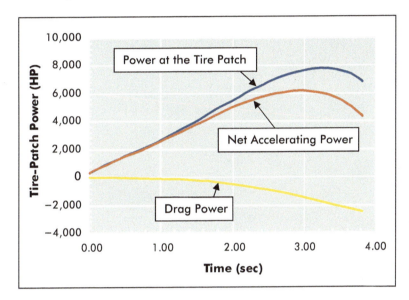

Figure 17 Power at the tire patch for McClenathan's dragster, also split into acceleration and aerodynamic drag components. As much as 7,800 HP is required at the tire patch.

The fact that the power going into acceleration drops off at the end isn't surprising, with more power going into overcoming drag. But I wouldn't have expected the *total* power to the tire patches to drop after 3.3 seconds. This could be for several reasons, including 1) the engine is so hot that it degrades combustion, 2) heat expands the pistons to where they "scrape" the cylinders, increasing friction, and 3) the clutch has locked, and engine speed is past its power peak.

MAJOR POINT

Don't jump to the conclusion that you need engine torque and engine speed to calculate horsepower. It's often much easier to work it the other way, from thrust force and vehicle speed.

11.2 Land-Speed-Record Power

How much power does it take to go faster than sound, while on the ground? We have what we need to calculate the power used by the Thrust SSC during its record run, based on the thrust (force) data from Chapter 6, and its velocity during the run. Multiplying the force by the velocity, and converting ft-lbf/sec to horsepower, we get the power traces in **Figure 18**.

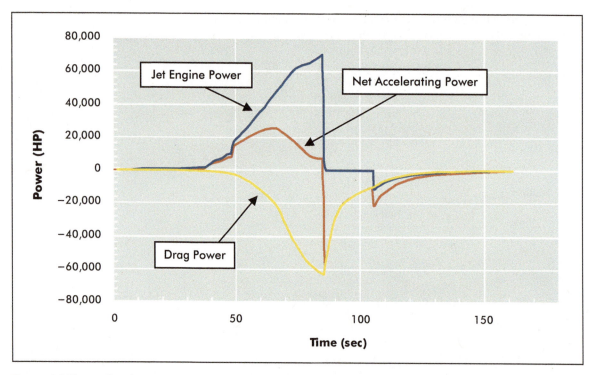

FIGURE 18 The engine thrust power, drag power, and accelerating power of the Thrust SSC when it broke the sound "barrier." Peak power was about 70,000 HP!

The power here is hard to picture. At 70,000 HP, the jet engines produce 140 times the power of the Corvette Z06 we'll study in Chapter 15, and require 5 gallons/second of jet fuel. Flying low isn't cheap.

12   Engines vs. Motors

What's the difference between engines and motors? Remember the rule that all squares are rectangles, but not all rectangles are squares? There's a similar distinction between engines and motors. All engines are motors, but not all motors are engines. Motors are devices that move something else:[15]

A **MOTOR** is any device that converts *supplied energy* into work.

The definition for an engine is more restricted:

An **ENGINE** is a device that converts *heat* into work.

Since converting thermal energy to mechanical energy runs counter to Nature's nature, it requires a complex device called an engine.[16] We've been calling them *heat engines* to emphasize it. The two main types are

---

Note 15: The Latin *motor* means a "mover" of something.

Note 16: The word "engine" comes from the Latin *ingenium*; roughly, an ingenious device. It takes ingenuity to turn heat into work. That's where the word "engineer" comes from, too.

piston engines and turbines. Either can use internal combustion or external combustion, the fuel either burned inside or outside the engine.

The typical automotive engine is an internal combustion piston engine, fuel being burned in the cylinders to create pressure and drive the pistons. Most external combustion piston engines heat steam in a boiler and then pipe it into the cylinder, letting the steam expand as it pushes on the piston. Steam engines were common until the mid-1940s in locomotives.

Turbine engines are very common in electric power plants and in jet airplanes. Power plants use external combustion to heat steam, which then drives the turbines. Turbines in planes use internal combustion, burning the air/fuel mixture in a combustion chamber after compressing it. Besides sending a fast jet of burned gases rearward, they often drive a propeller or fan with the turbine shaft.

KEY CONCEPT

Where mass is not an issue, external combustion engines are common. But where the engine moves itself (automotive and aeronautic use), mass and space are more important, and internal combustion is used. This avoids having the boiler.

We usually use the word "motor" to describe machines that are supplied some sort of mechanical energy and put out work. They include electric motors, compressed air motors, and hydraulic motors. These don't produce power, but only convert it to shaft power.

13 | Efficiency

Having discussed engines and motors, it's time to review energy efficiency. It's a pretty simple concept: the ratio of work out vs. energy in. So you'd think it wouldn't cause any confusion. But it does, because efficiency numbers are often thrown around without regard to what type of energy is being put in.

13.1 Thermal Efficiency

The energy input to an engine is heat, so when we rate its efficiency, we divide the engine output work W by the thermal energy supplied (the high-temperature heat Q_H from the fuel):

Physical Law Eq. 14-21   **Engine thermal efficiency**

$$\eta_{Thermal} = \frac{W}{Q_H}$$

An engine converting ⅓ of its supplied heat to work would have a thermal efficiency of 33.3%. Depending on driving conditions, the engine in your car could be between 5% and 40% efficient, once all the inefficiencies are taken into account. With gasoline engines, it depends strongly on how highly loaded the engine is. They get much more efficient near wide-open throttle, and are very inefficient at idle. Diesel engines don't change as much with load.

Thermal efficiency only applies to engines, since only they are fed with heat.

13.2 Mechanical Efficiency

We want to use motor/engine mechanical energy (and thus power) as completely as possible to drive loads. But as discussed, some work (or power) is lost to heat along the way. We define mechanical efficiency as the ratio, output vs. input, of either work or power:

Physical Law Eq. 14-22   **Mechanical efficiency**

$$\eta_{Mechanical} = \frac{W_{Out}}{W_{In}} = \frac{P_{Out}}{P_{In}}$$

Every device a car uses to transfer mechanical energy dissipates power as heat, reducing its mechanical efficiency. That's why many transmissions and some final drive gears need coolers.

In hybrid gasoline/electric vehicles, electrical motors/generators also transfer some of the work done by the engine. In electric vehicles, electric motors transfer mechanical energy from the battery to the drive axles. These processes can all be rated by their mechanical efficiency.

Because you could ideally convert one type of mechanical energy completely to another, the goal for mechanical efficiency should be 100%. The efficiency of manual transmissions is about 95% or so, and modern hypoid ring and pinion sets have efficiencies over 90%. To transmit power as efficiently as gears, electric motor/generator pairs should also be over 90%.

14 | 25% Efficient Engines vs. 95% Efficient Electric Motors

True or False: Gasoline/electric hybrids are more efficient because electric motors are 95% efficient, while gasoline engines are only 25% efficient.

The answer is "false." To see why, first let's look at the energy flow in a gasoline engine. The engine is shown in **FIGURE 19** as a black box, with high-temperature heat added as fuel.[17] For argument's sake, let's assume the engine is operating at 25% thermal efficiency, so 75% of the supplied energy is rejected as low-temperature heat.

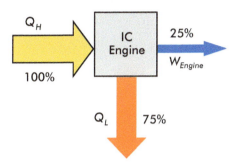

FIGURE 19 Energy flow in a gasoline engine. Being a heat engine, it receives heat Q_H and partially converts it to work W_{Engine}, the rest being rejected as heat Q_L.

Now for the hybrid powertrain in **FIGURE 20**, using the IC engine from **FIGURE 19**. Unless it's a plug-in hybrid, all the electrical energy driving the electric motor was originally generated by the engine, in driving the generator.[18] So the same thermal efficiency limits come into play. Assuming 100% efficiency of the generator/motor pair, the thermal efficiency of the powertrain is still 25%. This figure assumes that at this moment, we are not using energy from the battery.

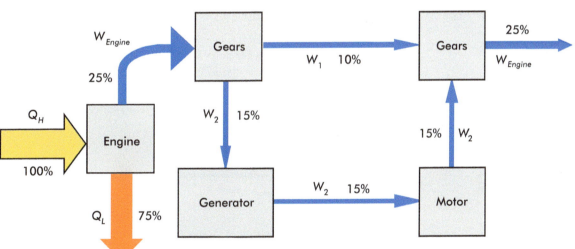

FIGURE 20 Energy flow in a gasoline/electric hybrid powertrain. The engine supplies work to the transmission, which splits it into two paths. Path 1 drives the output gears through a shaft, and Path 2 drives the generator, which drives the motor. This is called a "power split."

Note 17: Okay, the box isn't actually black, but a "body in white" coming out of a weld shop isn't white either.

Note 18: Within a few miles from the dealer's lot, even the energy in the battery is from the engine-driven generator.

There are two main things to learn from comparing FIGURE 19 to FIGURE 20:

1. The electric motor/generator pair doesn't generate work, it transmits it like a gear set.
2. If the motor/generator pair is less than 100% efficient, it can *reduce* efficiency.
3. There isn't much difference in overall thermal efficiency, fuel to wheels.

> **KEY CONCEPT**
>
> An electric motor/generator pair is a way of transferring power from one shaft to another, just like a transmission or gear set.

All right, why have a hybrid powertrain? Well, they do save fuel, mostly by shutting off the engine at "idle" and by using the motor/generators to brake (regenerative braking). The extra power from the electric motors during acceleration may also allow a smaller IC engine, which should improve its thermal efficiency. Most other improvements in a hybrid (such as lowering drag and rolling resistance) can be done on other vehicles.

So hybrids can certainly help fuel economy, but don't expect anything like 95% efficiency in converting fuel to power.

> **KEY CONCEPT**
>
> Comparing efficiencies of electric motors and IC engines is like comparing apples and oranges. Electric motors must be rated by mechanical efficiency, and IC engines by thermal efficiency.

15 Summary

How is power different than energy?

Energy can be in different forms, which we split into 1) mechanical and 2) thermal. But power is always about *mechanical energy*, specifically the rate at which work is put out or brought in (in either ft-lbf/sec or N-m/sec). The only time we directly relate power to thermal energy is when power is dissipated, such as in bearings, braking, or aerodynamic drag. Even then we should really calculate the "negative power" from the torque or force opposing motion.

But we also use the terms energy and power differently when it comes to their sources. By *power source* we mean anything you can hook up to drive a mechanism, such as electrical outlets, compressed air, and rotating shafts. But an *energy source* is whatever fed the device that produced the power, like gasoline fueling an engine, natural gas feeding an electric power plant, or sunlight feeding a solar cell. While most energy sources are thermal energy (like nuclear), some are types of mechanical energy (like water pressure in a hydroelectric plant).

In vehicle performance, we've seen that acceleration at *constant power* isn't possible at low speeds, and that after just a few seconds, *constant acceleration* isn't either. Chapter 15 will combine these facts to predict the performance of an "ideal" drag racer.

Tracking power is a powerful way (pun intended) to simplify your understanding of propulsion. When we followed the "power flow" through a driveline, the torque and speed changed greatly, but the power only slightly. Except for losses, the power at the engine equaled the power at the tires: $T_{Engine}\omega_{Engine} = F_{Thrust}v_{Vehicle}$. So, as **Figure 12** showed, a driveline will:

1. Increase or decrease torque,
2. Decrease or increase shaft speed (inversely to torque), and
3. Maintain power (at best).

Remember that while power can be calculated by multiplying torque by engine speed, it is often easier to multiply force by vehicle speed.

It's clear that electric motor/generator pairs in hybrid powertrains are equivalent to conventional transmissions in conveying power from one shaft to another. They do have more flexibility in the speed ratio between input and output, and so are common in diesel/electric locomotives.[19]

About that "power vs. torque" debate—the highest acceleration at a given vehicle speed comes with the right gearing to put the engine at its peak power. So more power increases max acceleration, and power is more important for overall performance. Then why do people like "torquey" engines, even with less power? First off, they're fun. They usually have a flatter torque curve than a smaller, high-winding engine, making them flexible—you can punch it and accelerate pretty well regardless of engine speed. They have more "grunt." But that also makes them good for road racing, where you want linear changes in torque vs. throttle position and engine speed, and you'd like to avoid having to shift in a curve to stay in a narrow power band. If all that doesn't settle the argument, sorry.

You can adapt the formulas here to many powertrain variants, including any power source or draw. A genuine understanding of power is a must for anyone who wants to appreciate the trade-offs in design decisions—hot rodders, engineers, designers, race enthusiasts—really any gearhead.

Note 19: The diesel engine drives a generator that drives the traction motor, making it a "series hybrid."

Major Formulas

Definition	Equation	Equation Number
Power in Straight-Line Motion	$P = Fv$	14-1
Power in Rotational Motion	$P = T\omega$	14-2
Driveline Power Out vs. Power In	$P_{Out} = \eta_{Mechanical} P_{In}$	14-3
Vehicle Speed as a Function of Engine Speed and Gear Ratios	$v = \dfrac{\omega_{Engine} r_{Tire}}{R_{Trans} R_{FD}}$	14-4
Traction Force Using a Real Driveline (If Not Traction-Limited)	$F_{Thrust} = \dfrac{T_{Engine} \eta_{Trans} R_{Trans} \eta_{FD} R_{FD}}{r_{Tire}}$	14-6
Power at the Tire Patch from Engine Power, No Slip	$P_{Thrust} = P_{Engine} \eta_{Trans} \eta_{FD}$	14-7
Power Going to Friction in a Slipping Clutch	$P_{Friction} = T_{Friction} \left(\omega_{In} - \omega_{Out} \right)$	14-8
End Time After Constant Power Acceleration	$t_2 = t_1 + \dfrac{m \left(v_2^{\,2} - v_1^{\,2} \right)}{2P}$	14-9a
End Speed After Constant Power Acceleration	$v_2 = \sqrt{\dfrac{2P \left(t_2 - t_1 \right)}{m} + v_1^{\,2}}$	14-9b

Definition	Equation	Equation Number
Power to Accelerate at a Given Speed, No Drag	$P_{Accel} = mav$	14-10
Power in Constant Acceleration a, Standing Start	$P_{Accel} = ma^2 t$	14-11
Power in Constant Acceleration a, Standing Start	$P_{Accel} = \left(ma^{\frac{3}{2}} \right) \sqrt{2s}$	14-12
Power to Overcome Rolling Resistance	$P_{Rolling} = F_{Rolling}\, v$	14-13
Power to Overcome Aerodynamic Drag	$P_{Aero} = \dfrac{1}{2} \rho v^3 C_D A_{Frontal}$	14-14
Top Speed, Considering Aerodynamic Drag Only, Power at the Tire Patches	$v_{Max} = \sqrt[3]{\dfrac{2 P_{Max}}{\rho C_D A_{Frontal}}}$	14-15
Thermal Efficiency	$\eta_{Thermal} = \dfrac{W}{Q_H}$	14-21
Mechanical Efficiency	$\eta_{Mechanical} = \dfrac{W_{Out}}{W_{In}} = \dfrac{P_{Out}}{P_{In}}$	14-22

15 Power Applications

Power Production and Utilization, and Energy Efficiency

As a 6.2-liter LS3 engine from a Cadillac CTS-V SCCA World Challenge car is tested, it puts out lots of power to the dyno, and lots of heat to the headers as a by-product. *Photo by Randy Beikmann with permission of Katech Inc.*

Power lingo and measurements are explained. The concept of the CVT is presented, and used to calculate the ET and top speed for an "ideal drag racer." Traction force at the driving tires is found for a Z06 in different gears, and then used to predict its acceleration performance. A close look is taken at power production and efficiency in an internal combustion engine. Internal combustion engines and electric propulsion are compared.

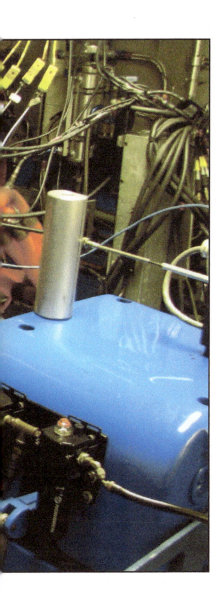

Contents

Key Symbols Introduced

Symbol	Quantity	SFS Units	MKS Units
a_{TL}	Traction-Limited Acceleration	ft/sec^2	m/sec^2
a_{PL}	Power-Limited Acceleration	ft/sec^2	m/sec^2

1 | Introduction

We've seen that power is critical in automotive physics, where every function of the car depends on how *quickly* work gets done. It isn't just important in racing, but also in merging onto a busy freeway or passing safely on a two-lane. Still, in between accelerations, you'd like to get good fuel economy.

So the first part of this chapter concentrates on the fun stuff: maximizing performance by making use of the available power and traction. We'll carry on with investigating components of the driveline, like a torque converter and a CVT (continuously variable transmission). We'll use some of these to do a "first shot" calculation of the performance of an ideal drag racer, and of a Corvette Z06 going through the gears.

The middle part of the chapter deals with converting a fuel's energy into power. In Chapter 13 we put a simplified 1-D gas in a cylinder, compressed it, and then let it expand. We got the compression work back during expansion, breaking even (work out vs. work in). Now we will *produce* work by adding heat to the gas at the right time, but also see the effect of the second law of thermodynamics. Although it's a very simplified engine, it illustrates the basics of converting heat to work.

We round out the chapter with a more serious side of driving: fuel economy. Filling your tank hits you in the pocketbook every time you pull into a gas station, and can put a crimp on your enjoyment of driving. To me this is part of the performance challenge. I'd like to get good fuel economy and still be able to have a little fun. Plus, bad memories of previous oil crises are still with me, and I don't want to get burned again.

Comparing today's performance cars to the muscle cars of the 1960s, we've already come a long way. A car that did 0–60 in 6 seconds back then would get 10 mpg—maybe. Now there are cars that go 0–60 in just over 4 seconds and get in the upper 20s on the highway. But the better things get, the harder it is to improve on them. How do we decide where to go from here?

In considering this, you need to fairly compare the efficiencies of different propulsion systems. This is relatively easy since we've grouped energy into mechanical and thermal types. Here, we'll compare electric propulsion to the typical internal combustion engine. You may be surprised by the outcome, and may not even believe it. That is fine. Science is all about being skeptical about a theory or claim, testing it, and either verifying, modifying, or scrapping it. Anyway, you're welcome to look up the data and prove it to yourself.

2 Brake Horsepower, Net and Gross

Horsepower is often poorly explained or "glossed over" when discussing it. To make matters worse, there isn't just one "type" of horsepower quoted; you see *brake* horsepower, as well as *SAE gross* and *SAE net*. Besides SAE, you may also see DIN (the Deutsches Institut für Normung—the German Institute for Standardization) and JSAE (the Japanese Society of Automotive Engineers) horsepower quoted. With all this, how can you compare engines, especially today's vs. the legends of the 1960s?

Brake horsepower is what the engine produces at the back of the crankshaft while turning an engine dynamometer, or "dyno" (an electric motor/generator). But it got the name "brake" horsepower because it was first measured by having the engine drive the shaft of a Prony brake (**FIGURE 1**). Here, friction with the driven shaft lifts a weight through a torque arm. It is essentially a drum brake built into a lever, where the bolts are tightened until the lever *just* lifts the weight mg. The torque T equals mgr, and power is calculated from $P = T\omega$. In the figure, the engine produces 400 ft-lbf of torque to lift the 200 lbf of weights. Turning at 6,000 rpm, the engine is producing 457 HP.

FIGURE 1 Prony brakes measure engine torque by balancing the torque in the engine-driven shaft with torque from the weights, through the friction material in the brake.

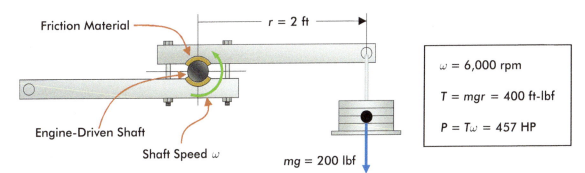

Friction Material

$r = 2$ ft

Engine-Driven Shaft

Shaft Speed ω

$mg = 200$ lbf

$\omega = 6{,}000$ rpm

$T = mgr = 400$ ft-lbf

$P = T\omega = 457$ HP

Water brakes do the same thing, and are often used in race shops because they are more durable and still have low inertia; you can measure torque during fast acceleration (see **FIGURE 2**). The engine's power is absorbed by a pump pressurizing and heating the water flowing through it. The force at the end of the torque arm is measured with a load sensor.

But whether measured against a brake or a dyno, it's still called brake horsepower. An electric dyno has the capability of driving the engine without it firing (called "motoring" the engine), to measure the torque required to overcome friction.

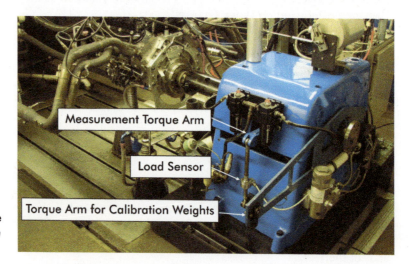

FIGURE 2 The shorter lever arm (blue) is attached to the pump housing, and measures force with a load sensor. That force is used to calculate torque. *Photo by Randy Beikmann with permission of Katech Inc.*

What about SAE gross HP vs. SAE net? Gross power is the brake power produced while the cooling fan and engine accessories are driven by something besides the engine, and with low restriction intake and exhaust systems (a best-case scenario). Net horsepower is measured with realistic accessory loads, and is a more accurate measure of how well the engine will perform in the car. Gross power overrated engines by about 20%. Car manufacturers advertised gross power in the 1960s, and net power was reported beginning in the 1970s.[1]

In 2005, the SAE updated their test procedure J1349 to stop the funny business, introducing more stringent conditions for testing. It also requires an SAE-approved witness to be present at the test, after which the engine is rated in "SAE J1349 Certified Power."

Note 1: Once insurance companies started basing insurance rates on horsepower ratings, manufacturers sometimes underrated horsepower to make ownership more affordable.

3 Driveline and Engine Characteristics

Let's take another look at a couple more factors in matching the driveline to the engine. We'll start with a quick discussion of torque converters, and a quick reality check to see if you've been paying attention. Then we'll go through engine torque and power curves, and the CVT.

3.1 How Torque Converters Work

Automatic transmissions are almost always driven by the engine through a torque converter, which can increase the driving torque. It's got a donut-shaped shell (a toroid), and is filled with automatic transmission fluid (ATF). As it spins, its impeller blades take fluid in near the center, accelerate it outward, and then aim it at the turbine. The fluid then pushes on the side of the turbine blades, producing torque to drive the transmission.

FIGURE 3 shows one sliced through, as seen from along the crankshaft axis, so you can see the inside of the (engine-driven) *impeller*, which drives the fluid. It rotates at speed ω_{Crank} and its blades force the fluid to go around with it. The figure shows one "chunk" of fluid, which enters the impeller (at 12:00) at the low speed v_{In}, but later exits (at 5:30) at the outer radius with the much faster speed v_{Out}.

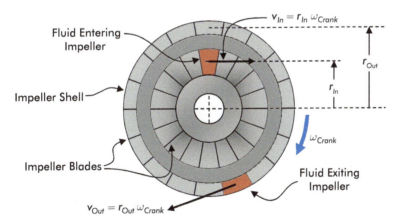

FIGURE 3 A slice through a torque converter, with an edge-on view of the impeller blades that push the fluid around with it. The fluid chunk (in "ATF red") enters the impeller slowly, but is flung out much faster. This high-speed fluid drives the turbine.

The fluid also rotates around the toroidal cross-section (**FIGURE 4**), which is how fluid enters and exits the impeller.[2] This is called the "toroidal flow." When the fluid is "slung out" by the impeller, it hits the side of the turbine blades. This assumes the turbine blades are "flat" just like the impeller blades. In reality the turbine blades are curved to accept the fluid more smoothly, increasing efficiency. The resulting torque drives the turbine, which in turn drives the transmission shaft.

Note 2: It's good to switch between **FIGURE 3** and **FIGURE 4** to visualize the action.

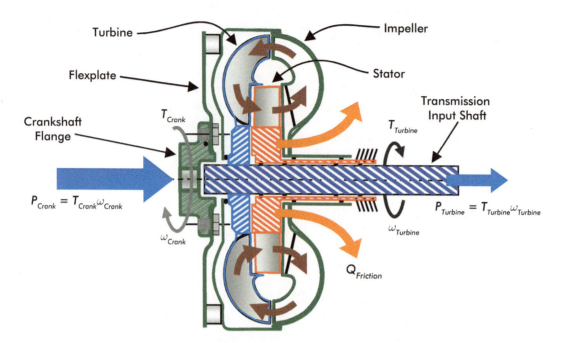

FIGURE 4 The engine drives the impeller (green border), which flings fluid into the turbine (blue border), exerting force on the side of the turbine blades. The resulting turbine torque drives the transmission. When the engine is putting out torque T_{Crank}, the input speed ω_{Crank} is greater than the output speed $\omega_{Turbine}$.

Unlike clutches, torque converters can *increase* torque, producing a turbine torque up to twice the engine torque, depending on the conditions. This is only because of the stator, which can exert torque on the transmission case. The stator steers the fluid flow in the direction of the impeller rotation, "assisting" the impeller in producing high-speed fluid. Torque converters trade speed for torque; they can't create power. For example, when multiplying torque by 1.5, the turbine speed must be ⅔ the crank speed or less.

3.2 A Torque Converter at Stall

Getting ready to launch a drag racer with an automatic transmission, you "stall" the engine against the torque converter. In gear, you hold the brakes to keep the axles stationary, throttle down. When stalled, what is the power output from the torque converter? Assume the engine speed is 5,000 rpm and putting out 600 ft-lbf of torque.

This sounds hard, because we don't know the converter output torque. But it's easy—in gear with the axles not turning, the transmission and turbine aren't either. Since power is torque times angular velocity, the power output from turbine to transmission is

$$P_{Turbine} = T_{Turbine}\omega_{Turbine} = \left(T_{Turbine} \text{ ft-lbf}\right)\times\left(0 \text{ rad/sec}\right) = 0 \text{ ft-lbf/sec}$$

Regardless of its output torque, if the turbine isn't turning, its power output is zero. The instant you release the brake and things start turning, power is produced. At stall, the input power (here 571 HP) is all being turned to heat, warming up the transmission fluid (very quickly). This is similar to how a water brake dissipates power.

3.3 Torque and Horsepower Curves, and Gear Selection

Chapter 14 examined torque and horsepower curves for a generic gasoline engine at wide-open throttle (WOT). Actual WOT curves for a 2008 Corvette Z06 engine (**FIGURE 5**) look very similar. The torque builds from 1,000 rpm, reaching its major torque band from about 2,000–6,500 rpm, with a peak of 475 ft-lbf at 4,800 rpm. Since the torque curve is almost flat, the power increases almost linearly from 2,000 to 5,500 rpm. It peaks at 6,500 with 505 HP.

In a given gear, the thrust at the tire patch depends on the engine torque, the overall gear ratio, and the tire diameter. When do you hit maximum acceleration in a given gear? Could you accelerate quicker in a different gear?

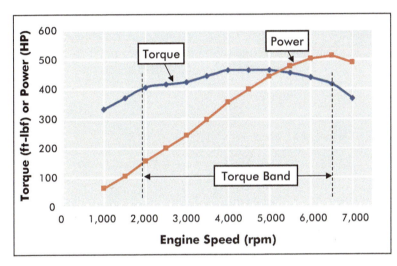

FIGURE 5 WOT torque and HP curves for a 2008 Corvette Z06 7.0-liter engine. The torque is relatively flat and peaks at 4,800 rpm, while power continues to build to a peak at 6,500 rpm.

To keep it simple, we'll ignore mechanical losses. Using $F = ma$ and equation 14-6, the maximum acceleration in any gear is at maximum engine torque T_{Max}:

Equation 15-1   **In a given gear**

$$a_{Max} = \frac{T_{Max} R_{Trans} R_{FD}}{r_{Tire} m}$$

So the peak acceleration *in a given gear* is at peak engine torque (here 4,800 rpm). But what if, at the same vehicle speed, you shift into a lower gear (more gear reduction)? With no losses, drag, or tire slip, the power from the engine is the same as at the tire patch:

$$P_{Engine} = P_{Tire}$$

$$T_{Engine} \omega_{Engine} = F_{Thrust} v$$

So:

$$F_{Thrust} = \frac{T_{Engine} \omega_{Engine}}{v}$$

Knowing that $a = F_{Thrust}/m$ gives

$$a = \frac{T_{Engine}\omega_{Engine}}{mv}$$

So, the acceleration at speed v is not just dependent on the engine torque, but engine torque times engine speed, i.e., power. If we downshift so that the rpm is at peak power (assuming the right gear ratio is available):

Equation 15-2   **At a given vehicle speed**

$$a_{Max} = \frac{P_{Max}}{mv}$$

Your maximum acceleration at a given vehicle speed would be with the engine at its power peak. Here, downshifting from peak torque (475 ft-lbf @ 4,800) to peak power (505 HP @ 6,500) could increase acceleration by 16%. It sounds like a contradiction, but when the engine feels its strongest in third gear, you'd accelerate quicker if you downshift to second. The extra torque multiplication makes up for the drop in engine torque.

3.4 The CVT

Given the traction, you could accelerate quickest if your transmission held the engine speed at peak power, regardless of vehicle speed, whenever your foot was to the floor. This is the idea behind the *continuously variable transmission*. A CVT can also adjust engine speed for maximum efficiency at cruise. The CVT typically runs a metal belt between two pulleys, the speed ratio controlled by the pulley sheave separation (see **FIGURE 6**).

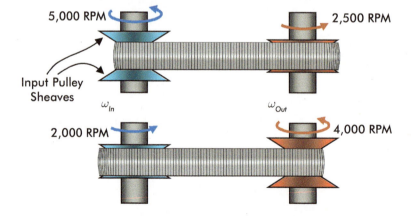

FIGURE 6 The drive mechanism of a CVT, using a metal belt. The pulley sheaves are pressed against the belt to prevent slip, and their separation adjusts to change the speed ratio.

The "constant power" acceleration of an ideal CVT (zero loss) is a good goal in maximizing performance. The next section will use that idea to calculate the ET and top speed of the ideal drag racer.

4 Traction, Power, and Gearing

We have been assuming that the tires can use all of the engine's power, but obviously they only have so much traction. The next couple of sections investigate the interplay between available power and traction.

4.1 ET and Top Speed for the Ideal Drag Racer

The ideal drag racer would use all its traction and power to make it down the track in the lowest elapsed time (ET). So given certain levels of power and traction, what is the best it could do in a quarter mile? As an example we'll use a 3,220-pound car with 300 HP available at the tire patch, and 1,500 lbf of traction. We'll ignore drag and whether we can actually get a transmission that does what we want. This is the ideal case to work toward.

First, what traction force is required to put a given power P to the ground? Power is force times velocity, so the traction force needed at a given speed is:

Equation 15-3   **Required traction to apply power P**

$$F_{Thrust} = \frac{P}{v}$$

What does the thrust look like at 300 HP? We divide power by speed, so at low speeds, the required traction is *very* large, as shown at the left end of **Figure 7**. Up to 40 ft/sec (27 mph), it would take over 4,000 lbf, *way* above the 1,500 lbf available! In fact, the tires can't use all 300 HP until 110 ft/sec (75 mph). Up to then acceleration is limited by the available traction.[3] There is no way a vehicle can use all its available power at launch.

So with that, how would we run the ideal drag racer, given its power and traction limit?

> **MAJOR POINT**
>
> The ideal drag racer would run at its peak acceleration, limited by traction, until its engine couldn't produce enough power to do so. After that it would accelerate at constant peak power, using a CVT to hold the engine speed on the power peak.

Note 3: Looking at the force required at say, 5 mph to produce 300 HP, it's probably best you don't have the traction to do it. That much force might produce enough acceleration to snap your neck!

So we split the run into two sections. The "traction-limited" section begins at the starting line (point 1 in **FIGURE 7**). It ends where the car first hits its power limit (point 2). The "power-limited" section runs from there to the end of the quarter-mile (point 3). Traction-limited acceleration is easy to calculate. With maximum thrust of F_{TL}:

Equation 15-4   **Traction-limited acceleration**

$$a_{TL} = \frac{F_{TL}}{m}$$

where "TL" stands for traction-limited. From a standing start, the velocity is:

Equation 15-5

$$v_{TL} = a_{TL}t$$

and the displacement is:

Equation 15-6

$$s_{TL} = \frac{1}{2}a_{TL}t^2$$

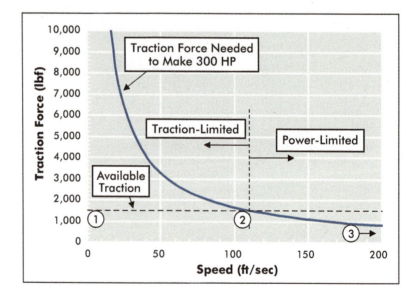

FIGURE 7 Limits on traction force. The solid curve is the required traction to use the engine's 300 HP, while the dashed line shows that only 1,500 lbf is available. The maximum thrust you can actually get, at any given vehicle speed, is the smaller of the two.

We don't need to plot the two curves in **FIGURE 7** to find the crossover from being traction-limited to power-limited (point 2). Taking equation 15-3, setting the power to P_{Max}, and thrust to the traction limit F_{TL}, the speed at point 2 is:

Equation 15-7   **Speed when reaching power limit**

$$v_2 = \frac{P_{Max}}{F_{TL}}$$

which here is 110 ft/sec. We can now find the time and distance to point 2. Using the fact that the velocity v_2 equals $a_{TL}t_2$, the traction-limited acceleration a_{TL} is F_{TL}/m, and using equation 15-7, the time t_2 is:

$$t_2 = \frac{v_2}{a_{TL}} = \frac{m}{F_{TL}}v_2 = \frac{m}{F_{TL}} \times \frac{P_{Max}}{F_{TL}}$$

That is:

Equation 15-8   **ET at reaching power limit**

$$t_2 = \frac{mP_{Max}}{F_{TL}^2}$$

Plugging this into equation 15-6, the distance traveled at point 2 is:

$$s_2 = \frac{1}{2}a_{TL}t_2^2 = \frac{1}{2}\frac{F_{TL}}{m}\left(\frac{mP_{Max}}{F_{TL}^2}\right)^2$$

or:

Equation 15-9   **Displacement when reaching power limit**

$$s_2 = \frac{mP_{Max}^2}{2F_{TL}^3}$$

So the larger the mass and the more power, the longer it takes before you can use all your power. But doubling the traction would let you use max power in one quarter the time, and in one eighth the distance! It's easy to see why traction is especially important to drag racers.

Now let's finish the run. We know everything at the crossover point: s_2, v_2, and t_2. We also know the ending displacement s_3 is ¼ mile, or 1,320 feet. After hitting the power limit, acceleration drops. The power-limited acceleration is found knowing that $a_{PL} = F_{PL}/m$, and $F_{PL} = P_{Max}/v$:

Equation 15-10   **Power-limited acceleration**

$$a_{PL} = \frac{P_{Max}}{mv}$$

The higher the speed, the less acceleration. Using this, and some math that is outside the scope of the main text, the top speed v_3 is:[4]

Equation 15-11   **Top speed**

$$v_3 = \sqrt[3]{v_2^3 + \frac{3P_{Max}(s_3 - s_2)}{m}}$$

Note 4: The derivations of this equation and the next are in Appendix 5.

Notice that is a *cube* root. This value can be used to find the time at the end of the quarter-mile, after we do a little crunching:

Equation 15-12   ET

$$t_3 = t_2 + \frac{m}{2P_{Max}}\left(v_3{}^2 - v_2{}^2\right)$$

Now let's see if it's in the ballpark, by finding the ET and top speed of our 300 HP (165,000 ft-lbf/sec), 3,220-lbm (100-slug) car, traction limited to 1,500 lbf. The time t_2 to point 2 is:

$$t_2 = \frac{mP_{Max}}{F_{TL}{}^2} = \frac{100 \text{ slugs} \times 165,000 \text{ ft-lbf/sec}}{(1,500 \text{ lbf})^2} = 7.3333 \text{ seconds}$$

The distance to point 2 is:

$$s_2 = \frac{mP_{Max}{}^2}{2F_{TL}{}^3} = \frac{100 \text{ slugs} \times (165,000 \text{ ft-lbf/sec})^2}{2(1,500 \text{ lbf})^3} = 403.33 \text{ ft}$$

As we'd calculated, the velocity at the crossover point (point 2) is 110 ft/sec. So, using equation 15-11, we calculate the top speed:

$$v_3 = \sqrt[3]{v_2{}^3 + \frac{3P_{Max}(s_3 - s_2)}{m}} = 180.4 \text{ ft/sec}$$

which is about 123 mph. The elapsed time (ET) for the quarter-mile is then:

$$t_3 = t_2 + \frac{m}{2P_{Max}}\left(v_3{}^2 - v_2{}^2\right) = 7.333 \text{ sec} + 6.192 \text{ sec} = 13.525 \text{ sec}$$

The results are pretty reasonable. They are a bit optimistic, because we neglected losses, and assumed we were always at the traction limit or using maximum power. On the other hand, we didn't include rollout, which would have reduced ET.

Speed is plotted vs. time in **Figure 8** for our 1,500 lbf traction limit, and for 3,000 lbf (because 1,500 is pretty low for drag slicks). Each trace ends when the car completes the quarter-mile. Looking closely at the 1,500 lbf run, it shows the speed increasing linearly (constant acceleration) for the first 7.33 seconds, then curving slightly once the power can't keep up.

With 3,000 lbf of traction, the crossover happens at 1.833 seconds, and is much easier to see. A major thing to note is that with twice the traction, the ET drops from 13.53 to 11.52 seconds, *without* an increase in power.

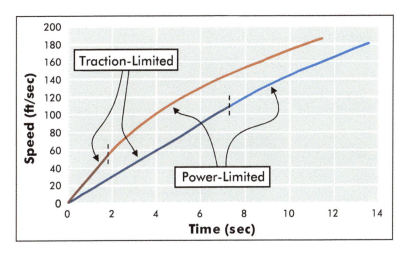

FIGURE 8 Speed vs. time for our ideal drag racer (blue line 1,500 lbf traction, red line 3,000 lbf). Doubling traction cuts the ET by about 2 seconds.

Now let's see speed vs. *distance* in FIGURE 9. It's harder to tell where the acceleration becomes power-limited, because the trace starts curved anyway. For the record, the 1,500-lbf traction car is traction limited until 403 feet, while with 3,000-lbf it's only 50 feet—it hooks up fast. It's interesting how much the speed picks up in the first ⅛ mile vs. the last ⅛. The car travels so fast during the second ⅛ mile, there isn't nearly the time to accelerate, or the power to do it.

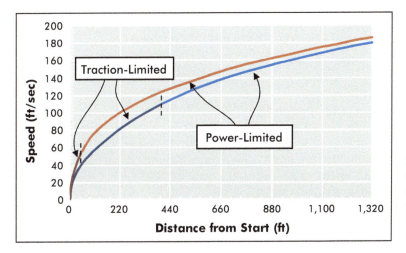

FIGURE 9 Speed vs. distance for the two versions of our ideal drag racer. Notice how close the top speeds are: 186.2 ft/sec vs. 180.4 ft/sec (127 mph vs. 123 mph).

The other major point is that although doubling the traction cuts the ET by 2 whole seconds, top speed is only 4 mph higher! Since it's low ET that wins races, it pays to get a quick start by having enough traction and the right gearing to use it. *Then* let power do its thing.

4.2 Maximum Acceleration for the Z06

If you know the torque curve for an engine, and the car's gear ratios, it's fairly straightforward to find its acceleration vs. speed. Let's use the torque curve from **FIGURE 5** to analyze the WOT acceleration of the 2008 Z06 Corvette.

First, let's find its traction limit. I looked up some *Road & Track* data, and from 0 to 60 mph, its speed increased in a straight line, a constant acceleration (**FIGURE 10**). By dividing the change in velocity by the change in time, the acceleration a_{TL} is 22.9 ft/sec², or 0.71 g's.

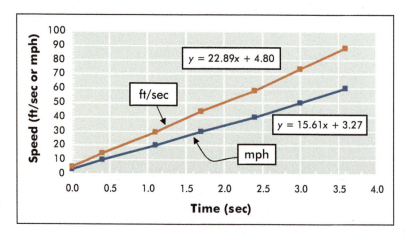

FIGURE 10 Speed vs. time for a 2008 Corvette Z06 over its 0–60 mph WOT acceleration. Data, from *Road & Track* magazine, are plotted in mph and ft/sec.

The constant acceleration indicates the car is at its traction limit. Knowing the acceleration a_{TL}, and assuming the car weighs 3,550 lbf (mass is 110.25 slugs), we can find max traction:

$$F_{TL} = ma_{TL} = 110.25 \text{ slugs} \times 22.89 \text{ ft/sec} = 2{,}520 \text{ lbf}$$

That's a good bit of traction for a 3,550-lbm car on street tires. Now let's see what tire force the engine could actually power in each gear.[5] We multiply the engine torque curve by the gear reductions and mechanical efficiency to get axle torque, and then get traction force by dividing by tire radius:

Equation 15-13   **Torque and gear-limited thrust**

$$F_{TGL} = \frac{T_{Engine}\eta_{Mech}R_{Trans}R_{FD}}{r_{Tire}}$$

This is the maximum thrust the engine can drive, after being "torque and gear-limited" by the available engine torque at your engine operating speed, and by the selected gear ratio. We can calculate this in every gear, vs. engine speed. The driveline data for this are in **TABLE 1**, and the tire radius is 1.07 feet.

Note 5: Using published acceleration data to find the traction limit, then calculating acceleration with that, may seem like circular logic. But it lets us use known data to combine the effects of weight distribution, load transfer, and tire properties. Plus, it gives you a way to check your results for the first part of the run.

Gear	First	Second	Third	Fourth	Fifth	Sixth
Transmission Ratio	2.66	1.78	1.30	1.00	0.74	0.50
Final Drive Ratio	3.42	3.42	3.42	3.42	3.42	3.42
O/A Ratio	9.10	6.09	4.45	3.42	2.53	1.71
MPH per 1,000 RPM	8.42	12.59	17.24	22.40	30.28	44.81

TABLE 1 Transmission and final drive gearing for a 2008 Corvette Z06. The vehicle speeds per 1,000 rpm are given in each gear, given its tire radius of 1.07 feet.

Using equation 15-13, we can calculate the traction force the engine could power in each gear, as plotted in FIGURE 11 vs. engine speed. An 85% mechanical efficiency of the driveline is assumed.

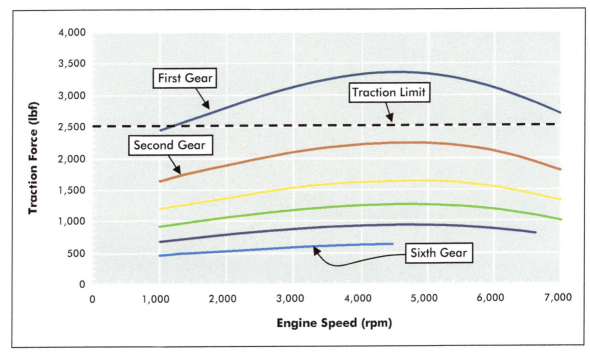

This plot is interesting, though not overly helpful. It does tell us that the early part of the acceleration was definitely traction limited, because the thrust capability all through first gear is at or above the 2,520-lbf traction limit. But it doesn't say where to shift to maximize vehicle acceleration. For that, we need to plot the thrust vs. vehicle speed (FIGURE 12).

FIGURE 11 The traction force the engine could power in each gear vs. engine speed. This is at or above the 2,520-lbf traction limit all the way through first gear.

Now we're talking. FIGURE 12 shows the whole picture of the thrust you could power in each gear, at any vehicle speed. It also plots the maximum thrust possible (in black) with the engine at its power peak at all vehicle speeds.

Nothing new in first gear—more than enough power to break traction throughout. A 1–2 shift at redline (about 60 mph) puts the engine right at its torque peak in second, and fairly close to the traction limit. Second gear produces more thrust than third all the way through, so there's no reason to shift until redline (about 85 mph). In third we start off slightly past the torque peak, and hold that to redline, shifting into fourth at 120 mph(!). That's good until about 155, when you shift to fifth. Note that for maximum acceleration, you never use sixth. It's only there for fuel economy at cruise.

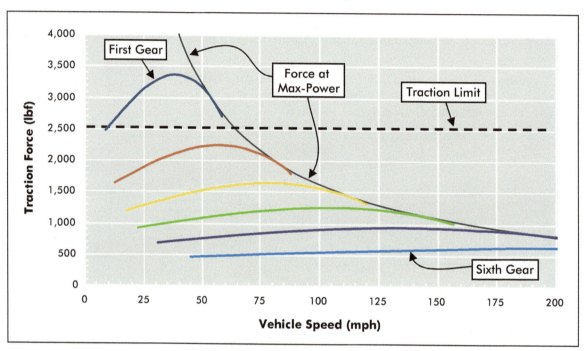

FIGURE 12 A more revealing picture of the thrust possible at the rear tires of the Z06, vs. vehicle speed. The fine black line plots the tire force with max power to the rear wheels. Each gear's trace skims it when the engine reaches its power peak.

To summarize, maximizing acceleration means maximizing the thrust, which for this example means always shifting at redline. Many cars would want closer gear ratio spacing than this. Because this engine produces so much rear-tire force over a broad speed range, wide spacing works fine. A heavier car (more traction and mass) with a less powerful engine or a narrow power band would benefit from having less rpm drop between gears, to keep the engine closer to its power peak. Many racing engines have had a "narrow power band," where power drops off sharply at speeds higher and lower than the power peak. These need closer transmission ratios to stay within that band.

How well do these calculations predict actual performance? If we take the thrust and subtract aero drag, we can estimate the resultant force on the car. Then it's just $F=ma$. FIGURE 13 compares the speed vs. time plots out to 150 mph, measured by R&T vs. calculated. A frontal area of 22.3 ft² and a drag coefficient of 0.34 were used. [6]

Note 6: The elapsed times were found in a spreadsheet, by taking small steps in speed. Calculate acceleration at one speed v_1 (say 90 ft/sec) and find the time to get to v_2 (say 91 ft/sec), as $\Delta t = a/(v_2 - v_1)$.

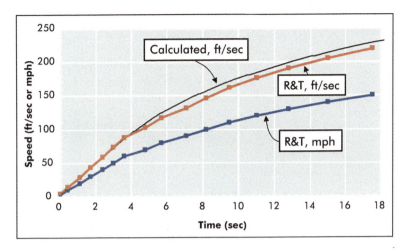

Figure 13 A comparison of the measured performance of the Z06 (*Road & Track*) vs. calculated from the vehicle specs. The calculation neglects shift times, so it leads actual performance past 60 mph.

The calculations match the tests almost exactly out to 60 mph (they should, since we based our traction limit on that). But why do the test results lag the calculations after 60 mph? My calculations didn't include the time it takes for each shift, when no power is delivered to the wheels.[7] Other than that, the results are quite good.

5   Harnessing Heat with an Engine

Once I read a list of supposed employee reviews circulating on the Internet. My favorite was "Works well when under constant supervision and cornered like a rat in a trap." Employing a fuel's heat to produce work is pretty much the same.[8] In Chapter 13 we saw that gas in a cylinder could store energy when being compressed, and return it by doing work during expansion. As an encore, we'll use the expansion stroke to harness heat.

The thermal energy in a hot gas is the random motion of its billions of gas particles, with no sense of direction; an "employee" with no focus. So we'll harness the fuel's heat by cornering the hot gas "like a rat" in a cylinder, and then "constantly supervise" its expansion, by only letting the piston move. The hot gas then has no choice but to do work on the moving piston.

Figure 14 shows the heat addition and expansion, using the 1-D gas from Chapter 13: a single particle bouncing back and forth between piston and cylinder head. From the left, it starts compressed at TDC, has heat added to it to increase its particle speed to v_3, and then expands without heat transfer as the piston moves. But the gas does cool

---

Note 7: Shift times could have been put into the calculations, but at some point I just had to quit monkeying around.

Note 8: First runner-up in the "Difficulty in Converting-Heat-to-Work Analogy" category was "Herding Cats."

during expansion, since each impact with the receding piston slows down the gas particle, while doing work on the piston.

Compressed, Before Burn

TDC

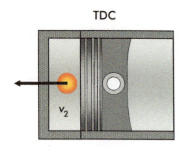

v_2

Burn

Q_H

TDC

v_3

Partially Expanded

45° ATDC

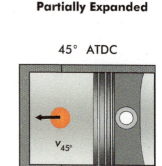

$v_{45°}$

FIGURE 14 A simple engine with our 1-D gas. At left, we have the compressed gas at TDC. In the middle, we add heat Q_H from fuel, speeding up the particle from v_2 to v_3. At the right is expansion, during which the particle speed drops (the gas cools) as work is done on the piston.

Now we have a simple internal combustion engine. We'll operate it using our 1-D gas, starting with the piston at BDC, the cylinder full of gas at atmospheric pressure:[9]

1. We compress the gas by pushing the piston to TDC, and find the work *input*.
2. We add heat (as from a burning fuel).
3. We allow the gas to expand and push the piston, and then find the work *output*.
4. We (reluctantly) reject heat at BDC.

Together, these four processes create a *thermodynamic cycle*, because we have a way to produce work *repetitively*. Afterward, we'll calculate the thermal efficiency. As in Chapter 13, we assume that the piston, cylinder wall, and cylinder head are insulated, so none of the gas's internal energy is lost to heat during compression or expansion.

5.1 Compression Stroke

We covered the compression stroke in Chapter 13.[10] As the piston moves from BDC to TDC:

1. The distance between the piston and cylinder head decreases.
2. The gas particle velocity increases from repeated impacts with the moving piston. Each does work on the gas.
3. The kinetic energy of the particle, the gas's internal energy, increases.
4. Particle/piston collisions become harder and more frequent, increasing the pressure.

At TDC, the distance from piston to head is ⅓ the original at BDC, and the pressure force goes up by a factor of 27 (from 184.7 lbf to 4,986

Note 9: This would be at the end of the intake stroke in an ideal internal combustion engine with its throttle wide open. Remember that BDC is bottom dead center.

Note 10: This process would be worth reviewing if necessary.

lbf). The internal energy increases by a factor of 9, the particle speed increasing from 980 ft/sec to 2,940 ft/sec. The work to compress the gas is 253.6 ft-lbf, equal to its increase in internal energy. We have a hot, high-pressure gas.

5.2 Heat Addition from a "Burning" Fuel

Now let's add fuel energy, in a simple way. In a real engine, combustion changes the gas composition, which we'll ignore. Combustion also takes time in a real engine, so it must start well before TDC, and finishes well after. Here we'll assume heat is added *instantly* at TDC, and that it doubles the gas's internal energy.[11] The square of the particle speed must then double. That is, the speed v_3 (after heat addition) must equal the speed v_2 (after compression) multiplied by the square root of 2. So our new particle speed is

$$v_3 = v_2 \cdot \sqrt{2} = (2{,}940 \text{ ft/sec}) \times 1.414 = 4{,}160 \text{ ft/sec}$$

As described in Chapter 13, the pressure force is also proportional to the square of the particle speed, so it doubles after combustion, to 9,970 lbf, as shown in **FIGURE 15**.

FIGURE 15 The pressure force increases from compression (Points 1 to 2) and from heat addition (Points 2 to 3) in our simplified engine.

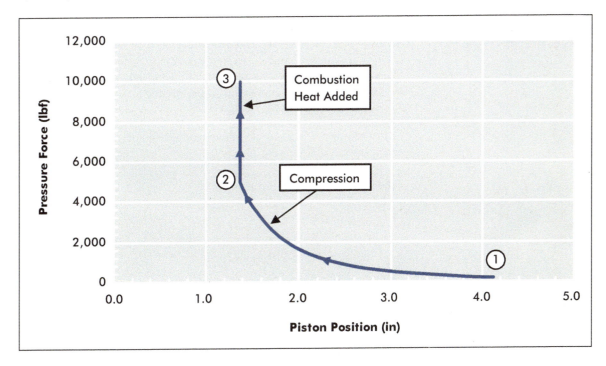

Note 11: By doing this, it is also doubling the absolute temperature of the gas.

Using the first law, the heat Q_H added by the fuel must equal the increase in particle energy:

$$Q_H = \frac{1}{2}mv_3^2 - \frac{1}{2}mv_2^2 = \frac{1}{2}m\left(v_3^2 - v_2^2\right)$$

$$= \frac{1}{2}(0.000066 \text{ slugs})\left[(4{,}160 \text{ ft/sec})^2 - (2{,}940 \text{ ft/sec})^2\right] = 285.3 \text{ ft-lbf}$$

Now we've invested 253.6 ft-lbf of compression work and 285.3 ft-lbf of heat to get the gas to this high pressure at TDC. Let's see how it pays off as the high-pressure gas expands.

5.3 Gas Expansion: The Power Stroke

If you've lifted a car on a hoist that uses compressed air, you know how an engine uses the power stroke. Fill the cylinder with high-pressure air, shut off the air supply, then release the hoist. The air in the cylinder will accelerate the car upward quickly while the pressure is high. As the hoist moves up, the air expands and its pressure drops. The car rises more and more slowly until the pressure *just* supports the car. As the hoist rises, the expanding air does work to lift the car while its own internal energy is reduced. It's the same in an engine.

KEY CONCEPT

Burning the fuel doesn't produce power; the resulting hot, high-pressure gas pushing the piston down the cylinder, while *expanding*, does it. There is no "combustion stroke."

To calculate the work done, we can adapt the equation we used for compression in Chapter 13:

Equation 15-14   **1-D expansion work**

$$W_{Exp} = \frac{1}{2}mv_3^2\left[1 - \left(\frac{x_{Clear}}{L}\right)^2\right] = \frac{1}{2}mv_3^2\left[1 - \left(\frac{1}{R_{Comp}}\right)^2\right]$$

Our expansion work is then:

$$W_{Exp} = \frac{1}{2}(0.000066 \text{ slugs})(4{,}160 \text{ ft/sec})^2\left[1 - \left(\frac{1}{3}\right)^2\right] = 507.2 \text{ ft-lbf}$$

and our net work W is

$$W = W_{Exp} - W_{Comp} = 507.2 \text{ ft-lbf} - 253.6 \text{ ft-lbf} = 253.6 \text{ ft-lbf}$$

A fair amount of work output. Next we make a tough decision.

5.4 Tough Choice

Everyone hates rejection, but at the end of the expansion stroke (point 4), we've run out of expansion room, while the gas hasn't run out of internal energy; it is still hotter than when brought in at point 1. This presents a dilemma; we need to operate the engine in a repeating cycle (otherwise this whole exercise would be very short-lived), so our only choice is to reject heat Q_L from the gas to get back to point 1. We're done "extracting" work.

Isn't this wasteful? Yes, but it's the second law at work. For our input of 285.3 ft-lbf of heat Q_H, we got out a net 253.6 ft-lbf of work W, so our thermal efficiency is:

$$\eta_{Thermal} = \frac{W}{Q_H} = \frac{253.6 \text{ ft-lbf}}{285.3 \text{ ft-lbf}} = 0.889 \text{ or } 88.9\%$$

It is lower than 100% because 11.1% was rejected as Q_L.[12] FIGURE 16 shows the entire cycle of compression (1–2), heat addition (2–3), expansion (3–4), and heat rejection (4–1).

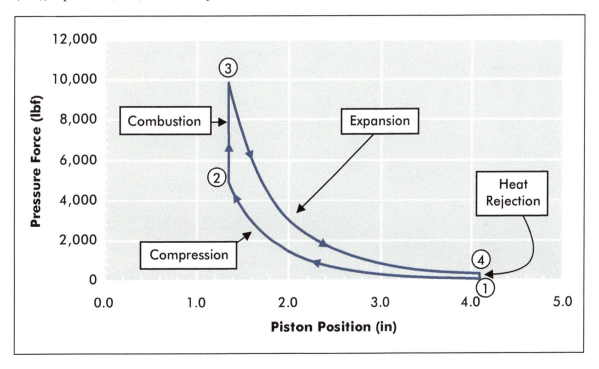

So heat leaves the engine unused after the power stroke, equal to 11.1% of the fuel energy in this case. But this engine with our fictional

FIGURE 16 The force on the piston during the full cycle, using our 1-D gas. The area inside this force-displacement curve is the net work done by the engine during the cycle (once we convert the x-axis from inches to feet).

Note 12: More work could have been squeezed out with a longer expansion, but that would require a longer stroke than for compression. That is called the Atkinson Cycle.

1-D gas still produced a much higher thermal efficiency than with a 3-D gas. The second law of thermodynamics will help explain why.[13]

5.5 Heat-Engine Summary

The main purpose of this was to mechanically explain the thermodynamics of the engine cycle. But we also saw less than 100% efficiency, even without "the usual suspects."

> **KEY CONCEPT**
>
> We assumed complete combustion of the fuel, no heat lost through the cylinder walls during compression or expansion, no leakage allowed past the rings, and no friction. Still, 11.1% of the energy from the fuel was not converted to work.

These factors decrease the efficiency of real engines, but they are not inherent. The main inefficiency has more to do with the lack of concentration of energy in the gas.[14]

In our fictional 1-D gas, all its internal energy was bound up in particle motion parallel to the cylinder walls, so it hit the piston squarely. But molecules in real air move in three directions, and are dumbbell shaped (see **FIGURE 17**). This means they carry kinetic energy in five different motions: the translational DOFs (degrees of freedom) and two rotational DOFs. The effect is that when we run an engine with a real gas, much of the fuel energy goes into motion in directions that do not push on the piston. In the end, this reduces thermal efficiency.

1-D Gas:

3-D Gas: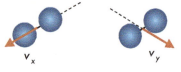

FIGURE 17 Our 1-D gas moves only along the cylinder axis (the x-direction, top), and so carries all its kinetic energy in that direction. But molecules in real air can move and carry kinetic energy, in five different DOFs. This lowers engine efficiency. Note that v_x is parallel to the cylinder centerline.

So let's compare results for our ideal 1-D gas engine to one using a 3-D gas (air). To make it fair we'll run the 3-D gas at a 10:1 compression

Note 13: Thanks to Prof. Emeritus Donald Patterson, of the University of Michigan, for reviewing this problem.

Note 14: For the record, about 98% of the fuel burns in modern engines, so increasing to 100% would barely change efficiency. If you see a product claiming a 20% fuel-economy increase by improving combustion (with a fuel-line magnet or a "turbulence inducer" in the air intake), don't be taken for a ride!

ratio, to produce a compression pressure similar to our 1-D gas engine at 3:1. But we'll keep the same stroke, and add the same heat Q_H.

The plots of pressure force vs. piston position are shown in **Figure 18** for both. One difference is that adding the same heat to the 3-D gas (from points 2′ to 3′ at TDC) doesn't give the same bump in pressure as with the 1-D. But also, the pressure drops faster during expansion. So the pressure during the expansion stroke is smaller, producing less work on the piston. The area inside the curve, the net work, is visibly less for the 3-D gas.

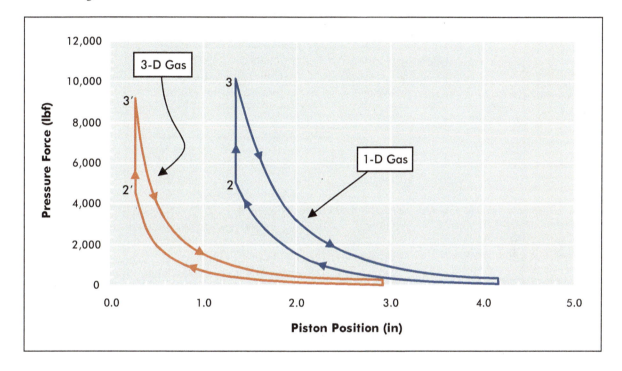

The major results for comparison are shown in **Table 2**. Note how the 1-D gas engine only rejects 31.51 ft-lbf of heat at BDC, while with the 3-D gas it's 113.59 ft-lbf. The 3-D gas had much more "leftover energy" at the end of expansion, that didn't get converted to work.

You can compare these results to theory by using the equation for efficiency of this cycle, the "Air Standard Otto Cycle," found in any thermodynamics book. For the 3-D gas, we use a specific-heat ratio of 1.4, and a 10:1 compression ratio. For the 1-D gas, the "specific-heat ratio" is 3.0, and the compression ratio is 3:1. You'll find very good agreement.

Figure 18 Comparing a full cycle of 1-D and 3-D gas engines, with the same heat addition. The area inside the curve, the net work done during the cycle, is much smaller with the 3-D gas.

Air Model:		1-D Gas	3-D Gas
Compression Ratio:		3:1	10:1
Pressures (psi):	Before Compression (at BDC)	14.7	14.7
	End of Compression (at TDC)	396.8	369.15
	End of Burn (at TDC)	793.6	622.76
	End of Expansion (at BDC)	29.4	24.8
Energy (ft-lbf):	Compression Work	253.62	177.37
	Heat Added	285.33	285.33
	Expansion Work	507.24	349.11
	Heat Rejected	31.51	113.59
	Net Work	253.82	171.74
Thermal Efficiency:		88.9%	60.2%

TABLE 2 The pressures and energy balances in different parts of the cycles for the 1-D and 3-D gas engines. With 285 ft-lbf of heat added, the 1-D gas produces about 254 ft-lbf of work ($\eta = 89\%$), while the 3-D gas only produces 172 ft-lbf ($\eta = 60.2\%$).

The thermal efficiency using a realistic gas, at 60.2%, is much lower than with our fictional gas. If we could keep the particles' motion parallel to the cylinder axis, and only add energy to that direction, we could greatly increase thermal efficiency. We just can't. The second law and statistics say that the fuel's heat will be split up evenly among the five ways (DOFs) the air molecules can move. And after expansion, some energy will be left over in each.

6   Is It Waste or Operating Expense?

We call the heat rejected by an engine "waste heat." But is that the right name? If a tailback runs 30 yards cross-field to gain ten yards downfield, were the 30 yards wasted? If a batter gets a hit in 3 out of 10 at bats, were the other 7 attempts wasted?[15]

Rejecting some heat is the price we pay for using the random motion in thermal energy to create the coordinated energy of work, the cost of doing business. It's like the phrase "It takes money to make money." You have to invest something to get something back. **FIGURE 19** compares the energy flow in a heat engine to the cash flow in a business. The engine takes in heat Q_H from a fuel, rejects Q_L, and nets the work W; the business takes in revenue from sales, rejects (pays) expenses,

Note 15: I considered including the waste in getting beef from a steer, but the visual was just offal.

and nets a profit (here a 25% profit margin, which is extremely high!). Just like work drives a machine, profit drives a company.

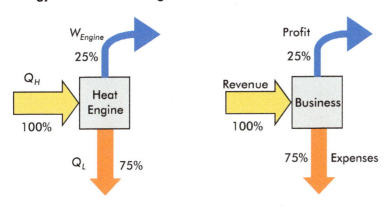

Energy Flow in a Heat Engine

W_{Engine}
25%

Q_H
100%

Heat Engine

Q_L 75%

Cash Flow in a Business

Profit
25%

Revenue
100%

Business

75%   Expenses

FIGURE 19 The flow of energy through a heat engine is much like cash through a business. Most businesses don't have a profit margin of 25%, though.

So whether it's called waste or operating expense, don't be too discouraged that we can't convert heat completely to work. We need to keep increasing thermal efficiency, but every other area of life involves "waste" too.

7   The Basics of Fuel Economy

Being born in 1960, fuel efficiency is near and dear to my heart. After gas prices reached inflation-adjusted historic lows in 1972, the first energy crisis caused skyrocketing prices, and a 55-mph speed limit, just before I got my learner's permit (see **FIGURE 20**). While my '69 Cougar had the standard 351 2-barrel engine, a manual transmission, and the "economy" 3.00:1 rear axle, it still only got 16 mpg, on a long trip, at 55 mph.[16] And paying $1.60/gallon for premium in 1981 when I made $3.35/hour left quite an impression.

Note 16: With a 55-mph speed limit on the Great Plains, every trip seemed long.

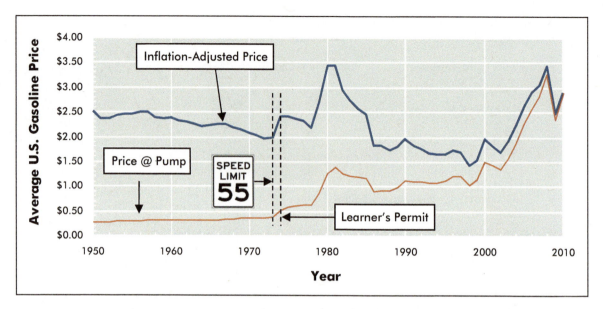

FIGURE 20 U.S. gasoline prices, at the pump, and adjusted for inflation (October 2010 dollars). Although 2008 saw "record prices" up to that time, the real price was just as high in 1981. Note that the all-time real-price low was in 1998. Source: U.S. Energy Information Administration, http://www.eia. gov/steo/fsheets/index.cfm.

So how do we design vehicles to reduce fuel consumption? Reducing mass, aerodynamic drag, and rolling resistance are all important factors. How much each factor helps depends on your driving conditions. Let's compare highway and city driving.

Highway driving is mainly cruising at a constant speed, so most of your energy goes to overcoming rolling resistance and aerodynamic drag. Let's compare two cars that weigh 2,000 and 4,000 pounds, but have the *same aerodynamic drag* of 120 lbf. This isn't as crazy at it sounds. The 4,000-pound car may not have much more frontal area, and a longer, more tapered body might give it a lower drag coefficient.

FIGURE 21 shows the energy flow for the 2,000-lbf and 4,000-lbf cars, each running at constant speed. This uses the kinetic energy "bucket" idea introduced earlier. The "water level" signifies the vehicle speed, and the "water volume" signifies its kinetic energy. Note that the bucket on the right for the heavier car is wider, to hold twice as much for the same height, just as the car has twice the kinetic energy for the same speed.

Let's assume that they have 20 lbf and 40 lbf of rolling resistance, respectively. The heavier car then has total drag of 160 lbf, vs. 140 lbf for the lighter one. This reduces its fuel economy, but doesn't cut it in half, assuming equal engine efficiency. (It is likely that the larger vehicle will have a larger engine, which will slightly reduce its efficiency.)

So heavier cars do okay on the highway, especially with a low drag coefficient. The amount of energy needed to accelerate to speed is twice as high, but you only do that occasionally.

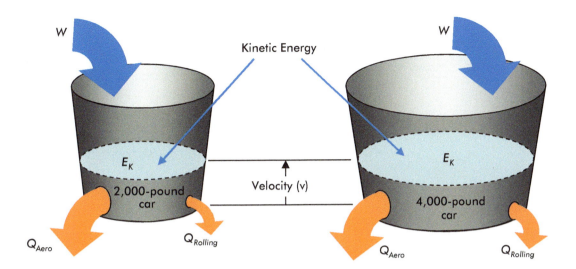

Now take the same cars in the city, driving away from a stoplight together (**FIGURE 22**). At this low speed there is little aerodynamic drag, and rolling resistance power is low, but since the car on the right weighs twice as much, it takes twice as much power to accelerate its inertia.

FIGURE 21 "Energy buckets" for two cars cruising at constant speed (highway driving), with the same aero drag but different weights. The effect of extra weight is just more rolling resistance, so the power and fuel needed aren't much higher.

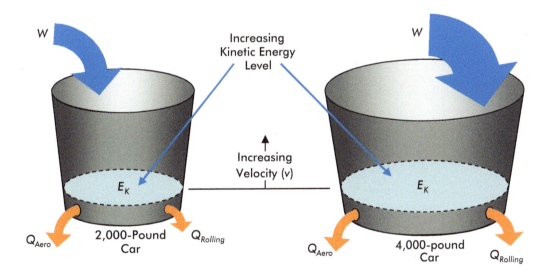

All right, you might say no mechanical energy is lost in acceleration, because the work goes into kinetic energy. True, but it all gets dumped out the brakes at the next stoplight (barring regenerative braking). So any heavy car begins at a disadvantage for city fuel economy.

FIGURE 22 The "energy bucket" concept for accelerating away from a stoplight. The same acceleration takes twice the power for the heavier car (right). This hurts fuel economy because all that mechanical energy soon gets lost to heat through the brakes.

8   Why Kilowatt-Hours?

The automotive industry isn't the only one where custom has led to odd ways to measure things. For instance, the electric meter on your building measures energy in kilowatt-hours, and the labels on electric cars have inherited the same units. If you invented the electric meter today, you'd probably measure the delivered energy in joules. A watt is 1 joule/sec, so a power of one watt, delivered for 1 second, delivers 1 joule of (mechanical) energy:

$$E = Pt$$

$$1 \text{ joule} = 1 \text{ watt} \times 1 \text{ second}$$

Since a joule of energy is pretty small, you'd have it read in megajoules (1 MJ being one million joules). Electric meters do measure energy by multiplying power by time. But instead of using a second for time, they use an hour. So the unit for electrical energy became the kilowatt-hour (1 kW-hr), which is 1,000 watts multiplied by an hour. Its units are:

$$1 \text{ kW-hr} = 1{,}000 \ \frac{\text{joules}}{\text{sec}} \times 1 \text{ hr} = 1{,}000 \ \frac{\text{joule-hr}}{\text{sec}}$$

This is a roundabout way of getting energy, with two units for time in the same calculation. It's easier to use battery storage energy in other calculations if we put it into consistent units:

$$1 \text{ kW-hour} = 1 \text{ kilowatt} \times 1 \text{ hour} = 1{,}000 \text{ joules/sec} \times 3{,}600 \text{ sec}$$
$$= 3{,}600{,}000 \text{ joules} = 3.6 \text{ megajoules}$$

So 1 kW-hr is equal to 3.6 MJ, or about 2.65 million ft-lbf. Electric-vehicle batteries are rated both by their energy storage capacity (in kW-hr or MJ) and maximum discharge power (in kW).

9   Which Should We Use: The IC Engine or the Electric Motor?

There is an ongoing controversy about whether vehicles should be powered by electric motors, or by internal combustion engines. But this is a false choice. This section will argue that each has its place. Of course, physics is not the whole picture. Some propulsion types are technically viable, but are not common because of cost or lack of infrastructure, and some energy sources are attractive because of which countries produce them, and their relationship with the U.S.[17]

9.1 Propulsion Efficiencies

Let's compare the efficiency of an IC engine powertrain to electric propulsion. We'll follow energy flow as before, considering the first and second laws. The figures will be for a constant-speed cruise, but we'll also briefly discuss the effects of stop-and-go driving.

FIGURE 23 shows power flow for a conventional gasoline powertrain. Vehicle kinetic energy remains constant, so all the power is used to overcome aero drag and rolling resistance. The engine receives high temperature heat Q_H from the fuel. Assuming a 28% thermal efficiency, 72% of this will go to the surrounding air as Q_L. The rest flows as power from the engine to the driveline. It transmits 90% of that to the wheels, 25% of the original fuel energy. Once it drives the vehicle down the road, that energy is dissipated through drag.

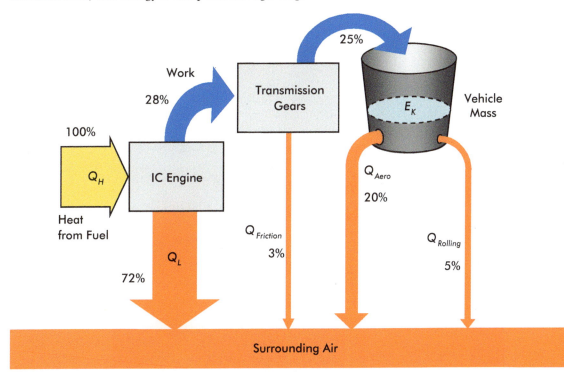

How about the electric vehicle? We start at the electric power plant (see FIGURE 24). Here, about 32.5% of the fuel energy is converted to electrical power, 67.5% rejected through the smokestack and cooling tower. Of the 32.5%, some is lost in the transmission lines, in

FIGURE 23 An IC engine driving a vehicle at constant speed. We assume the engine converts 28% of the fuel's heat into work, 90% of which makes it through the driveline to drive the vehicle, for a propulsion efficiency of 25%.

Note 17: The world is making important decisions about propulsion, so we should discuss things rationally. I'd suggest reading this entire section to get the whole picture, and avoid dwelling on any one point.

charging and discharging the vehicle battery, and in the electric motor.[18] The overall propulsion efficiency is 25%, about the same as for the IC engine!

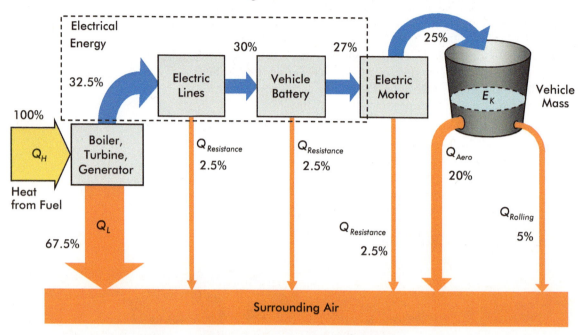

FIGURE 24 The process of converting heat to electric propulsion. Between turbine efficiency, generator efficiency, and losses along the way, the propulsion efficiency is about 25%.

Is this surprising? It really shouldn't be. At some point in either case, heat is converted to mechanical energy, which we know can't be done at high efficiency.[19] In city driving, the IC engine doesn't do as well; we'll cover that in a couple of sections.

NOTE

It doesn't matter whether heat is converted to power in the vehicle or at a power plant. The second law takes its cut in either case, so the overall propulsion efficiency is essentially the same.

Interestingly, the same conclusion is reached in *Energy and Climate Change: Creating a Sustainable Future*, by David Coley, 2008.

For a complete comparison, the energy used in mining/transporting/refining the energy should also be considered—a "well-to-wheels" approach. About 17% of the energy in crude oil is spent in refining gasoline. The percentage for electricity depends on the energy source.

Note 18: While other sources show similar efficiency values, these are from *Automotive Engineering International*, August 2009, page 46, published by SAE International.

Note 19: I do have to admit that I was surprised myself at how close they were. In fact, I felt a little sheepish that this wasn't obvious to me right off. I've gotten over it since.

9.2 Why Internal Combustion Engines?

Once while GM's EV1 was being developed, I stood underneath one on a hoist with a friend of mine, to check it out. At the time, the only reliable batteries available were lead-acid, so it took about 1,175 pounds of battery to give it a range of 80–100 miles, and they took up a lot of room. With a curb weight of about 3,000 pounds the vehicle, minus battery pack, only weighed 1,800 pounds. Further, it had a very low 0.19 drag coefficient.

We eyeballed the battery size and did some calculations in our heads to calculate the car's range if it had a small turbo-diesel engine. In the space occupied by the batteries, we figured you could put in a 50-gallon fuel tank. With fuel, that would weigh about 350 lbf. Replacing the electric powertrain and electronics with a turbo-diesel powertrain might add another 200 pounds.

All in all, this would produce a 2,350-pound vehicle with very low drag, so we figured an easy 60 mpg on the highway. With the 50-gallon tank, that's a range of 3,000 miles! [20]

While no one needs a 3,000-mile range, the point is that hydrocarbon fuels like diesel and gasoline provided 30–40 times the range of an EV of the time. **Table 3** supports this, showing energy densities for common storage batteries, and for diesel fuel and gasoline assuming 25% conversion to work. Higher numbers mean you can store more energy in the same mass. Gasoline and diesel fuel are still many times higher than batteries. Finding room for a fuel tank isn't quite an afterthought, but it's much easier than for a battery with the same range.

TABLE 3 Energy densities for various electrical batteries. A range is shown for each, as well as projected lithium-ion capability. For comparison, the same is shown for two common fuels, assuming a 25% thermal efficiency. (The battery storage capabilities are from *Automotive Engineering International*, December 2009, page 31, published by SAE International.)

| | Energy Density | | | | | |
| | kW-hr/kg | | J/kg | | ft-lbf/lbm | |
	Low	High	Low	High	Low	High
Pb-Acid	15	30	54,000	108,000	18,000	36,000
Ni-MH	50	80	180,000	288,000	60,000	96,000
Li-Ion	90	180	324,000	648,000	108,000	216,000
Li-Ion Target	240		864,000		289,000	
Gasoline @ 25% Efficiency	3,320		11,960,000		4,000,000	
Diesel Fuel @ 25% Efficiency	3,110		11,210,000		3,750,000	

Note 20: Using the 1,175-lbf battery weight, and a 125 lbm/ft³ density of lead-acid batteries, I actually get a 70-gallon tank. So the range would be more like 4,000 miles.

Remember our Porsche from Chapter 12 that required 792,000 ft-lbf of work per mile? Using a lead acid battery, you'd need around 22 pounds of battery per mile of range; with a Ni-MH battery, 8.25 pounds; with today's Li-Ion batteries, 3.7 pounds per mile; and with targeted Li-Ion capability, 2.75 pounds per mile. By comparison, you'd need about 0.20 pounds of gasoline or diesel per mile. Batteries have improved, but there's still a big gap.

The primary reason for the IC engine's success is energy storage; the ease of carrying a tremendous amount of propulsion energy in a small, light package. Since the 1960s it has also proven to be very adaptable, with greatly reduced emissions and increased fuel economy.

9.3 Why Electric Propulsion?

The main advantage of electric vehicles is that they don't emit local pollutants, like unburned hydrocarbons (HC), carbon monoxide (CO), or oxides of nitrogen (NO_x). Fuel is burned remotely at a power plant, so the pollutants are emitted there. It's easier to control pollutants at a large power plant than in vehicles.

Another *possible* advantage to EVs is reduced emissions of carbon dioxide (CO_2), a greenhouse gas that may adversely affect the global climate. It doesn't create smog, so it's not a local pollutant, but a global pollutant. To reduce the CO_2 effect produced by a vehicle, it doesn't matter whether it is emitted directly or remotely (at a power plant).

The CO_2 produced by an EV depends on the fuel at the plant. Nuclear plants produce no CO_2, natural gas produces a moderate amount, and coal emits quite a bit. Wind and solar power produce no CO_2, but the wind speed and sunshine are not always strong enough to be used. Hydroelectric plants do not produce CO_2 either, and are more reliable, but they only work in locations with the right terrain.

9.4 The Bottom Line

Both electric and IC engine powertrains are improving as we speak, so comparing them is a moving target. But we can summarize their strengths and weaknesses.

The conventional IC engine powertrain is tailor-made for automobiles, where the propulsion unit must move itself and its energy storage unit; they both "want" to be light. Its high horsepower per weight, its ability to use energy-dense fuels, and its quick starting and refueling make it a difficult device to unseat. The IC engine will be preferred for road trips for a very long time.[21] One of its main weaknesses is that it keeps running at 600 rpm or so when the vehicle is stopped, using fuel. Another is that it is very inefficient at low loads. At idle or in very slow traffic,

---

Note 21: An 18-wheeler with electric propulsion would use over a third of its 80,000-pound limit on batteries, to have the same range as a typical semi with 250 gallons of diesel.

a gasoline engine might be 5–10% efficient. (A diesel engine is more efficient at low loads and idle because it does not use a throttle.)

EVs are attractive for dense cities and/or heavy traffic for several reasons. Their lack of local pollution is the first. They also have two advantages in stop-and-go driving: 1) they don't have to "idle," and 2) they use regenerative braking to partially re-use otherwise dissipated kinetic energy. Electric propulsion makes a lot of sense for taxis and delivery trucks in cities. As explained, the electric vehicle's main weakness is range, requiring heavy batteries to store enough energy. (For people making short commutes, or who have more than one car, this may not be a factor.) The increased weight also requires a stronger (heavier) vehicle body and suspension, and the added mass must be propelled by the motor. This in turn requires more energy.

But not only is the choice between IC engines and EVs a false one, we also have the option of smoothing out their strengths and weaknesses by combining them, such as in gasoline/electric hybrids. At low speeds, when the IC engine is inefficient, they can use the electric motor(s). The electric motor(s) can also perform regenerative braking.

Further, we can balance the hybridization to different extents. Already we have the Toyota Prius, most of which exclusively use gasoline as their energy source and only use their electric motors to improve the engine's overall efficiency. We also have plug-in hybrids, and EREVs (extended range electric vehicles) like the Chevy Volt, which are pure electric for a certain range (about 40 miles for the Volt), and then switch to gasoline to keep going once they've used their batteries' energy.

Your "best choice" depends on your situation. But getting this all across to a customer, to the extent that they can make a logical decision, is no easy task. Based on what I read, it's hard enough for reporters, who should have more time to research the subject. Hopefully this improves.

Considering all these options, it is unlikely that electric propulsion will *completely* replace IC engines. The physics just doesn't work. Even with improved batteries, electric-only vehicles will not compete for range with IC engine vehicles using energy-dense fuels. I'm sure we'll always have vehicles powered by IC engines, electric motors, or both. Just don't ask me how the mix will play out.[22]

10 Summary

With the concepts and formulas in these last four chapters, you can get quite a ways in understanding how 1) power is related to performance, 2) energy usage is related to fuel economy, and 3) power is produced and transmitted from an energy source.

Note 22: As Niels Bohr, the ground-breaking quantum physicist said, "It is exceedingly difficult to make predictions, particularly about the future."

For performance, "power to the pavement" is what matters, which can be limited by the engine itself, the mechanical losses and gearing of the driveline, and the tires' traction. Maximizing performance means making best use of power and traction. Reduced mass and aerodynamic drag increase performance for the same power.

For fuel economy, you need an efficient engine, matched to a transmission that allows it to turn at its most efficient speed for the conditions. Hybridization can save fuel by using regenerative braking, and by stopping the engine when it's not needed. But again, low mechanical losses, reduced vehicle mass, and reduced aerodynamic drag are beneficial.

As for power production and transmission, you need to remember that all power comes from somewhere. With IC engines, we burn a fuel, "coerce" it into doing work, and transmit the work to the drive axles. With EVs, we (typically) burn a fuel to heat steam at a power plant, "coerce it" into doing work on a turbine to drive an electrical generator, and transmit the power through the lines, battery, motor, and drive axle.

What can we take away from this for vehicle design? For me the first step is low mass. As mentioned, it not only helps performance, but also helps city fuel economy. This is like having your cake and eating it too. From there I'd reduce aero drag, both the drag coefficient and frontal area.[23] This helps performance at higher speeds and improves highway fuel economy. And reducing parasitic losses from cooling fans, power steering systems, A/C systems, etc., helps both fuel economy and performance.

Reducing mass and drag lets you choose between keeping the increased performance and fuel economy, or reducing the engine size to maintain performance and further improve fuel economy.[24] My ideal sports car would weigh about 2,200 pounds and have 250 HP or so, and I miss the days of 3,000-pound V8 Mustangs and Camaros.

The perfect power source would propel our vehicles without causing any ill effect on our local community or Earth as a whole. Unfortunately, the second law of thermodynamics says that won't happen; we can only minimize our effect. All in all, we need to take a factual, rational approach to develop ways to propel our vehicles that are environmentally *and* economically sustainable.[25] In other words, we need to use physics and ingenuity, and keep our heads.

But still have some fun.

Note 23: My favorite cars use this combination to improve performance and fuel economy: low mass, low frontal area, and a low drag coefficient, like a Chevy Corvette or Lotus Elise. The ultimate, in my mind, was the McLaren F1 road car that had 600 HP but was very compact and only weighed 2,650 pounds (priced accordingly). In their time, the Ford GT40, Porsche 550 Spyder, and Lotus 23 were also great examples.

Note 24: I know a few people who would increase the engine size to increase performance and preserve fuel economy instead.

Note 25: Without a healthy economy, we can't pay for effective pollution controls or more efficient vehicle technology. Pollution in the former East Germany was a good example.

Major Formulas

Definition	Equation	Equation Number
Maximum Acceleration in a Given Gear, No Mechanical Losses	$$a_{Max} = \frac{T_{Max} R_{Trans} R_{FD}}{r_{Tire} m}$$	15-1
Acceleration in Terms of Power at the Tire Patch and Vehicle Speed; No Drag	$$a_{Max} = \frac{P_{Max}}{mv}$$	15-2
Thrust Force in Terms of Power at the Tire Patch, and Vehicle Speed	$$F_{Thrust} = \frac{P}{v}$$	15-3
Formulas for the "Ideal Drag Racer"		
Traction-Limited Acceleration Rate	$$a_{TL} = \frac{F_{TL}}{m}$$	15-4
Velocity During Traction-Limited Acceleration, Standing Start	$$v_{TL} = a_{TL} t$$	15-5
Displacement During Traction-Limited Acceleration, Standing Start	$$s_{TL} = \frac{1}{2} a_{TL} t^2$$	15-6
Speed at End of Traction-Limited Acceleration, with Power at the Tire Patch	$$v_2 = \frac{P_{Max}}{F_{TL}}$$	15-7
Time at End of Traction-Limited Acceleration	$$t_2 = \frac{m P_{Max}}{F_{TL}^2}$$	15-8
Displacement at End of Traction-Limited Acceleration	$$s_2 = \frac{m P_{Max}^2}{2 F_{TL}^3}$$	15-9
Power-Limited Acceleration Rate	$$a_{PL} = \frac{P_{Max}}{mv}$$	15-10
Top Speed for the Ideal Drag Racer	$$v_3 = \sqrt[3]{v_2^3 + \frac{3 P_{Max} (s_3 - s_2)}{m}}$$	15-11
ET (Elapsed Time t_3) for the Ideal Drag Racer	$$t_3 = t_2 + \frac{m}{2 P_{Max}} \left(v_3^2 - v_2^2 \right)$$	15-12
Torque and Gear-Limited Thrust in a Given Gear	$$F_{TGL} = \frac{T_{Engine} \eta_{Mech} R_{Trans} R_{FD}}{r_{Tire}}$$	15-13

16 Statics and Quasi-Statics Basics

Center of Gravity Location, Weight Distribution, and Load Transfer

Mike Malone uses his Lotus 26R to demonstrate complete load transfer to the outside tires, while going into Turn 8 at Road America. You can see the Porsche 911s of Ronnie Randall (#24) and David Bland (#231) also experiencing load transfer, but not so extreme. *Photo by Randy Beikmann with permission of the cars' owners.*

Chapter 16

The center of gravity (CG) of a rigid body is better defined, and tied to the weight distribution of a stationary vehicle, using statics. Quasi-statics is used to calculate a car's load transfer in straight-line acceleration and in cornering, as well as a motorcycle's roll angle in cornering. Forces in links are examined.

Contents

Key Symbols Introduced

Symbol	Quantity	SFS Units	MKS Units
F_z	Load on a Tire or Axle	lbf	newtons
M_A	Moment About a Point "A"	ft-lbf	N-m
M_{CG}	Moment About the CG	ft-lbf	N-m
L_{WB}	Length of Wheelbase	feet	meters
L_1	Distance from CG to Front Axle	feet	meters
L_2	Distance from CG to Rear Axle	feet	meters
$a_x,\ a_y,\ a_z$	Longitudinal, Lateral, Vertical Acceleration	ft/sec^2	m/sec^2
$\theta_x,\ \theta_y,\ \theta_z$	Roll, Pitch, Yaw Angles	radians	radians
W_{Track}	Track Width (Between Tire Centers)	feet	meters
h_{CG}	Height of CG Above Ground	feet	meters

1 Introduction

As we've seen, power does you no good if you don't have the traction to use it. Since traction is a friction force, you can use more power by pressing the driving tires harder against the pavement, and by installing stickier tires. The same is true for cornering; you want to have the right amount of load on each tire to make the best use of them as a set.

Managing the loads on the tires is so critical to vehicle performance that the first thing you hear when someone talks about a car's layout is weight distribution, the percentage of weight carried by the front vs. the rear axle. This, and how the tire loads *change* with acceleration, braking, and cornering, have a major effect on the car's capability in each maneuver.

This chapter will cover the details of statics and quasi-statics, developing the theory and equations to balance forces and moments in a plane. This will be used to find the tire loads, and which factors in the vehicle design control them.

We use **STATICS** to find the forces that are acting on an object that is moving with *constant velocity*, which of course includes *stationary* objects. In other words, if there is no net force acting on a system, meaning no acceleration, we can use statics.

To use statics for weight distribution, we know that the weight of the vehicle must be supported by the tires, and using Newton's laws, we can find what each axle must carry. We can also use statics to find the CG (center of gravity) location of the total vehicle if we know the mass and CG location of each individual component (engine, transmission, wheels, etc.).

Once the vehicle accelerates, it's a *dynamics* problem. When accelerating forward, the tire loads will be different than in the static case, as we get load transfer from the front axle to the rear. Depending on which wheels are driving, this will increase or reduce the available traction for thrust. Either way, we need to know the effect and understand it. But since the vehicle is accelerating, we can't really use statics, right?

Fortunately, a vehicle's dynamics are not always complicated. There are many times when the acceleration is nearly constant, as we saw in the traction-limited part of a 0–60 run. Soon after launch, a vehicle usually settles into a steady "attitude," where the nose and tail stay at constant heights. The loads on the tires will also be nearly constant, even though the vehicle is accelerating, and we can use **QUASI-STATICS** (almost statics) to find them. Now instead of setting acceleration to zero in our calculations, we set it to the steady acceleration. Otherwise, it's almost the same as statics.

We'll define why a body's center of gravity is also its center of mass, and therefore has been our center of attention all through the book. This will be especially true in quasi-statics.

Relating tire loads to the CG location isn't the only use for statics and quasi-statics of course, just the most obvious. Forces in suspension bushings and springs, clutches, brakes, and even piston pin and valve-train loads can be estimated using quasi-statics.

2   Overview and Historical Background

In ancient Greece, Archimedes made scientific history by understanding and explaining the lever. He calculated how, by applying a certain force to the longer arm, you could produce a much larger force with the shorter arm (see **FIGURE 1**, left). He was quoted as saying that given a long enough lever, and a proper fulcrum (pivot), he could move Earth.[1] But he only understood cases where the forces were perpendicular to the lever, since he knew geometry, but not trigonometry.

In the late 1500s Simon Stevin worked out the laws of statics for a lever with forces that *weren't* perpendicular to the lever (**FIGURE 1**, right). To do so, he introduced the force vector. He also realized that

---

Note 1: A lofty goal, but it indicates he had an eye toward *using* the knowledge he had, rather than just studying it like most Greek thinkers of his time. In fact, he also created some fearsome war machines.

the increase in force was accompanied by a decrease in motion, which we now know follows from the conservation of energy (work done *on* the lever equals work done *by* it).

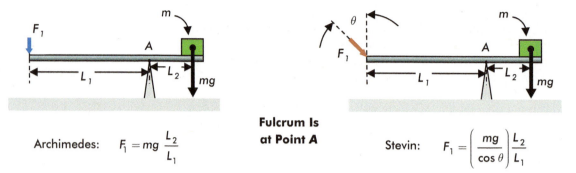

Fulcrum Is at Point A

Archimedes: $F_1 = mg \dfrac{L_2}{L_1}$

Stevin: $F_1 = \left(\dfrac{mg}{\cos\theta} \right) \dfrac{L_2}{L_1}$

FIGURE 1 Archimedes studied the forces in a lever with forces perpendicular to the arm (left). In the 1500s Stevin did the same for levers with angled forces (right).

Once Newton introduced his laws of motion in the late 1600s, statics was seen as dynamics with the acceleration set to zero. Remember that his first law said that if an object has no *resultant* force acting on it, it has a constant velocity. The forces acting on it balance out. The upward forces counteract the downward ones, and the forces acting left counteract those pushing right. Because this is a set of *equal forces* in opposite directions, this situation is called equilibrium. (In Latin, the word *aequi* means equal, and *libra* means force or balance.)

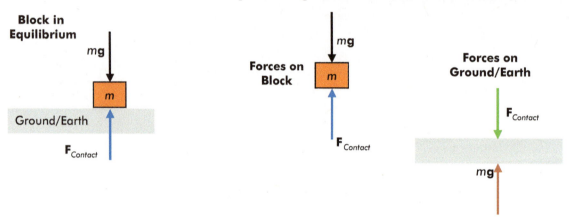

FIGURE 2 The weight of a stationary block sitting on the ground must be balanced by an upward contact force. The free-body diagrams on the right show there is one contact force pushing up on the block, another down on the ground. Also, one gravitational force pulls down on the mass, but a second pulls upward on the ground/Earth.

Equilibrium is illustrated in **FIGURE 2** (with free-body diagrams to the right), where a block with mass m sitting on the ground has its downward-acting weight $m\mathbf{g}$ balanced out by the support force $\mathbf{F}_{Contact}$ acting upward on it. Likewise, the block pulls upward on Earth with a gravitational force equal to mg (remember, Earth and the block are attracted toward each other), while it also pushes downward on the ground with an equal and opposite contact force. So the gravitational forces pull the block and the ground together, while the contact forces between the block and ground push them apart.

Additionally, Newton's laws for rotation must be met in statics. A simple example is two kids on a teeter-totter, which is really just a lever. If they're different weights, the heavier kid has to sit closer to the pivot than the lighter one, or the teeter-totter will rotate and drop him to the ground.

We can use the same facts to find a car's weight distribution, assuming we know its CG location and its wheelbase. **Figure 3** shows the free-body diagrams for a stationary car and the road/Earth that supports it. To keep the car in equilibrium, it must not accelerate; the forces on it must add up to zero. The upward support forces on the tires must equal the car's weight. As shown, the weight mg (3,650 lbf) is balanced by the support forces, 1,644 lbf at the rear axle and 2,006 lbf at the front.

Note also that a 3,650-lbf gravitational force acts upward on Earth, while the tires push down on it with a total of 3,650 lbf. Even the forces on Earth balance out, which should make us all rest easier.

The end effect is that gravitational forces, acting through the body and suspension, press the tire patches and road against each other, just like you might press two parts together in a vise.

KEY CONCEPT

Because we rely on it for traction, we should really think of vehicle weight as gravitational downforce.

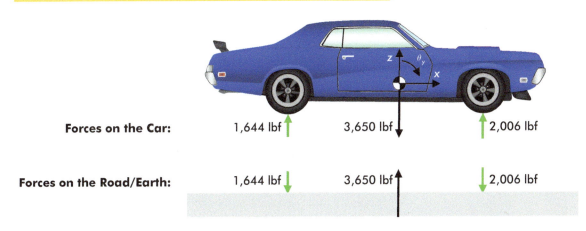

Forces on the Car: 1,644 lbf ↑   3,650 lbf ↓   ↑ 2,006 lbf

Forces on the Road/Earth: 1,644 lbf ↓   3,650 lbf ↑   ↓ 2,006 lbf

So far so good, but what about front tire forces vs. rear? **Figure 4** shows the car with the front tires on blocks. If that's all there were supporting the car, it would obviously rotate counterclockwise and fall. The moment of the car's weight about the front tire patch, 182,500 inch-lbf CCW, would accelerate it in rotation.[2] Counteracting this takes an upward force at the rear axle to create a 182,500 inch-lbf CW moment. Acting a distance of 111 inches from the front tires, the rear axle load must be 1,644 lbf.[3]

Figure 3 Free-body diagrams of a 1969 Cougar and the road it is sitting on, separated to show the forces acting on each. Each has a gravitational force acting on it, as well as contact forces between the road and tires, together balancing out.

Note 2: Remember that a moment is a torque on a body from an applied force, as covered in Chapters 7 and 10.

Note 3: Using inch-lbf is violating the "consistent-units" rule, but in statics it doesn't matter as long as we always use the *same* units. It matters more in quasi-statics.

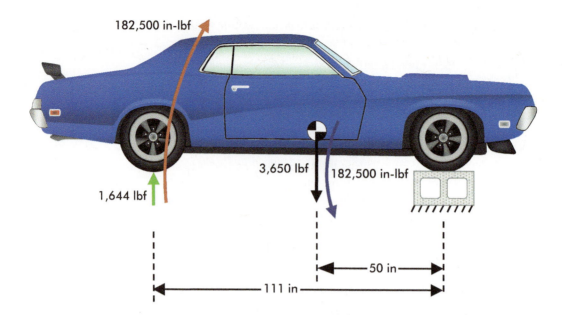

182,500 in-lbf

1,644 lbf

3,650 lbf

182,500 in-lbf

50 in

111 in

Figure 4 Calculating moments is necessary to determine how the weight is carried by the front and rear axles. Here, the CCW moment from the weight must be counteracted by the CW moment from the vertical rear tire force.

What about the load on the front tires? The loads on the axles must add up to the vehicle weight, so we subtract the rear axle load from the weight to get 2,006 lbf.

We use statics in many areas besides cars. Civil engineers use it in designing buildings and bridges. Pilots use it before every flight, checking their "weight and balance" to make sure the plane's CG will be lined up with the lift from the wings (not too far back or forward).

For all the work done on statics before Newton, without his discovery of dynamics we wouldn't have quasi-statics, finding the force balance in cases where there is constant acceleration. This opens up many other automotive uses, without adding much complication.

A famous example of quasi-statics was used by Einstein, and is one you've probably experienced yourself. Imagine you weigh 200 lbf, and you're standing in an elevator that is set to go up. Before accelerating (**Figure 5**, left), the floor pushes up on you with a force equal to your weight, mg. Suppose it then accelerates upward at 1 g. Now the upward force on you must increase by your mass times the 1 g acceleration (again mg), making a total force of twice your weight ($2mg$). Now you have 400 pounds acting upward on you from the floor.

Figure 5 Using quasi-statics to calculate the vertical force on a person standing in an elevator. When stationary or at constant velocity (left), it's a statics problem. During steady acceleration (right), the force increases proportional to the acceleration.

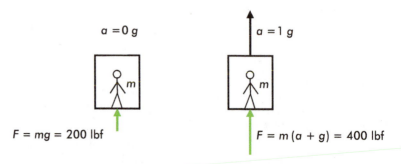

$a = 0\,g$

$a = 1\,g$

m

m

$F = mg = 200$ lbf

$F = m(a + g) = 400$ lbf

After the initial start-up "jerk," the forces on your knees, feet, and elsewhere in your body will stay the same, at twice what they are normally. It would be like living on a planet with 2 g's gravity, as long as the acceleration lasted.[4]

KEY CONCEPT

Quasi-statics is not quite statics because you aren't moving at a constant velocity. But the "situation" is constant, and you can calculate the steady forces by considering the steady acceleration. For us, statics will explain weight distribution, but quasi-statics will also explain load transfer.

3   The Center of Mass, Alias: The Center of Gravity

The **CENTER OF MASS** (or CM) of a rigid body is the weighted average location of all its mass particles. A force F acting on a free body, along a line of action through its CM, will cause an acceleration $a = F/m$, but will not cause rotation. If the force's line of action is to one side of the CM, it still causes an acceleration of F/m. But the moment M it also produces will cause an angular acceleration α equal to M/I.

Now for the CG. An object is made up of many small particles having their own weight, and these sum up to a resultant force mg, the body's weight. The **CENTER OF GRAVITY** of a rigid body is the point its weight mg acts upon it as a whole, as shown in **FIGURE 6**.

Individual Masses   **Masses Combined at CG**

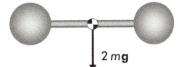

FIGURE 6 For translation, the combined effect of two bodies of equal mass m is twice that mass, placed halfway between them. This is their combined center of mass (CM). The weight from the two masses can also be thought of as being concentrated there, so their CM is also their CG.

So the CG is technically different than the CM, but for all objects in this book, the CM is at the CG.[5] Since the term CG is such a familiar term for the center of mass, we'll just go with it.

Note 4: Einstein used this example for his general theory of relativity, reasoning that all laws of physics should work the same whether you stand on a "1-g" planet, or accelerate at 1 g in space. A simple beginning, but it led him to great discoveries.

Note 5: Only rarely are the center of gravity and center of mass different. One example is when large objects like planets and moons are close to each other.

For a symmetric body like a car battery, the CG is at its geometric center. If the mass is unevenly distributed, the CG is closer to the heavier end. For complicated objects, we can find the CG mathematically, as we'll see in Chapter 17.

You can also do a test to find it. Suppose you have the engine out of your car. Since engines are pretty heavy, you're wondering how its mass affects the car's CG. You'd want to weigh it and get its own CG location. Weighing it is easy enough, but finding its CG isn't much harder.

To support something with one force and keep it from rotating, the force must act through the CG. So if you hang an object by a single flexible chain, the CG always hangs directly below the chain (see **FIGURE 7**).[6] If you hang it at several angles, and draw vertical lines in line with the chain for each, the CG is at the intersection of the lines. In Position A, the engine is hung rolled left; in the Position B, it is level. Position C is also level, but we look at it from the side.

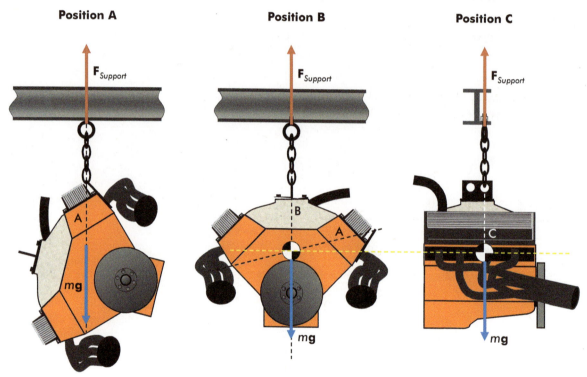

FIGURE 7 You can pin down a body's CG location in three steps, by hanging it from two or three attachment points by a chain. When still, its CG always hangs directly below the chain.

In Position A, assuming the back face of the engine is perfectly vertical and we view straight from the rear, we can sight down the chain to draw a line A down the back face, knowing that the CG lies somewhere straight ahead of this line (a plane through the engine). In Position B, we do the same. Now we know the CG is on a line straight forward of where lines A and B intersect. In Position C, we draw a third line from a side view, to find out how far forward of the rear face the CG is.

Note 6: The main thing is that the support must apply a vertical force and no moments. Using a rigidly connected stiff cable or steel rod that has bending stiffness would throw off the results.

Once you know its center of gravity, you can treat a complicated but rigid object (like a car or motorcycle) as if all its mass particles were concentrated at one point. Concentrating the mass to one point doesn't consider rotational inertia, but that doesn't matter in statics or quasistatics, since there's no angular acceleration.

4   The Mechanics of Statics in a Plane

With the CG understood, let's cover the details of statics. For us to use it, the object must move at constant velocity (including being stationary), and we must write and solve as many equations for the forces and moments acting on it as there are DOFs (degrees of freedom) involved.

In three dimensions there are three translations and three rotations to deal with, making three forces and three moments that could be important, and as many as six equations to find them. TABLE 1 spells out these degrees of freedom relative to the car.

Direction	Translational DOF	Rotational DOF
Fore/Aft	x (Longitudinal)	θ_x (Roll)
Left/Right	y (Lateral)	θ_y (Pitch)
Up/Down	z (Vertical)	θ_z (Yaw)

TABLE 1 The six degrees of freedom (DOFs) for motion, forces, and moments. Given in car-coordinates, forward is positive x, left is positive y, and up is positive z, as in FIGURE 8.

Since we will only work in a plane here, we only need to deal with two translational directions at once (say, longitudinal and vertical), and one rotation (say, pitch), for a total of three degrees of freedom. That means we'll need three equations, to balance the forces in two directions and the moments in one. FIGURE 8 shows the most likely sets of directions we would use, with one application for each. Tilted coordinate systems can also be used, if more convenient, but x, y, and z must still be perpendicular to each other. Note that the coordinates each originate at the vehicle CG.

Front/Rear Weight Distribution	**Left/Right Weight Distribution**	**Lateral Tire Forces in a Crosswind**
F/A, Vertical, Pitch	Lateral, Vertical, Roll	F/A, Lateral, Yaw

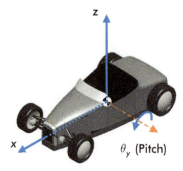

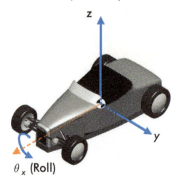

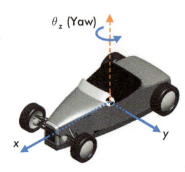

FIGURE 8 Sets of DOFs to use for statics problems in different cases, shown in blue. For front/rear weight distribution, we use the x-z plane and pitch; for side-to-side weight distribution, the y-z plane and roll; for crosswind forces, the x-y plane and yaw.

In each of these planes the three static conditions we can use to find the forces and moments involved are:

1. No acceleration in the first translational direction,
2. No acceleration in the second translational direction, and
3. No angular acceleration.

The next step is to make sure these are true, by requiring that the resultant forces and moment are zero. Now we'll spell out the specifics for each of these planes.

4.1 The Longitudinal/Vertical (x-z) Plane

With the longitudinal/vertical (x-z) plane, we have longitudinal forces, vertical forces, and pitch moments acting on the vehicle. Applying Newton's second law to statics:

1. The longitudinal forces (x-direction) add up to zero,
2. The vertical forces (z-direction) add up to zero, and
3. The pitch moments (θ_y-direction) add up to zero.

First we set the resultant force in the x-direction to zero:

Physical Law Eq. 16-1a

$$F_{Res,x} = 0$$

We do the same for the z-direction:

Physical Law Eq. 16-1b

$$F_{Res,z} = 0$$

Finally, we set the resultant pitch moment $M_{y,A}$ about some "point A" to zero:

Physical Law 16-1c

$$M_{y,A} = 0$$

KEY CONCEPT

In *statics*, you can choose any convenient "point A" to calculate moments about; the CG, a tire patch, etc. The best place is usually a point that has an unknown force acting on it, so that force produces a known moment (of zero).

Quasi-statics is more restrictive about where to take the moments.

4.2 The Lateral/Vertical (y-z) Plane

Here again, we set the resultant forces in the lateral/vertical (y-z) plane to zero, as well as the roll (x) moment:

1. The lateral forces (y-direction) add up to zero,
2. The vertical forces (z-direction) add up to zero, and
3. The roll moments (θ_x-direction) add up to zero.

So in the y-direction:

Physical Law Eq. 16-2a

$$F_{Res,y} = 0$$

in the z-direction:

Physical Law Eq. 16-2b

$$F_{Res,z} = 0$$

and for the roll moments:

Physical Law Eq. 16-2c

$$M_{x,A} = 0$$

Here, $M_{x,A}$ is the resultant roll moment, taken about point A.

4.3 The Longitudinal/Lateral (x-y) Plane

Using the same procedure, we have the following requirements:

1. The longitudinal forces (x-direction) add up to zero,
2. The lateral forces (y-direction) add up to zero, and
3. The yaw moments (θ_z-direction) add up to zero.

In other words, in the x-direction:

Physical Law Eq. 16-3a

$$F_{Res,x} = 0$$

in the y-direction:

Physical Law Eq. 16-3b

$$F_{Res,y} = 0$$

and in yaw:

Physical Law Eq. 16-3c

$$M_{z,A} = 0$$

Here, $M_{z,A}$ is the resultant yaw moment about point A.

4.4 Finding the Static Load on Each Axle, Knowing the CG Location

An axle load is the total contact force between the road and the tires on one axle. When the vehicle is stationary on a level surface, the axle loads together just support the weight of the car. The weight carried by each axle is called its **STATIC AXLE LOAD**.

Let's make this more concrete with a Porsche 911 weighing 3,365 lbf with driver, having a 92.9-inch wheelbase, and a longitudinal CG location 56.7 inches rearward of the front axle (see **FIGURE 9**). What are the static axle loads?

KEY CONCEPT

Remember that at the contact between tire and road, we have equal and opposite vertical forces. We should refer to the downward force of the tire on the road as the *tire load* (or *axle load* when combined), and the upward force of the road on the tire as the *support force*. Being equal and opposite, finding one finds the other.

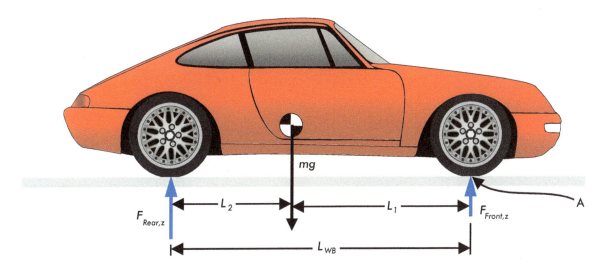

We did this numerically for **FIGURE 3** and **FIGURE 4**. Now let's do this with variables and plug numbers in at the end. First let's sum the vertical forces, taking upward as positive. Referring to **FIGURE 9**:

$$F_{Res,z} = -mg + F_{Front,z} + F_{Rear,z} = 0$$

or:

Equation 16-4

$$F_{Front,z} + F_{Rear,z} = mg$$

As expected, the static axle loads add up to the weight of the car. As for *x*-direction forces, they don't matter here because there aren't any; the car is on a level road and stationary.

We need to take *moments* from the support forces to find the individual axle loads. Let's take moments about the front tire patches (point A). Note that pitch is positive nose-down, clockwise in **FIGURE 9**:

$$M_{y,A} = -mgL_1 + F_{Rear,z}L_{WB} = 0$$

This can immediately be solved as:

Equation 16-5a   **Static Rear-Axle Load**

$$F_{Rear,z} = mg\,\frac{L_1}{L_{WB}}$$

FIGURE 9 A Porsche 911 that weighs 3,365 lbf, has a wheelbase L_{WB} of 92.9 inches, and has its CG 56.7 inches behind the front axle ($L_1 = 56.7$ inches).

Putting this into equation 16-4 and solving for the front-axle load, knowing that $L_{WB} = L_1 + L_2$:

Equation 16-5b   **Static Front-Axle Load**

$$F_{Front,z} = mg\frac{L_2}{L_{WB}}$$

Now let's plug the numbers in. With $mg=3{,}365$ lbf, a wheelbase of 92.9 inches, and L_1 of 56.7 inches, equation 16-5a gives the rear axle load:

$$F_{Rear,z} = mg\frac{L_1}{L_{WB}} = (3{,}365\ \text{lbf})\times\frac{56.7\ \text{inches}}{92.9\ \text{inches}} = 2{,}054\ \text{lbf}$$

And since $L_2 = L_{WB} - L_1 = 36.2$ inches, the front axle load is, using equation 16-5b:

$$F_{Front,z} = mg\frac{L_2}{L_{WB}} = 3{,}365\ \text{lbf}\times\frac{36.2\ \text{inches}}{92.9\ \text{inches}} = 1{,}311\ \text{lbf}$$

Dividing each axle load by the weight of the car yields a weight distribution of 39% front, 61% rear. Logically, biasing the CG rearward biases the weight distribution to the rear axle.

4.5 Finding the CG Location, Knowing the Static Load on Each Axle

FIGURE 10 A 2008 Honda Civic coupe that has a 1,870-lbf load on the front wheels, 1,196 lbf on the rear wheels, and a 104.3-inch wheelbase. What is the distance L_1?

It's more likely you'll know the axle loads than the CG location, so let's work the problem the other way. Suppose you weigh a 2008 Honda Civic Si, including a 180-pound driver, with a scale underneath each tire. You total 1,870 lbf on the front axle and 1,196 lbf on the rear (**FIGURE 10**). Where is the CG, given the Civic's 104.3-inch wheelbase?

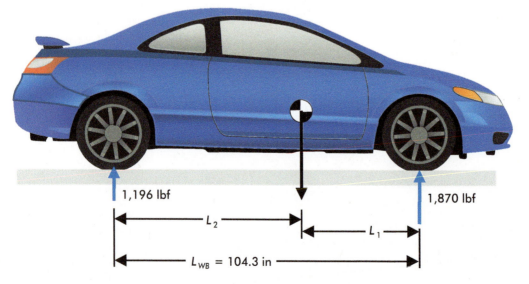

Using equation 16-4, we find the weight:

$$mg = F_{Front,z} + F_{Rear,z} = 1{,}870 \text{ lbf} + 1{,}196 \text{ lbf} = 3{,}066 \text{ lbf}$$

which incidentally gives it a weight distribution of 61% front, 39% rear. Now, to find the longitudinal CG location, let's use equation 16-5a:

$$L_1 = \frac{F_{Rear,z}}{mg} L_{WB} = \frac{1{,}196 \text{ lbf}}{3{,}066 \text{ lbf}} \times 104.3 \text{ in} = 40.7 \text{ in}$$

This means that the CG is 40.7 inches behind the front axle, 39% of the wheelbase. Note that the portion of weight carried on the rear axle is also 39%.

TABLE 2 shows a group of cars having different engine and drivetrain placements, going from rear-biased to front-biased. In each, the distance between the CG and *one* axle is always the same percentage of the wheelbase as the percentage of load on the *other* axle.

Car	Engine Position	Drive Axle	% Front Load	% Rear Load	L_1/L_{WB}, %	L_2/L_{WB}, %
2008 Porsche 911	Rear	Rear	39	61	61	39
2008 Porsche Cayman	Rear Mid	Rear	45	55	55	45
2008 Chevy Corvette	Front Mid	Rear	51	49	49	51
2008 Ford Mustang	Front	Rear	54	46	46	54
2008 Honda Civic Si	Front	Front	61	39	39	61
2008 Chevy Malibu	Front	Front	61	39	39	61

TABLE 2 Weight (static load) distributions of a range of road cars, starting with rear-engine RWD, then rear mid-engine RWD, front mid-engine RWD, front-engine RWD, and front-engine FWD.

Production cars generally lie in the 40/60 to 60/40 range, which is covered by the cars in **TABLE 2**. The 911, with its engine behind the rear axle and being RWD (rear-wheel drive), has a lot of mass toward the rear. The Honda and Malibu, with engines ahead of the front wheels, and FWD (front-wheel drive), have it concentrated toward the front.

In between is the Cayman, with the engine ahead of the rear axle (rear mid-engine), and a slightly rearward bias. The Corvette, with the engine just behind the front axle (front mid-engine), and its transmission in a rear-mid position, has its weight split almost evenly.

The front-engine rear-drive Mustang is front biased, but less so than the front-drive cars, because its drive-axle mass is in back.

Weight distributions are important and interesting, but are only exact when the car is stationary. Not the most fun part of driving. So let's move on to quasi-statics.

5   The Mechanics of Quasi-Statics

While static load distribution is important, we're more interested in what the tire loads become during acceleration, making it a dynamics problem, not statics. Once into steady acceleration, the loads don't change, or change "slowly." The resulting quasi-static condition is very similar to a static one, and applies in many situations. We'll concentrate here on weight and traction, although we could easily add aerodynamic drag and downforce.

As in our statics problems, we will work in a plane, so there are always two translations and one rotation to consider. **FIGURE 11** shows the same three possible sets of directions as for statics (**FIGURE 8**), but note that allowing acceleration creates different applications.

Longitudinal Load Transfer

F/A, Vertical, Pitch

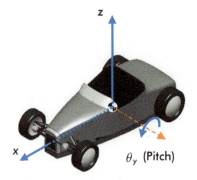

Lateral Load Transfer

Lateral, Vertical, Roll

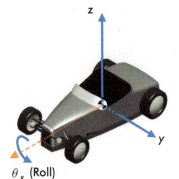

Lateral Axle Forces in Cornering

F/A, Lateral, Yaw

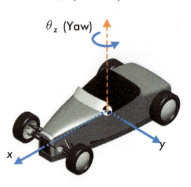

FIGURE 11 Sets of DOFs to use for quasi-statics problems; the x-z plane for longitudinal load transfer, the y-z plane for lateral load transfer in cornering, and the x-y plane for cornering forces at each axle.

You'll see that although quasi-statics is very similar to statics, there is one major difference to watch out for when taking moments.

5.1   Developing an Attitude

When you first stab the throttle pedal, the tires dig in, the front end noses up, and the rear end squats down. The car might pitch upward 2° at first. Then the front will drop down some, bounce a little, and settle into a "constant" attitude, say 1° nose-up.[7] The first part, the pitching, is transient motion. Quasi-statics cannot be used yet, because the tire and suspension forces are changing too quickly. But once this transient motion is over, and the car settles into its steady acceleration and pitch attitude, it's "steady enough" to use quasi-statics.

Note 7: In statics or quasi-statics, the attitude is settled out by damping, which here is from the shocks.

KEY CONCEPT

We can consider the forces acting on a vehicle as steady if they are changing "slowly." For example, when a typical car accelerates, the aerodynamic forces on it increase and its traction force decreases, but the forces change slowly enough to calculate the tire loads as if they were steady.

Now the next time you drop throttle, shift, or brake, you again have transient motion, and can't use a quasi-static approach until things settle down.

The assumption of constant acceleration and constant vehicle attitude hardly ever happens exactly. A purist could argue that quasi-statics *never* holds, but in many situations it is close enough to give you valuable insight.

5.2 Quasi-Statics in Longitudinal Acceleration

The conditions where we can use quasi-statics in a plane are only slightly different than for statics. In longitudinal acceleration:

1. The object's longitudinal acceleration (x-direction) must be constant.
2. The vertical acceleration (z-direction) must be constant.
3. The pitch acceleration (θ_y direction) must be zero.

So in straight-line motion, we have a new set of three conditions:

1. The longitudinal forces (x-direction) add up to the mass times the x-acceleration of the CG.
2. The vertical forces (z-direction) add up to the mass times the z-acceleration of the CG.
3. The pitch moments (θ_y-direction) about the CG add up to zero.

Putting this into equations, we have in the longitudinal direction:

Physical Law Eq. 16-6a
$$F_{Res,x} = ma_x$$

and in the vertical direction:

Physical Law Eq. 16-6b
$$F_{Res,z} = ma_z$$

Note that in purely longitudinal acceleration, we will set a_z equal to zero, so the resultant vertical force will be zero. Also, for pitch:

Physical Law Eq. 16-6c
$$M_{y,CG} = 0$$

That moment equation, 16-6c, is a subtle but critical difference between statics and quasi-statics. In statics, the mass merely leads to a *force* (its weight mg), which produces moments. So we completely take the mass's effect into account by its weight and moments.

In quasi-statics, the mass's weight effect is tracked in the same way. In addition, the mass is accelerating, so its *inertia* effect must be tracked while summing forces and moments. But its accelerating does not produce any additional moments to track, because a mass does not produce a force as a result of being accelerated (it *requires* a net force *to be accelerated*).

So we can't pick just any "point A" to take moments about; we can only use the CG. (This becomes very clear if you try to do it wrong.)

CAUTION

In quasi-statics, we must take into account the inertia effects of the body's mass. In a plane, we set each resultant force equal to its corresponding acceleration times the mass. But in addition, *we must take the moments about the CG* and set them to zero.

Earlier, we used statics to calculate the axle loads for the stationary Cougar in **FIGURE 3** and **FIGURE 4**. Now let's use quasi-statics to find them during constant straight-line acceleration. **FIGURE 12** shows the forces acting on the car. The thrust from the rear tires, being at ground level, acts on a horizontal line below the CG. So besides accelerating the car, the thrust also puts a CCW moment on it, tending to accelerate it nose-up in pitch, until something stops it.

As a result, the car does rotate slightly, the rear springs compressing and the front springs relaxing, until the forces from the tires adjust to counteract the nose-up moment from the thrust. This whole process is called **LOAD TRANSFER**. If this is hard to imagine, think about what would happen if the car were "floating" in space, so the only forces acting on it were pushing forward at the bottom of the rear tires. The car would flip.

KEY CONCEPT

Note that although load transfer is commonly called weight transfer, it is not: the weight of the car is still acting at the CG, straight downward.

KEY CONCEPT

The thrust force produces the same (nose-up) pitch moment whether the vehicle is FWD, RWD, or AWD (all-wheel drive). Therefore, with the same acceleration (same thrust), the same load transfer would occur.

The Cougar in **FIGURE 12** has a 111-inch wheelbase, and the CG is 50 inches rearward of the front axle, 16 inches off the ground. It weighs 3,650 lbf. If we drive with 1,825 lbf of thrust, what will the load on the rear tires be in steady acceleration?

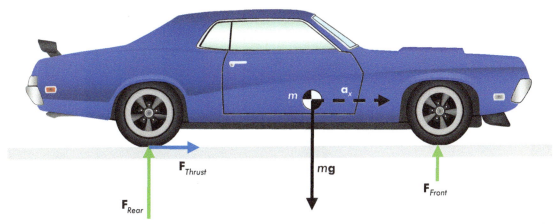

We can start off by summing the forces in the vertical direction. Using equation 16-6b:

$$F_{Front,z} + F_{Rear,z} = mg = 3,650 \text{ lbf}$$

FIGURE 12 Axle loads in a vehicle with constant acceleration a_x due to traction force F_{Thrust}. The thrust tends to pitch the nose up, while the axle loads adjust to counteract it.

So far it's like statics; the axle loads add up to the weight of the car. Next we sum fore-aft forces, which only consist of the thrust. It doesn't help much with loads, but it does tell us the straight-line acceleration (note that the mass is 113.35 slugs):

$$a_x = \frac{F_{Thrust}}{m} = \frac{1,825 \text{ lbf}}{113.35 \text{ slugs}} = 16.1 \text{ ft/sec}^2 = 0.5 \text{ g's}$$

Now let's look at the moments on the car. Remember that in quasi-statics, we must take them about the CG. Using equation 16-6c:

$$M_{y,CG} = -F_{Thrust} h_{CG} + F_{Rear,z} L_2 - F_{Front,z} L_1 = 0$$

Filling in the known values, and knowing that the front axle load is the car's weight minus the rear axle load:

$$-\left(1,825 \text{ lbf} \times 16 \text{ in}\right) + \left(F_{Rear,z} \times 61 \text{ in}\right) - \left(\left(3,650 \text{ lbf} - F_{Rear,z}\right) \times 50 \text{ in}\right) = 0$$

We solve this for the rear axle load:

$$F_{Rear,z} = 1{,}907 \text{ lbf}$$

Compared to the static loads, the forward thrust of 1,825 lbf increases the rear axle load by 263 lbf, and reduces the front axle load by the same amount, to 1,743 lbf. With a bit of work, we can get general formulas to find the front and rear axle loads during steady acceleration:

Equation 16-7a   **Front-axle load**

$$F_{Front,z} = mg \frac{L_2}{L_{WB}} - F_{Thrust} \frac{h_{CG}}{L_{WB}}$$

Equation 16-7b   **Rear-axle load**

$$F_{Rear,z} = mg \frac{L_1}{L_{WB}} + F_{Thrust} \frac{h_{CG}}{L_{WB}}$$

Let's think about this. If there is no traction force ($F_{Thrust} = 0$), then the front and rear axle loads are mgL_2/L_{WB} and mgL_1/L_{WB}, exactly the static loads from equations 16-5b and 16-5a. Adding thrust increases the rear-tire load and reduces the front, by the thrust times the CG height-to-wheelbase ratio h_{CG}/L_{WB}. For a given vehicle then, the load transfer depends on the amount of thrust, but not directly on vehicle weight or the original weight distribution. But we can also substitute ma_x in for the thrust in equations 16-7, to find directly how the tire loads depend on acceleration:

Equation 16-8a   **Front-axle load**

$$F_{Front,z} = mg \frac{L_2}{L_{WB}} - ma_x \frac{h_{CG}}{L_{WB}}$$

Equation 16-8b   **Rear-axle load**

$$F_{Rear,z} = mg \frac{L_1}{L_{WB}} + ma_x \frac{h_{CG}}{L_{WB}}$$

Equations 16-8 state the loads in terms of the longitudinal acceleration, vehicle mass, wheelbase, and CG height. Here you can see that the load transfer of a given car depends on its acceleration.

KEY CONCEPT

An obvious conclusion is that if forward acceleration *increases* the load on the drive wheels for rear drive, it *reduces* it for front drive. This will limit FWD traction and acceleration, as we'll see in Chapter 17.

5.3 Limitations in Using Statics and Quasi-Statics

There are several more conditions controlling whether we can use statics or quasi-statics to solve problems for motion *in a plane*:

1. There must be no more than a total of three unknown forces and moments, because we only have three equations to find them with.
2. The object must be rigid, or its flexibility must not affect the problem.
3. Each force must have a clearly defined line of action.

To the first point, there are four tires on a typical car, so we cannot solve for all four individual tire loads with the three equations. We've been using axle loads instead, because there are only two. We could also assume that the tire forces on one axle are equal in forward acceleration, but that depends on mass symmetry in the vehicle, and preloads in springs and antiroll bars. And with a solid rear axle, propshaft torque during acceleration increases the load on the left rear tire and reduces the load on the right rear.

5.4 Quasi-Statics in Lateral Acceleration

You're entering a left turn, which you will take at constant speed. Your car must "settle into it," just like at the start of straight-line acceleration. You steer into the curve, and the front tires bite. The vehicle starts to rotate CCW in yaw, and to accelerate left. The rotation causes the rear tires to bite, and the vehicle's leftward acceleration increases. It also rolls to the right as the right side springs compress and the left side springs extend. In a half second or so, the roll and slip angles settle, the vehicle follows a constant radius, and the yaw speed and lateral acceleration are constant, as in **FIGURE 13**.

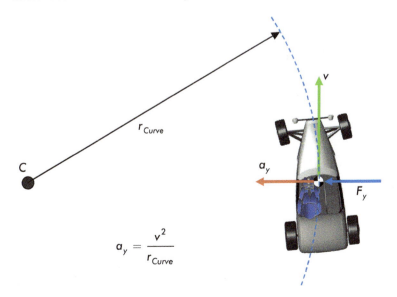

$$a_y = \frac{v^2}{r_{Curve}}$$

FIGURE 13 A car in a constant-radius, constant-speed curve, showing the lateral acceleration a_y and the resultant lateral force F_y causing it, resulting from the tires' cornering forces.

Once we're in the curve we can use quasi-statics to find the lateral load transfer. The forces and acceleration are shown in **Figure 14**. The tires' lateral forces are acting to the left to cause the lateral acceleration, and also apply a moment tending to cause CW roll. The loads on the tires become greater on the right side, to counteract the roll moment from the lateral forces.

Here, the lateral acceleration a_y is the centripetal acceleration v^2/r_{Curve}. But assuming we're on a flat road, the vertical acceleration a_z is zero. Let's write our three conditions for the forces and moments. In the lateral direction,

Equation 16-9a

$$F_{Res,y} = ma_y$$

in the vertical direction:

Equation 16-9b

$$F_{Res,z} = 0$$

and the roll moment:

Equation 16-9c

$$M_{x,CG} = 0$$

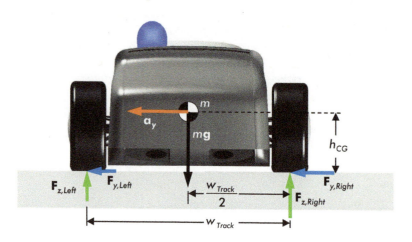

Figure 14 Vertical and lateral tire forces produced when turning left. The load on the right side tires is higher than for the left, due to lateral load transfer.

Just as in straight-line motion, lateral acceleration causes load transfer, but side-to-side. Using a similar approach to that for longitudinal acceleration, and assuming lateral symmetry, the tire loads are:

Equation 16-10a   **Left-side load, left turn**

$$F_{Left,z} = \frac{mg}{2} - ma_y \frac{h_{CG}}{w_{Track}}$$

and:

Equation 16-10b   **Right-side load, left turn**

$$F_{Right,z} = \frac{mg}{2} + ma_y \frac{h_{CG}}{w_{Track}}$$

Remember that the track width w_{Track} is the distance between the centers of the tires' contact patches (not the outside). In words, the total left-side load is half the car's weight *minus* the load transfer, while the right side is half the car's weight *plus* the load transfer. What do these equations mean? The load transfer will be greater if:

1. You accelerate harder (a tighter curve or higher speed),
2. The track width is narrower, or
3. The CG is higher.

> ### MAJOR POINT
>
> The major takeaway is that the outside tires have a higher load and more traction, and so supply most of the lateral forces in hard cornering. This effect is reduced with a lower CG and wider track.

You have no doubt sensed one effect of load transfer while cornering, especially if you've been in older vehicles that had much softer springs and antiroll bars—they leaned plenty.

6   Motorcycle Cornering

As you know, motorcycles have a different force balance when cornering. Lateral tire forces still try to roll the bike outward, but that can't be counteracted by transferring load to the "outside" tires; there aren't any.[8] Instead, moments are balanced by rolling the bike inward.

FIGURE 15 is a front view of a motorcycle taking a curve. Its centerline is angled 51°, but the tires contact the pavement slightly off-center, so the bike/rider CG is on a line angled 48° from the tire patches; the "effective lean angle" θ_x is 48°. What is its lateral acceleration a_y?

We can make quick work of this by first setting the resultant vertical force equal to zero, which says the total vertical tire force F_z (from both tires) equals the weight mg. We also know that the lateral forces must equal the mass times lateral acceleration, so F_y equals ma_y.

Note 8: Well, it could have a sidecar, which is a whole 'nother thing.

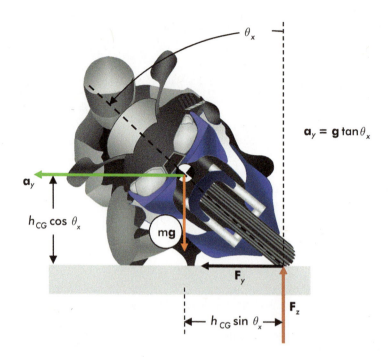

FIGURE 15 The lateral tire forces and weight each produce a roll moment on a bike, here in a steady curve. The steady roll angle θ_x must balance these, and only depends on the lateral acceleration.

Only one more step: we sum the moments about the CG from the lateral and vertical tire forces, then use that to find the acceleration. First, in terms of the tire forces:

$$F_z h_{CG} \sin \theta_x - F_y h_{CG} \cos \theta_x = 0$$

Then, substituting in the tire lateral and vertical forces we found above:

$$mgh_{CG} \sin \theta_x - ma_y h_{CG} \cos \theta_x = 0$$

Solving this for the acceleration:

Equation 16-11   **Motorcycle lateral acceleration**

$$a_y = g \tan \theta_x$$

Here, the 48° roll angle θ_x has a tangent of 1.11, for a lateral acceleration of 35.76 ft/sec², or 1.11 g's.

7 | Two-Force Members

Here's an important concept used in mechanisms like suspensions and crank trains. A **TWO-FORCE MEMBER** is a link with a "friction free" bearing at each end, so it can rotate freely and a moment can't be applied through it. One example is the suspension link in **FIGURE 16**, a straight rod with a rubber bushing at each end.[9] As shown, it has forces pulling on each end, and is inclined from horizontal by the angle θ.

On the left side of the figure, the y-forces (lateral) and z-forces (vertical) are shown. It's easy to see from statics that for equilibrium, the lateral forces on the link at the left and right ends must be equal and opposite; same for the vertical forces.[10]

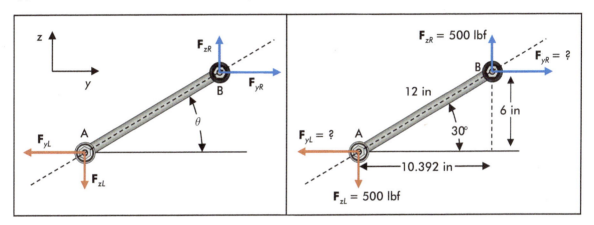

But what is the relationship between the vertical and lateral forces? To find out, we need to sum moments about one of the ends. As shown on the right of **FIGURE 16**, we know that the vertical force on each end is 500 lbf, that the length of the link is 12 inches, and that the angle is 30°. We find that their lateral offset Δy is 10.392 inches, and that their vertical offset Δz is 6 inches. Taking the moments about point B, CCW positive,

$$M_{xB} = 0 = F_{zL}\Delta y - F_{yL}\Delta z = 500 \text{ lbf} \times 10.392 \text{ in} - F_{yL} \times 6 \text{ in}$$

Solving this, F_{yL}= 866 lbf. Using this, **FIGURE 17** (left side) shows the resultant force on each end of the link.

Importantly, this shows we have two 1,000-lbf forces, one on each end of the link, pulling in opposite directions. Both act along the line between the two attachment points. This is true for any hinged link, and explains why a chain's links line up straight when you pull on it.

FIGURE 16 Force components on a suspension link with frictionless attachments at each end. On the left is shown the general geometry, with specific values on the right.

Note 9: Each end could also have a ball joint, or a greased shaft... some sort of hinge.

Note 10: If the rod is accelerating, the forces at each end will still be pretty close to equal, as long as the acceleration isn't too large, or the rod is "light."

KEY CONCEPT

Because only two forces act on a link, and no moments, it is called a two-force member. Because the hinge points can carry no moments, the two forces are equal, opposite, and along the line that goes between the attachment points.

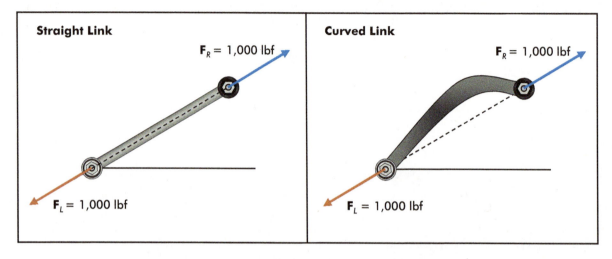

Straight Link

F_R = 1,000 lbf

F_L = 1,000 lbf

Curved Link

F_R = 1,000 lbf

F_L = 1,000 lbf

FIGURE 17 The opposing forces acting at each end of a two-force member are always equal and lie along the line through their attachment points. This is true whether the link is a straight rod (left) or curved bar (right).

If that seemed obvious, the point next may not be. We didn't require the rod to be straight when we took the moments, so it could have any shape. A curved link (**FIGURE 17**, right side) is one example. Sometimes a suspension link must be bent to clear a tire, frame, or another link. It can still do its job if it's curved, and the end forces will still be along the line connecting them. But the forces on each end (as shown) will tend to straighten out a curved link, so it must be made stronger (and heavier) to withstand this.

8   Summary

It may seem like a step backwards to take Newton's second law, set the acceleration to zero and call it statics. But as you can see, statics gives us a lot to work with in determining the design of a vehicle, and quasi-statics opens up further design aspects.

You use statics every day. When you tighten a lug nut on a car that's up on a jack, you need the brakes on to hold the wheel still. While you apply torque to the lug nut, it exerts a moment on the wheel. It takes an opposing moment (from the brake) to hold it still. Whatever you're applying a force or torque to, you hold it with a vise, a stand, a foundation, etc. So usually when you use statics, you don't perform calculations. You use common sense.

Using quasi-statics makes it plain how a vehicle's fore-aft CG location and CG height, as well as its wheelbase and track width, are crucial to all-around performance. Besides putting the right tires on the car, you have to put the right loads on the tires.

If you look back, you'll also see that the load transfer in steady straight-line acceleration is very similar to that for cornering. Compare equations 16-8 to 16-10, for example. Switch the fore-aft acceleration and vehicle dimensions to lateral ones, and one set of formulas becomes the other (no surprise, right?). We're not done with that part yet. Chapter 17 will use these concepts to investigate other influences of vehicle layout on its performance.

The same techniques can be used to calculate the forces within a valve-train, crank-train, or suspension linkage. **Figure 18** shows the forces in an SLA (short/long arm) suspension that has a torsion bar attached to the inner pivot of the LCA (lower control arm), and 1,000-lbf lateral and vertical forces at the tire patch. Given the tire patch forces, all the interface forces and moments were found using statics.

FIGURE 18 Statics can be used to find the forces (shown in lbf) between parts of a mechanism like this torsion bar suspension. Note that from this front view, the UCA (upper control arm) is a two-force member. But the LCA is not, because a moment from the torsion bar acts on it.

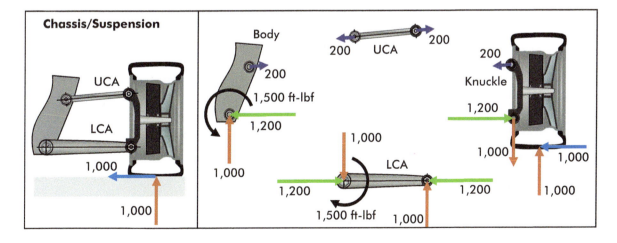

Just remember that not all situations are static, or even quasi-static. If you design a suspension to just withstand the loads on a smooth road, even a fully loaded car in hard cornering, it will not handle the additional loads from hitting bumps. If you design a propshaft to just withstand the maximum steady torque the transmission will produce, it will not handle the torque pulse during shifts, or the torsional vibration from the engine's firing pulses. These impact and vibration loads are more complicated, and we won't handle them here.

You now have the foundation you need for Chapter 17, which will begin by reinforcing these concepts. But mainly you'll get a chance to examine situations that are a bit more complex, investigating the combined influence of vehicle properties on its performance.

But as usual, the more intricate aspects of vehicle design are the most interesting and rewarding to learn. And sometimes they're surprising.

Major Formulas

Definition	Equation	Equation Number
Static Equilibrium in x-z Plane		
Resultant x-Force (Longitudinal)	$F_{Res,x} = 0$	16-1a
Resultant z-Force (Vertical)	$F_{Res,z} = 0$	16-1b
Resultant y-Moment (Pitch)	$M_{y,A} = 0$	16-1c
Quasi-Statics in x-z Plane		
Resultant x-Force (Longitudinal)	$F_{Res,x} = ma_x$	16-6a
Resultant z-Force (Vertical)	$F_{Res,z} = ma_z$	16-6b
Resultant y-Moment (Pitch)	$M_{y,CG} = 0$	16-6c
Front-Axle Load in Forward Acceleration	$F_{Front,z} = mg\dfrac{L_2}{L_{WB}} - F_{Thrust}\dfrac{h_{CG}}{L_{WB}}$	16-7a
Rear-Axle Load in Forward Acceleration	$F_{Rear,z} = mg\dfrac{L_1}{L_{WB}} + F_{Thrust}\dfrac{h_{CG}}{L_{WB}}$	16-7b
Front-Axle Load in Forward Acceleration	$F_{Front,z} = mg\dfrac{L_2}{L_{WB}} - ma_x\dfrac{h_{CG}}{L_{WB}}$	16-8a
Rear-Axle Load in Forward Acceleration	$F_{Rear,z} = mg\dfrac{L_1}{L_{WB}} + ma_x\dfrac{h_{CG}}{L_{WB}}$	16-8b
Load on Left Tires, Left Turn	$F_{Left,z} = \dfrac{mg}{2} - ma_y\dfrac{h_{CG}}{w_{Track}}$	16-10a
Load on Right Tires, Left Turn	$F_{Right,z} = \dfrac{mg}{2} + ma_y\dfrac{h_{CG}}{w_{Track}}$	16-10b
Lateral Acceleration vs. Lean Angle, Motorcycle	$a_y = g\tan\theta_x$	16-11

17 Statics and Quasi-Statics Applications

Contact Forces, CGs of Assemblies, and Vehicle Design for Maximum Performance

These are the two Don Garlits cars that turned drag racing on its ear in 1970–71: Swamp Rat XIII, his last front-engined car, and Swamp Rat XIV, the first successful rear-engined one. We'll explain why with quasi-statics. *Illustration 'Transformation — Don Garlits' Wynn's Chargers, 1969-1971' © 2004 by Kane Rogers.*

Chapter 17

We use statics and quasi-statics to find the CG of an assembly of components, and to find bearing forces in an engine. The factors limiting straight-line acceleration are found, and the results are used to analyze a quantum shift in dragster design in the early 1970s. Design features for maximizing performance of different types of vehicle are discussed.

Swamp Rat XIII, 1969-1970

Swamp Rat XIV, 1971

Contents

Key Symbols Introduced

Symbol	Quantity	SFS Units	MKS Units
a_{WS}	Wheel-Spin-Limited Acceleration	ft/sec^2	m/sec^2
a_{WL}	Wheelstand-Limited Acceleration	ft/sec^2	m/sec^2
SSF	Static Stability Factor	dimensionless	dimensionless

1 Introduction

Statics and its cousin quasi-statics have given us a good look at how a vehicle's design influences its dynamics. It's one thing to equip a vehicle with a powerful engine and performance tires and hope for the best. It's another to set the vehicle up to take advantage of them, or design it that way the first place. With ingenuity, even underpowered cars can over-perform. The CG location, wheelbase, and track width determine a lot about a car's ability to accelerate in straight-line motion or in cornering. A vehicle's layout and its design proportions help set today's cars apart from those 100, 50, or even 20 years ago.

Even though we'd talked about the importance of the CG several times in this book, it wasn't until the previous chapter that we had a way to find it. This chapter will show how to calculate the CG location of a simple object made of a few different parts, like a set of weights, or a more complicated one, like a car.

We'll spend a fair amount of time discussing the effects of vehicle design on straight-line acceleration limits, given a level of tire friction coefficient. These results explain a very significant (and famous) change in dragster design in the early 1970s, and the trade-offs involved. But the analysis also shows why you need to keep your eye on the math you use and make sure it accounts for all the physics of the problem.

We'll also investigate the desired layout for cars that hit the oval tracks and road courses. The technique is the same; only the goals are different. In doing this we'll typically assume tires with constant friction coefficient, but we'll briefly consider the effects of more realistic tires.

Because we use statics and quasi-statics in so many common ways, we'll start by analyzing forces used in lifting some everyday objects, like a barbell and a heavy flywheel. You use the same ideas when using pliers or stacking boxes of parts.

Then we go on to an example of forces in a crank train. It is a simple one, but its ideas can be used to find forces in more involved mechanisms, like an independent suspension, traction bars, or gears and bearings in a transmission. So let's go.

2 The Center of Gravity

A vehicle isn't the only object where the CG location matters. Anytime you exert a force that is not acting through the CG, you will have to watch out for unbalanced moments. This applies to using tools, balancing tires, and simply lifting objects.

2.1 Knowing Where the CG Is When You Lift

If you try to lift a heavy flywheel by grabbing it at one edge (point A in **FIGURE 1**, left), it can easily twist itself out of your grip. Its CG is offset from your hand by the radius r, so its weight exerts a CCW moment $M_A = mgr$, on the point of contact (your hand). Your hand must counteract that with a CW moment $M = mgr$ of its own. This is difficult because your fingers don't have much leverage; just the distance from their tips to the root of your thumb.

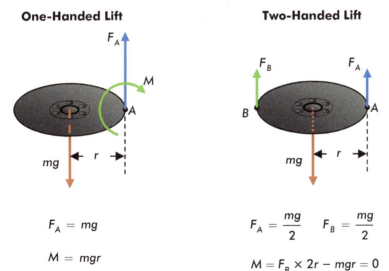

One-Handed Lift

$$F_A = mg$$

$$M = mgr$$

Two-Handed Lift

$$F_A = \frac{mg}{2} \qquad F_B = \frac{mg}{2}$$

$$M = F_R \times 2r - mgr = 0$$

FIGURE 1 Holding a flywheel by the right edge only (left figure) requires one large force F_A and a moment M from your hand. Holding it by both sides is easier, largely because the support forces from your hands cancel each other's moments, leaving no twisting moment for you to resist.

On the other hand, if you include the other hand and lift the flywheel from both sides, you can do it much more easily (**FIGURE 1**, right side). It's partly because you carry half the weight in each hand, but the main advantage is that neither hand needs to exert a moment. The CCW moment on your right hand from the weight, $-mgr$, is now cancelled by the CW moment $2F_B r = mgr$ from the support force F_B at the left hand.

2.2 The CG of an Off-Center Barbell

In **Figure 2**, a barbell has 60 lbf on one side and 40 lbf on the other, mounted on a solid 20-lbf bar, with the centers of the weights 50 inches apart.[1] To lift this with an equal force from each hand, you'd want them to be the same distance from the CG. So where is it?

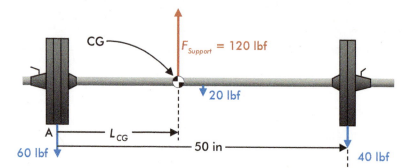

Figure 2 To lift a barbell with unequal weights, where would you need to center your hands to lift it evenly? In other words, where is its center of gravity?

Let's tackle this in an organized way, tracking each weight in **Table 1**. Their forces and distances from point A (see **Figure 2**) are shown in columns 2 and 3, with the moments about point A in the fourth column. Upward vertical forces and CW moments are positive.

Because the total weight is a 120-lbf downward force (-120 lbf), the support force must be a 120-lbf upward force; $F_{Support} = 120$ lbf. A force applied at the CG causes no rotation. The point where we could apply the support force, so that the moments cancel, is the CG.

Why are we using weight instead of mass when calculating the CG location? Weight is proportional to mass, so we get the same answer.

The CW moments about point A from the three barbell parts total 2,500 in-lbf (**Table 1**). We find the distance L_{CG} from point A to the CG by requiring the 120-lbf support force there to cancel the other moments:

$$M_A = 0 = 2,500 \text{ in-lbf} - \left(120 \text{ lbf} \times L_{CG}\right)$$

$$L_{CG} = \frac{2,500 \text{ in-lbf}}{120 \text{ lbf}} = 20.8 \text{ in}$$

The CG is closer to the heavier end, of course, 20.8 inches in.

Note 1: I don't know why anyone would do this unless they only had 20-lbf weights and wanted to lift 120 lbf.

	Force (lbf)	Distance (x) from A (in)	Moment (in-lbf)
Left weight	−60	0	0
Bar	−20	25	500
Right weight	−40	50	2,000
Total from Loads:	−120	N/A	2,500
Support Force	120	20.8	−2,500

TABLE 1 Forces, locations, and moments for each weight, and the support force, of **FIGURE 2**. The contributions from each of the weights are totaled, then the support force location is chosen to cancel their moments.

In this case we added the effects of three masses to find their combined CG, but you could combine five, twenty, or more together. It's just more entries in the table. This is how we'll "assemble" a dragster to get its CG location, but it will be more interesting because we'll find both the fore-aft and vertical CG positions (and because it's on a car!).

2.3 CGs of Simple Shapes

For a typical flywheel, its CG is easy to locate. It is symmetric, so the CG is at its geometric center. And if it is ¾-inch thick throughout, its CG is ⅜ inch below either surface.

Some other shapes are also obvious because of symmetry, like the rectangle in **FIGURE 3**. But what about a half-disc or a triangle? Like many other common geometric shapes, formulas for their CGs have been calculated, as shown.

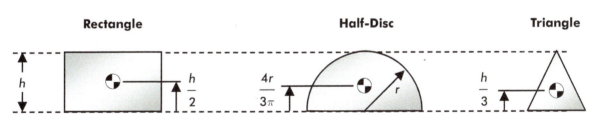

Even if the exact shape you're after isn't in a standard table of formulas, you can usually split the shape up into shapes with known formulas, locate each of their CGs, and add their effects like we did for the barbell.

Note that drilling/punching an off-center hole in the shape will move its CG. Flywheels are usually balanced by drilling them near the edge, at the right spot to *make sure* the CG is at the geometric center (on its spin axis).

FIGURE 3 Simple geometric shapes and their CG locations, assuming a constant thickness material. Note that a hole drilled in the shape would move the CG.

3   Bearing Forces in a Crank Train

Many automotive mechanisms use two-force members, links or rods with "friction-free" hinges at each end, as introduced in Chapter 16. We can use statics or quasi-statics to understand these. **Figure 4** is a free-body diagram of a crank train in an F1 engine. Several forces act on the connecting rod, mainly from cylinder pressure, and that necessary to accelerate the piston.

Here we'll just consider a 6,000-lbf cylinder pressure force. By not including acceleration, this becomes a statics problem. So the piston must push on the connecting rod with 6,000 lbf, through the piston pin. The connecting rod then pushes against the crank pin, and the engine block also pushes back on the crank. Per Newton's third law, contact force pairs of 6,000 lbf develop at each of the interfaces, in opposite directions.

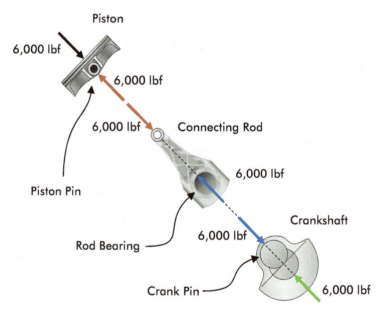

Figure 4 The pressure force (black) acting on the piston leads to bearing forces between the piston pin and connecting rod (red), the rod and the crank pin (blue), and the crankshaft and engine block (green, block not shown).

At each of the interfaces, the connecting parts must *bear* the loads on them, so these are called **BEARING FORCES** or **BEARING LOADS**. Bearing loads are very important in any design. In this case, the rod/crank and crank/block interfaces are even called bearings. But the piston pin also is a bearing.

With an angled rod (past TDC) the idea would be the same, except the forces would act in 2-D. The rod, being a two-force member, would cause a side-load on the piston, and actually create a moment (torque) on the engine block. An example was illustrated back in Chapter 9, **Figure 3**.

Of course the forces to *accelerate* the piston are also extremely important, especially at high speeds (as we investigated in Chapter 6). You can just as easily use quasi-statics to calculate those additional forces, and add them to the loads from the pressure. As always, divide and conquer.

4 Traction Limits in Longitudinal Acceleration

In Chapter 15, much was made of the traction-limited acceleration a_{TL} and power-limited acceleration a_{PL}, and how an increased traction limit lowers quarter-mile times drastically. But we didn't fully cover what defines the traction limit; and it isn't just traction! Let's look closer at how the vehicle design limits acceleration.

4.1 The Traction Limit for Rear-Wheel Drive—Take One

Increased load on the rear axle of a rear-drive vehicle increases its drive traction. Rearward weight bias is one way to increase it. The other is to encourage load transfer during acceleration. So the load transfer produces more traction, which allows more acceleration, which produces more load transfer, etc....Where does this all end, and how much does it affect performance? Let's find out.

To make it simple, we'll assume a constant friction coefficient for the tires, and no aerodynamic downforce. The rear-axle load, which per equation 16-7b is the static load plus the load transfer, is[2]

$$F_{Rear,z} = mg\left(L_1/L_{WB}\right) + F_{Thrust}\left(h_{CG}/L_{WB}\right)$$

So increased rear weight bias would result from either increased CG distance behind the front axle, or from a shorter wheelbase: both produce a larger ratio L_1/L_{WB}.

Increased load transfer would result from a higher CG or shorter wheelbase: a larger ratio h_{CG}/L_{WB}.[3]

Wheel spin would start when the thrust F_{Thrust} equals the friction coefficient times the rear axle load, $\mu_{Static}F_{Rear,z}$, at which point we have reached maximum traction. Substituting F_{Thrust}/μ_{Static} for the rear axle load above, and multiplying by μ_{Static}, we get the thrust at wheel spin:

$$F_{Thrust,WS} = \mu_{Static}mg\left(L_1/L_{WB}\right) + \mu_{Static}F_{Thrust,WS}\left(h_{CG}/L_{WB}\right)$$

Note 2: If you need to, look ahead to **Figure 8** for a refresher on the definitions of these variables. Also, remember that the wheelbase is $L_{WB} = L_1 + L_2$.

Note 3: Many early drag racers took advantage of this, by lifting the entire car high up on its springs, and moving the CG rearward by moving the axles forward, altering the wheelbase. This gave them the name "altereds."

Here, the "*WS*" subscript stands for "wheel-spin" limit. Notice that the wheel-spin thrust we're after appeared in two places. By rearranging we can isolate it, to get:

Equation 17-1   **RWD wheel-spin limit**

$$F_{Thrust,WS} = \mu_{Static} \frac{mg\left(L_1/L_{WB}\right)}{1 - \mu_{Static}\left(h_{CG}/L_{WB}\right)}$$

Now we can use Newton's second law, as $F_{Thrust,WS} = ma_{WS}$. Substituting ma_{WS} for thrust in equation 17-1, and dividing by the mass, we get the wheel-spin-limited acceleration a_{WS}:

Equation 17-2a   **RWD wheel-spin limit**

$$a_{WS} = \frac{\mu_{Static}g\left(L_1/L_{WB}\right)}{1 - \mu_{Static}\left(h_{CG}/L_{WB}\right)}$$

In *g*'s this would be:

Equation 17-2b   **RWD wheel-spin limit**

$$\frac{a_{WS}}{g} = \frac{\mu_{Static}\left(L_1/L_{WB}\right)}{1 - \mu_{Static}\left(h_{CG}/L_{WB}\right)}$$

Note how the top of this equation has L_1/L_{WB} which affects weight distribution, and the bottom has h_{CG}/L_{WB} which affects load transfer. Without load transfer, the wheel-spin limit in *g*'s would then be the rear weight fraction times the friction coefficient: $\mu_{Static}(L_1/L_{WB})$. But the bottom of equation 17-2b divides that by $1-\mu_{Static}(h_{CG}/L_{WB})$ to comprehend the extra traction from load transfer. Call it a "rear-drive bonus," that depends on the wheelbase, CG height, and friction coefficient. Let's see how this plays out.

Suppose the CG height is 20% of the wheelbase ($h_{CG}/L_{WB} = 0.2$), four tenths of the weight is on the rear axle ($L_1/L_{WB} = 0.4$), and the friction coefficient is 1.20. The top of equation 17-2b says the wheel-spin limit using the rear tires' *static* load would be 0.48 *g*'s. But the bottom of equation 17-2b is 0.76. Dividing by 0.76 is the same as multiplying by 1.316, so there is a 31.6% jump in the wheel-spin limit from load transfer, increasing it from 0.48 *g*'s to 0.63 *g*'s.

If you were a drag racer, you'd also want to scope out the effect of changing rear weight bias. So let's find the wheel spin limit with the CG placed *anywhere* between the axles (we vary L_1/L_{WB} in equation 17-2b). Again, h_{CG}/L_{WB} is 0.2, and $\mu_{Static} = 1.20$. The results are in **Figure 5**, plotted vs. rear weight bias

The results present a good-news/questionable-news story. The good news is that we do see the extra 31.6% in the wheel-spin limit, resulting from load transfer; so with "only" 76% of the weight on the rear axle, the wheel-spin limit (in *g*'s) is equal to the tires' friction coefficient.

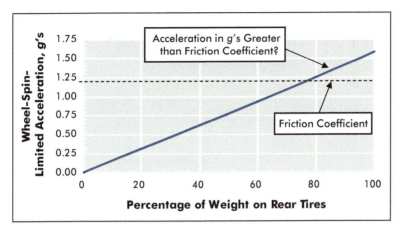

FIGURE 5 Wheel-spin-limited acceleration of a rear-drive car vs. its rear-weight bias, with $h_{CG}/L_{WB} = 0.2$ and μ_{Static} =1.20. With 76% rear bias, the wheel-spin limit (in *g*'s) equals the friction coefficient. Disturbingly, with greater rear-weight bias it predicts acceleration greater than the friction coefficient, which is impossible without downforce.

The questionable news? The graph also shows that with rear bias over 76%, the wheel-spin limit is greater than 1.20 *g*'s. That isn't physically possible: the acceleration in *g*'s cannot be larger than the friction coefficient (we have no aerodynamic downforce). When something sounds too good to be true, it probably is, so let's take a step back.

4.2 The Traction Limit for Rear-Wheel Drive—Take Two

There's obviously a problem with equation 17-2b, at least by itself. So let's look at the pieces that went into it. Remember that the calculated wheel-spin limit was based on the load on the rear tires, and their friction coefficient. Let's check the rear axle load at the wheel-spin limit by multiplying the acceleration from equation 17-2a by the mass of the car (to find thrust), then dividing by the friction coefficient:

Equation 17-3   **At wheel-spin limit, RWD**

$$F_{Rear,z,WS} = \frac{mg\left(L_1/L_{WB}\right)}{1 - \mu_{Static}\left(h_{CG}/L_{WB}\right)}$$

FIGURE 6 plots the rear axle load for our car at the wheel-spin limit, assuming a 3,000-lbf weight. Note that it is equal to the car's weight with 76% rear weight bias, but becomes greater than the weight with more bias. This would explain the high acceleration limit, but an axle load more than the vehicle weight isn't physically possible without downforce. What's going on?

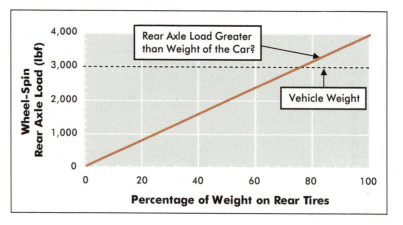

Figure 6 Rear axle load at wheel-spin limit vs. rear bias for the car from **Figure 5**, with a 3,000-lbf weight. Once the rear weight bias exceeds 76%, it says the rear axle load is more than the weight of the car! Not buying that.

Let's check out the *front* axle load with say, 85% rear bias. With that, equation 17-3 predicts a rear axle load of 3,355 lbf. Then the front axle load is:

$$F_{Front,z} = mg - F_{Rear,y} = 3{,}000 \text{ lbf} - 3{,}355 \text{ lbf} = -355 \text{ lbf}$$

When the rear tire load is larger than the car's weight, the front tire load is *negative*: the road is supposedly pulling the tire *down*. The road can hold the tire up, but it can't pull it down!

CAUTION

This is why we need to be careful with the math, and know the physics. The equations don't "know" that axle loads can't be negative. Once the front axle load goes to zero, the tire is lifting, and we have wheelstand. At that point we need to stop using equations 17-1–17-3.

But it's hardly a total loss. We just found another limit to acceleration: wheelstand. And we can still use our wheel-spin limit formulas unless we hit wheelstand before wheel spin.[4] Fortunately it's easy to predict when wheelstand will occur.

Figure 7 shows the Cougar with the front tires *just* coming off the pavement (that is, $F_{Front,z} = 0$). The thrust is trying to pitch the car CCW, while the vertical force on the rear tire patch (now equal to mg) is trying to pitch the car CW.

With no front tire load, these are the only two forces producing moments on the car's CG. So in steady acceleration at imminent wheelstand, their moments are equal and opposite:

$$F_{Thrust,WL} h_{CG} = mg L_2$$

Note 4: Just like a pet cat, you can only push equations so far.

Solving, the thrust at the wheelstand limit is:

Equation 17-4   **Wheelstand limit**

$$F_{Thrust,WL} = mg\left(L_2/h_{CG}\right)$$

Here, the subscript "*WL*" stands for "wheelstand limit."

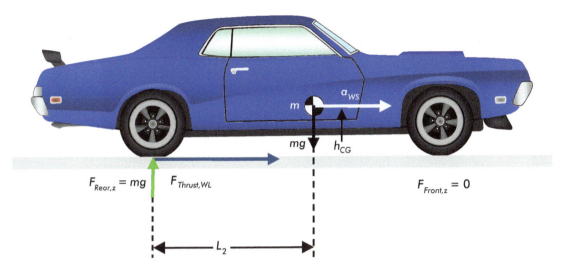

We'll call the acceleration at the wheelstand limit a_{WL}, so $F_{Thrust,WL} = ma_{WL}$. Substituting this into equation 17-4, and dividing by the mass, we get a_{WL}:

FIGURE 7 At the wheelstand limit, the load on the front axle is zero, and the rear axle load equals the weight of the car. The thrust force can't get any larger than $F_{Thrust,WL}$. Well, it couldn't for long....

Equation 17-5a   **Wheelstand limit**

$$a_{WL} = g\left(L_2/h_{CG}\right)$$

In *g*'s, this looks even simpler:

Equation 17-5b   **Wheelstand limit (g's)**

$$\frac{a_{WL}}{g} = \frac{L_2}{h_{CG}}$$

MAJOR POINT

The wheelstand limit for acceleration (in *g*'s) is simply the forward distance from the rear axle to the CG, divided by the height of the CG. Note that the wheelbase, in itself, doesn't matter.

Why do we call this the wheelstand *limit*? Because any increase in thrust would lift the front end, raising the CG, so h_{CG} increases. Once this happens, the thrust has a larger "lever arm" about the CG, so it pitches the nose up *faster*. You either let off or flip the car. Granted, once the front end lifts, the car's pitch acceleration means we don't have quasi-statics anymore. But quasi-statics has identified the edge of where we can operate.

Equation 17-5b can be plotted as a line from the rear tire patch, to find CG placements with different wheelstand limits. If the CG of the Porsche in **FIGURE 8** were on the solid line, it could accelerate at up to 1 g before wheelstand; on the dashed line, up to 2 g's. But the Porsche's CG is below the 2-g line, and could accelerate at about 2.3 g's before lifting the front. Even performance street tires wouldn't have the traction to do that, which is why most road cars can't come close to wheelstand, unless they're on a drag strip with slicks. Maybe.

Motorcycles are a different story, though. Their short wheelbase and high CG make it easy for many production motorcycles to pull a wheelie without much trouble.

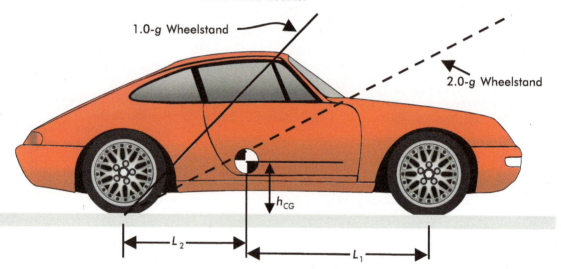

1.0-g Wheelstand

2.0-g Wheelstand

h_{CG}

L_2

L_1

FIGURE 8 The CG location for the Porsche 911, along with lines the CG must be below to avoid wheelstand at 1.0- and 2.0-g's acceleration. The 911 could accelerate about 2.3 g's without lifting the front end.

The downside of hitting the wheelstand limit is that, well, it limits your acceleration. But the upside is that if you plan for it, you can use this to your advantage. At wheelstand you have the largest possible rear axle load, equal to the car's weight:

Equation 17-6 At wheelstand limit

$$F_{Rear,WL,z} = mg$$

With it comes the most possible *traction*, equal to $\mu_{Static}mg$. But if that's more than $F_{Thrust,WS}$ in equation 17-1, you'd spin the wheels before reaching this thrust. The solution is to design the car so that the wheelstand limit thrust $F_{Thrust,WL}$ is equal to the maximum traction, $\mu_{Static}mg$:

$$F_{Thrust,WL} = mg\left(L_2/h_{CG}\right) = \mu_{Static}mg$$

Or simply:

Equation 17-7   **Max acceleration design**

$$\frac{L_2}{h_{CG}} = \mu_{Static}$$

Now the wheel-spin limit is equal to the wheelstand limit. So let's say your tires have a static coefficient of 2, and you want to get all the acceleration you can out of them: 2 *g*'s. Then the CG needs to be 2 times as far forward of the rear tire patch as it is high; if the CG is 48 inches forward of the rear axle, it needs to be 24 inches high. Any higher and it will wheelstand before 2 *g*'s. Any lower and the rear wheels will spin before 2 *g*'s. In practice, you might put the CG 22–23 inches high, to keep the front wheels down. There is a fine line between maximizing acceleration and losing the race (or worse) with a wheelstand.

4.3 The Traction Limit for Front-Wheel Drive

Letting the dust settle for the RWD case, what about FWD cars? After all, they start off with more load on the drive wheels, 61% in the case of our Honda Civic. At low traction values (low μ_{Static}), this means FWD cars have better drive traction, so they're popular in snowy climates.

But what if you want to make it into a performance car, putting stickier tires on it? The higher μ will increase the load transfer to the rear axle, which in this case takes load off the drive wheels. If we make the same assumptions for FWD as for RWD, its maximum traction force is:

Equation 17-8   **FWD wheel-spin limit**

$$F_{Thrust,WS} = mg \frac{\mu_{Static}\left(L_2/L_{WB}\right)}{1 + \mu_{Static}\left(h_{CG}/L_{WB}\right)}$$

Unlike the RWD case in equation 17-1, the load transfer now makes the bottom part of the equation *larger*, reducing the traction. It's a FWD *penalty*. The maximum acceleration is now, in *g*'s:

Equation 17-9   **FWD wheel-spin limit**

$$\frac{a_{WS}}{g} = \frac{\mu_{Static}\left(L_2/L_{WB}\right)}{1 + \mu_{Static}\left(h_{CG}/L_{WB}\right)}$$

So to increase its max acceleration, you'd not only want to move the CG forward, but also minimize the load transfer by lowering the CG and increasing the wheelbase. The wheelstand limit obviously doesn't apply to a FWD car, because the very attempt to lift the front end would eliminate the traction to do it with.

4.4 The Real Traction Limit: The Combined Effect of Wheel Spin and Wheelstand

To summarize, the effects of load transfer on traction-limited acceleration are shown in **Figure 9**, from both wheel-spin and wheelstand limits. It shows cars similar to a Porsche 911 (rear-engine rear-wheel drive, or RERWD), a Ford Mustang (FERWD), and a Honda Civic (FE-FWD), using the same tires (same μ_{Static}). For fairness, the ratio of CG height to wheelbase is assumed 0.1633 for each.

On slick surfaces (say μ_{Static} around 0.2), the RERWD and FEFWD cars are superior for acceleration, because each carries 60% of its weight on its drive axle. But with performance tires on dry pavement (μ_{Static} around 1.2), the FERWD car can accelerate at 0.69 g's, while the FE-FWD can only manage 0.60 g's. The load transfer has helped the rear-drive car and hurt the front-drive car. Meanwhile the RERWD car can accelerate at 0.90 g's, because its initial traction advantage has only been helped by the load transfer.

But with drag slicks that have $\mu_{Static} = 3.4$ or so, the situation changes again. The RERWD car hit its wheelstand limit at 2.45 g's, so it flat-lines, even with slicks. But the FERWD car's forward weight bias keeps the front tires down up to 3.3 g's. So the RERWD car can take full advantage of tires up to $\mu_{Static} = 2.45$, while for the FERWD car it's good up to 3.30.

The FEFWD car is never limited by wheelstand, but by load transfer, so $a_{TL} = a_{WS}$. With drag slicks having $\mu_{Static} = 3.40$, the FEFWD car can only reach 1.31 g's, 40% as much as the FERWD. You don't see too many front-drive drag racers.

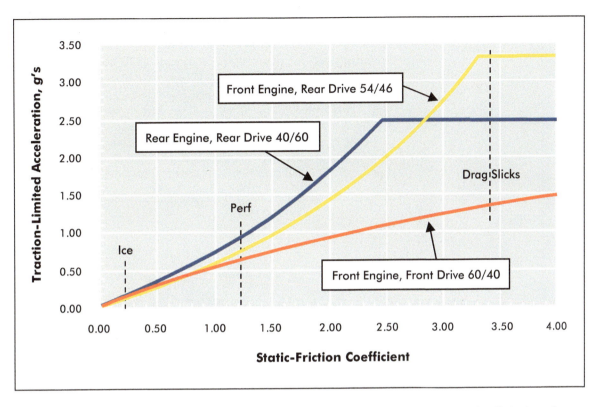

As **FIGURE 9** shows for the rear-drive cars, the usable traction can be limited by wheel spin or wheelstand, depending on the available traction μ_{Static}.

> **MAJOR POINT**
>
> The traction-limited acceleration a_{TL} is the smaller of the wheel-spin limit a_{WS} and the wheelstand limit a_{WL}.

FIGURE 9 Traction-limited acceleration for FERWD, RERWD, and FEFWD cars, vs. the tire friction coefficient. At low friction levels the FERWD is the slowest, but with high grip, on the strip, it can accelerate the fastest. The flat lines indicate wheelstand limits.

5   Garlits, μ, and the End of the Slingshot Era

Don Garlits was probably the most influential drag racer in history. Throughout his career, he took chances on developing new ways to make power and get it hooked to the track. He is best known for his Top Fuel dragsters, especially with making the rear-engined dragster (rear-mid, really) the machine to beat. Up to 1970, all Top Fuelers were front-engined and called slingshots. They were beautiful machines, but the drivers were seated precariously behind and over the rear axle. Wrapped around the transmission and differential, and sitting behind the engine, they had to contend with hot-oil baths and shrapnel from blower, clutch, and transmission explosions.

Unfortunately, Don was seriously injured in 1970 when his last slingshot, Swamp Rat XIII (shown in **Figure 10**) blew its transmission. Besides cutting his car in half, it also claimed a good part of his right foot. He returned within a year with the first successful rear-engined dragster, Swamp Rat XIV, shown with Swamp Rat XIII at the front of the chapter. The press reported the reason for switching to the rear-engine design was safety, getting his body ahead of all the powertrain mayhem. Not a bad reason, but is that the whole story?

Figure 10 Garlits in Swamp Rat XIII racing "Kansas John" Wiebe in 1970, at Beeline Dragway near Phoenix. *Photo by permission of Don Garlits Museum of Drag Racing.*

In writing this chapter, the equations for maximum acceleration made me think there was more to it. They say that the stickier the rear tires are (higher friction coefficient, μ_{Static}), *the less rearward* you want the CG. Otherwise the resulting wheelstand will limit your acceleration before you use the full traction of your rear tires. Load on the rear tires is still good, but you need to leave a little on the front.

More to the point, after introducing the rear-engined dragster (in 1971) Garlits was once again at the top of the drag racing world, and within two years had made the slingshot dragster obsolete.[5] Would someone as successful as Garlits "luck into" such an effective design just because it happened to be safe? I didn't think so either, so I asked him to shed some light on it.

Note 5: I told Don that I had been sad to see the front-engined dragsters disappear (of course, I didn't have to drive them). He said that when he brought out the rear-engine car, one woman told him she hoped that it wouldn't be successful, because it was the ugliest car she'd ever seen!

As mentioned, the slingshots were heavily rear-weight biased. But by 1970, drag tires had become sticky enough that this was no longer the right thing to do. In fact, Don said that in 1970, he had to run 150 pounds of lead ballast on the front of the car just to keep the front wheels on the track. Don said that it was hard to explain to people how adding weight to the *front* of the car could help you use the traction of the *rear* tires. It isn't your first choice. But if acceleration is limited by wheelstand (equation 17-5b) before the wheel-spin limit (equation 17-2b) is reached, it's the better choice.

The top of **Figure 11** shows the original CG for Swamp Rat XIII (black), and how ballast would move the CG forward by 15.6 inches, and slightly down (red). That kept the front wheels down, but added extra mass to accelerate (there was no minimum weight at the time).

With the rear-engined dragster, Garlits had found a better way to achieve the right CG placement. By moving the driver ahead of the engine in Swamp Rat XIV (bottom of **Figure 11**), *the driver* acted as ballast to keep the front end down instead of the 150 pounds of lead. Moving the driver in front of the engine also reduced the mass added by the roll cage, since it could be built into the existing frame. He was able to bring the car's weight (with driver) down from 1,760 to 1,510 pounds (50 pounds was from removing the transmission). That was over 14% lighter than the slingshots!

Figure 11 Outlines of Swamp Rat XIII, and Swamp Rat XIV with no wing, as it first ran. Each major component is shown as a solid steel sphere scaled to its weight. The CG of Swamp Rat XIII is shown without (black) and with ballast (red). The curved arrows show the major changes in component CG locations between the two cars.

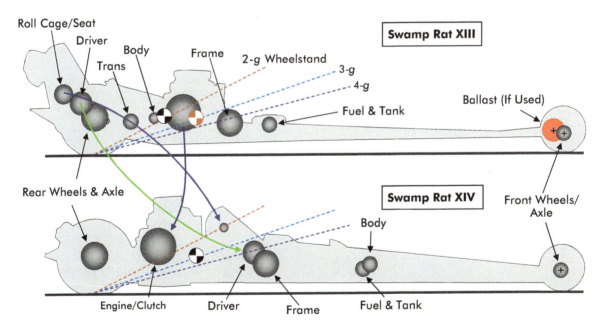

The CG calculations for Swamp Rats XIII and XIV, both with the driver's weight (assumed 160 pounds) added, are shown in **Table 2** and **Table 3**.[6] Swamp Rat XIII, without ballast, had its CG 33.5 inches

---

Note 6: Don was kind enough to estimate the weights of the major components of his dragsters, as summarized in **Table 2** and **Table 3**. I used the drawings Kane Rogers provided for the head of the chapter to estimate the component CG locations.

forward and 17.4 inches up from the rear tire patch, and had a wheelbase of 215 inches, for a 14.5/84.4 weight distribution and wheelstand limit of 1.92 g's. Swamp Rat XIV had its CG 49 inches forward and 17.2 inches up, and the same 215-inch wheelbase, for a 22.8/77.2 weight distribution and wheelstand limit of 2.84 g's.

So while it may have had 84.4% of its weight on the rear tires without ballast, Swamp Rat XIII couldn't use all of its traction once the tires' friction coefficient got near 2, because its acceleration was limited by wheelstand to 1.92 g's. Swamp Rat XIV could keep the front tires on the ground up to 2.84 g's without the performance-robbing ballast.

Component	Mass m (lbm)	x (in)	mx (in-lbm)	y (in)	my (in-lbm)
Engine & Clutch	650	40	26,000	18	11,700
Transmission	50	16.5	825	14.5	725
Driveshaft	5	8.5	42.5	15	75
Rear Axle	200	0	0	15	3,000
Rear Wheels, Tires	75	0	0	15	1,125
Driver	160	−7	−1,120	23	3,680
Driver Suit	10	−7	−70	23	230
Seat, Roll Cage	95	−14.5	−1,377.5	27.5	2,612.5
Fuel & Tank	50	80	4,000	13	650
Frame	250	61.5	15,375	14	3,500
Body	20	27.5	550	16	320
Front Axle	25	215	5,375	8.5	212.5
Front Wheels, Tires	20	215	4,300	10	200
Car without Ballast:	1,610 lbm	N/A	53,900 in-lbm	N/A	28,030 in-lbm
		CG, x:	**33.48 in**	**CG, y:**	**17.41 in**
Ballast	150	210	31,500	8.5	1,275
Car with Ballast:	1,760 lbm	N/A	85,400 in-lbm	N/A	29,305 in-lbm
		CG, x:	**48.52 in**	**CG, y:**	**16.65 in**

TABLE 2 Component masses and CG locations relative to the rear tire patch for Swamp Rat XIII, and the vehicle CG position without and with ballast. Ballast moved the CG forward by 15 inches and changed the weight distribution from 15.6/84.4 to 22.6/77.4.

Notice the effect of ballast in TABLE 2. With 150 pounds 5 inches behind the front axle, the CG of Swamp Rat XIII moved forward by 15 inches, to be 48.5 inches ahead of the rear axle, making a 22.6/77.4 weight distribution. That gave it a wheelstand limit of 2.91 g's, *almost identical* to Swamp Rat XIV without ballast, but with much more mass.

Note that the tables use mass instead of weight to calculate the CG, which is more proper than using weight, especially in a plane.

Component	Mass m (lbm)	x (in)	mx (in-lbm)	y (in)	my (in-lbm)
Engine & Clutch	650	29	18,850	20.5	11,700
Transmission	N/A	N/A	N/A	N/A	N/A
Driveshaft	5	8.5	42.5	16	75
Rear Axle	200	0	0	16	3,000
Rear Wheels, Tires	75	0	0	16	1,125
Driver	160	73.5	11,760	17.5	3,680
Driver Suit	10	73.5	735	17.5	230
Seat, Roll Cage	15	59.5	892.5	29	2,612.5
Fuel & Tank	50	123	6,150	10.5	650
Frame	250	78.5	19,625	13	3,500
Body	50	125.5	6,275	12.5	320
Front Axle	25	215	5,375	8.5	212.5
Front Wheels, Tires	20	215	4,300	10	200
Ballast	N/A	N/A	N/A	N/A	N/A
Total Vehicle:	1,510 lbm	N/A	74,005 in-lbm	N/A	26,028 in-lbm
		CG, x:	49.01 in	CG, y:	17.24 in

Don said that they paid close attention to the CG location, to maximize traction but still keep the front wheels down. Their rule was that however many g's you want to accelerate at, the CG needs to be that many times forward of the rear tire patch as it was high. He said in words what equation 17-5b says in math: the CG must be at or below the line going through the rear tire patch that has a slope equal to the acceleration in g's (check **FIGURE 11**).

All in all, Swamp Rat XIV could run with the same weight distribution and wheelstand limit as Swamp Rat XIII, but 250 pounds lighter. This was such a huge advantage that Don was able to run a de-tuned engine throughout 1971 and race at three meets between pulling the cylinder heads off the engine.[7] As tires got better (say μ_{Static} = 3.5), they could move components forward or add a bit of ballast to the front, and still be much lighter than the slingshots.

For me, the profound lesson here is the domino effect technology can trigger. *Better tires* not only made the cars faster, but also required a different vehicle layout to take full advantage of them. They made the slingshot dragster, king of the strip for years, a thing of the past.

TABLE 3 CG location calculations for Swamp Rat XIV. Its CG was 49.01 inches ahead of the rear axle and 17.24 inches up, giving it a wheelstand limit of 2.84 g's. This is almost exactly the same as Swamp Rat XIII with 150 pounds front ballast.

Note 7: Today Top Fuel engines are completely torn down and rebuilt after every pass.

6   Load Transfer in Braking

Everyone has seen a motorcycle pull a "brakey," where the forward load transfer is enough to unload the rear tire and lift it off the ground. Tipping a car over its nose from braking isn't a big concern, but load transfer still causes brake dive, and changes the necessary proportioning of the braking force front to rear, to fully use the available traction. So let's see what affects it.

First, we take equations 16-7 for forward acceleration, and replace the thrust by the braking forces shown in **Figure 12**. The front and rear axle loads are then:

Equation 17-10a   **Braking load, front**

$$F_{Front,z} = \frac{mgL_2}{L_{WB}} + \left(F_{Braking,Front} + F_{Braking,Rear} \right) \frac{h_{CG}}{L_{WB}}$$

and:

Equation 17-10b   **Braking load, rear**

$$F_{Rear,z} = \frac{mgL_1}{L_{WB}} - \left(F_{Braking,Front} + F_{Braking,Rear} \right) \frac{h_{CG}}{L_{WB}}$$

Just this once, we'll take braking forces and acceleration positive rearward. Notice that the *total* braking force determines load transfer, regardless of the front/rear braking proportions.

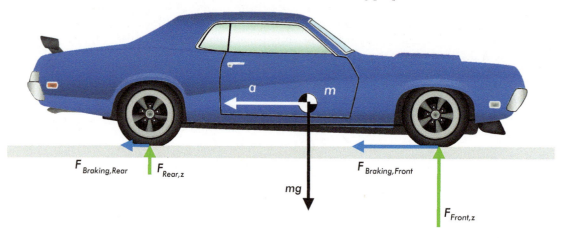

Figure 12 Vertical and longitudinal tire forces during braking. Braking forces and acceleration are taken positive rearward. Note that the load transfer is to the front.

We also need to know the axle loads in terms of braking *acceleration*. Since the total braking force $F_{Braking}$ equals $ma_{Braking}$, we can substitute that for $F_{Braking,Front} + F_{Braking,Rear}$ in equations 17-10. This produces the axle loads as:

Equation 17-11a   **Front axle load**

$$F_{Front,z} = mg\frac{L_2}{L_{WB}} + ma_{Braking}\frac{h_{CG}}{L_{WB}}$$

and:

Equation 17-11b   **Rear axle load**

$$F_{Rear,z} = mg\,\frac{L_1}{L_{WB}} - ma_{Braking}\,\frac{h_{CG}}{L_{WB}}$$

On a slick surface (μ_{Static} of 0.1 or so), where the braking won't be enough to cause much load transfer, the front/rear brake proportioning should be almost equal to the weight distribution. With 60/40 weight distribution, about 60% of the braking force should be at the front. But as you brake harder, the front axle load increases, and the fronts should produce a higher proportion of the braking. Let's examine a couple of examples.

First, let's look at the Honda Civic with 61/39 weight distribution. FIGURE 13 shows the axle loads at braking levels from zero to 1.6 g's, assuming the CG height is 20% of the wheelbase.

As expected when braking on ice and snow, the axle loads are straight from the weight distribution. But braking very hard on a dry surface, the front axle loads become a very large fraction of the car's weight. With good performance tires that could produce a braking deceleration of 1.2 g's, the front axle load is 85% of the weight. Assuming the traction goes up with axle load, around 85% of the braking force should be applied by the front brakes.

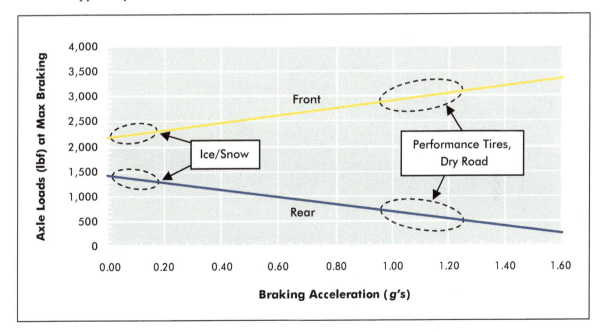

This explains why front brakes are so much bigger than the rear. Producing most of the braking force, they have to absorb and dissipate most of the heat. Under the hardest braking, the rear brakes are used very little.

FIGURE 13 Axle loads during braking in a Honda Civic with 61/39 weight distribution. At 1.2 g's the percentage of the load on the front axle has gone up from 61% to 85%.

FIGURE 14 shows a different picture for a Porsche 911, which has 39/61 weight distribution. For lack of better data I am assuming $h_{CG}/L_{WB} = 0.2$, same as the Honda. At 0.55-g braking, the axle loads become equal. At 1.2 g's, the front axle load is 63% of the car's weight. Most of the load ends up on the front axle, but the rear axle load is still 37% of the car's weight at the hardest braking the car would see on performance street tires.

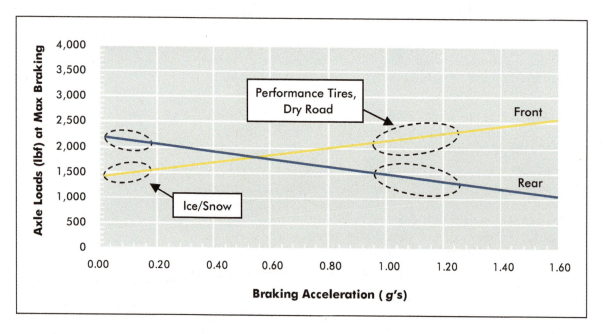

FIGURE 14 Axle loads during braking in a Porsche 911 with 39/61 weight distribution. At 1.2 g's, the percentage of the load on the front axle has gone up from 39% to 63%. The front and rear axle loads are more equal than with the front bias of FEFWD.

To summarize, the brake proportioning at low levels should be about 61/39 for the Civic and about 39/61 for the Porsche. But in hard braking, say 1.2 g's, the brake proportioning should move toward 85/15 and 63/37, respectively. With its rear bias, the 911's braking forces will be more balanced front to rear, especially in hard braking when the most heat is produced. The rear brakes can help out the fronts by dissipating over a third of the total braking heat.

7   Another Look at Cornering

We usually think of the tires' friction coefficient as determining a car's maximum cornering capability: its *traction-limited lateral acceleration* $a_{y,TL}$ in g's being equal to μ_{Static}. But the car's CG height is another crucial factor. Raising the CG will cause more load transfer from the inside to outside tires for any given curve and vehicle speed (refer back to equations 16-10) and will affect the cornering limit in two ways.

The first is a reduced traction-limited acceleration $a_{y,TL}$ from a given set of tires. With a higher CG causing a greater left/right difference in tire load, their total lateral grip is reduced (as discussed in Chapter 11, Section 7.2). With less grip, your cornering capability is reduced.

The second is a *rollover limited acceleration* $a_{y,RL}$. With enough lateral force at the tires, whether from traction or "tripping" by a curb or pothole, a car or truck can roll over.

Your actual cornering limit is the lower of the two; to negotiate the curve, you need traction and to keep all four wheels on the ground (well, three anyway). So how would you estimate your road car's CG height? Fortunately, the federal government has cataloged an approximate resistance to rollover for many cars, and we can use the data to back-calculate their CG height. While its value is not listed for every car, you may be able to find it for one similar to yours.

7.1 Rollover Limit and Finding Your Car's CG Height

Just as fore-aft load transfer can cause a wheelstand, excessive lateral load transfer can cause a rollover. This begins when the combined load on the inside tires goes to zero. On a flat, smooth road in a steady curve, the inside tire load is given by equation 16-10a. Setting that load to zero and solving for acceleration gives us a *first approximation* of rollover limit, in *g*'s, of:[8]

Equation 17-12   **Approximate rollover limit**

$$\frac{a_{y,RL}}{g} = \frac{w_{Track}}{2h_{CG}}$$

This assumes enough lateral force at the tire contact patches to reach this acceleration. The federal government has recorded this value for various vehicles, from their measured CG heights and track widths. Since they use (quasi) statics to calculate it, they call it the static stability factor, or *SSF*:[9]

Equation 17-13a   **Static stability factor**

$$SSF = \frac{w_{Track}}{2h_{CG}}$$

Note 8: This calculation is only approximate since it does assume a flat, smooth road, with constant curve radius and vehicle speed. There is more to it, such as tire and suspension deflection, tire properties, suspension geometry/roll center, wheelbase, the actual road surface and camber, longitudinal acceleration, electronic stability control, and the driver's ability. Any vehicle can be rolled under the right (wrong?) conditions, but an off-road vehicle isn't likely to resist rollover as strongly as a sports car.

Note 9: The Department of Transportation keeps reports of the *SSF* at www.regulations.gov, under the docket NHTSA-2001-9663. The data are a bit spotty, but they show a definite increase in the *SSF* for like cars from about 1985 to 2010. Note that they call the track width *T* and the CG height *h*, so they write the *SSF* as *T*/2*h*.

Besides its intended purpose, you can use a vehicle's *SSF* to find its CG height, which isn't typically listed in car magazines (maybe it *should* be).[10] Rearranging equation 17-13a, the CG height is:

Equation 17-13b   **CG height**

$$h_{CG} = \frac{w_{Track}}{2 \times SSF}$$

For example, a 2008 Nissan 350Z has an average track width of 60.7 inches and its *SSF* is 1.57 (see **TABLE 4**), so its CG height from equation 17-13b is:

$$h_{CG} = \frac{60.7 \text{ in}}{2 \times 1.57} = 19.33 \text{ in}$$

The *SSF* and CG height for a wide range of vehicles are listed in **TABLE 4**. Note that sports cars like a Corvette might have their CG lower than 18 inches off the ground, while a delivery van might have it twice that high. Accordingly, the maximum lateral *g*'s you'd want to pull in the delivery van is only about half as much as in the Corvette (I don't think that will surprise or disappoint anyone). Further, the *SSF* is measured with only a driver and a full fuel tank. Any additional passengers, luggage, or cargo will change the actual CG height.

TABLE 4 Track width, *SSF* (static stability factor), and CG height for a range of 2008 vehicles. The *SSF* is a first approximation of the onset of rollover, in lateral *g*'s. (The *SSF* for the 2008 C6 Corvette wasn't available, so the value for the C5 1997–2004 Corvette was used.)

Vehicle	Ave. Track (in)	SSF	CG Height (in)
Chevy Corvette	61.4	1.75	17.54
Nissan 350Z Coupe	60.7	1.57	19.33
Ford Mustang	62.5	1.53	20.42
Honda Civic Coupe	59.5	1.44	20.66
Chevy Malibu	60.3	1.41	21.38
Dodge Grand Caravan	65.2	1.24	26.29
Jeep Commander 4WD	62.6	1.09	28.72
Ford E350 Van	68.1	1.09	31.24
Dodge Sprinter 2500	67.7	0.92	36.79

Note that the *SSFs* in **TABLE 4** approximate the rollover limit in *g*'s, and the lateral traction limit is roughly the tires' friction coefficient. Since tires have a friction coefficient in the range of 0.8 to 1.2, you can see that on a flat, smooth curve, the lateral acceleration for most road vehicles would be limited by traction, rather than rollover. But note that

---

Note 10: As important as it is to handling, it makes sense to list the *vertical* CG position. After all, weight distribution is listed, which pinpoints the *longitudinal* position. But CG height measurement does require specialized equipment, to tilt the vehicle. As of this writing, at least one magazine *is* reporting it for some cars.

because of their high CGs, large trucks may tip before the tires reach their maximum lateral force, at maybe 0.5 g's or less. Caution signs on curves are usually set with them in mind.

7.2 Lateral Forces in a Steady Turn

We've found the load transfer in a steady turn, but how much of the cornering force is produced by the front vs. rear tires? After all, this is part of what goes into picking front and rear tire sizes. Just as for front/rear weight distribution, we can find the axle loads in a steady turn. But now it's quasi-statics, because the car is in steady lateral acceleration. Take a look at FIGURE 15.

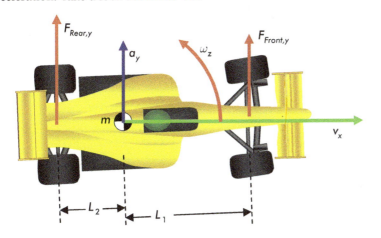

FIGURE 15 To find the lateral forces on the front and rear axles, we need to know the vehicle mass, lateral acceleration, and CG location.

Here we have loads in a horizontal plane, and the longitudinal acceleration is zero. Then we have, in the lateral direction:

Equation 17-14a

$$F_{Res,y} = ma_y$$

and for yaw:

Equation 17-14b

$$M_{z,CG} = 0$$

The lateral acceleration a_y is just the centripetal acceleration v^2/r_{Curve}. Without going through the whole process, the moments about the CG from the lateral tire forces must balance out, producing the following results:

Equation 17-15a   **Front cornering force**

$$F_{Front,y} = ma_y \left(L_2 / L_{WB} \right)$$

Equation 17-15b   **Rear cornering force**

$$F_{Rear,y} = ma_y \left(L_1 / L_{WB} \right)$$

Note the similarity to equations 16-5. It requires the same front/rear force proportioning to accelerate the mass laterally in cornering, as it does to support the vehicle weight statically.

8   Maximizing Performance

How would you design a purpose-built race car from a clean sheet of paper? First you need to know the type of racing. We've looked at straight-line acceleration in some detail, and we'll top it off here. But then we'll consider performance on a banked oval and a road course.

The amount of power available would affect some of these choices, so we'll speak in general terms.

8.1 Drag Racing: Pedal to the Metal

First off, a drag racer should be rear drive. With the traction available on a drag strip, there is too much load transfer to the rear for a front-drive car to compete (review **FIGURE 9**).[11] Next, you maximize the acceleration at launch and down track. We know from equation 17-7 that down-track traction (which includes load transfer) is greatest with the CG height so that L_2/h_{CG} is equal to the friction coefficient μ_{Static}. But at launch there's been no load transfer; the rear axle load comes straight from the weight distribution. Using that to calculate traction, the maximum launch acceleration is:

Equation 17-16   **Max launch capability (RWD)**

$$\frac{a_{Launch,Max}}{g} = \mu_{Static}\left(1 - \frac{L_2}{L_{WB}}\right)$$

which is highest with a rearward CG and long wheelbase. Ideally you'd like the same traction at launch and down track, which isn't possible here.[12] We can come closest by reducing the load transfer to a minimum by lowering the CG location and lengthening the wheelbase, per equation 16-8b—long wheelbase again!

Taken together, the ideal dragster design would be 1) long wheelbase, 2) low CG, and 3) CG just far enough forward to make $L_2/h_{CG} = \mu_{Static}$.

Note 11: On street tires, all-wheel drive would be an advantage, but on drag slicks there is so much rearward load transfer that rear drive is nearly as good, without the extra mass.

Note 12: This ignores any effect of traction bars and chassis flex. Drag racers use these to push the tires into the pavement at launch to increase launch traction.

This makes the down-track acceleration equal to the static friction coefficient, and the launch acceleration slightly less.[13]

Top Fuelers, being the ultimate drag racers, follow this formula; long wheelbase, with the heavy parts way back and low. While Garlits's Swamp Rat XIV had a wheelbase of 215 inches, Top Fuelers eventually grew to 300 inches, at which point the NHRA (the National Hot Rod Association) limited them.

8.2 Banked Ovals: Left Turn Only

Oval-track racing is similar to drag racing in that there is one major task the car needs to do well, in this case turn left. Most ovals are banked, which increases cornering capability because 1) the loads on the tires are higher than on a flat track, increasing their traction, and 2) the normal force from the track is providing centripetal force.

Let's take the 1970 Plymouth Superbird in **FIGURE 16** as an example, and calculate the forces at the tires. We'll assume it's taking a curve at Daytona, with a 31° bank angle and 1,000-foot radius, at an even 200 mph (293.3 ft/sec). We'll assume a weight of 3,500 pounds (mass = 108.70 slugs), track width of 60 inches (5 feet), and a centered CG with a height of 21 inches (1.75 feet). We'll ignore downforce.

With this speed and curve radius, the centripetal acceleration a is 86.04 ft/sec², or about 2.67 g's, horizontally left. From $F = ma$, we need a centripetal force of 9,353 lbf to the left.

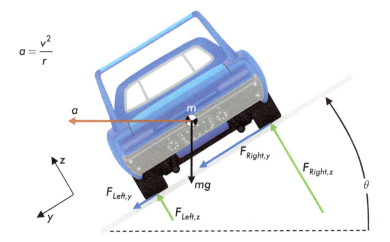

FIGURE 16 A banked curve increases loading on the tires. Besides increased cornering traction, the banked surface itself pushes the car into the curve. Note that there would be aerodynamic downforce or lift at the 200-mph speed, which we are ignoring.

Notice that **FIGURE 16** defines the y- and z-directions relative to the car; y is lateral to the car, parallel to the track surface; z is "vertical"

Note 13: With $\mu_{Static} = 4.0$, a 16-inch CG height, and a 300-inch wheelbase, L_2 would need to be 64 inches, per equation 17-7. This would predict a down-track acceleration of 4.0 g's, while equation 17-16 gives a launch acceleration of 3.15 g's.

to the car, and perpendicular to the pavement.[14] Defining components along these directions makes it easy to solve for the forces, as in FIGURE 17.

Note that in this slanted coordinate system, there are weight and acceleration components in both the y- and z-directions. In the y-direction, we have lateral tire forces $F_{Left,y}$ and $F_{Right,y}$, plus the lateral components of the acceleration and weight:

$$a_y = a\cos\theta \qquad mg_y = mg\sin\theta$$

The equation of motion in the y-direction is then, from $F_y = ma_y$,

$$ma_y = mg_y + F_{Left,y} + F_{Right,y}$$

or, substituting the above expressions for a_y and mg_y,

$$ma\cos\theta = mg\sin\theta + F_{Left,y} + F_{Right,y}$$

Solving for the sum of the lateral tire forces:

Equation 17-17a

$$F_{Left,y} + F_{Right,y} = m\left(a\cos\theta - g\sin\theta\right)$$

Plugging in the mass and acceleration for our case produces combined lateral tire forces of 6,214 lbf. You can follow the same process to find the "vertical" tire forces:

Equation 17-17b

$$F_{Left,z} + F_{Right,z} = m\left(a\sin\theta + g\cos\theta\right)$$

which here produces a combined "vertical" tire force (load) of 7,817 lbf. We can find how much of this load is carried by the left- and right-side tires, by taking moments about the CG:

$$\left(F_{Left,y} + F_{Right,y}\right)h_{CG} + \left(F_{Left,z} - F_{Right,z}\right)w_{Track}\big/2 = 0$$

Combining this with equation 17-17b and solving for the combined right-side tire load in a left-hand turn,

Equation 17-18a

$$F_{Right,z} = \frac{m}{2}\left(a\sin\theta + g\cos\theta\right) + m\left(a\cos\theta - g\sin\theta\right)h_{CG}\big/w_{Track}$$

Note 14: The "vertical" direction should probably be called the "normal" direction, since it is normal to the track surface.

Alright, so this one got a bit involved. But it is a banked curve. The combined left tire load is, for a left-hand turn,

Equation 17-18b

$$F_{Left,z} = \frac{m}{2}\left(a\sin\theta + g\cos\theta\right) - m\left(a\cos\theta - g\sin\theta\right)h_{CG}\big/w_{Track}$$

In our problem, this makes our right tire load 6,083 lbf and the left 1,734 lbf. Notice that on a flat track, where the bank angle is zero, these two equations are the same as equations 16-10.

Dividing our *lateral* force of 6,214 lbf between left and right tires requires an assumption, as we've used up our three quasi-static conditions. Let's keep it simple and proportion it to the tire loads. This produces a lateral tire force of 4,836 lbf on the right, and 1,378 lbf on the left.

The tires produce 6,214 lbf lateral force with 7,817 lbf load, so they only need a 0.80 friction coefficient to do it (speed at Daytona is not limited by cornering, but by drag). At 200 mph, the tires "think" they are on a 3,500-lbf car with 4,317 lbf downforce, in a 1.78-g curve. Of course the track does not cause downforce. But when you roll 31° into a banked curve, and then look ahead of you, it starts to feel like the track is "curving up," becoming like a loop-the-loop (at 90°, it would be). It's sort of like **FIGURE 2** in Chapter 13, but you're at the *bottom* of the loop.

Car-Lateral Components

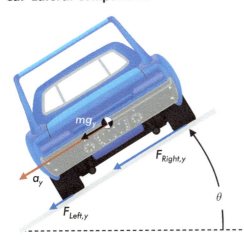

Car-Vertical Components

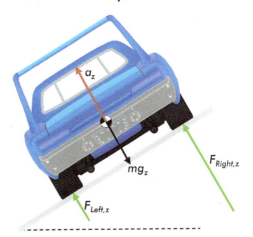

You can see why many NASCAR pit stops only involve right-side tires. Here, the rights had to bear about 6,100 lbf load, vs. only 1,700 on the left. With over three times the load, the rights wear much faster and their traction "goes away." It pays to even out the tire loads if you can, because the tire degradation increases faster than the load on the tire.

FIGURE 17 Writing the equations of motion parallel and perpendicular to the track surface makes it easier to solve for the forces. Note that, even ignoring downforce, the tires only need a 0.80 friction coefficient to produce the lateral forces.

To balance the loads, the car itself was historically designed to be as light as possible, and then ballast was added to shift the CG to the left and down as far as possible and reach the minimum weight. NASCAR cars have traditionally used lead "bricks" in the left rocker below the door. The rules were changed to require a certain minimum weight on the right-side tires, to limit the CG shifting. But it still helps to *lower* the CG.

Another approach, once used in Indycars, was to offset the heavy parts of the car to the left. In the early 1960s, the engines were often shifted to the left, with the driver to the right. An extreme example of leftward shift is shown by the Lotus 38 in FIGURE 18, whose left-side suspension control arms are much shorter than the right, offsetting the whole body to the left.

FIGURE 18 Notice the short left-side control arms on Jim Clark's Lotus Indycar, shifting the car's CG to its left and reducing loads on the right-side tires in turns. This was the first rear-mid-engine car to win Indy, in 1965. *Photo by permission of Ford Motor Company.*

The original reason for minimum weight rules was to equalize competition, knowing that everyone will run at the minimum. I believe it's also meant to reduce costs, since no matter how light you make your car, you'll have to add ballast to it anyway to make the minimum weight. But you can see that it doesn't take away all the advantage of beginning with a light car.

8.3 Road Racing: Finding Balance

In road racing you have to accelerate, brake, and turn left and right, so you can't skew your car's performance to one aspect at the expense of the others. How do you balance them out?

Because they need to accelerate out of corners, and some race from a standing start, road racers need to have a lot of traction on the drive wheels. This suggests rear-wheel drive and rearward weight bias, with bigger tires in the rear to take advantage of it. In braking we've seen that the front to rear balance of forces, and of heat generation, are closest with weight rearward. Again, rear bias wins. And to minimize load transfer to the front on braking, you want a long wheelbase and a low CG.

What about cornering? There are a couple of aspects to this. One is the effect of load transfer to the outside tires, and the other is maneuverability. Remember that the tires as a set will corner best when they have most nearly equal loads.

The effects of load transfer are easiest to see in cars that are *not* purpose-built race cars. The Mini Cooper at the top of **Figure 19** has a high CG and relatively narrow track, so in a tight corner has nearly all the load on the outside tires. The outside front in particular is deflected in the extreme, while the inside tires are nearly undeformed or completely off the ground. The Porsche 911 in the lower part of the figure is accelerating out of a left-hand curve. In other words, it's simultaneously producing lateral and longitudinal acceleration. The result is load transfer to the right and the rear, to the point of lifting the left front wheel.

Figure 19 You can see the effect of lateral load transfer of the Mini Cooper and the 911, both in lifted wheels and tire deflection. The Mini shows more effect because of its higher CG and narrower track. *Photos by Randy Beikmann with permission of Najeeb Khan (top) and Steve Grundahl (bottom).*

There is always load transfer to the outside tires, so the loads will never all be equal. But we can minimize their difference by having 1) a low CG, and 2) a wider track.

As for maneuverability, you want a low polar moment of inertia, so the car will turn in quickly when steered. As shown in Chapter 10 (FIGURE 18), the polar moment is smallest when the heavy parts of the car are closest to the CG. This gives the nod to a mid-engine design.

So it's not surprising that the top classes in road racing are rear-mid-engine, running as close as possible to the ground, components down low, with a wide track and long wheelbase. F1 cars like the Benettons in FIGURE 20 are so composed it's hard to tell by looking whether they are cornering or not. No drama.

FIGURE 20 The Benetton-Fords of Michael Schumacher and Martin Brundle show the typical road racing layout: long wheelbase relative to the body, low, wide, rear-mid engine, rear-wheel drive. This is at the Belgian Grand Prix in Spa in 1992, Schumacher's first F1 victory. *Photo by permission of Ford Motor Company.*

F1 cars have had a minimum weight rule since the early 1960s. In the years leading up to 2009, the cars were rumored to be so light compared to their roughly 1,300-pound minimum (including driver), that over a hundred pounds of ballast could be placed to advantage. For road racers like F1 cars, this would typically be low and centered, to minimize load transfer and keep the polar moment small.

The increasingly complex energy recovery systems introduced since are thought to have claimed most of that margin, despite increases in the minimum weight. Although it's difficult to say how true rumors are (by definition), it is known that in 2009 drivers were heavily dieting, to the point of reducing their muscle strength and endurance.

9 Summary

This chapter showed how important a vehicle's CG location is to performance, and how it's affected by the mass and location of each part. Statics and quasi-statics took our intuition of the effects of vehicle design and put numbers to them.

Both are indispensable to the vehicle design and tuning process. It begins immediately with placement of its large/heavy components: engine, transmission, wheels/suspension, fuel tank, occupant(s), body structure, etc. This in concert with which wheels drive. It continues in the geometry and detailed design of the suspension, the engine and drive mechanisms, and structures.

You take it further in modifying a vehicle for performance, selecting which components need to be moved or lightened, or where to add ballast. Those decisions depend on the tires' traction capabilities, engine power, and driveline gearing. And vice versa. On a rear-drive car, lightening the car is good, but lightening it in front is most important, increasing the ratio of traction to mass. So buy a carbon-fiber hood before a carbon-fiber trunk. Moving an underhood battery to the trunk is another tried-and-true method. Fascinating.

If you look back through the book, you'll see we've been using statics quite a bit, even in Chapter 1 with F1 suspension forces. Section 3 here found static loads in a crank train at TDC, but **Figure 3** in Chapter 9 showed it being used to find the forces at 45° ATDC in making crank torque. It could also have found the side loads on the cylinder wall and main bearing.

Introduce quasi-statics, and we could add the cornering forces in that F1 suspension. Section 5 in Chapter 7 basically used it in force resolution. We can take it further, like with the SLA suspension illustrated in the Chapter 16 summary, to get the loads in the pushrod, knuckle, and ball joints. That would be good to try out.

Some factors were *not* so intuitive, such as improving a Top Fueler's performance by shifting the CG *forward* to match the car's design to the tire capability. We found this by finishing the discussion on factors limiting straight-line acceleration: traction and power. Here we found the traction-limited acceleration a_{TL} was actually the smaller of:

1. Wheel-spin-limited acceleration a_{WS} and
2. Wheelstand-limited acceleration a_{WL}.

So traction wasn't necessarily limiting our "traction-limited" acceleration. Putting this together with Chapter 14, the acceleration limit is then the smaller of:

1. Traction-limited acceleration a_{TL} and
2. Power-limited acceleration a_{PL}.

Only by understanding what limits vehicle performance can you hope to improve it. You need to change what matters, and not waste effort on what's okay.

So enjoy your driving, and know your limits!

Major Formulas

Definition	Equation	Equation Number
Thrust Force at Wheel-Spin Limit (RWD)	$F_{Thrust,WS} = \mu_{Static} \dfrac{mg\left(L_1/L_{WB}\right)}{1 - \mu_{Static}\left(h_{CG}/L_{WB}\right)}$	17-1
Acceleration at Wheel-Spin Limit, g's (RWD)	$\dfrac{a_{WS}}{g} = \dfrac{\mu_{Static}\left(L_1/L_{WB}\right)}{1 - \mu_{Static}\left(h_{CG}/L_{WB}\right)}$	17-2b
Thrust Force at Wheelstand Limit (RWD)	$F_{Thrust,WL} = mg\left(L_2/h_{CG}\right)$	17-4
Acceleration at Wheelstand Limit, g's (RWD)	$\dfrac{a_{WL}}{g} = \dfrac{L_2}{h_{CG}}$	17-5b
CG Placement for Maximum RWD Acceleration, with Tire Friction Coefficient μ_{Static}	$\dfrac{L_2}{h_{CG}} = \mu_{Static}$	17-7
Acceleration at Wheel-Spin Limit, g's (FWD)	$\dfrac{a_{WS}}{g} = \dfrac{\mu_{Static}\left(L_2/L_{WB}\right)}{1 + \mu_{Static}\left(h_{CG}/L_{WB}\right)}$	17-9
Load on Front Tires During Braking, as a Function of Total Braking Force	$F_{Front,z} = \dfrac{mgL_2}{L_{WB}} + \left(F_{Braking,Front} + F_{Braking,Rear}\right)\dfrac{h_{CG}}{L_{WB}}$	17-10a
Load on Rear Tires During Braking, as a Function of Total Braking Force	$F_{Rear,z} = \dfrac{mgL_1}{L_{WB}} - \left(F_{Braking,Front} + F_{Braking,Rear}\right)\dfrac{h_{CG}}{L_{WB}}$	17-10b

Definition	Equation	Equation Number
Load on Front Tires During Braking, as a Function of Braking Acceleration	$F_{Front,z} = mg\dfrac{L_2}{L_{WB}} + ma_{Braking}\dfrac{h_{CG}}{L_{WB}}$	17-11a
Load on Rear Tires During Braking, as a Function of Braking Acceleration	$F_{Rear,z} = mg\dfrac{L_1}{L_{WB}} - ma_{Braking}\dfrac{h_{CG}}{L_{WB}}$	17-11b
Static Stability Factor, or Approximate Rollover Limit in g's, from Quasi-Statics	$SSF = \dfrac{w_{Track}}{2h_{CG}}$	17-13a
CG Height Calculated from the SSF	$h_{CG} = \dfrac{w_{Track}}{2\times SSF}$	17-13b
Cornering Force at Front Axle	$F_{Front,y} = ma_y\left(L_2/L_{WB}\right)$	17-15a
Cornering Force at Rear Axle	$F_{Rear,y} = ma_y\left(L_1/L_{WB}\right)$	17-15b
Combined Right-Side Tire Load in Left-Hand Curve with Bank Angle θ	$F_{Right,z} = \dfrac{m}{2}\left(a\sin\theta + g\cos\theta\right)$ $+ m\left(a\cos\theta - g\sin\theta\right)h_{CG}/w_{Track}$	17-18a
Combined Left-Side Tire Load in Left-Hand Curve with Bank Angle θ	$F_{Left,z} = \dfrac{m}{2}\left(a\sin\theta + g\cos\theta\right)$ $- m\left(a\cos\theta - g\sin\theta\right)h_{CG}/w_{Track}$	17-18b

Appendix 1
Units and Conversions

All approximate conversions are given to five significant digits except for those starting with 1, which are given to six. Exact conversions are given to only as many digits as necessary to be exact (Example: 25.4 millimeters = 1 inch). Exact conversions in the tables are written in blue.

When possible, it reduces round-off error to divide by an exact conversion, such as 1 ft/(0.3048 meters) rather than multiplying by an approximate one, such as 3.2808 feet/meter.

Source: Haynes, William M., ed., *CRC Handbook of Chemistry and Physics, 95th Edition*. London: CRC Press, 2014.

Metric Prefixes

Multiplier	Number	Prefix	Symbol
10^1	ten	deka-	da
10^2	hundred	hecto-	h
10^3	thousand	kilo-	k
10^6	million	mega-	M
10^9	billion	giga-	G
10^{12}	trillion	tera-	T
10^{15}	quadrillion	peta-	P
10^{18}	quintillion	exa-	E
10^{21}	sextillion	zetta-	Z
10^{24}	septillion	yotta-	Y

Multiplier	Number	Prefix	Symbol
10^{-1}	tenth	deci-	d
10^{-2}	hundredth	centi-	c
10^{-3}	thousandth	milli-	m
10^{-6}	millionth	micro-	μ
10^{-9}	billionth	nano-	n
10^{-12}	trillionth	pico-	p
10^{-15}	quadrillionth	femto-	f
10^{-18}	quintillionth	atto-	a
10^{-21}	sextillionth	zepto-	z
10^{-24}	septillionth	yocto-	y

Length

SFS Unit: 1 foot (ft)
MKS Unit: 1 meter (m)

Key Unit Conversions

- 1 inch is defined as exactly 25.4 millimeters (mm) = 2.54 centimeters (cm).
- 1 mile is exactly 1.609344 kilometers (km).

- 1 meter is approximately 39.370 inches (in) = 3.2808 feet (ft).
- 1 kilometer is approximately 0.62137 miles (mi).

Conversions Between British Units

		Multiply the Number in These Units by			
		inches	feet	yards	miles
To Get a Value in These Units	inches	1	12	36	63,360
	feet	0.083333	1	3	5,280
	yards	0.027778	0.33333	1	1,760
	miles	0.0000157828	0.000189394	0.00056818	1

Conversions Between Metric Units

		Multiply the Number in These Units by			
		millimeters	centimeters	meters	kilometers
To Get a Value in These Units	millimeters	1	10	1,000	1,000,000
	centimeters	0.1	1	100	100,000
	meters	0.001	0.01	1	1,000
	kilometers	0.000001	0.00001	0.001	1

Conversions Between Common British and Metric Units

		Multiply the Number in These Units by			
		feet	meters	kilometers	miles
To Get a Value in These Units	feet	1	3.2808	3,280.8	5,280
	meters	0.3048	1	1,000	1,609.344
	kilometers	0.0003048	0.001	1	1.609344
	miles	0.000189394	0.00062137	0.62137	1

Time

SFS Unit: 1 second (sec)
MKS Unit: 1 second (sec)

Conversions Between Common Units

		Multiply the Number in These Units by		
		sec	minute	hour
To Get a Value in These Units	sec	1	60	3,600
	minute	0.0166667	1	60
	hour	0.00027778	0.0166667	1

Note: 1 millisecond (msec) is one-thousandth of one second.

Mass

SFS Unit: 1 slug
MKS Unit: 1 kilogram (kg)

Key Unit Conversions

- 1 kilogram is approximately 0.068522 slugs, and 1 slug is approximately 14.5939 kilograms.

- 1 pound-mass (lbm) is defined to be 0.45359237 kilograms.

Conversions Between Common Units

		Multiply the Number in These Units by		
		lbm	kg	slugs
To Get a Value in These Units	lbm	1	2.2046	32.174
	kg	0.45359237	1	14.5939
	slugs	0.031081	0.068522	1

Note: The pound-mass (lbm) is not the consistent SFS mass unit, but in common U.S. usage, mass is expressed in pounds (and usually called "weight"). This makes 1 slug approximately equal to 32.174 lbm.

Velocity or Speed

SFS Unit: 1 foot/second (ft/sec)
MKS Unit: 1 meter/second (m/sec)

Key Unit Conversions

- 1 ft/sec is exactly 0.3048 m/sec.
- 1 mi/hr (or mph) is exactly 1.609344 km/hr (or kph).
- 1 m/sec is approximately 3.2808 ft/sec = 39.370 in/sec.

- 1 km/hr is approximately 0.62137 mi/hr.
- 1 mi/hr is approximately 1.46667 ft/sec; 60 mi/hr is exactly 88 ft/sec.

Conversions Between British Units

		Multiply the Number in These Units by		
		in/sec	ft/sec	mi/hr
To Get a Value in These Units	in/sec	1	12	17.6
	ft/sec	0.083333	1	1.46667
	mi/hr	0.056818	0.68182	1

Conversions Between Metric Units

		Multiply the Number in These Units by			
		mm/sec	cm/sec	km/hr	m/sec
To Get a Value in These Units	mm/sec	1	10	277.78	1,000
	cm/sec	0.1	1	27.778	100
	km/hr	0.0036	0.036	1	3.6
	m/sec	0.001	0.01	0.27778	1

Conversions Between Common British and Metric Units

		Multiply the Number in These Units by			
		km/hr	ft/sec	mi/hr	m/sec
To Get a Value in These Units	km/hr	1	1.09728	1.609344	3.6
	ft/sec	0.91134	1	1.46667	3.2808
	mi/hr	0.62137	0.68182	1	2.2369
	m/sec	0.27778	0.3048	0.44704	1

Acceleration

SFS Unit: 1 foot/second^2 (ft/sec^2)
MKS Unit: 1 meter/second^2 (m/sec^2)

Key Unit Conversions

- 1 ft/sec^2 is exactly 0.3048 m/sec^2.
- 1 m/sec^2 is approximately 3.2808 ft/sec^2.
- 1 g (standard) is defined as 9.80665 m/sec^2.
- 1 g (standard) is approximately 32.174 ft/sec^2.

Conversions Between Common Units

		Multiply the Number in These Units by		
		ft/sec^2	m/sec^2	g's
To Get a Value in These Units	ft/sec^2	1	3.2808	32.174
	m/sec^2	0.3048	1	9.80665
	g's	0.031081	0.101972	1

Note: While a standard value for g is taken to be 9.80665 m/sec^2, the actual free-fall acceleration varies with location.

Force

SFS Unit: 1 pound-force (lbf)
MKS Unit: 1 newton (N)

Key Unit Conversions

- 1 pound-force (lbf) is defined as the weight of 0.45359237 kg in standard gravity.
- 1 pound-force (lbf) is exactly 4.448221615152605 newtons (N).
- 1 newton is approximately 0.22481 lbf.

Conversions Between Common British and Metric Units

		Multiply the Number in These Units by			
		N	lbf	ton	metric ton
To Get a Value in These Units	N	1	4.4482	8,896.4	9,806.7
	lbf	0.22481	1	2,000	2,204.6
	ton	0.000112404	0.0005	1	1.10231
	metric ton	0.000101972	0.00045359	0.90718	1

Note: Given that the pound-force (lbf) is defined as the weight of 0.45359237 kg in standard gravity, the relationship "weight equals mg" produces the rest of the conversions.

A metric ton is the weight of a mass of 1,000 kg in standard gravity.

Torque

SFS Unit: 1 foot-pound-force (ft-lbf)
MKS Unit: 1 newton-meter (N-m)

Key Unit Conversions

- 1 ft-lbf is approximately 1.35882 N-m.
- 1 N-m is approximately 0.73756 ft-lbf.
- 1 ft-lbf is exactly 12 in-lbf.

Conversions Between Common Units

		Multiply the Number in These Units by		
		in-lbf	N-m	ft-lbf
To Get a Value in These Units	in-lbf	1	8.8507	12
	N-m	0.112985	1	1.35582
	ft-lbf	0.083333	0.73756	1

Rotation

SFS Unit: 1 radian (rad)
MKS Unit: 1 radian (rad)

Key Unit Conversions

- 1 revolution is exactly 360 degrees (360°).
- 1 revolution is exactly 2π radians (approximately 6.2832 radians).
- 1 radian is approximately 57.296°.

Conversions Between Common Units

		Multiply the Number in These Units by		
		degrees	radians	revolutions
To Get a Value in These Units	degrees	1	57.296	360
	radians	0.0174533	1	6.2832
	revolutions	0.0027778	0.159155	1

Rotational Velocity

SFS Unit: 1 rad/sec
MKS Unit: 1 rad/sec

Key Unit Conversions

- 1 rad/sec is approximately 9.5493 rpm.
- 1 rpm is approximately 0.104720 rad/sec.

Conversions Between Common Units

		Multiply the Number in These Units by			
		deg/sec	rpm	rad/sec	rev/sec
To Get a Value in These Units	deg/sec	1	6	57.296	360
	rpm	0.166667	1	9.5493	60
	rad/sec	0.0174533	0.104720	1	6.2832
	rev/sec	0.0027778	0.0166667	0.159155	1

Rotational Acceleration

SFS Unit: 1 rad/sec^2
MKS Unit: 1 rad/sec^2

- 1 rad/sec^2 is approximately 9.5493 rpm/sec.

Pressure

SFS Unit: 1 lbf/ft^2
MKS Unit: 1 N/m^2 (1 pascal, or Pa)

Key Unit Conversions

- 1 atmosphere (atm) is defined as 101,325 pascals (Pa), or 101.325 kilopascals (kPa).
- 1 atmosphere is about 14.6959 lbf/in^2 (psi), or 2,116.2 lbf/ft^2.
- 1 lbf/in^2 is exactly 144 lbf/ft^2.
- Note that pressure is sometimes given in inches of mercury (inches Hg) or millimeters of mercury (mm Hg), corresponding to the height of a mercury column produced in a manometer by that pressure.

Conversions Between British Units

		Multiply the Number in These Units by			
		lbf/ft^2	inches Hg	lbf/in^2	atm
To Get a Value in These Units	lbf/ft^2	1	70.726	144	2,116.2
	inches Hg	0.0141390	1	2.0360	29.921
	lbf/in^2	0.0069444	0.49115	1	14.6959
	atm	0.00047254	0.033422	0.068046	1

Conversions Between Metric Units

		Multiply the Number in These Units by			
		Pa	mm Hg	kPa	atm
To Get a Value in These Units	Pa	1	133.322	1,000	101,325
	mm Hg	0.0075006	1	7.5006	760.00
	kPa	0.001	0.133322	1	101.325
	atm	0.0000098692	0.00131579	0.0098692	1

Conversions Between Common British and Metric Units

		Multiply the Number in These Units by			
		Pa	lbf/ft^2	kPa	lbf/in^2
To Get a Value in These Units	Pa	1	47.880	1,000	6,894.8
	lbf/ft^2	0.020885	1	20.885	144
	kPa	0.001	0.047880	1	6.8948
	lbf/in^2	0.000145038	0.0069444	0.145038	1

Area

SFS Unit: 1 ft^2
MKS Unit: 1 m^2

Key Unit Conversions

- 1 square inch (1 in^2) is exactly 645.16 square millimeters (mm^2), or 6.4516 square centimeters (cm^2).
- 1 square foot (1 ft^2) is exactly 144 in^2.
- 1 square meter (1 m^2) is exactly 10,000 cm^2, or 1,000,000 mm^2.

Conversions Between Common British and Metric Units

		Multiply the Number in These Units by			
		cm^2	in^2	ft^2	m^2
To Get a Value in These Units	cm^2	1	6.4516	929.0304	10,000
	in^2	0.155000	1	144	1,550.00
	ft^2	0.00107639	0.0069444	1	10.7639
	m^2	0.0001	0.00064516	0.09290304	1

Volume

SFS Unit: 1 ft^3
MKS Unit: 1 m^3

Key Unit Conversions

- 1 cubic inch (1 in^3) is exactly 16,387.064 cubic millimeters (mm^3), or 16.387064 cubic centimeters (cm^3).
- 1 cubic foot (1 ft^3) is exactly 1,728 in^3.
- 1 cubic meter (1 m^3) is exactly 1,000,000 cm^3, or 1,000,000,000 mm^3.
- 1 ft^3 is approximately 7.4805 gallons, making 1 gallon approximately 231.00 in^3.
- 1 liter is exactly 1,000 cm^3.

Conversions Between Common British and Metric Units

		Multiply the Number in These Units by			
		cm^3	in^3	ft^3	m^3
To Get a Value in These Units	cm^3	1	16.387064	28,317.846592	1,000,000
	in^3	0.061024	1	1,728	61,024
	ft^3	0.000035315	0.00057870	1	35.315
	m^3	0.000001	0.000016387064	0.028316846592	1

Energy

SFS Unit: 1 ft-lbf
MKS Unit: 1 N-m (1 joule, or J)

Key Unit Conversions

- 1 joule (J) equals 1 N-m.
- 1 BTU (a British Thermal Unit) is the approximate amount of heat required to raise the temperature of 1 pound of pure water by 1 degree Fahrenheit.
- 1 kilocalorie (1 kcal) is the approximate amount of heat required to raise the temperature of 1 kilogram of pure water by 1 degree Celsius.

- 1 BTU is approximately 1,055.06 joules, based on having defined the kilocalorie (kcal) as exactly 4,186.8 joules at the Fifth International Conference on the Properties of Steam, 1956. There are several other standards, but all are close to this value.
- 1 kilowatt-hour (kW-hr) is 3,600,000 joules, or 3.6 megajoules (MJ).

Conversions Between Common British and Metric Units

		Multiply the Number in These Units by			
		joule	ft-lbf	BTU	kcal
To Get a Value in These Units	joule	1	1.35582	1,055.06	4,186.8
	ft-lbf	0.73756	1	778.17	3,088.0
	BTU	0.00094782	0.00128507	1	3.9683
	kcal	0.00023885	0.00032383	0.25199	1

Power

SFS Unit: 1 ft-lbf/sec
MKS Unit: 1 watt (W)

Key Unit Conversions

- 1 watt (W) is defined as the power in doing 1 joule of work per second, so 1 W = 1 J/sec, or 1 N-m/sec.

- 1 horsepower (HP) is defined as the power in doing 550 ft-lbf of work per second, so 1 HP = 550 ft-lbf/sec.
- 1 kW (1,000 watts) is equal to about 1.34102 HP.

Conversions Between Common British and Metric Units

		Multiply the Number in These Units by			
		watt	ft-lbf/sec	HP	kW
To Get a Value in These Units	watt	1	1.35582	745.70	1,000
	ft-lbf/sec	0.73756	1	550	737.56
	HP	0.00134102	0.00181818	1	1.34102
	kW	0.001	0.00135582	0.74570	1

Temperature

Key Reference Information

- The Celsius temperature scale was originally defined by setting 0 °C at the freezing point of pure water and 100 °C at the boiling point (at atmospheric pressure), with equal divisions in between. This is approximately the scale today, but more repeatable reference points are now used.

- The Fahrenheit temperature scale is divided into 180 equal parts, with 32 °F set to 0 °C, and 212 °F at 100 °C. Therefore a change in temperature of 1 °C is a change of exactly 1.8 °F.

- Absolute temperature is defined with zero being the temperature of minimum entropy, called absolute zero (roughly the lowest temperature possible). The Celsius scale is now defined so that absolute zero occurs at exactly −273.15 °C. This also puts absolute zero at −459.67 °F.

- The absolute temperature in metric units is measured in degrees kelvin (more properly "kelvins," or K without a degree symbol). The kelvin-scale degree is the same size as for Celsius, so add 273.15° to the temperature in Celsius to get its equivalent in kelvins. The kelvin is the international standard for temperature. Absolute zero is at exactly 0 K.

- The absolute temperature in British units is measured in degrees Rankine, or °R. The Rankine scale degree is the same size as for Fahrenheit, so add 459.67° to the temperature in Fahrenheit to get Rankine.

Conversion Formulas for Temperature

	Fahrenheit	**Celsius**	**Rankine**	**Kelvin**
Fahrenheit	N/A	$°F = °C \times \dfrac{9}{5} + 32°$	$°F = °R - 459.67°$	N/A
Celsius	$°C = \dfrac{5}{9}\left(°F - 32\right)$	N/A	N/A	$°C = °K - 273.15°$
Rankine	$°R = °F + 459.67°$	N/A	N/A	$°R = °K \times \dfrac{9}{5}$
Kelvin	N/A	$°K = °C + 273.15°$	$°K = °R \times \dfrac{5}{9}$	N/A

Appendix 2
Greek Letters and Pronunciations

Uppercase	Lowercase	Name	Pronunciation
A	α	alpha	AL-fuh
B	β	beta	BAY-tuh
Γ	γ	gamma	GA-muh
Δ	δ	delta	DEL-tuh
E	ε	epsilon	EP-sil-on
Z	ζ	zeta	ZAY-tuh
H	η	eta	AY-tuh
Θ	θ	theta	THAY-tuh
I	ι	iota	eye-OH-tuh
K	κ	kappa	KAP-uh
Λ	λ	lambda	LAM-duh
M	μ	mu	mYOO
N	ν	nu	NOO
Ξ	ξ	xi	KS-EYE
O	o	omikron	OH-mi-kron
Π	π	pi	PIE
P	ρ	rho	ROW
Σ	σ	sigma	SIG-muh
T	τ	tau	TOW[1]
Y	υ	upsilon	UP-sil-on
Φ	φ	phi	FEE
X	χ	chi	KIE[2]
Ψ	ψ	psi	SIGH
Ω	ω	omega	oh-MAY-guh

Note 1:   rhymes with "cow"

Note 2:   rhymes with "pie"

Appendix 3
Math Reference

Order of Math Operations

				Name of Each Term		
Operation	Symbol(s)	Order	Example	b	c	a
Exponentiation	(superscript)	1st	$b^c = a$	Base	Exponent	Power
Multiplication	$\times$, $\cdot$, position	2nd	$bc = a$	Factor	Factor	Product
Division	$\div$, $/$	3rd	$b/c = a$	Dividend	Divisor	Quotient
Addition	$+$	4th	$b+c = a$	Augend	Addend	Sum
Subtraction	$-$	5th	$b-c = a$	Minuend	Subtrahend	Difference

Algebraic Manipulations Used to Solve Equations

Add Same Quantity to Both Sides:	If	$a = b$	then	$a+c = b+c$
Subtract Same Quantity from Both Sides:	If	$a = b$	then	$a-c = b-c$
Multiply Both Sides by Same Quantity:	If	$a = b$	then	$ac = bc$
Divide Both Sides by Same Quantity:	If	$a = b$	then	$a/c = b/c$
Exponent—Base to the Second Power:	If	$a = b \times b$	then	$a = b^2$
Exponent—Base to the Third Power:	If	$a = b \times b \times b$	then	$a = b^3$
Exponent—Square Root:	If	$a = \sqrt{b}$	then	$a = b^{1/2}$
Exponent—Cube Root:	If	$a = \sqrt[3]{b}$	then	$a = b^{1/3}$
Commutative Law for Addition:	$a+b = b+a$			
Commutative Law for Multiplication:	$ab = ba$			
Associative Law for Multiplication:	$a(b+c) = ab + ac$			
Associative Law for Division:	$(a+b)/c = a/c + b/c$			

The Quadratic Equation

Given an equation with variable x, of the form:

$$ax^2 + bx + c = 0$$

The roots of the equation are at:

$$x = \frac{-b \pm \sqrt{b^2 - 4ac}}{2a}$$

For example, if:

$$2x^2 - 16x + 30 = 0$$

then the roots are at:

$$x = \frac{16 \pm \sqrt{256 - 4 \cdot 2 \cdot 30}}{4} = 4 \pm 1$$

So one root is at $x = (4-1) = 3$, and the other is at $x = (4+1) = 5$.

Special Symbols

Name	Symbol(s)	Explanation/Example
Absolute Value	| |	Magnitude of the number, regardless of its sign: $\|25\| = 25$; $\|-25\| = 25$
Angle	$\angle$	"Angle A" is written $\angle A$. Also, an angular measure of say 45°, may be stated as $\angle 45°$.
Delta	Δ	Change in a quantity: Δt is a change in t.

Appendix 4
Symbols

Uppercase Roman

Symbol	Name	SFS Units	MKS Units
A	Area	ft^2	m^2
B	Cylinder Bore	feet (ft)	meters (m)
C_D, C_L	Drag and Lift Coefficients	dimensionless	dimensionless
C_λ, C_γ	Cornering or Camber Stiffness	lbf/rad	N/rad
D	Diameter	feet	meters
E	Energy	ft-lbf	joules (J)
F	Force	pounds (lbf)	newtons (N)
G	Universal Gravitational Constant	lbf-ft^2/slug^2	N-m^2/kg^2
I	Rotational Inertia	slug-ft^2	kg-m^2
K	Radius of Gyration	ft	m
L	Length	ft	m
M	Moment of a Force	ft-lbf	N-m
N	Number (of Gear Teeth, Etc.)	dimensionless	dimensionless
P	Power	ft-lbf/sec	watt (W)
Q	Heat	ft-lbf	joules (J)
R	Ratio	dimensionless	dimensionless
S	Speed	ft/sec	m/sec
S	Entropy	ft-lbf/°R	joules/K
T	Torque	ft-lbf	N-m
T	Absolute Temperature	°Rankine (°R)	kelvins (K)
U	Internal Energy	ft-lbf	joules (J)
V	Volume	ft^3	m^3
W	Work	ft-lbf	joules (J)

Lowercase Roman

Symbol	Name	SFS Units	MKS Units
a	Translational Acceleration	ft/sec^2	m/sec^2
a_g	Translational Acceleration in g's	dimensionless	dimensionless
c	Speed of Light	ft/sec	m/sec
d	Distance	ft	m
f	Frequency	cycles/sec (hertz)	cycles/sec (hertz)
g	Downward Acceleration from Gravity	ft/sec^2	m/sec^2
h	Height	ft	m
k	Spring Stiffness	lbf/ft	N/m
m	Mass	slugs	kg
p	Pressure	lbf/ft^2	pascal (Pa)
q	Electric Charge	N/A	coulomb
r	Radius	ft	m
s	Translational Displacement	ft	m
t	Time	seconds	seconds
v	Translational Velocity	ft/sec	m/sec
w	Width	ft	m
x	Position in Reference Frame Direction x	ft	m
y	Position in Reference Frame Direction y	ft	m
z	Position in Reference Frame Direction z	ft	m

Uppercase Greek

The only uppercase Greek character used is Δ, which is defined in the Special Symbols table in Appendix 3.

Lowercase Greek

Symbol	Name	SFS Units	MKS Units
α	Angular Acceleration	rad/sec^2	rad/sec^2
β	Not Used	N/A	N/A
γ	Camber Angle	radians (rad)	radians (rad)
δ	Not Used	N/A	N/A
ε	Not Used	N/A	N/A
ζ	Not Used	N/A	N/A
η	Efficiency	dimensionless	dimensionless
θ	Angle, Especially Angular Displacement	radians	radians
ι	Not Used	N/A	N/A
κ	Not Used	N/A	N/A
λ	Tire Slip Angle	radians	radians
μ	Coefficient of Friction	dimensionless	dimensionless
ν	Not Used	N/A	N/A
ξ	Not Used	N/A	N/A
o	Not Used	N/A	N/A
π	Circumference $\div$ Diameter	dimensionless	dimensionless
ρ	Mass Density	slugs/ft^3	kg/m^3
σ	Not Used	N/A	N/A
τ	Not Used	N/A	N/A
υ	Not Used	N/A	N/A
ϕ	Not Used	N/A	N/A
χ	Not Used	N/A	N/A
ψ	Not Used	N/A	N/A
ω	Angular Velocity	rad/sec	rad/sec

Appendix 5
Selected Derivations

Chapter 2

Derivation of Equation 2-10: Velocity vs. Time with Constant Acceleration, Calculus-Based

To derive equation 2-10 using calculus, first note that by taking the limit of equation 2-8 as Δt tends to zero, it becomes $a = \mathrm{d}v/\mathrm{d}t$. Then $\mathrm{d}v = a\,\mathrm{d}t$, and knowing that a is constant, we can integrate from $t = t_0$ to t, to find

$$\int_{v_0}^{v} \mathrm{d}v = a \int_{t_0}^{t} \mathrm{d}t$$

$$v - v_0 = a\left(t - t_0\right)$$

where v_0 is the beginning value of the velocity, at time t_0. Then the velocity v at any time t is simply

Equation 2-10

$$v = v_0 + a\left(t - t_0\right)$$

Derivation of Equation 2-15: Displacement vs. Time with Constant Acceleration, Calculus-Based

To derive equation 2-15 using calculus, we can take the differential of each side of equation 2-5b to find that $\mathrm{d}s = v\,\mathrm{d}t$. Taking a to be constant, and knowing from equation 2-10 that $v = v_0 + a(t\text{-}t_0)$, we can integrate from $t = t_0$ to t to find

$$\int_{s_0}^{s} \mathrm{d}s = \int_{t_0}^{t} v\,\mathrm{d}t = \int_{t_0}^{t} \left[v_0 + a\left(t - t_0\right)\right]\,\mathrm{d}t$$

$$s - s_0 = v_0\left(t - t_0\right) + \frac{1}{2}a\left(t^2 - t_0^{\,2}\right) - at_0\left(t - t_0\right)$$

where s_0 is the beginning value of the displacement, at time t_0. Grouping terms involving a, completing a square, and solving for the displacement s simplifies the equation to

Equation 2-15

$$s = s_0 + v_0\left(t - t_0\right) + \frac{1}{2}a\left(t - t_0\right)^2$$

Derivation of Equation 2-16: Velocity vs. Displacement in Constant Acceleration

We want to directly find the relationship between velocity and displacement during constant acceleration, without calculating each versus time. We start with equation 2-14:

Equation 2-14

$$s = s_0 + v_0 t + \frac{1}{2} a t^2$$

To remove time from this, we use the relationship between velocity and time during constant acceleration, $v = v_0 + at$, equation 2-10. Solving it for time:

$$t = \frac{v - v_0}{a}$$

Putting this value of t into 2-14:

$$s = s_0 + v_0 \frac{(v - v_0)}{a} + \frac{1}{2} a \frac{(v - v_0)^2}{a^2}$$

After multiplying through, combining the fractions, and canceling terms, we get our result:

Equation 2-16

$$s = s_0 + \frac{v^2 - v_0{}^2}{2a}$$

Derivation of Equation 2-16: Displacement vs. Velocity with Constant Acceleration, Calculus-Based

We can derive equation 2-16 by knowing taking $ds = v\ dt$, and $dv = a\ dt$, solving for the two values of dt, and setting them equal. We then find that $v\ dv = a\ ds$. Integrating both sides, we find

$$\int_{v_0}^{v} v\ dv = \int_{s_0}^{s} a\ ds$$

$$\frac{1}{2}\left(v^2 - v_0{}^2\right) = a\left(s - s_0\right)$$

From this we can solve for the displacement:

Equation 2-16

$$s = s_0 + \frac{\left(v^2 - v_0{}^2\right)}{2a}$$

Chapter 3

Derivation of Equation 3-5: Acceleration from 60-Foot Time

Taking the equation from the text,

$$s_{60} = s_{Rollout} + \sqrt{2a_{Launch}s_{Rollout}}\; t_{60} + \frac{1}{2}a_{Launch}\left(t_{60}\right)^2$$

we then group terms, based on powers of a_{Launch}:

Equation A3-1

$$\frac{t_{60}^{\;2}}{2}a_{Launch} + t_{60}\sqrt{2s_{Rollout}}\;\sqrt{a_{Launch}} + \left(s_{Rollout} - s_{60}\right) = 0$$

We can't use the quadratic equation on this directly, but let's define a variable H so that:

$$H = \sqrt{a_{Launch}}$$

We can only do this assuming $a_{Launch} > 0$ (if not, you're going the wrong way anyway!). Now we can substitute $H^2 = a_{Launch}$ into equation A3-1, which becomes:

$$\frac{t_{60}^{\;2}}{2}H^2 + t_{60}\sqrt{2s_{Rollout}}\;H + \left(s_{Rollout} - s_{60}\right) = 0$$

Now we can use the quadratic equation, and solve for H as:

$$H = \frac{-\sqrt{2s_{Rollout}}\;t_{60}}{t_{60}^{\;2}} \pm \frac{\sqrt{2s_{Rollout}t_{60}^{\;2} - 2t_{60}^{\;2}\left(s_{Rollout} - s_{60}\right)}}{t_{60}^{\;2}}$$

Simplify, cancel terms, and choose the positive square root to get:

$$H = \frac{-\sqrt{2s_{Rollout}} + \sqrt{2s_{60}}}{t_{60}}$$

Now using $a_{Launch} = H^2$, this is:

Equation 3-5

$$a_{Launch} = \frac{2}{t_{60}^{\;2}}\left(\sqrt{s_{60}} - \sqrt{s_{Rollout}}\right)^2$$

So we have the launch acceleration. As always, care must be taken when taking square roots. Usually the positive square root is what we're after, but not always.

Derivation of Equation 3-7: Going Faster by Stopping Faster

We start with the equations for the displacement during acceleration:

Equation 3-6a

$$s_{1,F} - s_{1,0} = \frac{v_{1,F}^2 - v_{1,0}^2}{2a_1}$$

and during braking:

Equation 3-6b

$$s_{2,F} - s_{2,0} = \frac{v_{2,F}^2 - v_{2,0}^2}{2a_2}$$

Remember that the velocity $v_{1,F}$ at the end of acceleration equals the velocity $v_{2,0}$ at the beginning of braking, and that the displacement $s_{1,F}$ at the end of acceleration equals the displacement $s_{2,0}$ at the beginning of braking. Further, the displacement $s_{1,0} = 0$. That simplifies equation 3-6a to

Equation A3-2

$$s_{1,F} = \frac{v_{1,F}^2 - v_{1,0}^2}{2a_1}$$

and changes 3-6b to

Equation A3-3

$$s_{2,F} - s_{1,F} = \frac{v_{2,F}^2 - v_{1,F}^2}{2a_2}$$

Next, we solve for $v_{1,F}^2$ in each of these two equations. From equation A3-2 we get:

Equation A3-4

$$2a_1 s_{1,F} + v_{1,0}^2 = v_{1,F}^2$$

And from equation A3-3 we get:

Equation A3-5

$$-2a_2 s_{2,F} + 2a_2 s_{1,F} + v_{2,F}^2 = v_{1,F}^2$$

Since the right sides of equations A3-4 and A3-5 are equal, the left sides must also be equal:

$$2a_1 s_{1,F} + v_{1,0}^2 = -2a_2 s_{2,F} + 2a_2 s_{1,F} + v_{2,F}^2$$

Now we have one equation in which the only unknown quantity is $s_{1,F}$, the point where braking begins. Solving for $s_{1,F}$:

Equation 3-7

$$s_{1,F} = \frac{-2a_2 s_{2,F} + v_{2,F}{}^2 - v_{1,0}{}^2}{2a_1 - 2a_2}$$

Once we use this to find the braking point, we can plug that into equation 3-8 in the main text, to find the peak speed.

Chapter 7

Finding the Direction Angle from Vector Components

The tangent of a vector's direction angle θ in Cartesian coordinates is equal to the y-component divided by the x-component. So θ is found by taking the inverse tangent of their ratio, $\tan^{-1}(y/x)$, also called arctangent(y/x). The arctangent is found in most computer applications, using the command "atan" or "ATAN," as in "theta=ATAN(y/x)." This will produce the angle in radians.

The problem is that the ATAN command only produces angles between $-\pi/2$ to $\pi/2$ radians ($-90°$ to $90°$). This is because the tangent of a vector (x, y) is the same as for the vector $(-x, -y)$; that is, $x/y = -x/-y$. For example, the vector $(1,1)$ is directed at $45°$, and the vector $(-1, -1)$ at $225°$, but both have a tangent of 1.

To get around this, there is a function called "atan2" or "ATAN2." In it, you put the x- and y-components in separately, and the direction angle is found uniquely. Be careful though. In Excel, the format is "theta = ATAN2(x,y)," while in Matlab, the format is reversed: "theta = atan2(y,x)."

Chapter 10

Derivation of Equations 10-14 and 10-16: Centripetal Acceleration vs. Velocity and Curve Radius, Calculus-Based

We start by writing the equation for the velocity vector in Figure 10 of Chapter 10, and assume that $t = 0$ at this moment, so that $\theta = \omega t$:

$$\mathbf{v} = v_x \mathbf{i} + v_y \mathbf{j}$$
$$= r\omega \left(\cos \omega t \; \mathbf{i} + \sin \omega t \; \mathbf{j} \right)$$

where **i** and **j** are unit vectors in the x- and y-directions. To find acceleration, we take the derivative of velocity with respect to time:

$$\mathbf{a} = \frac{d\mathbf{v}}{dt} = r\omega\left(-\omega \sin \omega t \ \mathbf{i} + \omega \cos \omega t \ \mathbf{j}\right)$$

Knowing the time $t = 0$, and evaluating this, the acceleration is

$$\mathbf{a} = \frac{d\mathbf{v}}{dt} = r\omega^2 \ \mathbf{j}$$

So, at this moment, there is a y-acceleration with magnitude of $r\omega^2$, and thus directed at the center C of the curve. Therefore, this is the centripetal acceleration:

Equation 10-16

$$a_{Centripetal} = r\omega^2$$

Knowing that $\omega = v/r$, we can substitute this into equation 10-16, and get

Equation 10-14

$$a_{Centripetal} = \frac{v^2}{r}$$

Chapter 11

Derivation of Equation 11-5: The Required Motorcycle Jump Velocity

To find the required ramp velocity v_0 to make the jump, in terms of the ramp angle θ and jump distance L, we start with one equation from the text involving the ramp speed and time at landing:

$$\left(v_0 \sin \theta_{Ramp}\right) t_F - \frac{1}{2} g t_F^2 = 0$$

Solving this for t_F,

Equation A11-1

$$t_F = \frac{2 v_0 \sin \theta_{Ramp}}{g}$$

We then take a second relationship for t_F:

Equation 11-4

$$t_F = \frac{L}{v_0 \cos \theta_{Ramp}}$$

Both values for t_F must be equal, so setting equation A11-1 equal to equation 11-4,

$$\frac{2v_0 \sin\theta_{Ramp}}{g} = \frac{L}{v_0 \cos\theta_{Ramp}}$$

Solving for the ramp speed v_0:

Equation 11-5

$$v_0 = \sqrt{\frac{gL}{2\cos\theta_{Ramp}\sin\theta_{Ramp}}}$$

Chapter 12

Derivation of Equation 12-2, Translational Kinetic Energy, Calculus-Based

We find the kinetic energy by calculating the work done on a mass by a force in the same direction, in accelerating the mass from a speed of zero to its current speed v. Assuming that we have straight-line motion with constant acceleration, and knowing that $F = ma$, we can find the work done by integrating the force with the displacement. Utilizing $v\,dv = a\,ds$, multiplying by m, and integrating,

$$m\int_0^v v\,dv = ma\int_0^s ds = F\int_0^s ds$$

$$\frac{1}{2}mv^2 = Fs = W$$

Assuming no drag, friction, etc., we know that the work $W=Fs$ went completely into the motion of the mass, into its kinetic energy E_K:

Equation 12-2

$$E_K = \frac{1}{2}mv^2$$

Derivation of Equation 12-6, Elastic Strain Energy in a Translational Spring, Calculus-Based

We find the elastic strain energy by calculating the work done on a spring in deflecting it from its free length ($x = 0$) to a deflection x. Knowing that $F = kx$, and that $dW = F\,ds$, we can find the work done by integrating the force with the deflection:

$$W = \int_0^x F\,dx = \int_0^x kx\,dx$$

$$= \frac{1}{2}kx^2$$

Since the work went into the deflection of the spring, it is equal to the strain energy E_{Strain}:

Equation 12-6

$$E_{Strain} = \frac{1}{2}kx^2$$

Chapter 13

Derivation of 13-8: Pressure Force of a 1-D Gas

As noted in Chapter 13, these 1-D gas equations are *not valid for real 3-D gases*! They are here to show how the transformation of gas compression work, into internal energy of the gas, follows directly from Newton's laws. Real gases have the same qualities, but use slightly different mathematical relationships between pressure, volume, work, etc.

In Section 9 on momentum, we saw that the change in a body's momentum, $m\Delta v$, is equal to the impulse $F_{Impact}\Delta t$ applied, equation 13-7. Since our gas-particle velocity v changes by $-2v$ (where v is the particular speed) when impacting the piston, the impulse *on the particle* is, with a stationary piston,

$$F_{Impact}\Delta t_{Impact} = m \cdot \Delta v = -2mv$$

where F_{Impact} is the *average* force during impact, and Δt_{Impact} is the impact duration. This makes each impulse *on the piston* the equal and opposite amount:

Equation A13-1

$$F_{Impact}\Delta t_{Impact} = 2mv$$

To find the pressure force, the average force F_{Press} on the piston, we need to know the time between impacts in the piston. Since the distance between the piston and head at any given time is x, there is a delay time Δt_{Delay} of $2x/v$ between impacts; the time for the particle to make one round trip. The impulse from the (average) pressure force between each impact (as seen in **FIGURE 12** in the text) must equal the impulse from each impact:

Equation A13-2

$$F_{Press}\Delta t_{Delay} = F_{Impact}\Delta t_{Impact}$$

From this, we can find the pressure force F_{Press} as $F_{Impact}\Delta t_{Impact}/\Delta t_{Delay}$. But we also know from equation A13-1 that $F_{Impact}\Delta t_{Impact} = 2mv$. Using all this, the pressure force is:

$$F_{Press} = \frac{2mv}{\Delta t_{Delay}} = \frac{2mv}{\left(2x\middle/v\right)}$$

or simply:

Equation A13-3   **Pressure force in general**

$$F_{Press} = \frac{mv^2}{x}$$

For example, at BDC, $v = v_1$, and $x = L$. There the pressure force becomes:

Equation 13-8   **Pressure force at BDC**

$$F_{Press,1} = \frac{mv_1^2}{L}$$

So a little physics and simple algebra gave us a compact formula for 1-D gas pressure.

Derivation of 13-9: Pressure During Compression of the 1-D Gas

Finding the pressure during compression takes some calculus. First we'll set things up. Assuming a *very low* piston velocity v_P (positive *toward* the head):[1]

Equation A13-4
$$x = x_1 - v_P t$$

Taking the differential of each side:

$$dx = -v_P dt$$

from which:

Equation A13-5
$$\frac{dt}{dx} = -\frac{1}{v_P}$$

With each impact against the moving piston, the particle speed v increases by $2v_P$. The time between impacts is $2x/v$, so impacts happen

---

Note 1: Instead of assuming a very low piston velocity, we could have split the single particle back into many. Either way makes the pressure a smooth function of piston position.

with a frequency f of $v/2x$ per second. Then the rate of increase in particle speed, dv/dt, is the impact frequency times the speed increase per impact:

$$\frac{dv}{dt} = f \cdot 2v_P = \frac{2v}{2x}v_P = \frac{v}{x}v_P$$

or simply:

Equation A13-6

$$\frac{dv}{dt} = \frac{v}{x}v_P$$

Now let's put this together to find how the particle speed v depends on the piston position x. Using equations A13-5 and A13-6,

$$\frac{dv}{dx} = \frac{dv}{dt}\frac{dt}{dx} = \frac{v}{x}v_P \cdot \left(-\frac{1}{v_P}\right)$$

giving:

Equation A13-7

$$\frac{dv}{dx} = -\frac{v}{x}$$

Cross-multiplying this and integrating between a beginning piston Position 1 and an ending Position 2:

$$\int_{v_1}^{v_2} \frac{1}{v}dv = -\int_{x_1}^{x_2} \frac{1}{x}dx$$

$$\ln v \Big|_{v_1}^{v_2} = -\ln v \Big|_{v_1}^{v_2}$$

$$\ln\left(\frac{v_2}{v_1}\right) = \ln\left(\frac{x_1}{x_2}\right)$$

where "ln" is the natural logarithm function. But this last relationship simply requires that:

$$\frac{v_2}{v_1} = \frac{x_1}{x_2}$$

So the new particle speed, from moving the piston from Position 1 to Position 2, is:

Equation A13-8 $$v_2 = \frac{x_1}{x_2}v_1$$ **Particle speed at x_2**

Now let's set x_2 and v_2 to their instantaneous values x and v. Also, let's set x_1 to L, the piston position at BDC, and let v_1 be the particle velocity at BDC, which we already called v_1. This gives us the particle speed v as a function of piston position x during compression:

Equation A13-9   **Particle speed during compression**

$$v = \left(\frac{L}{x}\right)v_1$$

The particle speed is thus inversely proportional to the distance between the piston and the cylinder head, so as the distance is cut in half, the speed is doubled. The increasing speed and decreasing distance also increase the pressure force. Using equation A13-3 with A13-8, the pressure force is

Equation A13-10   **1-D gas pressure force**

$$F_{Press} = \frac{m v_1^2}{x_1}\left(\frac{x_1}{x}\right)^3$$

Which is just the pressure at an initial piston position x_1 divided by the cube of the volume ratio, from Position 1 to Position 2. Setting this to have started at BDC (so $x_1 = L$), the pressure force during compression is

Equation 13-9   **1-D compression force**

$$F_{Press} = \frac{m v_1^2}{L}\left(\frac{L}{x}\right)^3 = F_{Press,1}\left(\frac{L}{x}\right)^3$$

Note that we removed the piston velocity from the problem; the particle speed and pressure force depend on the piston's position x, but not how fast it's moving when it gets there.

Derivation of 13-10: Compression Work for the 1-D Gas

Now that we know how the pressure force varies during compression, we can calculate the work the piston does on the gas. In integral form, the work done by a varying force applied through a distance is:

$$W_{12} = \int_{x_1}^{x_2} F\,dx$$

Inserting the pressure force from equation A13-9:

                    1-D gas work

$$W_{12} = \int_{x_1}^{x_2} \frac{-m v_1^2}{x_1}\left(\frac{x_1}{x}\right)^3 dx = -m v_1^2 x_1^2 \left(\frac{-1}{2x^2}\right)\Bigg|_{x_1}^{x_2}$$

So the work done on the gas between any two piston positions is:

Equation A13-11   **Work on a 1-D gas**

$$W_{12} = \frac{1}{2}mv_1^2\left(\left(\frac{x_1}{x_2}\right)^2 - 1\right)$$

To find the total compression work, we specify that $x_1 = L$, that $x_2 = x_{Clear}$, and rearrange:

Equation 13-10   **1-D compression work**

$$W_{Comp} = \frac{1}{2}mv_1^2\left(\left(\frac{L}{x_{Clear}}\right)^2 - 1\right) = U_1\left(R_{Comp}^2 - 1\right)$$

Note that $R_{Comp} = L/x_{Clear}$ is the compression ratio, and U_1 is the gas's internal energy at BDC, $\frac{1}{2}mv_1^2$.

Chapter 15

Derivation of 15-11 and 15-12: The Ideal Drag Racer

From Chapter 15 the velocity, time, and displacement of the car at point 2 (the end of the traction-limited section of the run) are v_2, t_2, and s_2; and the power-limited acceleration a_{PL} from point 2 to point 3 (the end of the quarter-mile) is:

Equation 15-10

$$a_{PL} = \frac{P_{Max}}{mv}$$

To find the velocity v_3 at point 3, we use calculus, first stating the power-limited acceleration a_{PL} as the derivative of velocity, and then manipulating the result:

$$a_{PL} = \frac{dv}{dt} = \frac{dv}{ds}\frac{ds}{dt} = \frac{dv}{ds}v$$

or:

Equation A15-1

$$v\frac{dv}{ds} = a_{PL}$$

Substituting the expression for a_{PL} from equation 15-10 into A15-1:

$$v\frac{dv}{ds} = \frac{P_{Max}}{mv}$$

Cross-multiplying and integrating from point 2 to point 3:

$$m \int_{v_2}^{v_3} v^2 dv = P_{Max} \int_{s_2}^{s_3} ds$$

which yields:

$$\frac{m}{3}\left(v_3^{\,3} - v_2^{\,3}\right) = P_{Max}\left(s_3 - s_2\right)$$

Solving for the top speed v_3:

Equation 15-11

$$v_3 = \sqrt[3]{v_2^{\,3} + \frac{3P_{Max}\left(s_3 - s_2\right)}{m}}$$

Now for the ET, which is t_3. We again use 15-10, but substitute in dv/dt for a_{PL}:

$$\frac{dv}{dt} = \frac{P_{Max}}{mv}$$

Cross-multiplying this and integrating from point 2 to point 3:

$$m \int_{v_2}^{v_3} v dv = P_{Max} \int_{t_2}^{t_3} dt$$

Integrating this yields:

$$\frac{m}{2}\left(v_3^{\,2} - v_2^{\,2}\right) = P_{Max}\left(t_3 - t_2\right)$$

Now, solving for the elapsed time t_3, we get:

Equation 15-12

$$t_3 = t_2 + \frac{m}{2P_{Max}}\left(v_3^{\,2} - v_2^{\,2}\right)$$

Derivation of 15-14: Expansion Work for the 1-D Gas

To find the expansion work, we do the opposite that we did for compression work. We adapt equation A13-11 to calculate the work done on the gas between positions 3 and 4:

Equation A13-11   **Work on a 1-D gas**

$$W_{34} = \frac{1}{2}mv_3^{\,2}\left(\left(\frac{x_3}{x_4}\right)^2 - 1\right)$$

Now, the expansion will start with the piston at the clearance distance $x = x_{Clear}$, and last until it reaches the end of its travel, at $x = L$. Plus now the piston is moving away from the head instead of toward it, so work is being done on the piston, not on the gas. We'll have to flip the sign on the equation to reflect this.

So to get the expansion work, we specify that $x_3 = x_{Clear}$, and that $x_4 = L$. Finally, v_3 in the equation will be the particle speed at TDC after heat addition, which we have already called v_3. We also flip the sign. This produces

Equation 15-14   **1-D expansion work**

$$W_{Exp} = \frac{1}{2}mv_3^2\left(1 - \left(\frac{x_{Clear}}{L}\right)^2\right) = \frac{1}{2}mv_3^2\left(1 - \left(\frac{1}{R_{Comp}}\right)^2\right)$$

Note again that $R_{Comp} = L/x_{Clear}$ is the compression ratio.

Appendix 6
Glossary

Term	Definition	Symbol
A		
Absolute Temperature	See *Temperature, Absolute.*	
Absolute Zero	The lowest temperature possible, where all thermal motion theoretically stops, equal to $-459.67\ °F$ or $-273.15\ °C$, 0 kelvins or 0 °Rankine.	
Acceleration	Rate of change in velocity, or how quickly velocity is increasing. Applies to linear and angular motion.	a, or α
Action/Reaction	The equal and opposite forces that act between two bodies, per Newton's third law that forces always act in pairs.	
Angle	A measure of rotation, or a difference in direction.	θ
Angular Acceleration	How quickly the angular velocity is increasing.	α
Angular Displacement	How much an object is rotated in the positive direction, relative to a reference angle.	θ
Angular Motion	See *angular displacement, angular velocity,* and *angular acceleration.*	
Angular Velocity	How quickly the angular displacement is increasing.	ω
Applied Force	A force applied to a body by some action outside the body. Examples are gravity and drag. Same as an external force.	$F_{Applied}$
Area	The two-dimensional measure of the extent of a surface or a cross-section.	A
B		
Bearing Load	A contact force that develops between two surfaces, as one supports the other.	
C		
Center of Gravity	As used here, it is the same as the center of mass.	CG

Term	Definition	Symbol
Center of Mass	Weighted average position of all the mass in a body. A force acting through it does not change the body's angular velocity.	CM or CG
Centripetal Acceleration	Acceleration toward the center of a turn. In cornering, this is roughly equal to lateral acceleration.	$a_{Centripetal}$
Centripetal Force	Force applied on a body, directed toward the center of a curve, causing centripetal acceleration.	$F_{Centripetal}$
Coefficient of Kinetic Friction	Ratio of the kinetic friction force magnitude to the normal force between two sliding surfaces.	$\mu_{Kinetic}$
Coefficient of Static Friction	Ratio of the maximum static friction force magnitude, to the normal force between two non-sliding surfaces.	μ_{Static}
Combustion	A chemical reaction between a fuel and an oxidizer, which produces heat. Burning.	
Conduction	Heat transfer between contacting materials having no relative motion.	
Convection	Heat transfer involving at least one moving fluid, such as air through an oil cooler.	
Cornering Stiffness	The ratio of a tire's lateral force $F_{Lateral}$ to its slip angle λ, in the tire's linear operating range.	C_λ
Cosine	Cosine of an angle; the length of the adjacent side of the right triangle, divided by the length of the hypotenuse.	cos
Couple	A "pure torque" produced by two equal and opposite, but offset forces.	
Curvature	A curve's "tightness," equal to the reciprocal of the radius.	$1/r$
D		
Degree	Angular measure, equal to 1/360 revolutions.	° or **deg**
Degrees of Freedom	The minimum number of quantities necessary to describe a body's motion. For example, a point mass moving on a flat track has 2 degrees of freedom.	DOF
Density	The amount of mass a material contains per unit volume. Example: 1,000 kilogram/meter³.	ρ
Diameter	Distance across a circle through its center.	D
Dimension	A general type of measurement. Examples: length, mass, time, temperature.	
Displacement	The change in position (or angle) of a body from some reference point (or angle).	s, or θ

Term	Definition	Symbol
Dissipation	The "spreading out" of a quantity, such as energy, over some area or volume.	
Distance	The accumulated path length of a body's travel, without regard to its direction during each instant.	d
Dynamics	The study of motion, and the forces or torques causing it.	
E		
Elastic Material	A material that can be stretched, compressed, or twisted and then return to its original state without loss of mechanical energy.	
Elastic Strain Energy	Potential energy stored in the material of a deflected body.	E_{Strain}
Electric Charge	Typically refers to a net amount of charged particles, either positive (as protons have) or negative (as electrons have).	q
Electric Current	The flow of electric charge through a conductor.	
Electromagnetism	The group of electrical and magnetic phenomena that always occur together. It involves, forces, currents, waves, fields, etc.	
Energy	The motion of mass, or the capability to produce such motion. This may be either coordinated or random motion, corresponding to mechanical or thermal energy, respectively.	E
Engine	See *heat engine*.	
Entropy	A measurement of the lack of concentration of energy. It increases as energy spreads out.	S
Equation	A formula in which one quantity is set equal to another.	
External Force	Force applied to a body by another body. Same as an applied force.	$F_{External}$
F		
Fluid	A material that can "flow," either a gas or a liquid.	
Fluid Dynamics	The interaction between forces and motion of a fluid.	
Fluid Statics	The balance of forces in a stationary fluid.	
Force	An action that pushes or pulls on a body. It has a magnitude and direction.	F

Term	Definition	Symbol
Free-Body Diagram	A diagram in which each interacting body is drawn separately, along with the forces acting on it.	
Frequency	The number of times per second an event happens, such as an impact, vibration, or sound wave.	f
Friction	The forces acting between, and parallel to, the contacting surfaces of two bodies. It can be between two nonsliding surfaces (static friction) or two sliding surfaces (kinetic friction).	
Friction Coefficient	See coefficient of *kinetic* friction, or coefficient of *static* friction.	μ
Fulcrum	The pivot point of a lever.	

G

Term	Definition	Symbol
Gas	A fluid in which the molecules have enough energy to overcome their molecular bonds, and so can move far apart.	
Gear Ratio	The ratio of the angular motion of the input gear to the angular motion of the output gear. Also equal to the ratio of output torque to input torque.	R
Gravitational Force	A force between two bodies caused by gravity. Most often refers to the downward force on a body from Earth's gravity.	

H

Term	Definition	Symbol
Heat	The movement of thermal energy from one body or fluid to another.	Q
Heat Engine	A device that (partially) transforms thermal energy into work, using a series of thermodynamic processes. Loosely, it converts heat to power.	
Hypotenuse	Longest side in a right triangle.	

I

Term	Definition	Symbol
Inequality	A formula in which one side is set to be larger or smaller than the other.	
Inertia	In translation, the tendency for a body to resist being accelerated by a force, and called mass. In rotation, the tendency for a body to resist being accelerated by a moment, and called rotational inertia.	m, or I
Internal Energy	The thermal energy consisting of the random molecular motion within a body.	U

Term	Definition	Symbol
Irreversible Process	Thermodynamically, a process that cannot be "played backwards" to restore a system to its original state. Examples: sliding friction, and heat transfer from hot to cold. See also *reversible process*.	

J

Term	Definition	Symbol
Jerk	The rate of increase of acceleration.	

K

Term	Definition	Symbol
Kinematics	The study of motion, without regard to the cause of it.	
Kinetic Energy	The energy carried in the motion of a body, proportional to its mass times the square of the velocity.	E_K
Kinetics	The study of motion and the forces causing it (dynamics).	

L

Term	Definition	Symbol
Lateral	Left/right relative to the vehicle. Sideways.	
Lateral Acceleration	The acceleration directed roughly toward the center of an arc, causing the vehicle to move along the arc.	a_y
Lateral Force	A force acting sideways on a vehicle.	F_y
Law	In science, a generally accepted rule describing some type of event or state, based on being supported by experimental evidence. It may have limits on its conditions for use.	
Length	A basic dimension, measured in feet, meters, etc.	L
Lever	A device in which a bar rotates about a fulcrum, and has a force applied to one point on it in order to apply another force in another direction and/or a different magnitude.	
Liquid	A material whose molecules are bound tightly enough by intermolecular forces to stay in contact, but not tight enough to be held in place.	
Load Transfer	The "shift" in normal load from one tire to another as a result of acceleration or driving on an angled surface.	
Longitudinal	Forward/rearward relative to the vehicle.	
Longitudinal Acceleration	Acceleration in line with a vehicle, usually considered positive forward.	a_x

Term	Definition	Symbol
Longitudinal Force	A force acting in line with a vehicle.	F_x
M		
Magnetic Force	A force that develops from the interaction of magnetic fields, and can either attract or repel two bodies.	
Magnitude	The "size" of a measurement. Example: speed is the magnitude of velocity.	
Mass	A property of a material, which makes up its inertia.	m
Mechanical Energy	Energy that either is coordinated motion of mass, or could ideally be converted completely into it.	
Mechanics	The science determining the effects of energy and forces on solids, liquids, and gases.	
Moment, General	The product of a quantity and some power of the radial distance from a given point. Usually means "moment of a force," unless specified otherwise.	
Moment of a Force	A force times its offset/radius from its line of action, essentially a torque.	M
Moment of Inertia	The inertial resistance of a body to angular acceleration from an applied torque. Also called *rotational inertia*.	I
Momentum	The product of a body's mass and velocity.	mv
Motor	A device which utilizes supplied mechanical or thermal energy, to provide mechanical energy (work/power).	
N		
Normal Force	A contact force perpendicular to the surface between two contacting bodies.	F_N
O		
P		
Perpendicular	Two lines having a 90-degree angle between them.	
Pitch	Rotation of a vehicle about its lateral axis, such as its nose diving during braking.	θ_y
Polar Moment of Inertia	Moment of inertia of a vehicle about the vertical (z-) axis through its CG.	I_z
Potential Energy	Energy stored in forms such as gravitational, strain, chemical, electromagnetic, and nuclear.	

Term	Definition	Symbol
Power	The rate at which work is done.	P
Pressure	For a distributed force applied to a surface, the amount of force per unit surface area, usually by a fluid.	p
Q		
Quasi-Statics	The use of Newton's laws to calculate forces on a body in "constant" acceleration.	
R		
Radial	A vector directed along a radius of a curve, such as a radial force.	
Radian	An angular measure in which a circular arc length spanning the angle is equal to the radius of the arc.	rad
Radiation	Heat transfer from a body through electromagnetic radiation, most commonly infrared.	Q_{Rad}
Radius	The distance from the center of a circle/arc, out to the curve itself. It is perpendicular to the curve/arc at their intersection. Also, the distance from the center of rotation to the "working edge" of a tire, gear, pulley, etc.	r
Ratio	One quantity divided by another.	
Reversible Process	A thermodynamic process which can be "undone" without leaving any physical sign of having occurred. Example: the bob of a frictionless pendulum swinging back to its starting point. See also *irreversible process*.	
Revolution	One full rotation of a body, back to its original angle.	rev
Right Angle	An angle of 90 degrees.	
Right Triangle	A triangle containing one right angle.	
Rigid Body	A body that flexes little enough that the flexing can be ignored for the problem being worked.	
Roll	Rotation of a vehicle about its longitudinal axis, as in a motorcycle leaning into a corner.	θ_x
Rotation	The turning of a body about some point, such as its CG, or a fixed point. Also see *angular displacement*, *angular velocity*, and *angular acceleration*.	
Rotational Inertia	The inertial resistance of a body to angular acceleration from an applied torque. Also called *moment of inertia*.	I

Term	Definition	Symbol
S		
Scalar	A physical quantity having no implied direction, such as energy.	
Science	The practice of proposing theories to explain observations, then proving/disproving/updating them, based on new data.	
Sine	Sine of an angle; the length of the opposite side of the right triangle, divided by the length of the hypotenuse.	$\sin$
Slip Angle	The difference in angle between a tire's heading and its actual direction of travel. Necessary for a tire to produce lateral force.	λ
Slip Ratio	The ratio of a tire's slip velocity to its "no-slip" wheel center velocity.	R_{Slip}
Slip Velocity	The difference between the "no-slip" contact patch velocity (based on the tire's radius and angular velocity) and its actual wheel center velocity. Necessary to produce longitudinal force.	v_{Slip}
Slope	The increase in "height" of a line or curve per unit of horizontal travel.	
Solid	A material whose particle energy is low enough that its molecular bonds can hold its molecules in place, except for their thermal vibrations.	
Speed	The rate at which distance increases. The magnitude of velocity. Unlike velocity, it has no specified direction.	v
Statics	The use of Newton's laws to calculate forces on a body at constant velocity.	
Steady State	Conditions that are either constant, or nearly so. For example, cruising at a (nearly) constant speed on a (nearly) level road.	
Stiffness	The force required to deflect a spring or other body by a unit amount.	k
T		
Tangent	Tangent of an angle; the length of the opposite side of the right triangle, divided by the length of the adjacent side.	$\tan$
Tangential	A vector directed along the instantaneous path of motion, such as tangential velocity in a curve.	
Temperature	A measure of the average thermal energy in a material.	T
Temperature, Absolute	The temperature given relative to absolute zero, the point where molecular motion theoretically stops.	T

Term	Definition	Symbol
Thermal Energy	Energy associated with random particle motion.	
Thermodynamics	The science of the transformation between thermal and mechanical energy.	
Thermodynamic Cycle	A series of thermodynamic processes that eventually return to the same condition. Usually intended to produce work. For example, the Air Standard Otto Cycle.	
Thermodynamic Process	A single process that transforms energy. Using friction to brake is a thermodynamic process converting kinetic energy to heat.	
Thermodynamic System	A physical system that we mentally bound so that we can assess the energy transfer within it, and across its boundary.	
Time	A physical quantity that progresses as physical processes do, from past to present to future. It is measured in seconds, minutes, etc.	t
Torque	An action that tends to cause rotation, caused by a force acting on an effective lever arm.	T
Transient	A situation where conditions are changing. Launching a car in a drag race is a transient event.	
Translational Motion	Motion that moves the center of gravity of a body from one place to another.	
Two-Force Member	A link, bar, etc., with "frictionless" joints at each end, and thus only capable of producing two forces, equal and opposite, along the line between the joints.	

U

Term	Definition	Symbol
Unit	A specific amount of some measurement, as in "one meter is a unit of length."	

V

Term	Definition	Symbol
Vector	A physical quantity having magnitude and direction.	**bold type** or → arrow
Velocity	The speed of a body, with direction specified. The rate of change of displacement. Applies to translation and rotation.	v, or ω
Volume	The three-dimensional space occupied by an object, or by space.	V

W

Term	Definition	Symbol
Weight	The gravitational force exerted by a large planetary body (like Earth or the moon) on a body near it. For instance, the weight of a car on Earth's surface.	F_{Weight} or mg

Term	Definition	Symbol
Weight Transfer	A common but inaccurate term for load transfer. See *load transfer*.	
Work	The effort expended when a force acts through a distance.	W
X		
Y		
Yaw	Rotation of a vehicle about its vertical axis.	θ_z
Z		

Index

References
and Recommended Reading

Bentley, Ross. *Ultimate Speed Secrets — The Complete Guide to High-Performance and Race Driving.* Minneapolis, MN: Motorbooks, 2011.

Brown, Sanborn C. *The Collected Works of Count Rumford.* Volume I, *The Nature of Heat.* Cambridge, MA: The Belknap Press of Harvard University Press, 1968.

Carnot, Sadi. *Reflections on the Motive Power of Fire.* Mineola, NY: Dover Publications, Inc., 1960.

Çengel, Yunus A., and Michael A. Boles. *Thermodynamics: An Engineering Approach, Sixth Edition (SI Units).* New York, NY: McGraw-Hill, 2007.

Donohue, Mark, and Paul Van Valkenburgh. *The Unfair Advantage.* Cambridge, MA: Bentley Publishers, 2000.

Gillespie, Thomas D. *Fundamentals of Vehicle Dynamics.* Warrendale, PA: Society of Automotive Engineers, 1992.

Gillispie, Charles Coulston. *The Edge of Objectivity — An Essay in the History of Scientific Ideas.* Princeton, NJ: Princeton University Press, 1960.

Guillen, Michael. *Five Equations that Changed the World — The Power and Poetry of Mathematics.* New York, NY: Hyperion, 1995.

Lightman, Alan. *Great Ideas in Physics.* New York, NY: McGraw-Hill, 2000.

Lopez, Carl. *Going Faster! Mastering the Art of Race Driving.* Cambridge, MA: Bentley Publishers, 1997.

Milliken, William F., and Douglas L. Milliken. *Race Car Vehicle Dynamics.* Warrendale, PA: Society of Automotive Engineers, 1995.

Wong, J.Y. *Theory of Ground Vehicles, Third Edition.* New York, NY: Wiley Interscience, 2001.

Zafiratos, Chris. *Physics.* New York, NY: John Wiley & Sons, 1976.

Acknowledgements

Paul Potrykus helped me realize that I had to do this book. We've always talked cars, and our talks shaped the discussions here. His son Scott reviewed a recent version and gave me great feedback.

Dave Freiman provided valuable input and encouragement while reading my early (and literally rough) drafts.

Don Garlits' discussions with me about his shift to rear-engine dragsters gave me a story I hadn't seen anywhere. It helped transform the book, making it part "detective story."

Richard Noble and Ron Ayers provided data from the Thrust SSC land speed record run, creating one of the recurring themes in the book. Ron also shared learnings that greatly improved the discussion.

Quite a few people helped by reviewing chapters at various stages: John Beardmore, Lori Ertell-Tower, Mark Gehringer, Ron Godlewski, Dean Hauersperger, Bernie Kuschel, Bill Logsdon, Devin Marshall, Michael Mueller, Sheldon Plaxton, Theresa Prior, Don Straitiff, Mahkameh Salehi, Thom Timmons, Rich Waters, Steve Wolfe, and Jim Woods.

John Kyros patiently guided my image searches at the GM Heritage Center, and Al Scott spent a Friday after Thanksgiving helping me dig through racing photos at the Ford Archives.

Richard Lammert, as my high school physics teacher, got me started in the right direction.

Bentley Publishers bought into my idea and allowed me to carry it out. Their commitment to quality, starting with Michael Bentley, was stellar. Janet Barnes has seen it all the way through as editor, and saw to its technical review. Andrea Corbin was instrumental in producing a page design that presents the material in a clear and orderly way. She and colleague Audrey Saunders then fit my content to the design — no small task.

My son Aric provided material by being my garage buddy, "grudgingly" accompanying me on trips to Road America, and racing me in go-karts there.

My wife Tara stayed supportive, despite hearing more mentions of "the book" than a wife should have to, and living with a do-it-yourselfer who didn't have time to do it himself.

My colleagues at General Motors, the majority at the proving ground, have taught me more about cars than I thought there was to know. I feel blessed to have rubbed shoulders with so many incredibly knowledgeable yet down to earth people.

Many others may not think they contributed, but kept me going with their encouragement. It really did help!

About the Author

Randy Beikmann holds a Ph.D. in mechanical engineering from the University of Michigan. He received his bachelor's degree in ME from Kansas State University. An automotive engineer at General Motors for over 30 years, he is a technical specialist in noise and vibration (N&V) at their proving ground in Milford, MI. He has published numerous papers on powertrain N&V, and has helped design and teach classes in it at GM. He currently holds six patents.

His first assignment was the enviable task of developing Corvettes for minimized "rough road shake." At first he based his recommendations on physics-based intuition, evaluation rides, and measurements. Gaining experience, and learning from other engineers and technicians, he began to spend less time testing and more time applying physics-based analysis to develop a better design.

Randy's greatest job satisfaction is in dissecting a complex problem, stripping it down to its basics, and formulating a viewpoint that makes the solution look simple. He sees his role at GM as one of collaboration and synthesis: gathering knowledge from the experts around him, adding his own perspective, and putting it all together to improve his group's abilities. Although noise and vibration may seem like a narrow focus, it actually involves every area of the vehicle. Think about how many things can produce a drone or whine in your car! He especially enjoys working on the powertrain, suspension, and body structure.

He learned physics from the ground up, first by storing up many questions. When he was six he knew that the engine made noise, then the car ran. But what about it made the car move? "That's just how it works" was no answer. He had to understand it for himself. Over time he answered many of his own questions by reading car magazines, driving and working on his 1969 Cougar, and hanging around his uncle's garage.

With high school physics, things really connected. Randy viewed every new topic in terms of where he could use it in a car. Although the textbook seemed boring, learning physics was exciting! He soon began thinking that the ideal physics book would use cars to illustrate every area of physics, because cars are so interesting and complex.

So after 35 years, here is that book, one he really could have used as a teenager.

Selected Books and Repair Information From Bentley Publishers

Engineering

Bosch Fuel Injection and Engine Management *Charles O. Probst, SAE* ISBN 978-0-8376-0300-1

Maximum Boost: Designing, Testing, and Installing Turbocharger Systems *Corky Bell* ISBN 978-0-8376-0160-1

Supercharged! Design, Testing and Installation of Supercharger Systems *Corky Bell* ISBN 978-0-8376-0168-7

Scientific Design of Exhaust and Intake Systems *Phillip H. Smith & John C. Morrison* ISBN 978-0-8376-0309-4

The Hack Mechanic Guide to European Automotive Electrical Systems *Rob Siegel* ISBN 978-0-8376-1751-0

Motorsports

Alex Zanardi: My Sweetest Victory *Alex Zanardi and Gianluca Gasparini* ISBN 978-0-8376-1249-2

The Unfair Advantage *Mark Donohue and Paul van Valkenburgh* ISBN 978-0-8376-0069-7

Equations of Motion - Adventure, Risk and Innovation *William F. Milliken* ISBN 978-0-8376-1570-7

Audi Repair Manuals

Audi A4 Service Manual: 2002-2008, 1.8L Turbo, 2.0L Turbo, 3.0L, 3.2L *Bentley Publishers* ISBN 978-0-8376-1574-5

Audi A4 Service Manual: 1996-2001, 1.8L Turbo, 2.8L *Bentley Publishers* ISBN 978-0-8376-1675-9

Audi TT Service Manual: 2000-2006, 1.8L turbo, 3.2 L *Bentley Publishers* ISBN 978-0-8376-1625-4

Audi A6 Service Manual: 1998-2004 *Bentley Publishers* ISBN 978-0-8376-1670-4

BMW

Memoirs of a Hack Mechanic *Rob Siegel* ISBN 978-0-8376-1720-6

BMW X3 (E83) Service Manual: 2004-2010 *Bentley Publishers* ISBN 978-0-8376-1731-2

BMW 5 Series (E60, E61) Service Manual: 2004-2010 *Bentley Publishers* ISBN 978-0-8376-1689-6

BMW 5 Series (E39) Service Manual: 1997-2003 *Bentley Publishers* ISBN 978-0-8376-1672-8

BMW X5 (E53) Service Manual: 2000-2006 *Bentley Publishers* ISBN 978-0-8376-1643-8

BMW 3 Series (F30, F31, F34) Service Manual: 2012-2015 *Bentley Publishers* ISBN 978-0-8376-1752-7

BMW 3 Series (E90, E91, E92, E93) Service Manual: 2006-2011 *Bentley Publishers* ISBN 978-0-8376-1723-7

BMW 3 Series (E46) Service Manual: 1999-2005 *Bentley Publishers* ISBN 978-0-8376-1657-5

BMW Z3 (E36/7) Service Manual: 1996-2002 *Bentley Publishers* ISBN 978-0-8376-1617-9

Porsche

Porsche 911 (996) Service Manual: 1999-2005 *Bentley Publishers* ISBN 978-0-8376-1710-7

Porsche 911 (993) Service Manual: 1995-1998 *Bentley Publishers* ISBN 978-0-8376-1719-0

Porsche Boxster. Boxster S (986) Service Manual: 1997-2004 *Bentley Publishers* ISBN 978-0-8376-1645-2

Porsche 911 Carrera Service Manual: 1984-1989 *Bentley Publishers* ISBN 978-0-8376-1696-4

Porsche 911 (964): Enthusiast's Companion *Bentley Publishers* ISBN 978-0-8376-0293-6

Porsche: Excellence Was Expected *Karl Ludvigsen* ISBN 978-0-8376-0235-6

Ferdinand Porsche – Genesis of Genius *Karl Ludvigsen* ISBN 978-0-8376-1557-8

Porsche – Origin of the Species *Karl Ludvigsen* ISBN 978-0-8376-1331-4

Volkswagen

Volkswagen Rabbit, GTI Service Manual: 2006-2009 *Bentley Publishers* ISBN 978-0-8376-1664-3

Volkswagen Jetta, Golf, GTI Service Manual: 1999-2005 *Bentley Publishers* ISBN 978-0-8376-1678-0

Volkswagen Jetta Service Manual: 2005-2010 *Bentley Publishers* ISBN 978-0-8376-1616-2

Volkswagen Jetta, Golf, GTI: 1993-1999, Cabrio: 1995-2002 Service Manual *Bentley Publishers* ISBN 978-0-8376-1660-5

MINI Repair Manuals

MINI Cooper Service Manual: 2007-2013 *Bentley Publishers* ISBN 978-0-8376-1730-5

MINI Cooper Service Manual: 2002-2006 *Bentley Publishers* ISBN 978-0-8376-1639-1

Mercedes-Benz

Mercedes-Benz C-Class (W202) Service Manual 1994-2000 *Bentley Publishers* ISBN 978-0-8376-1692-6

Mercedes Benz E-Class (W124) Owner's Bible: 1986-1995 *Bentley Publishers* ISBN 978-0-8376-0230-1

B BentleyPublishers®
Automotive Reference

Bentley Publishers has published service manuals and automobile books since 1950. For more information, please contact Bentley Publishers at 1734 Massachusetts Avenue, Cambridge, MA 02138, or visit our web site at BentleyPublishers.com